Nanosensing and Bioanalytical Technologies in Food Quality Control

Pranjal Chandra • Parmjit S. Panesar
Editors

Nanosensing and Bioanalytical Technologies in Food Quality Control

Springer

Editors
Pranjal Chandra
Laboratory of Bio-Physio Sensors
and Nanobioengineering,
School of Biochemical Engineering
Indian Institute of Technology (BHU)
Varanasi
Varanasi, Uttar Pradesh, India

Parmjit S. Panesar
Food Biotechnology Research Laboratory,
Department of Food Engineering and
Technology
Sant Longowal Institute of Engineering and
Technology
Longowal, Punjab, India

ISBN 978-981-16-7028-2 ISBN 978-981-16-7029-9 (eBook)
https://doi.org/10.1007/978-981-16-7029-9

This Springer imprint is published by the registered company Springer Nature Singapore Pte Ltd.
The registered company address is: 152 Beach Road, #21-01/04 Gateway East, Singapore 189721, Singapore

Preface

This book attempts to address several recent advancements in food engineering based on biosensing and bioanalytical technologies. Developments related to classical unit operations in biosensing-engineering practised for food manufacturing are discussed extensively. Additionally, topics involving sensor-concentrated progress in the quality control and storage of liquid and solid foods; processing, i.e. heating, chilling, and freezing of foods; mass transfer in foods; chemical and biochemical aspects of food engineering and the use of kinetic analysis; thermal and nonthermal processing, extrusion, membrane processes, shelf life, electronic indicators in inventory management; sustainable technologies in food processing; and packaging, cleaning, and sanitation have also been conferred methodically. The book also presents the advances in electrode modification techniques and nanoengineered manipulations of biosensors. These advanced sensing systems are being fascinatingly projected in food technology to address their maximum analytical performances in recent times. Primarily, it aimed at food industries, professional food engineers, academics researchers addressing food-related findings, and graduate-level students.

In a nutshell, the book provides an accessible and readable summary of the uses of nanomaterial and modern electrochemical techniques deployed to food technology. The editors are leading engineers and scientists, widely acclaimed in their respective fields. They have presented this book according to the market need and pinpoint the cutting-edge technologies in food technology. All contributions are inscribed by world-renowned experts who have both academic and professional substantiation. All authors have attempted to provide critical, comprehensive, and readily accessible information on the art and science of a relevant topic in each chapter, with reference lists for further details. Therefore, this book can serve as an essential reference source to students and researchers in universities and research institutions.

Varanasi, India — Pranjal Chandra
Longowal, India — Parmjit S. Panesar

Contents

About the Editors

Pranjal Chandra is an Assistant Professor at the School of Biochemical Engineering, Indian Institute of Technology (BHU) Varanasi, India. He earned his Ph.D. from Pusan National University, South Korea and did post-doctoral training at Technion-Israel Institute of Technology, Israel. His research focus is highly interdisciplinary, spanning a wide range in biotechnology, nanobiosensors, material engineering, and nanomedicine. He has designed several commercially viable biosensing prototypes that can be operated for onsite analysis for biomedical diagnostics. He is a guest editor and an editorial board member of various international journals. He has authored over 110 high-impact publications including research/reviews papers and invited book chapter. He has published 11 books on various aspects of biosensors/medical diagnostics. Dr. Chandra is the recipient of many prestigious awards, coveted honors, and fellowships such as the DST Ramanujan fellowship (Government of India); Early Career Research Award (ECRA) (DST, Government of India); ICMR Shakuntala Amir Chand Prize 2020; BK-21 and NRF fellowship, South Korea; Technion Post-Doctoral Fellowship, Israel; Nano Molecular Society India Young Scientist Award; Biotech Research Society India (BRSI) Young Scientist Award; Young Engineers Award 2018; Highly Cited Corresponding authors in The Royal Society of Chemistry (RSC), Cambridge, London; Top 10% cited article in the General Chemistry Section RSC Journal, Cambridge, London, Gandhian Young Technology Innovation Award (GYTI) 2020, etc. Dr. Chandra has been listed among the world's top 2% scientists for the consecutive two years (2020 & 2021) in the report prepared by Stanford University, USA.

Prof. Parmjit S. Panesar is working as Dean (Planning & Development), Sant Longowal Institute of Engineering and Technology (Deemed to be University: Established by Govt. of India), Longowal, Punjab, India. Earlier Prof. Panesar has also served as Dean (Research & Consultancy) and Head, Department of Food Engineering & Technology, SLIET Longowal, Punjab. In 2005, he was awarded BOYSCAST (Better Opportunities for Young Scientists in Chosen Areas of Science and Technology) fellowship by the Department of Science and Technology (DST), Ministry of Science and Technology, Govt. of India. Prof. Panesar was awarded the Young Scientist Fellowship by Punjab State Council for Science and Technology, India. His research is focused in the area of value addition of food industry by-products, green extraction of bioactive from food industry by-products, food fermentation, food enzymes, prebiotics, biopigments, and biodegradable films. He has published more than 150 research articles in peer-reviewed journals of high repute, 32 chapters, and has authored/edited 07 books. He is a member of the editorial advisory boards of several journals, including the International Journal of Biological Macromolecules, Journal of Food Science and Technology, Carbohydrate Polymer Technologies and Applications. He acted as a former member of the National level Scientific Panel on "Genetically modified organisms and foods" by the Food Safety and Standards Authority of India (FSSAI) and Member in the international collective of experts of the Foundation for Science and Technology (FCT), Portugal.

In recognition of his work, Prof. Panesar was awarded "Fellow Award 2018" by The Biotech Research Society of India (BRSI) and "Fellow Award 2019" by National Academy of Dairy Science India (NADSI). He has been selected for the award of prestigious "INSA Teachers Award (2020)" by Indian National Science Academy (INSA). Recently, Prof. Panesar has also been listed in the most coveted list of "World Ranking of Top 2% Scientists" published by Stanford University, USA.

Application of Nanomaterials in Food Quality Assessment

1

Milad Torabfam, Qandeel Saleem, Prabir Kumar Kulabhusan, Mustafa Kemal Bayazıt, and Meral Yüce

Abstract

The food industry has a significant role to play in governing local economies all over the world. This sector includes some processes such as storage of raw materials, food production, and preservation. Food processing, food quality, and safety are vital to protect public health. Thus, food safety monitoring, for example, early detection of food pathogens, food-related toxins, allergens, chemicals, and enterotoxins, is of great significance. Nanoscience-based sensing platforms have become alternatives to conventional food safety monitoring techniques. Over the past few years, scientists have designed various novel nanosensors with high sensitivity and selectivity to detect a wide variety of hazardous substances. The nanomaterials, such as carbon-based nanoparticles, plasmonic/metallic nanoparticles, and inorganic fluorescent nanomaterials, have been extensively used to develop various detection platforms over the past few decades. The surface functionalization of nanoparticles using target-specific biological agents, such as aptamers and antibodies, has contributed to improving the efficiency of those nanoparticle-based diagnostic tools. In this chapter, general structural, physicochemical, and optical features of the nanoparticles were described, and their applications in food safety monitoring were reviewed. Following this, affinity agents and fundamental sensing principles employed in developing food-related hazardous substance detection tools were elaborated

M. Torabfam · Q. Saleem
Sabanci University, Faculty of Engineering and Natural Sciences, Istanbul, Turkey

P. K. Kulabhusan
The Institute for Global Food Security, Queen's University, Belfast, UK

M. K. Bayazıt · M. Yüce (✉)
SUNUM Nanotechnology Research and Application Centre, Sabanci University, Istanbul, Turkey
e-mail: meralyuce@sabanciuniv.edu

P. Chandra, P. S. Panesar (eds.), *Nanosensing and Bioanalytical Technologies in Food Quality Control*, https://doi.org/10.1007/978-981-16-7029-9_1

based on the recent publications in the literature. Finally, we expect to pave the way for enhancing the efficiency and applicability of nanosensors in the initial sensing of food-related targets that cause a significant risk for humankind worldwide.

Keywords

Nanoparticles · Aptamers · Monoclonal antibodies · Food safety · Biosensors · Toxins · Allergens · Pathogens · Enterotoxins

1.1 Introduction

Human health and well-being largely depend on the quality of food that we consume. The simple yet essential idea of "you are what you eat" is well-acknowledged and adequately established. However, environmental degradation caused by rapid industrialization, infiltration of hazardous chemical and biological contaminants into food through soil and water, novel foodborne diseases and health complications, and swiftly changing eating habits have made food quality assessment and safety more critical now than ever (Pan et al. 2019). Moreover, the advances in the food and agriculture industry, stringent guidelines set by global regulatory agencies, and the growing competition between local catering businesses as well as international food supply chains have also made food quality assurance inevitable.

Food quality assessment encompasses the identification and quantification of food contaminants like colorants and additives, chemicals, toxins and enterotoxins, allergens, pathogens, heavy metal ions, residual pesticides, and other pollutants. In a typical food production process, quality control is often performed at the end with one or a combination of standard screening approaches like chromatography, mass spectrometry, ultraviolet detection, or fluorescence-based screening. However, this reduces the efficiency considerably as the poor-quality food has already passed the costly and tedious production process. These analytical techniques are also limited by large test sample requirements, intricate pretreatments, and the need for well-trained staff (Mishra et al. 2018). Hence, it is crucial to develop new and improved techniques and sensing devices that can detect traces of impurities within the sophisticated matrix of edible products with high sensitivity and great precision in less amount of time (Liu et al. 2018a, b).

The unprecedented advancement in nanotechnology has led to extraordinary breakthroughs in the field of food inspection, diagnostics, and environmental sensing. In recent years, nanomaterials, organic as well as inorganic, are utilized in their pristine and functionalized forms in nanosensors. Owing to their small size (less than 100 nm) and unique physicochemical characteristics, nanomaterial-based food monitoring systems are capable of effectively detecting tiny molecules of contaminants. They offer a large surface area which can be further fine-tuned by controlling the size, shape, and size distribution. To take their exclusive surface-dependent properties one notch up, surfaces of nanomaterials are decorated with various functional groups and/or biological and chemical receptors, which enhances the affinity as well as stability of resulting nanosensors. Nanosensors exhibit outstanding biocompatibility, high sensitivity, improved selectivity, low detection limit, faster response time, thus boosting its quality and efficiency (Chandra et al. 2012; Choudhary et al. 2016; Deka et al. 2018; Mahato et al. 2018; Verma et al. 2019). A tremendous amount of research work has been focused on the synthesis and application of nanomaterials in food quality assessment. Some common nanomaterials used in sensing applications are: carbon-based nanomaterials (Gupta et al. 2019; Nehra et al. 2019; Pan et al. 2019; Yang et al. 2018), like graphene (Liu et al. 2017a, b; Parate et al. 2020; Vanegas et al. 2018; Wang et al. 2018), graphene oxide (GO) and reduced graphene oxide (rGO) (Karthik et al. 2018; Zhou et al. 2018), MXenes (Khan and Andreescu 2020; Szuplewska et al. 2020; Wu et al. 2018a, b; Xie et al. 2019; Zhou et al. 2017; Zhu et al. 2020), graphitic carbon nitride (g-C_3N_4) (Hua et al. 2018; Ramalingam et al. 2019; Tabrizi et al. 2017; Zhou et al. 2016), plasmonic metallic nanoparticles (Khateb et al. 2020; Loiseau et al. 2019a; Oh et al. 2017; Shams et al. 2019; Song et al. 2017), quantum dots (QDs) (Duan et al. 2019; Hu et al. 2017, 2021; Na et al. 2019; Nsibande and Forbes 2016; Yüce et al. 2018), and upconversion nanoparticles (UCNPs) (Annavaram et al. 2019; Li et al. 2020a; Na et al. 2019; Yin et al. 2019). These materials are effectively incorporated as active constituents (detector/transducer) to build small-sized sensing devices capable of signal enhancement and detection of different toxins and microbes in the smallest of food samples.

Food contaminants are detected using a range of methodologies, such as surface-enhanced Raman scattering (SERS), electrochemical sensing, fluorescence methods, colorimetric methods, molecular imprinting technology (MIT), and the latest chromatographic approach (Liu et al. 2018a, b). Each class of nanomaterials offers unique advantages when used with the appropriate inspection technique. For instance, precious metal-based plasmonic nanoparticles (Ag, Au, Pt, and Pd) provide excellent optical properties, commendable extinction cross-section, and surface plasmon resonance (SPR) that can be modified with the chemical entities that surround them. This makes them ideal signal converters as well as candidates for SERS substrates. The grouping of metallic nanoparticles and Raman spectroscopy makes SERS a rapid, easy-to-use, and high sensitivity detection technique (Liu et al. 2018a, b).

On the other hand, low-cost and highly precise biosensors with carbon-based nanostructures are designed for real-time detection of harmful elements in edibles and beverages. Most of these sensors work on electrochemical, optical, and piezoelectric principles. Moreover, luminescent nanoparticle-based sensors have gained immense popularity due to the diversity of the sensing mechanisms that can be combined with them, such as direct fluorescence, energy transfer based on fluorescence resonance, bioluminescence resonance or chemiluminescence, photon induction, and electrochemiluminescence (Yüce and Kurt 2017).

This chapter focuses on the application of a range of carbonaceous, metallic, and fluorescent nanoparticles in the field of food quality assessment. The most recent studies, published within the past 4 years, have are included in this chapter, with a particular emphasis on the role of nanomaterials in the development of inexpensive and sensitive sensors for food quality assessment and enhancement of their overall performance.

1.2 Carbon-based Nanoparticles and Their Applications in Food Safety Monitoring

Carbon-based nanomaterials are extensively used in diverse fields due to their attractive structural properties, surface chemistry, and electrical and optical characteristics. Based on the size, layers, and shapes, these materials exhibit metal-like conductivity and semiconducting behavior. Different synthetic routes and surface modification procedures can be followed to produce a range of carbonaceous nanomaterials (Yüce and Kurt 2017). Thus, the significance of carbon-based nanomaterials in sensing platforms cannot be overemphasized. The contribution of these materials in biosensing, bioimaging, and food safety and quality monitoring applications is quite noteworthy. Carbon-nanomaterials are used as platforms in biosensors working on electrochemical, optical, and piezoelectric principles. Mainly, in electrochemical sensors, these nanomaterials serve as adsorbents or transducers due to their geometrical benefits, rapid electron transfer capabilities, wide potential range, large surface area, inert nature, and excellent electrocatalytic abilities. Moreover, they provide support and surface for biological affinity agents, such as enzymes, proteins, antibodies, DNA, which enhances their sensitivity toward organic and inorganic targets (Nehra et al. 2019). Herein, the applications of three promising classes of carbonaceous nanomaterials in food quality assessment were detailed. A summary of the reviewed publications is presented in Table 1.1.

1.2.1 MXene Nanoparticles

MXene is a significant group of nanomaterials comprising two-dimensional transition metal carbides, nitrides, and carbonitrides. Since their discovery by Gogotsi and co-workers (Naguib et al. 2011) in 2011, MXenes have garnered considerable attention due to their 2D structure that can be easily modified, a large surface area with fantastic potential for functionalization, broad interlayer spacing, and optimum blend of physicochemical properties, such as high electrical conductivity, hydrophilic nature, biocompatibility, and chemical and thermal stability (Khan and Andreescu 2020). They are labeled as MXenes due to their general formula $M_{n+1}X_n$ ($n = 1, 2, 3$), where M represents an initial transition metal (e.g., Sc, Ti, V, Cr, Zr, Nb, Mo, Hf, Ta) and X represents carbon, nitrogen, or both, and graphene-like structure. The wet chemical synthesis of MXenes involves selective etching of a parent 3D MAX phase, in which the sandwiched layer A (an element belonging to group 13 or 14) is removed, and M_2X, M_3X_2, or M_4X_3 structures are obtained (Alhabeb et al. 2017). While hydrofluoric acid (HF) is a popular etchant to produce $Ti_3C_2T_x$ MXenes, other more sophisticated synthetic routes have been developed

Table 1.1 Carbon-based nanomaterials in food safety monitoring

Nanoparticle	Affinity agent	Target/s	Sensing principle	Limit of Detection	Real sample measurements	Ref
MXene (Graphite/ TiC/Ti_3C_2/ Chitosan)	Horse Radish Peroxidase (HRP) enzyme	H_2O_2	Electrochemical	0.74 μmol/L	Yes, in milk and dried scallop	Bao-Kai et al. (2020)
MXene/Au-Pd	Acetylcholinesterase (AChE)	Paraoxon	Electrochemical	1.75 ng/L	Yes, in pear and cucumber	Zhao et al. (2020)
MXene-$Ti_3C_2T_x$	Acetylcholinesterase (AChE)	Malathion	Electrochemical	0.3×10^{-14} M	Yes, in tap water	Zhou et al. (2017)
MXene-Ti_3C_2	Tyrosinase enzyme	Phenol	Electrochemical	12 nmol/L	Yes, in tap water spiked with phenol	Wu et al. (2018a, b)
MXene/AuNR	–	R6G, CV, MG, thiram, and diquat	SERS	1×10^{-12} (for R6G and CV), 1×10^{-10} M (for MG), 1×10^{-10} M (for thiram), and 1×10^{-8} M (For diquat)	No	Xie et al. (2019)
MXene-Ti_2C /Au–Ag NSs	–	Carbendazim (CBZ)	Electrochemical and SERS	0.002 μM (electrochemical) and 0.01 μM (SERS)	Yes, in rice and tea	Zhu et al. (2020)
MXene/Au@Pt nanoflower	5′-Nucleotidase-xanthine oxidase and bovine serum albumin (BSA)	Inosine monophosphate (IMP)	Electrochemical	2.73 ng/mL	Yes, in chicken, pork, beef, and lamb	Wang et al. (2020a, b)
P-g-C_3N_4/O-MWCNT	–	Cd(II), Hg(II), Pb (II), and Zn(II)	Electrochemical	8 and 60 ng/L	Yes, in cabbage, capsicum, and noodles	Ramalingam et al. (2019)

(continued)

Table 1.1 (continued)

Nanoparticle	Affinity agent	Target/s	Sensing principle	Limit of Detection	Real sample measurements	Ref
g- C_3N_4-TiO_2	Aptamer, 5′-NH_2-(CH_2)6–5′ TACTA ACGGTACAAGCTACCAG GCCGCCAACGTTGACCTA GAAGCACTGCCAGACCCGA ACGTTGACCTAGAAGC-3?	Tropomyosin (TROP)	Photoelectrochemical	0.23 ng/mL	Yes, in human serum	Tabrizi et al. (2017)
g-C_3N_4	–	Vanillin	Electrochemical	4 nM	Yes, in milk, tea, and biscuits	Fu et al. (2020)
g-C_3N_4 and C-dots/ 3DGH	Aptamer, 5′-GCA ATG GTA CGG TAC TTC CCC ATG AGT GTT GTG AAA TGT TGG GAC ACT AGG TGG CAT AGA GCC GCA AAA GTG CAC GCT ACT TTG CTA A-3′-NH_2-$(CH_2)_6$	*E. coli*	Photoelectrochemical	0.66 CFU/mL	Yes, in diluted milk	Hua et al. (2018)
C_3N_4 NTs/ Pt NPs	Molecular Imprinted Polymer (MIP)	Atrazine (ATR)	Electrochemical	1.5×10^{-13} M	Yes, in wastewater samples	Yola and Atar (2017)
Graphene	Antibody	Histamine	Electrochemical	30.7 μM	Yes, in fish broth	Parate et al. (2020)
Cu-plated Graphene	Diamine oxidase (DAO) enzyme	Biogenic Amines (BA)	Electrochemical	11.6 μM	Yes, in fresh and fermented fish paste	Vanegas et al. (2018)
NiFeSP@Graphene	–	Paraoxon Ethyl (PE)	Electrochemical	3.7 nmol/L	Yes, in tap water, tomato juice, and cucumber juice	Aghaie et al. (2019)
Graphene-AuNPs	CRD of hGal-3	Lactose	FET	200aM	No	Danielson et al. (2020)

Au-ZrO_2-Graphene Nanosheets	–	Methyl parathion (MP)	Electrochemical	1 ng/mL	Yes, in Chinese cabbage	Gao et al. (2019)
Ag@GO Nanoribbons	–	Methyl parathion (MP)	Electrochemical	0.5 nM	Yes, in fresh cabbage, green beans, strawberry, and nectarine fruit	Govindasamy et al. (2017)
GO-wrapped Fe_3O_4@Au particles	Aptamers sequence: 5′-SH-TCTAAAAATGGGCAAAGAAA CAGTGAC TCGTTGAGATACT-3′ (apt 1) and 5′-TAMRA-TCTAAAAATGGGCAAAGAAA CAGTGACT CGTTGAGATACT-3′ (apt 2).	*V. parahaemolyticus*	SERS	14 CFU/mL	Yes, in fresh salmon	Duan et al. (2017)
PrM/rGO	–	Methyl parathion (MP)	Electrochemical	1.8 nM	Yes, in water samples (tap and lake) and vegetables/fruit	Karthik et al. (2018)
Cu_2O/rGO	–	Sunset yellow (azo food colorant)	Electrochemical	6.0×10^{-9} mol/L	Yes, in carbonated drinks, orange juice, and candies	He et al. (2018)

over recent years. The etching is often followed by ultrasonic delamination to acquire a stack of MXene sheets that are single layered or only a few layers thick. As the last step, intercalation via huge cations is performed to increase the available surface area and improve efficiency (Szuplewska et al. 2020). Unlike other 2D nanomaterials, the intercalation capacity of MXenes is excellent.

While MXenes are attractive materials for catalytic and photocatalytic processes, energy storage, and biotechnological applications, they are particularly useful for biosensing and food quality assessment. The unsaturated, hydrophilic, and conductive surface of MXene-based platforms can be readily decorated with a range of biomolecules (proteins/enzymes) (Bao-Kai et al. 2020; Zhou et al. 2017) and nanoparticles (Zhao et al. 2020; Zhu et al. 2020) to predominantly enhance the selectivity, sensitivity, and stability of sensors and/or one-time tests using them. A recent study explored the effect of tailoring the surface of pristine Ti3C2-MXene with noble-metal NPs and ceramic oxide on its ecotoxicity, phytotoxicity, and antimicrobial characteristics. It was reported that surface modification could alter the toxic properties of pure MXene, making some tailored materials showing undesirable toxicity; however, the antimicrobial response against both Gram-positive and Gram-negative bacteria was positively affected (Rozmysłowska-Wojciechowska et al. 2019). Many MXene biosensors with low detection limits and fast response times have been proposed to detect toxins, microbes, pesticides, and pollutants found in food and drinks (Bao-Kai et al. 2020). To detect phenol in drinking water, Wu et al. (2018a, b) produced an ultrasensitive biosensor featuring an MXene-Ti_3C_2 platform. MXene nanosheets were obtained by etching its parent MAX, i.e., Ti_3AlC_2 in HF, which were then casted on glassy carbon electrode (GCE), and tyrosinase enzyme was immobilized on its surface. Tyr-MXene-Chi/GC biosensor exhibited outstanding bioanalytical tendency and detected phenol in real water samples without any mediator. It was also found to be highly sensitive with low response time and detection limits. In another study, Zhou et al. (2017) similarly fabricated MXene-Ti_3C_2 nanofilms and cast them onto GCE with chitosan acetylcholinesterase (AChE) enzyme for the detection of a toxic organophosphorus pesticide-malathion. Spike recovery tests in tap water proved the reliability of the proposed amperometric sensor with a low LOD and broad linear range. Another enzymatic biosensor was designed by Zhao et al. (2020) for the detection of highly toxic organophosphorus pesticides found in edible plants and agricultural produce. The disposable sensor consisted of a screen-printed electrode (SPE) covered with ultra-thin MXene nanosheets (1.5 nm thickness), which were synthesized by etching Ti_3AlC_2 in lithium fluoride (LiF) and hydrogen chloride solution (HCl) for 1 day at 35 °C, followed by sonication in water and centrifugation at 3500 rpm for 60 min each. MXene-coated SPE was immersed in a solution of $HAuCl_4$ and $PdCl_2$, which allowed Au and Pd particles of size 30–80 nm to grow on its surface within 5 s because of self-reduction. Finally, 10 μL of AChE enzyme was immobilized on the electrode surface, and the SPE/MXene/Au-Pd/GA/AChE electrode was oven-dried for 30 min. Electrochemical impedance spectroscopy (EIS) proved that the produced MXene-based nanocomposite had sufficient conductivity, increased electrocatalytic activity, and excellent transfer of electrons. Through amperometric analysis and

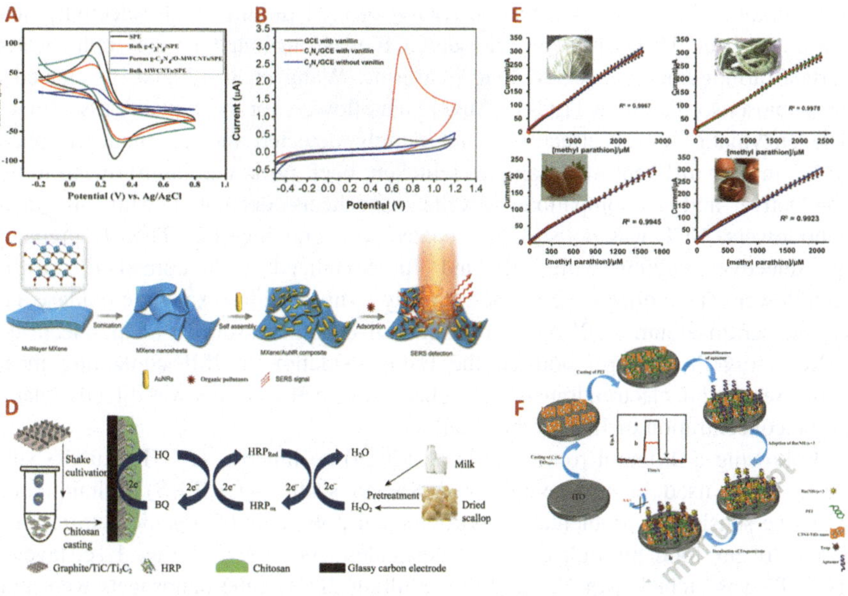

Fig. 1.1 (**a**) Electrochemical behavior of different modified electrodes in 5 mM $[Fe(CN)_6]^{3-/4-}$ in 0.1 M KCl as supporting electrolyte at a scan rate of 50 mVs^{-1} [16]; (**b**) Cyclic voltammograms of bare GCE and C_3N_4/GCE toward 10 μM vanillin in 0.1 M phosphate-buffered saline (PBS) (pH 7) [15]; (**c**) Schematic illustration of the fabrication of the MXene/AuNR composite and application as a substrate to SERS detection [11]; (**d**) Schematic illustration for the fabrication of HRP@MXene (Graphite/TiC/Ti_3C_2)/chitosan/GCE and H_2O_2 sensing principle of HRP@MXene/chitosan/GCE [6]; (**e**) Calibration plots for real sample analysis; [methyl parathion] (μM) vs. current (μA). (**a**) Cabbage, (**b**) green beans, (**c**) strawberry, and (**d**) nectarine fruit; (**f**) Schematic illustration of proposed PEC aptasensor fabrication and the sensing mechanisms employed by [17]

testing on real pears and cucumbers spiked with paraoxon, good stability and selectivity, and reliable performance of the sensor were confirmed.

Bao-Kai and co-workers (2020) synthesized an MXene-based electrochemical sensor to detect hydrogen peroxide (H_2O_2) in food (Fig. 1.1d). Firstly, 2D MXene nanosheets (graphite/TiC/Ti_3AlC_2) were prepared by mixing graphite, Ti, and Al powders with NaCl and KCl in a molar ratio 4 : 4: 1: 10 : 10. The mixture was heated and kept at 800 °C for 5 h and then at 1100 °C for 3 h. Subsequently, the mixture was cooled, washed, and etched with HF to obtain MXene nanosheets with a large specific surface area, high electronic conductivity, and desirable dispersion in an aqueous medium. To further enhance the electromagnetic absorption, enzyme entrapment, and electron transferability of the MXene-based platform, these sheets were arranged perpendicular to the graphite plane, forming a unique vertical junction structure. Next, GCE was uniformly coated with the fabricated MXene and Chitosan film. Finally, Horse Radish Peroxidase (HRP) enzyme was entrapped on MXene/chitosan/GCE, and no conformational alteration was observed in its secondary structure as per the FT-IR analysis. The biosensor successfully sensed the H_2O_2 traces in milk and contaminated dried scallops. The wide linear range from 5 to

1650 μmol/L, low limit of detection (LOD) of 0.74 μmol/L, high selectivity, and stability of the HRP@MXene/chitosan/GCE was attributed to the sophisticated vertical arrangement of the MXene platforms. Wang et al. (2020a, b) created a novel biosensor based on Ti_3C_2TXAu@Pt nanoflowers for meat quality assessment. The enzymatic sensor effectively and rapidly detected Inosine monophosphate (IMP) in four different meat samples (chicken, beef, pork, and lamb) owing to the conductive and biocompatible MXene nanosheets decorated with bimetallic nanoflowers. GCE was polished and coated with graphene-like $Ti_3C_2T_x$ prepared by selective etching with LiF and functionalized with core-shell Au@Pt nanoflowers. Two different enzymes, namely 5′-nucleotidase-xanthine oxidase and bovine serum albumin (BSA), were immobilized on the surface of the electrode. MXene/Au@Pt platform boosted the H_2O_2 oxidation in IMP-containing meat, which resulted in electron transfer. The change in electric current was directly related to the concentration of IMP in the meat.

Following a different route, Xie et al. (2019) designed a $Ti_3C_2T_x$-AuNRs substrate to be used in surface-enhanced Raman scattering (SERS) technique for effective sensing of problematic environmental pollutants of organic nature transferred to our food through dyes and pesticides. As shown in Fig. 1.1c, layered $Ti_3C_2T_x$ was etched with HF, and the resulting 2D MXene nanosheets were uniformly coated with Au nanorods prepared with a seedless procedure. The composite assembled on its own due to the electrostatic forces and carried abundant hot spots for SERS. The substrate detected typical organic dyes, like rhodamine 6G (R6G), crystal violet (CV), and malachite green (MG) with excellent sensitivity, reliability, and reproducibility. It even proved its efficiency as an ultrasensitive food safety sensor for more sophisticated pollutants like thiram and diquat. Furthermore, Zhu and co-researchers (2020) constructed an advanced MXene-based smart sensor with the help of machine learning techniques and proposed it to monitor harmful carbendazim (CBZ) traces in liquid and solid food items. The fabricated MXene/Au–Ag nano shuttles (NSs) can be used as electrochemical sensors as well as SERS substrate. 2D Ti_2C MXene powder was fabricated by HF etching of Ti_2AlC powder at 25 °C for 1 day, followed by centrifugation, washing, and drying. The nanohybrid was designed through ultrasonic dispersion of pristine MXene in Au–Ag NSs solution. This bifunctional platform was drop-coated on GCE for electrochemical analysis and on glass matrix for SERS analysis. Characterization revealed that the synthesized nano shuttles had a rough surface with a large area, high conductivity, commendable electrochemical behavior, and outstanding Raman enhancement. The nanosensor was highly stable and offered a broad linear range and low LOD for both electrochemical and Raman sensing. The successful tests on real samples of rice and tea showed the efficacy of these MXene-based nano shuttles in food safety and monitoring.

1.2.2 Carbon Nitride Nanoparticles

Among all the allotropes of naturally existing carbon nitride (CN), graphitic carbon nitride (g-C_3N_4) is lauded for its stability at room temperature, planar geometry, large surface area, unique physical and electrocatalytic properties, and metal-free polymeric nature (Chen and Song 2017). It features a layered, graphene-like 2D structure with sp^2 bonded C and N atoms arranged as a tri-s-triazine ring linked with tertiary amines. This unique structure gives it high thermal stability, chemical resistance, and a bandgap like semiconductors. It comprises nitrogen atoms and defects that act as active spots and boosts its electric conductivity (Magesa et al. 2019). Traditionally, g-C_3N_4 is fabricated in bulk via polymerization of typical precursors (like cyanamide, dicyandiamide, and melamine) by physical vapor deposition (PVD), thermal nitridation, solvothermal, chemical vapor deposition (CVD), and solid-state reaction. To obtain ultrathin single-lamellar g-C_3N_4 films, bulk material is exfoliated thermally, chemically, or ultrasonically. What makes it even more appealing for a range of applications is the resulting large surface ready to be embellished with metallic, non-metallic, and carbonaceous nanomaterials (Chen and Song 2017).

Several sensors have been designed by using g-C_3N_4 in its pristine or functionalized form for the detection of food colorants, antimicrobial remnants, metal ions, and toxic pesticides. In research by Fu et al. (2020), layered C_3N_4 nanosheets were fabricated by thermal oxidation of melamine precursor and coated as an absorbent unto the GCE. This sensor offered fast and simple monitoring of vanillin in edibles. A comparison of GCE with and without the C_3N_4 layer is shown in Fig. 1.1b in terms of vanillin's electrochemical behavior. The C_3N_4/GCE showed a decline in background current when no vanillin was present, and the oxidation signal was enhanced at a lower overpotential. This further confirms that vanillin receives an electrocatalytic response from C_3N_4. Hence, it can be deduced that C_3N_4 reportedly increased the active-specific area and electrochemical activity of the electrode. The resultant vanillin sensor had an LOD of 4 nM and a wide linear range of 20 nM–10 μM and 15 μM–200 μM, when tested on real milk, tea, and biscuits. Ramalingam and co-workers (2019) designed a highly selective sensing matrix created from porous g-C_3N_4 nanosheets and oxidized multiwalled carbon nanotubes, named p- g-C_3N_4-NSs/O-MWCNTs. Fabrication of the platform involved a single-step oxidation process in which bulk carbon nitride was mixed with bulk MWCNTs in potassium dichromate and sulfuric acid. As surfaces of both materials underwent acid functionalization, 1D O-MWCNTs was firmly embedded into the pores of 2D g-C_3N_4, and a 3D nanocomposite platform was obtained. After ATR-IR spectroscopy, XRD, and TEM analysis, the platform was drop-casted on SPE. The cyclic voltammograms of different electrodes shown in Fig. 1.1a reveal that P-g-C_3N_4/O-MWCNTs/SPE has a greater ΔEp value, which can be attributed to the carboxylic acid and hydroxyl groups (carrying a negative charge) attached to the surface. Owing to the uninterrupted electron transfer and excellent electrocatalytic activity of the tailored SPE, the biosensor detected four heavy metal ions Zn(II) Pb (II), Hg(II), and Cd(II) in spiked vegetables and noodles simultaneously by using anodic stripping voltammetry technique. The biosensor showed great accuracy, sensitivity, and selectivity.

In a pioneering study, Tabrizi and group (2017) built a photoelectrochemical aptasensor driven with visible light and made the most of the high photoactivity of g-C_3N_4/TiO_2, as shown in Fig. 1.1f. The produced sensor was used to detect tropomyosin (an allergen found in seafood) and showed a promising range up to 400 ng/mL and a low LOD of 0.23 ng/mL. They synthesized g-C_3N_4 by heating melamine at a rate of 3 °C/min up to 520 °C and kept the product at the same temperature under argon flow for 4 h. Consequently, it was dissolved with TiO_2 and deposited over the ITO electrode, followed by a coating of polyethyleneimine (PEI). Finally, the amine terminal TROP probe was attached to the electrode surface, and the cleaning procedure was followed. This PEC aptasensor was tested on diluted human serum samples and demonstrated a decreasing photocurrent intensity with a corresponding increase in the concentration of tropomyosin. Hua et al. (2018) followed the synthetic route detailed in (Tabrizi et al. 2017) to prepare g-C_3N_4 and modified one side of the ITO electrode with it, whereas the adjacent side was functionalized with carbon quantum dots embedded into 3D graphene hydrogel (C-dots/3DGH) to build an ultrasensitive ratiometric PEC biosensor. The aptamer was coated unto the C-dots/3DGH, while g-C_3N_4 acted as a reference providing stable anodic current. The difference in bias voltage helped distinguish the current from the anode and cathode, and the ratio between the two was used for *E. coli* concentration monitoring. Milk samples containing different amounts of *E. coli* confirmed recovery and stability. Yola and Atar (2017) adopted an emerging technique called molecular imprinting to produce a unique electrochemical sensor that featured a C_3N_4 NTs/Pt NPs nanocomposite. The molecularly imprinted polymer (MIP) contained active spots for the determination of atrazine (ATR)—a pesticide released in water sources from where it travels into soil and agricultural products. Nanotubes of C_3N_4 embellished with Pt NPs were produced via a single-step hydrothermal method and deposited onto a clean GCE. ATR was then imprinted onto the electrode via cyclic voltammetry (CV). The prepared electrode was then used for ATR detection in contaminated water fractions. According to the sample tests, the voltammetric sensor exhibited good recovery, low LOD, high stability and selectivity, and low response time.

1.2.3 Graphene and Derivatives

The discovery of graphene in 2004 (Novoselov et al. 2004) marked a new era in nanotechnology and opened doors for a myriad of applications, one of which is food quality monitoring and sensing. Flaunting a honeycomb-like rigid framework made up of a single layer of sp^2-hybridized carbons, graphene has served as a stepping stone for the fabrication of 0D (nanoparticles and fullerenes), 1D (single-walled nanotubes, nanorods, and nanowires), and 3D (graphite, nanocomposites) structures. Its exceptional mechanical strength, attractive chemical and thermal properties, unparalleled electron transfer capability, and generous surface area is bound to spike every scientist's interest (Magesa et al. 2019). Some conventional methods for graphene synthesis are chemical vapor deposition (CVD), epitaxial growth, laser ablation approach, and silicon carbide decomposition. However, a recent study by Parate et al. (2020) proposed the aerosol jet printing (AJP) technique as a low-cost method to produce a graphene-based histamine sensor that could find application in identifying contaminated seafood and prevent consequent allergic reactions. Using graphene-nitrocellulose ink, an interdigitated electrode (IDE) was printed on a polyimide substrate and modified with oxygenated moieties via annealing. Histamine antibody was then covalently embedded on graphene, and the final product was used to detect histamine in real fish broth electrochemically. The sensor showed a broad detection range with a low LOD, and the detection was completed in around half an hour in liquified food samples.

Another low-cost amperometric food safety biosensor was designed by Vanegas et al. (2018) by reagents and materials procured from local stores. The proposed sensor showcased a system of electrodes prepared with a laser scribed graphene (LSG) technique. This synthetic approach transformed the sp3 carbon of polyimide into an sp^2-hybridized allotrope of graphene. The electrode possessed a multilayer porous graphene structure, plated with copper nanocubes, and functionalized with diamine oxidase (DAO). The biofunctionalized, Cu-coated LSG electrodes were designed to offer selectivity and electrochemically detect high concentrations of biogenic amines (BA), such as histamine, cadaverine, tyramine, and/or putrescine, in food to prevent food poisoning and food intolerance. The tests performed on real fish samples, before and after fermentation, confirmed that the LSG-Cu-DAO biosensor could detect histamine with an LOD of 11.6 μM, and a response time of 7.3 s.

Rouhani (2019), in his computational analysis, discussed the electronic efficiency of fluorographene and its role in the detection of ammonia and typical amines (like methylamine, dimethylamine, and trimethylamine) that can help sense contaminated seafood. Since inadequate adsorption of gas molecules limits the sensing ability of pristine graphene, fluorine-doping could be an effective method to overcome this limitation. With the help of the density functional theory (DFT) model, the adsorption of ammonia and amine molecules was tested for pristine and fluorine-doped graphene sheets. The theoretical findings confirmed that gas molecules showed better adsorption on fluorinated sheets as more precise signals were received due to high electron transfer and electrical conductivity. Hence, such a sensor could be used to assess the quality of seafood and detect fish spoilage. Aghaie and co-workers (2019) followed a sophisticated fabrication route to produce an enzyme-free sensor featuring a graphene-based bimetallic (NiFe) phosphosulfate nanocomposite to boost the voltammetric signals received from paraoxon ethyl (PE: an organic pesticide found in food items) and facilitate its detection. Since the target OP has an aromatic structure, it experiences desirable π-π stacking interactions with graphene. The NiFe phosphosulfate further ensures that PE sticks to the surface of GCE. PE was adsorbed completely into the electrode within 5 min, and a tremendous increase in maximum square wave voltammetric (SWV) signals was noted, which was ascribed to the properties of the graphene-based electrode. The sensor was also tested on contaminated water and fruit and vegetable juices and was found to be efficient. A non-enzymatic, ultrasensitive graphene field-effect transistors (G-FETs) biosensor was proposed by Danielson and his group (2020) for lactose sensing. They produced a single-layer graphene sheet enriched with uniformly distributed Au NPs and immobilized carbohydrate recognition domain (CRD) of the human galectin-3 (hGal-3) protein on its surface. Minimum voltage shifted in the negative direction during liquid gate measurement, which is characteristic for lactose, whereas a positive shift was observed for other carbohydrates. This indicates the suitability of this biosensor in detecting lactose in food.

Gao et al. (2019) used a GCE covered with graphene nanosheets decorated with gold and zirconia particles to detect methyl parathion (MP), a harmful organophosphorus pesticide. The prepared electrode provided excellent electrocatalytic activity and sensed MP with great accuracy. This was backed by the results of square wave voltammetry that showed a wide linear range of 1–100 ng/mL and 100–2400 ng/mL. For the same target pesticide, Govindasamy et al. (2017) used graphene oxide (GO) nanoribbons considering its various structural features, like tunable large surface areas, solubility in water, and high adsorption capacities for biological materials like enzymes, proteins, peptides, and nucleic acids (Chen and Nugen 2019). GO NRs were embellished with Ag NPs and loaded on an SPE, which exhibited commendable electrocatalytic activity as MP underwent reduction. Good linearity with MP concentration was observed in the calibration plots for real food items, as shown in Fig. 1.1e. In another food sensing-related assay, the authors explored the application of graphene oxide wrapped around Fe_3O_4@Au particles to construct an SERS-based sensor. The aptasensor's surface carried two different aptamers separated by a magnet to serve as capture probe and SERS sensing probe, respectively. This sensor was used to detect a notorious food pathogen, *Vibrio parahaemolyticus*, by measuring the SERS intensity of TAMRA aptamer. It showed a wide linear range, i.e., between 1.4×10^2 to 1.4×10^6 CFU/mL and a low LOD of 14 CFU/mL. Tests with contaminated fish yielded impressive recoveries in the range of 98.5–105% (Duan et al. 2017).

The reduced form of graphene oxide (rGO) is also suitable for electrode fabrication (Yang et al. 2018). Karthik and co-workers (2018) designed an electrochemical sensor for MP monitoring but used reduced graphene oxide (rGO) and functionalized it with 3D praseodymium molybdate (PrM) to create a novel nanocomposite. The voltammetric findings revealed lower and higher peaks for potential and cathodic current, respectively. Also, good linearity and sensitivity deemed this rGO-based composite fit for MP detection. He et al. (2018) reduced graphene oxide electrochemically and combined it with Cu_2O to produce a nanocomposite, tagged as Cu_2O-ErGO. The fabricated nanocomposite was then deposited on GCE to detect a popular food colorant—sunset yellow. Owing to the use of Cu_2O-ErGO, the dye produced 25 times greater anodic peak. The LOD of the sensor was as low as 6.0×10^{-9} mol/L, which is even better than electrodes prepared with precious metals. Real samples were used to establish the performance of the sensor.

1.3 Metallic Nanoparticles and Their Applications in Food Safety Monitoring

Over the years, noble-metallic nanoparticles have drawn considerable attention due to their fascinating physical and chemical properties. The metallic nanoparticles can be easily synthesized and modified with various chemical and biological functional moieties such as antibodies, peptides, DNA, and drugs. Therefore, they have myriads of applications in biotechnology, drug delivery, biosensing, and imaging. Furthermore, the unique optical properties of plasmonic nanoparticles lead to the development of various imaging modalities such as MRI, PET, CT, and SERS. The optical properties of these noble-metal nanoparticles have attractive aspects under investigation. The metallic nanoparticles exhibit plasmonic properties, i.e., surface plasmon resonance (SPR) and localized surface plasmon resonance (LSPR). The SPR phenomenon occurs when a beam of light is incident on the metal nanoparticles. The free surface electrons collectively oscillate in a phase with the incident light. When the frequency of the collective oscillations is approximately the same as the incident light, or the resonance condition occurs, such a phenomenon is known as LSPR. In brief, plasmonic properties deal with the interaction of light with metallic nanostructures to enable optical measurements that are related to the binding events. These help in deriving precious information about the character of the molecules (Kahraman et al. 2017; Liu et al. 2020a, b). SPR and LSPR property generally are governed by several factors such as the size, shape, geometry of the nanoparticles. Further, dielectric properties of the surrounding medium and the interparticle-coupling interactions play an essential role in the SPR and LSPR properties of nanoparticles (Yang et al. 2016).

Over the past decades, plasmonic nanoparticle-based detection sensors have gained considerable attention due to their tuned absorption wavelength, electromagnetic control, and single-molecule detection capability. Plasmonic sensors belong to a class of optical affinity biosensors. When the recognition element captures the target analytes, it provides a measurable signal (Kurt et al. 2019). The commonly used plasmonic nanoparticles are gold (AuNPs) and silver colloidal nanoparticles (AgNPs) employed to immobilization several recognition elements (antibodies, aptamers, and peptides). Due to the unique optical and physical properties, plasmonic sensors are indispensable for real-time and label-free analyte detection from various complex matrices. These can also be employed for the ultrasensitive detection of targeted substances from clinical samples and food matrices (Zhan et al. 2020). The biosensing schematic based on the plasmonic nanomaterials and various detection methods are presented in Fig. 1.2a.

1.3.1 Gold Nanoparticles (AuNPs)

Gold nanoparticles (AuNPs) are the most used plasmonic nanomaterials owing to their well-characterized LSPR phenomenon. The LSPR of AuNPs yields a high absorption coefficient and scattering properties within the visible wavelength to the

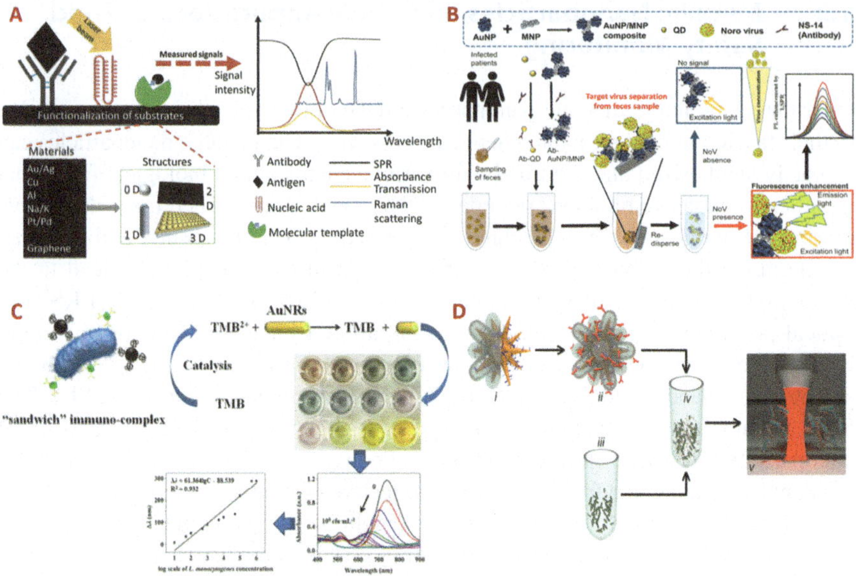

Fig. 1.2 (**a**) The schematic of the biosensing system by using plasmonic nanomaterial. (**b**) The human norovirus detection using LSPR-amplified magneto-fluoroimmunoassay. (**c**) A multi-colorimetric assay based on the plasmonic properties of AuNRs for the detection of *L. monocytogenes*. (**d**) Schematics of the SERS-based *L. monocytogenes* detection. (i) Fabrication of SERS-encoded GNSs (ii) Conjugation of the mAb C11E9 (iii) Sample bacteria at specific concentration (iv) incubation with SERS tag for recognition (iv) SERS detection in a microfluidic channel. Figures A, B, C, and D are adapted from the references (Liu et al. 2019, 2020a, b; Rodríguez-Lorenzo et al. 2019; Takemura et al. 2019), respectively

near-infrared range (Caucheteur et al. 2015). The spectral characteristics of LSPR produced by AuNPs are independent of size, shape, and the local dielectric environment. Therefore, after binding the target analyte, the refractive index shift could be used to monitor the molecular binding events. Several studies on the implication of LSPR of AuNPs for detecting pathogens or toxins associated with food safety have been demonstrated.

Oh et al. (2017) developed a label-free portable LSPR platform for the selective detection of *Salmonella typhimurium* from the pork meat sample (Oh et al. 2017). The *S. typhimurium* is considered as the "zero tolerance" microorganism in food samples and is the cause of acute gastroenteritis in humans. The sensor involved the self-assembly of AuNPs (20 nm) on a glass substrate of 5 cm $\times$ 0.8 cm (length $\times$ width. The aptamers were used as a recognition probe to capture the target bacteria. The sensor exhibited an LOD of around 10^4 CFU/mL in half an hour. Further, the developed sensor detected *S. typhimurium* from the spiked pork meat sample with an LOD of 1.0×10^4 CFU/mL without any pre-enrichment steps. The study also showed that the AuNPs immobilized plasmonic chip did not affect the food matrix or background contaminant of microflora. In another study, Yaghubi et al. 2020, developed a high-resolution optical method by employing the LSPR property of spherical AuNPs for the detection of *E. coli* O157:H7 (Yaghubi et al. 2020). The *E. coli* O157:H7 is one of the most important foodborne pathogens, which causes developing hemorrhagic colitis and hemolytic uremic syndrome. The sensor was developed by conjugating the specific anti-*E. coli* O157:H7 chicken antibody (IgY) under the specific pH of the nanoparticles. The characterization was performed by several analytical methods such as UC-VIS and DLS. The sensitivity was found to be 10 CFU/mL, i.e., the LSPR of AuNPs (λ max) redshifted from 530 nm to 543 nm in the presence of 10 bacteria. *E. coli* O157:H7 was also detected in complex food matrices within 2 h.

Apart from detecting bacteria from the food matrices, plasmonic sensors have also been adopted to detect foodborne viruses. Norovirus is a foodborne virus that causes gastroenteritis, non-bloody diarrhea, vomiting, stomach pain, and a significant public health concern worldwide. Su Heo et al. 2019, developed a peptide-guided plasmonic biosensor to detect the human norovirus as well as the capsid proteins (Heo et al. 2019). The LSPR chip was prepared by using colloidal AuNPs on a glass substrate treated with APTES. During this process, AuNPs (16–18 nm) were immobilized on the APTES-treated glass substrate ($\approx$50 particles/ 200 $\times$ 200 nm), and the peptide (recognition probe) was immobilized using cysteine-gold conjugation chemistry. The performance of the sensor was examined by LSPR signals, and the analyte-specific binding event was proportional to the absorbance of the LSPR signal. The LOD was found to be 0.1 ng/mL for noroviral capsid protein and around 10 copies/mL for the norovirus. The plasmonic sensor was able to detect the norovirus capsid protein from complex tissue culture media such as MEM and FBS. Takemura et al. (2019), developed another LSPR-amplified magneto-fluoroimmunoassay platform to detect the norovirus from the complex media (Takemura et al. 2019). They synthesized AuNPs with magnetic nanoparticles composite, and the anti-norovirus antibody was conjugated with CdSe quantum dots (QDs). This nanohybrid composite performed by the enhanced magnetic field and the high LSPR effects of AuNPs, which separates the norovirus from the matrices. The target norovirus-like particles were found in a linear range between 1 pg and 5 ng/mL with an LOD of 0.48 pg/mL in faces. The LOD was found to be 84 copies of RNA/mL from the clinical samples without any interference from the matrices. The detailed schematic of the plasmonic biosensor for norovirus was presented in Fig. 1.2b.

Raman spectroscopy is a sensitive characterization technique constructed on the inelastic scattering of photons from the targeting molecules. It provides essential information about chemical bond formation. However, the weak signal intensity is the major drawback of its application in biotechnology. The LSPR phenomenon of AuNPs has been utilized to enhance the light scattering signal and Raman spectroscopy. The metallic nanoparticles enhance the Raman scattering of molecules immobilized at their surface, constituting the surface-enhanced Raman spectroscopy (SERS). Here, it is important to mention that the SERS phenomenon is entirely different from the SPR, which occurs at the plane surface (Loiseau et al. 2019a). Several studies have recently reported the SERS-based biosensors utilizing AuNPs by detecting specific signals from the target analytes. Duan et al. (2020), developed an SERS aptasensor to detect multiple foodborne pathogens by SERS tags from complex food matrices (Duan et al. 2020). The sensor consists of polydimethylsiloxane (PDMS) coated with AuNPs acting as a substrate for Raman scattering. The Au-PDMS chip was chemically modified with target-specific aptamer oligonucleotides and Raman reporter, i.e., 4-mercaptobenzoic acid (4-MBA)/nile blue A (NBA) as a recognition probe to capture the target pathogens and as pathogen-specific SERS probe, respectively. The sensor facilitated sandwich assay formation by capturing the target pathogen and then binding with the SERS probe. The sensor was validated with two foodborne pathogens, *Vibrio parahaemolyticus* and *S. typhimurium,* which demonstrated an LOD of 18 CFU/mL and 27 CFU/mL, respectively.

The point-of-care detection of the foodborne pathogens was also carried out by using lateral flow immunochromatographic assays (LFIA), which is fast, easy to operate, and cost-effective. However, the low sensitivity of the assay in the presence of various matrices hinders its further applications. The limitations of the LFA are addressed by employing the LSPR properties of the AuNPs. Wu et al. (2019), developed an LFIA to detect *L. monocytogenes* and *S. typhimurium* membrane based on the principles of SERS (Wu 2019). The monoclonal antibodies for *L. monocytogenes* and *S. typhimurium* were used as the recognition probe and DTNB as the SERS tag. The SERS-based LFIA exhibited good linear response in the range of 10^2–10^7 CFU/mL. The LOD was found to be 75 CFU/mL for the foodborne bacteria from the milk sample.

Accumulated evidence shows the extensive use of spherical AuNPs in the plasmonic biosensor; however, recently, gold nanorods (AuNRs) have gained interest for this purpose. The LSPR of AuNRs has two absorption bands; the first one is located at 520 nm due to the transverse localized surface plasmon resonance (t-LSPR). The second one is at a higher wavelength due to the longitudinal localized surface plasmon resonance (l-LSPR) (Chen et al. 2013). The position of the LSPR can be altered by modulating the aspect ratio of AuNRs rather than their length. Hence, AuNRs have become the ideal candidates for a wide range of biomedical applications. Shams et al. 2019, developed an AuNRs-based plasmonic detection method to detect *Campylobacter jejuni* and *Campylobacter coli,* the two most critical foodborne pathogens (Shams et al. 2019). The AuNRs were modified by the specific ssDNA probes of the *cadF* gene of *Campylobacter.* The assay's sensitivity was validated by comparing the results with conventional culture methods, PCR, and RT-PCR. The sensitivity of the assay was found to be 88%, and the specificity was 100%. The LOD was found to be 10^2 copy number/mL, which is equivalent to the RT-PCR-based detection. Liu et al. 2019, developed a multi-calorimetric assay to detect *L. monocytogenes*-based on the etching of AuNRs (Liu et al. 2019). The assay is based on the TMB^{2+} etching of AuNRs, and the aptamer-modified magnetic nanoparticles were employed to concentrate on the target organism. Furthermore, to oxidize the TMB to TMB^{2+}, IgY-BSA-MnO_2 NPs were chosen as an oxidase-like nano-artificial enzyme. The longitudinal shift of LSPR of AuNRs was linearly correlated with the concentrations of

L. monocytogenes (10–10^6 CFU/mL) under the optimal condition with an LOD of 10 CFU/mL. The schematics of the detection platform is illustrated in Fig. 1.2c.

Liu et al. (2020a, b), reported a detection platform based on AuNRs and SERS tags conjugated with target-specific aptamer molecules and Raman reporters for food pathogen sensing (Li et al. 2020a, b, c, d). The aptamers were used as a recognition probe and induced the AuNRs into different geometrical shapes, which enhanced the Raman signal of the sensing platform. The AuNRs-based plasmonic sensor detected *E. coli O157:H7* and *S. typhimurium* with a concentration of 10^1–10^6 CFU/mL, with an LOD of 8 CFU/mL from spiked food samples. Loiseau et al. (2019a, b), developed a homogenous plasmonic sensor for staphylococcal enterotoxin A (SEA) sensing by the naked eye (Loiseau et al. 2019b). SEA causes food poisoning and toxic shock syndrome (TSS) produced by *S. aureus*. The principle lies in the metallic nanoparticles' LSPR properties, where the small redshift (~2–3 nm) could be observed and visually detectable after binding of the target. Two types of nanoparticle systems were synthesized, i.e., Au inside Ag nano-shells (Au@AgNPs) and Ag inside Au nano-shells (Ag@AuNPs). The nanoparticles' thickness and surface chemistry were controlled by anti-SEA antibody, and the LSPR band was tuned near 495 (Ag@AuNPs) and 520 nm (Au@AgNPs), respectively. After the target's binding, SEA, a large redshift of the LSPR band, was observed, and visual detection was enabled. The LODs were found to be 0.2 and 0.4 nM for Au@AgNPs and Ag@AuNPs, respectively. The visual color changed from orange to red, which was visible by the naked eye. Thus, the plasmonic sensor showed potential for medical diagnostics and environmental screening.

Apart from AuNPs and AuNRs, gold nanostars are (GNSs) also used as a plasmonic nanomaterial for the development of biosensor. Rodríguez-Lorenzo et al. (2019), demonstrated the use of GNSs paired to the antibody to detect *L. monocytogenes* from the food samples (Rodríguez-Lorenzo et al. 2019). The detection was achieved in a microfluidic cartridge under a flow in real-time. The assay was also able to distinguish from *L. monocytogenes* and *Listeria innocua* in 100 s. The schematic of the GNSs-based sensor was represented in Fig. 1.2d. Khateb et al. (2020), also described a label-free aptasensor for *Staphylococcus aureus* detection (Khateb et al. 2020). The plasmonic sensor exploited gold nano-disks' LSPR properties, and the aptamer was used as a recognition element. The gold nano-disks were fabricated on a glass substrate. The sensing device was used to detect *S. aureus* from pure culture and artificially contaminated milk samples with an LOD of 10^3 CFU/mL. Furthermore, the sensor required no pre-concentration steps, and the total turn-around time of detection was 30 min. From the above sections, it could be concluded that the application of gold nanoparticles, gold nanorods, and gold nanostars could be used in plasmonic biosensors for the food safety assessment. Apart from antibodies and aptamers, sugar receptors (Kaushal et al. 2019) can also be utilized for foodborne pathogen detection. The quantification of the sensor can also be achieved by different methods such as spectroscopic method, dark field microscope, and micro-spectroscopy system.

1.3.2 Silver Nanoparticles (AgNPs)

Recently, an investigation on anisotropic morphologies of metallic NPs have been carried under large scale due to the structural, optical, electronic properties and are superior to spherical NPs. Notably, the most striking feature of anisotropic lies "in the appearance of plasmon band at a longer wavelength (near-infrared region) than that of spherical NPs"(Loiseau et al. 2019a). Inspired by the AuNRs, silver nanorods (AgNRs) were synthesized by using ascorbic acid to reduce the silver nitrate ($AgNO_3$) with $NaBH_4$ and cetyltrimethylammonium bromide (CTAB). Researchers also demonstrated the synthesis of flower-like AgNPs used as SERS substrate. Another interesting morphology of Ag is silver nanoplates (AgNPLs), where the lateral dimensions are more extensive than their heights. These anisotropic silver nanoplates have vastly been used in SERS, photovoltaics, and plasmonic biosensing. Like the AuNPs, AgNPs also demonstrate unique LSPR properties and utilize it for the biosensing application. Colloidal AgNPs are yellow and display an absorption band around 380 nm in the visible range of the electromagnetic spectrum (Lee and Jun 2019). This facilitates the colorimetric detection of the target analytes by inducing the LSPR band changes.

The application of the plasmonic nature of the AgNPs was used in a surface plasmon resonance based on a fiber-optic (FOSPR) sensor for the detection of *E. coli* O157:H7 in water and juice (Zhou et al. 2018). The antimicrobial peptide, i.e., Magainin I, was used as a recognition element to capture the target pathogen. The signal was amplified by AgNPs-reduced graphene oxide nanocomposites (AgNPs-rGO) covered with a gold film. The SPR resonance wavelength presented a linear relationship with the target bacteria at a concentration from 1.0×10^3 to 5.0×10^7 CFU/mL with an LOD of 5.0×10^2 CFU/mL, under the optimized experimental conditions. The FOSPR displayed 1.5 times higher sensitivity as compared to the sensor fabricated with only AgNPs. Furthermore, the sensor was directed to the real-time sensing of *E. coli* O157:H7 in food samples such as water, fruit, and vegetable juice. Zhao et al. (2016), developed a colorimetric method based on the LSPR of silver nano prisms (Zhao et al. 2016). In the presence of enzyme catalase, the redox balance of silver nano prism disrupted, leading to the change in the particles' size and a color shift from blue to purple, red, orange, and yellow. The color transition of Ag colloidal solutions provided a quantification of *E. coli* (10^6–10^7 CFU/mL). The ability of this method to detect *E. Coli* was also confirmed from the contaminated lettuce leaf. Hassan et al. (2021), developed a SERS-based detection method using the flower-like silver nanoparticles to detect pesticides (methomyl, acetamiprid-(A.C.) and 2,4-dichlorophenoxyacetic acid-(2,4-D) residue) residue from the foodstuff (Hassan et al. 2021). The results exhibited a linear relation between the SERS signal and the pesticide concentration. The LOD of the sensor was found to be 5.58×10^{-4}, $1.88 \times 10^{-}4$, and 4.72×10^{-3} μg/mL. Table 1.2 represents various plasmonic nanoparticles, their target analytes, and the LOD of each developed sensor. From the preceding sections, it can be concluded that the applications of plasmonic nanoparticles for biosensing are enormous. Current challenges include accurate control of the size, shape, and functionalization of

Table 1.2 Different plasmonic nanoparticles, sensing principle, and limit of detection

Nanoparticle	Affinity agent	Target	Sensing Principle	LOD (CFU/mL)	Real sample measurement	Ref.
AuNPs	Antibody	*Campylobacter jejuni*	SPR	4×10^4	NA	Masdor et al. (2017)
AuNPs	Antibody	*E. coli O157:H7*	Plasmonic lateral flow assay	100–600	Liquid food system	Ren et al. (2019)
AuNPs	Antibody	*E. coli O157:H7*	Colorimetric	50	Chicken food sample	Zheng et al. (2019)
Au nano bones	Aptamer	*E. coli O157:H7*	SERS	3	NA	Zhou et al. (2020a, b)
AuNPs with starch magnetic beads (AuNP@SMBs)	Antibody	*E. coli O157:H7*	SERS	10	NA	You et al. (2020)
AuNRs@SiO_2	Antibody	*E. coli O157:H7*	SERS	10	NA	Song et al. (2017)
Flower-shaped AuNPs	ssDNA	*Listeria monocyotogenes*	Colorimetric	48–4 ng (*hly A* gene) 100.4 (genomic DNA)	NA	Du et al. (2018)
AuNRs	Antibody	*Staphylococcus aureus*	Colorimetric detection	476	Chinese cabbage and beef samples	Pang et al. (2019)
AuNRs	Peptide	*E. coli* *S. aureus*	LSPR	46 (*E. coli*) and 89 (*S. aureus*)	NA	Chen et al. (2018)
AuNR@Pt	Antibody	*Campylobacter jejuni*	SERS	50	Milk sample	He et al. (2019)
AuNRs	Antibody	Aflatoxin B1	Plasmonic ELISA	22.3 pg/mL	NA	Xiong et al. (2018)
AuNPs	Single-stranded oligonucleotide	*Salmonella spp.*	Colorimetric	10	Blue berried and chicken meat	Quintela et al. (2019)

(continued)

Table 1.2 (continued)

Nanoparticle	Affinity agent	Target	Sensing Principle	LOD (CFU/mL)	Real sample measurement	Ref.
AuNRs	Antibody	*E. coli*	SERS	10	NA	Bozkurt et al. (2018)
Plasmonic nanorod array	Antibody	*S. aureus*	SERS	17.8	NA	Zhang et al. (2018)
AgNPs	Antibody	*E. coli* O157:H7, *Staphylococcus aureus, and Salmonella*	SERS	NA	NA	Wei et al. (2018)
Au nanoflower	Antibody	*E. coli* O157:H7	Scattering	2.7	Pasteurized milk	Zhan et al. (2019)
Au nanodimers	Aptamer	*S. typhimurium*	Raman Spectroscopy	35	NA	Xu et al. (2018)
Spiny AuNPs	Aptamer	*S. typhimurium*	SERS	4	Pork sample	Ma et al. (2018)
Fe_3O_4@Au	Antibody	*Vibrio parahaemolyticus*	SERS	14	Salmon sample	Duan et al. (2017)
AuNPs	Antibody	*Salmonella*	SPR	7.6×10^6	Chicken meat	Wang and Park (2020)

NPs. However, the large-scale and low-cost synthesis with uniform size and shape will promote plasmonic nanoparticles for its wider application.

1.4 Fluorescent Nanoparticles and Their Applications in Food Safety Monitoring

The working principle of the methods that depends on fluorescent systems is the emission of fluorescent nanomaterial signals after exposure to radiation with a predefined level of energy, called excitation wavelength. In other words, absorption and emission of fluorescence radiation are considered critical factors in these fluorescence-based techniques. Various fluorescent nanomaterials, including quantum dots, upconverting nanoparticles, graphene, and carbon quantum dots, have been utilized to detect toxins, allergens, and other food-related pathogens. This section describes such nanoparticles (NPs), their general structural and optical features, and their applications in food safety assessment (Pehlivan et al. 2019).

1.4.1 Quantum dots (QDs)

As a semiconductor crystalline nanostructure, QDs consist of components that belong to groups II–VI or III–V of the periodic table (Matea et al. 2017). Their color is influenced by the particle size, which is smaller than 10 nm, benefiting from quantum confinement. Considering this phenomenon, along with other tunable features, they have been more frequently used in sensing assays in comparison with other NPs (Nsibande and Forbes 2016). Structurally, these inorganic fluorophores are made up of a core element and covering shell. Regarding the core part, it is frequently composed of heavy metals, such as cadmium telluride (CdTe) as well as cadmium selenide (CdSe). Due to the higher quantum yield (the ratio of emitted photons divided by photons absorbed) of CdSe, it has been more commonly used in QDs synthesis. In stark contrast, however, the toxicity of this element and its damaging impact on nature cannot be neglected. Hopefully, scientists found that passivating the core with an inorganic shell such as zinc sulfide (ZnS) has a significant effect on increasing quantum efficiency and decreasing toxicity by blocking cadmium leakage (Bonilla et al. 2016). Besides being bandgap-tunable, and their light emission can be easily adjusted along the part of the electromagnetic spectrum ranging from 400 to 4000 nm by tuning the particle size. Reducing the particle size leads to the emission of energy at higher levels, followed by an increase in the energy gap (Wagner et al. 2019). It should be noted that the chemical composition of the core part has an impact on the emission wavelength of synthesized quantum dots. Following this, they benefit from a broader spectrum in excitation and a narrower spectrum of emission while being compared with other conventional fluorophores. The narrow emission spectrum paves the way for multiple sensing assays. Moreover, they can quench the light more effectively at a specific wavelength, which makes these NPs effective probe even under a low level of light exposure. Furthermore, these fluorophore nanostructures have high robustness against degradation from the chemical aspect. QDs demonstrate a high spectral shift, namely Stokes shift, which is of great importance. This feature paves the way for developing fluorescence-based sensing methods using signals that are not high (Reshma and Mohanan 2019). These outstanding features of QDs make these nanostructures the best candidates for biological applications such as in vitro and in vivo imaging (Chen et al. 2008), drug delivery (Al-Nahain et al. 2013), gene therapy (Mansoori et al. 2014), and food science, including sensing of foodborne pathogens and toxins (Bonilla et al. 2016).

As an example, zoonoses, which can spread widely among human being all over the world, have become one of the main health issues over the past few years. These pathogens lead to agricultural losses and endanger the existence of billions of humans (McElwain and Thumbi 2017). Some of these zooneos with high infection risk are *Escherichia coli* (*E. coli*), *Listeria monocytogenes* (*L. monocytogenes*), and *Brucella melitensis* (*B. melitensis*) with numerous diseases. As a result, the fabrication of biosensors for the detection of these pathogens is important in food safety. Liu et al. (2020a, b, c, d) provided peptide modified magnetic beads (MBs) by binding biotin-modified peptide to streptavidin MBs through avidin-biotin interaction, a labeling technique, and enriched the pathogens in half an hour. Following this, polyclonal antibodies were coupled with various QDs to provide detection probes. Combining peptide-functionalized MBs with multicolor QDs, fluorescence-based multiple detections of bacteria was developed successfully.

Another primary concern in food safety and ecological contamination is the remaining pesticides in food after being disposed to the water and soil. Thus, developing a sensitive and consistent technique for the detection of these residues is of great significance. For instance, adding diazinon, a liquid pesticide in agriculture, to the soil is frequently detected in the food crops, which is a severe concern for human health (Sullivan and Goh 2000). Arvand et al. (2019) provided a fluorescence resonance energy transfer (FRET)-based sensor with high selectivity for sensing this pesticide, utilizing QDs and graphene oxide (GO) as donor and quencher, respectively. They modified L-cysteine-functionalized QDs with aptamer with a high affinity for diazinon. The fluorescence emission of this conjugation was quenched by adding GO. After adding target to the system, the growth in the donor's fluorescence intensity was noticed due to the separation of aptamer from acceptor and binding to diazinon pesticide, which is a target in this sensing system (Fig. 1.3a). In this developed aptasensor, the limit of detection (LOD) achieved by their group was 0.13 nM.

Acetamiprid is an insecticide widely utilized as an alternative to other conventional ones (Jin et al. 2016). Although this compound is organic, several health risks can be provided by this insecticide for humans, causing severe contamination in the environment. Therefore, the accurate tracking of this component is vital. For this aim, an aptasensor working based on fluorescence was designed by Guo et al. (2016). In this assay, QDs and the inner filter effect of gold nanoparticles (AuNPs) fluorescent probes and quenchers. It should be noted that they used aptamers with binding affinity for acetamiprid, which can attach to the AuNPs with a negative

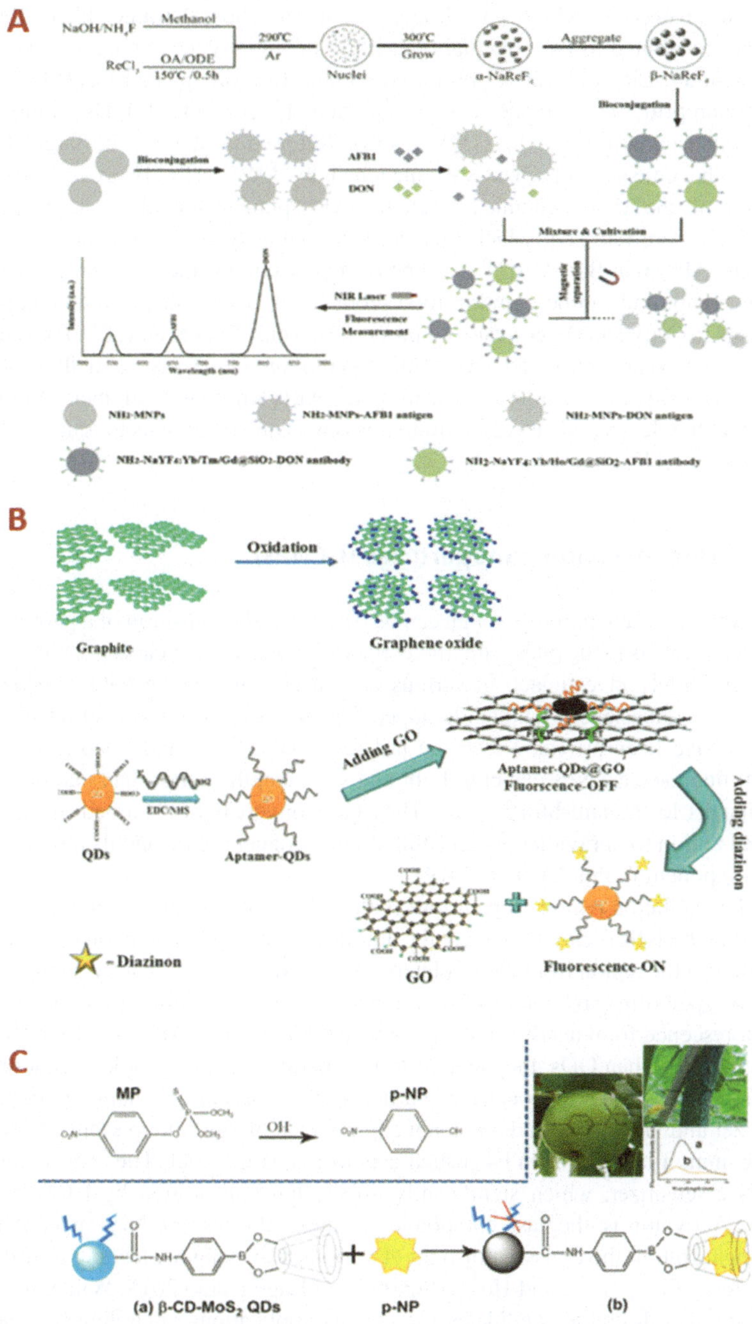

Fig. 1.3 (**a**) Schematic demonstration of the developing aptamer-based sensing platform for diazinon detection (Chen et al. 2016a, b), (**b**) Graphical illustration of UCNPs-based immunosensor designed for the detection of mycotoxins (Arvand and Mirroshandel 2019), (**c**) Graphical representation of MP hydrolysis along with fluorescent-based detection of MP using β-CD-MoS2 QDs conjugation (Yi et al. 2021)

charge to protect the NPs from salt agglomeration. Thus, the inner filter effect of AuNPs contributed to the effectual quenching of fluorescence emitted by QDs. In another study, Lin et al. (2016) proposed another turn-on aptamer-based biosensor for acetamiprid. They made use of aptamer functionalized QDs along with multiwalled carbon nanotubes (MWCNTs). Before introducing the target to the system, the emission of QDs was quenched by MWCNTs. However, after the addition of the target chemical, acetamiprid, aptamer bound to it. Therefore, MWCNTs were released, which increased the intensity of fluorescence related to the QDs. Aflatoxin B1 (AFB1) is a type of mycotoxin synthesized by *Aspergillus*. As this compound can be found in food products, it poses a threat to human health due to its toxicity and cancer-causing properties. Thus, Guo et al. (2019) provided a fluorescence-based sensor to detect this mycotoxin. In this assay, a fluorescence probe was formed by coating QDs with molecularly imprinted polymers. Noteworthy is that the developed signal-on immunoassay is an effective technique for AFB1 detection in real samples.

1.4.2 Upconversion nanoparticles (UCNPs)

Upconversion is a phenomenon that can be defined as the transition of a photon from low energy levels to the ones with higher energy. Since many years ago, this process has been widely investigated in various optical designs. Along with considerable growth in nanoscience, lanthanide-doped UCNPs with enhanced efficiency and features have been provided over the last few years (Chen and Zhao 2012). This type of fluorescence NPs can emit light at a wavelength shorter than that of excitation, thanks to the anti-Stokes shift. They can convert light absorbed in the near-infrared region to ultraviolet by adjusting doping concentration and its ratio with the host component (Pehlivan et al. 2019).

In the excitation region, dopant ions with considerably high concentrations were utilized in the NIR region to prevent the autofluorescence phenomenon of biological structures in the region from ultraviolet to visible. Equally significantly, though, their emission spectrum can be improved in terms of the signal to noise ratio by reducing the fluorescence from nearby entities (Yüce and Kurt 2017). Although UCNPs exist in fewer colors than QDs, they benefit from separate and changeable excitation and emission peaks. UCNPs consist of particles at nanoscale and lattice structure; a sensitizer and an activator are the main components of these fluorescence NPs. The most common lattice is NaYF4, which acts as a host material. The second component is a sensitizer, which significantly affects light absorption by UCNPs (e.g., Yb^{3+}). Activator is the last compound of these fluorescent NPs, which has a prominent role in the emission of light at various wavelengths, leading to different colors (e.g., Er^{3+}, Tm^{3+}, and Ho^{3+}) (Sharma and Raghavarao 2018; Wilhelm 2017). They are also desirable candidates for various applications, including bioimaging (Wen et al. 2018), solar cells (Haase and Schäfer 2011), display equipment (Park et al. 2017), biosensing, and food safety (Annavaram et al. 2019). Even though UCNPs possess remarkable capabilities for various applications, several limitations are required to be tackled (Wilhelm 2017). For example, sulfonamides (SAs) as a type of synthetic drug are effective in the treatment of most of the animal-related microbial diseases (Shen et al. 2016). As the residues of this medicine that cause a severe risk to human health can be noticed in animal-derived foods, Hu et al. [27] developed a new method for detecting this drug. Their provided system was a FRET-based sensor utilizing erbium-doped UCNPs and colloidal AuNPs as donor and quencher, respectively. It should be noted that achieving results from low fluorescence signals, being highly sensitive and selective, and low priced are salient advantages of this turn-on sensing technique.

Molds are forming many types of mycotoxins as secondary metabolites through food settlement. Like some of the insecticides, they can cause severe harmful impacts on public health due to their cancer-causing features on the kidney and other parts of the body (Chauhan et al. 2016). AFB1 and deoxynivalenol (DON) are salient examples of which their detection is of great significance. Chen et al. (2016a, b) developed a UCNPs-based aptasensor for multiple detections of mycotoxins. In this fluorescence-based assay, they functionalized magnetic nanoparticles (MNPs). They enhanced UCNPs ($NaYF^4$: Yb/Ho/Gd and $NaYF^4$: Yb/Tm/Gd) with antigen and antibody to develop capture probe and signal probe. Introducing their target mycotoxins to the mixture containing the probes mentioned above led to the formation of UCNPs-antibody-antigen MNPs, and UCNPs-antibody-targets based on antibody affinity for each target. Following this, a magnetic field was applied for separating the part, which includes MNPs, and fluorescence measurements were performed for the residue part (Fig. 1.3c).

In another food safety-related assay, a biosensor working based on fluorescence was provided by Zhang et al. (2020a, b) for dual sensing of histamine and tyramine, two types of biological amines. Considering the diseases such as heart rhythm disorders and high blood pressure caused by immoderate consumption of these amines in foods, developing a useful technique for their detection is significant for human health (Erdogan et al. 2018). They functionalized $NaYF^4$Yb, Tm, and $NaYF^4$Yb, Er with antibodies with an affinity for tyramine and histamine. Moreover, magnetic microspheres were examined as binding couples of target antigens to develop a biosensing probe. The tendency of this probe for binding to a signal probe takes it into the competition with targets. Finally, the amount of tyramine and histamine was measured by fluorescence intensity obtained at 483 and 550 nm wavelengths. The LOD obtained by this biosensor for tyramine and histamine was 0.1 mg L^{-1} and 0.01 mg L^{-1}, respectively. In the case of evaluation of the provided fluorescence-based sensor specificity, nonspecific targets including phenylethylamine, serotonin, histidine, octopamine, spermidine, tryptamine, spermine, and tyrosine were tested. There was no effective interaction between antibody and other nonspecific analytes which indicates high specificity of provided sensor.

Atrazine is a type of herbicide useful for preventing undesirable vegetation. Due to its resistance in nature, it can be noticed in soil and water. This compound in food crops considering its high toxicity is cause for concern regarding human health. In one recent research, Sheng et al. (2019) modified Er-doped UCNPs with a specific antibody of atrazine to form a signal probe. In the case of the biosensing probe, antigen-modified polystyrene magnetic microspheres were utilized. There is a competition between antigen located on biosensing probe and atrazine for being attached to the antibody to develop the immune conjugate. This step was followed by applying a magnetic field to separate conjugates and measuring fluorescence intensity. Noteworthy is that the LOD for atrazine was 2 ng L^{-1} in water samples taken from the river.

1.4.3 Other Fluorescent Nanoparticles

It has been a decade passed from the first synthesis of carbon QDs which was initially nominated as carbon NPs. However, these days they are entitled as carbon dots (CDs) due to resemblances in their physicochemical and optical features with QDs (Jelinek, 2017). Possessing a broad absorption spectrum and being chemically stable are striking features related to CDs as fluorescent NP of carbon with size between 1 and 10 nm, which have made the center of attention in recent years (Li et al. 2019). Same as other fluorescent NPs, emitting light at different wavelengths bring them to the center of attention for a variety of applications, including biosensing, drug delivery, cancer therapy, and food safety. The feature that makes these NPs stand out among others is being made up of carbons that are not toxic, and plenty of them can be found in nature. As a result, they are great candidates for application in which toxicity matters. Following this, being biodegradable due to its components is another important factor that makes these NPs an ideal alternative to other fluorescent NPs (Jelinek 2017).

Furthermore, carbons are mostly bound to other carbon atoms through sp^2 and sp^3 hybridization. They frequently have the shape of a crystal or do not have any distinct structure. It should be pointed out that the solubility of CDs in frequently used solvents can be improved by their surface modification with functional groups such as carboxyl (Molaei 2019). In food safety, shellfish allergy has been diagnosed as a severe threat to human health over the past few years. Therefore, the detection of arginine kinase, known as the primary cause of shellfish, is vital for public health. For this aim, Zhou et al. (2020a, b) provided a fluorescence probe using fluorescence emission of synthesized carboxyl functionalized CDs and the quenching feature of GO and successfully achieved a detection limit of 0.14 ng/mL for sensing arginine kinase. In another food safety-related assay, Hu et al. (2021) used CDs provided by a one-pot green method in fluorescence-based sensing of *E. coli* in milk. As a donor fluorophore, CDs were modified with aptamer with specificity for *E. coli*. Additionally, they functionalized MNPs with DNA, which has a complementary strand of the *E. coli* aptamer. Incubating the target with CDs-MNPs conjugates led to reducing fluorescence intensity obtained from the fluorescent spectra of CDs, signifying signal-off sensing. In one recent experiment, Hu et al. (2019) developed AuNPs-CDs conjugate to detect melamine in milk. In the food industry, this nitrogen-based target is added illegally to raise apparent protein percentage in milk, which poses a threat to human health due to its combination with cyanuric acid. The detection limit achieved in this assay was 3.6 nM. It is worth noting that increasing the amount of melamine incubated with the conjugation resulted in fluorescence intensity.

Another fluorescent NPs which have been widely utilized in providing fluorescence-based detection methods over recent years are silicon quantum dots (SiQDs). Besides being eco-friendly, highly soluble in common solvents, and biodegradable, they can be easily functionalized with other components that enhance their usage in sensing systems. In one experiment, SiQDs were produced by Wei et al. (2021) through a one-step method for sensing nitrite, a chemical compound which its immoderate amounts in food are damaging for human health. The capability of nitrite in quenching the fluorescence radiation emitted by these quantum dots owing to photoinduced electron transfer paves the way for target detection. With regard to the detection limit, they reached 25 nM in food samples. Moreover, molybdenum disulfide quantum dots (MoS_2 QDs) as fluorescent NPs are suitable for designing of detection platform in recent years. They exhibit fluorescent emission thanks to quantum confinement, and they are surface modifiable. To improve the fluorescence probe's sensitivity, the above-mentioned QDs can be modified with β-cyclodextrin (β-CD). This compound has some features such as hydrophilic exterior and hydrophobic internal space, which makes it a potential candidate for being used as a host in host-guest bindings. In one assay, Yi et al. (2021) provided a fluorescence probe by functionalizing MoS_2 QDs with β-CD to detect the analyte parathion-methyl (MP), an insecticide which its accumulation in food can cause harmful effects on human health. As can be seen in Fig. 1.3d, MoS_2 QDs were functionalized by 3-aminophenyl boronic acid (APBA) through amidation. It was followed by conjugating with β-CD to form a fluorescence probe. Additionally, the MP turned to p-nitrophenol through a hydrolysis reaction. Following this, p-nitrophenol developed complex with β-CD through host-guest binding, which led to effectual quenching of the fluorescence emitted from β-CD-MoS_2 QDs. Because the increase in MP concentration leads to a reduction in the intensity, target concentration can be accurately determined using this turn-off method.

Pieces of graphene at the nanoscale are named graphene quantum dots (GQDs). During the past decade, they have become a center of attention in developing sensing platforms, thanks to their characteristics, such as photoluminescence features. However, their quantum yields are noticeably lower than other fluorescent NPs. To undertake this barrier, scientists doped them with heteroatoms, including sulfur and nitrogen (Yang et al. 2015). As an example, a FRET-based detection approach was designed and provided by Nemati et al. (2018) for ethion sensing in the existence of Hg^{2+}. This sensing system's target is a pesticide which its residue in food crops should be monitored to improve the safety of food. Mercury ions bind to the surface of S and N doped GQDs with a negative charge (act as a donor). Before the addition of ethion, Hg^{2+} quenched the fluorescence emission of the donor. After introducing the target, Hg^{2+} bound to ethion, which results in the recovery of N and S doped GQDs. The concentration of ethion could be accurately determined using the intensity change because of the target added to the system. It is worth mentioning that 8 mg/L's detection limit was achieved in this sensing-based assay. In Table 1.3,

Table 1.3 Application of sensing platforms in food safety utilizing fluorescent nanomaterials

Nanoparticle	Affinity agent	Target/S	Sensing principle	LOD	Real sample measurements	Ref
QDs and colloidal AuNPs	Monoclonal antibody	Sulfa quinoxaline	FRET	1 ng/mL	Yes, in foods	Hu et al. (2017)
H_2O_2-sensitive mercaptopropionic acid-modified CdTe QDs	Polyclonal antibody	*E. coli*	Fluorescence	5 × 102 CFU/mL	Yes, in real milk	Chen et al. (2016a, b)
CdSe/ZnS QDs	Molecularly imprinted polymer	NƐ-carboxymethyl lysine	Fluorescence	2.6 μg/L	Yes, in foods	Liu et al. (2016a, b, c)
QDs	Molecularly imprinted optopolymer	Tyramine	Fluorescence	7.0 μg/kg	Yes, in fermented meat products	Zheng et al. (2019)
MNPs and QDs	Antibody	*E. coli*	Fluorescence	14 CFU/mL	Yes, in the spiked milk samples	Xue et al. (2018)
QD beads	Antibody	Zearalenone, ochratoxin A, and Fumonisin B1	Fluorescence	5, 20, and 10 ng/mL	Yes, in maize	Duan et al. (2019)
QDs and UCNPs	Monoclonal antibody	Norfloxacin	Fluorescence	2.5 μg/L	Yes, in real milk	Hu et al. (2020a, b)
CdSe/ZnS QDs and AuNPs	Egg yolk antibody	Vibrio parahaemolyticus	Fluorescence	10 CFU/mL	Yes, in real marine food samples	Liu et al. (2017a, b)
QDs	Biomimetic antibody	Methyl parathion, chlorpyrifos, and trichlorfon	Fluorescence	0.21 ± 0.021, 0.44 ± 0.069, and 0.32 ± 0.033 μg/L	Yes, in pear, carrot, kiwifruit, and banana samples	Jiang et al. (2019)
UCNPs	Monoclonal antibody	Sibutramine	Fluorescence	0.02 μg/mL	Yes, in diet food samples	Zhang et al. (2020a, b)
UCNPs	Antibodies	*E. coli* and *Staphylococcus aureus*	Fluorescence	13 and 15 CFU/mL	Yes, in adulterated meat and milk samples	Zhang et al. (2017)

$NaYF_4$: Yb/Tm UCNPs	Antibody	Chloramphenicol	Fluorescence	0.01 pg/mL	Yes, in the muscle tissue, milk, and honey samples	Sheng et al. (2020)
Gold nanorods and UCNPs	Aptamer	Zearalenone and Fumonisin B1	Fluorescence	0.01 μg/L and 0.003 ng/L	Yes, in spiked corn samples	He et al. (2020)
UCNPs and magnetite-modified AuNPs	Aptamer	Lead (II)	FRET	5.7 nM	Yes, in tea and wastewater	Chen et al. (2020a, b)
Gold nanorods and UCNPs	Aptamer	Enterotoxin B	Fluorescence	0.9 pg/mL	Yes, in spiked milk samples	Wu et al. (2018a, b)
UCNPs and AuNPs	Imidacloprid antibody	Imidacloprid	Fluorescence	0.79 ng/mL	Yes, in spiked water, Chinese cabbage, and honey samples	Si et al. (2018)
UCNPs	Aptamer	Zearalenone	Fluorescence	0.126 μg/kg for corn and 0.007 μg/L for beer	Yes, in corn and beer	Wu et al. (2017)
MNPs conjugated with UCNPs	Aptamer	Enrofloxacin	Fluorescence	0.06 ng/mL	Yes, in food samples	Liu et al. (2016a, b)
UCNPs	Aptamer	Mercury ions, ochratoxin A, and *Salmonella*	Fluorescence	5 ppb, 3 ng/mL, and 85 CFU/mL	Yes, in real water samples	Jin et al. (2018)
UCNPs	Aptamer	*E. coli*	FRET	17 CFU/mL	Yes, in tap water and green tea powder	Wang et al. (2020a, b)
UCNPs	Aptamer	*E. coli*	FRET	3 CFU/mL	Yes, in tap/pond water, milk	Jin et al. (2017)
Magnetic and UCNPs	Aptamer	*E. coli*	Fluorescence	10 CFU/mL	Yes, in food	Li et al. (2020a)
Rare-earth doped UCNPs	Aptamer	Diazinon	FRET	0.023 ng/mL	Yes, in food	Rong et al. (2020)
UCNPs	Aptamer	Malathion	FRET	1.42 nM.	Yes, in food	Chen et al. (2020a, b)

(continued)

Table 1.3 (continued)

Nanoparticle	Affinity agent	Target/S	Sensing principle	LOD	Real sample measurements	Ref
UCNPs	Aptamer	AFB1	LRET	0.17 ng/mL	Yes, in food samples	Wang et al. (2019)
UCNPs	Anti-sulfa quinoxaline monoclonal antibody	Sulfa quinoxaline	Fluorescence	0.5 μg/kg	Yes, in animal-derived foods	Hu et al. (2016)
UCNPs	Polyacrylic acid	Fe^{3+}	Near-infrared luminescence	1 μM	Yes, in infant formula, milk powder	Zhang et al. (2019a, b)
NH_2–$NaYF_4$: Yb, Er/$NaYF_4$@SiO_2 UCNPs	AuNPs	Cadmium ions	FRET	0.059 μM	Yes, in drinking water	Sun et al. (2020)
Au@Ag/graphene upconversion	Aptamer	Hg^{2+}	Fluorescence	1 ppb	Yes, in food	Li et al. (2020b)
CDs		Acrylamide	Fluorescence	8.1 × 10–7 M	Yes, in artificially spiked white bread crust samples	Wei et al. (2020)
SiQDs		Potassium ferrocyanide	Fluorescence	30 ng/mL	Yes, in table salt and salted food samples	Na et al. (2019)
ZnO QDs	Vitamin B6 cofactors	Histamine	Fluorescence	0.59 μM and 0.97 μM	No	Yadav et al. (2020)
CDs	AuNPs	Aldicarb	Fluorescence	3.02 μg/L	Yes, in fruits and vegetables spiked sample	Sajwan et al. (2021)
Sulfur, chlorine, and nitrogen co-doped CDs		Manganese (VII)	Fluorescence	12.8 nM	Yes, in vegetable, cereal, and tea samples	Hu et al. (2020a, b)
Bluish green-emitting CDs		Curcumin	Fluorescence	28.7 nM	Yes, in dietary food samples	Liu et al. (2020a, b)
CDs		Ascorbic acid	Fluorescence	42 nM	Yes, in fresh fruits, vegetables, and commercial fruit juices samples	Liu et al. (2016a, b, c)

fluorescent NPs, affinity agents, target analytes, and the detection limit of each provided fluorescence-based sensor are represented.

1.5 Conclusion

The impact of nanoscience on the food industry cannot be neglected that it fulfills various requirements and contributes to undertaking the obstacles. Many articles in this field indicates the effective use of NPs in developing cost-effective, environmentally friendly sensors with high sensitivity and selectivity. Carbon-based and plasmonic/metallic NPs and fluorescent NPs, including QDs and UCNPs, are among the NPs which are useful for sensing food-related toxins, allergens, chemicals, pathogens, and enterotoxins over the past few years. Furthermore, aptamers and antibodies are salient examples of frequently used affinity agents for developing sensing platforms. Moveable analytic sensors with high efficiency can be developed to detect food-related analytes with considerably lower LODs in real samples. Most of these platforms possess great potential for highly selective multiplex sensing of more than one target simultaneously, which leads to time-saving detection of targets. The tables prepared in this chapter summarize NP-based sensing principles, food-related targets, and their detection limits, affinity agents, and applicability of methods for real sample measurements. Though sensing techniques which are based on nanoscience have numerous pros in comparison with other techniques, from many aspects such as LOD, and selectivity, their efficacious application in the above-mentioned field for various sorts of real sample measurements is still considered as demanding issue because of the samples with complex nature and difficulties in sample separation steps. Developing robust and operational sample separation methods can significantly enhance the applicability of the detection techniques. Although these methods were assessed using real samples, only a few food samples were utilized for performance studies. However, using a diversity of real samples for sensing assays and comparing them with other conventional techniques can contribute to undertaking the issue as mentioned above. Other challenging issues are the intricate NP synthesis process along with costly precursors, which can be overcome by discovery of low-priced and efficient synthesis techniques. Additionally, some of the disease-causing pathogens cannot be noticed by available detection platforms due to the insufficiency of the detection agents. To conclude, the construction of novel, selective, multiplexed, and cost-effective sensing methods based on stable engineered NPs and target-specific affinity probes may revolutionize the field in near future.

References

Aghaie A, Khanmohammadi A, Hajian A, Schmid U, Bagheri H (2019) Nonenzymatic electrochemical determination of paraoxon ethyl in water and fruits by graphene-based NiFe bimetallic phosphosulfide nanocomposite as a superior sensing layer. Food Anal Methods 12(7): 1545–1555

Alhabeb M, Maleski K, Anasori B, Lelyukh P, Clark L, Sin S, Gogotsi Y (2017) Guidelines for synthesis and processing of two-dimensional titanium carbide (Ti3C2Tx MXene). Chem Mater 29:7633–7644. https://doi.org/10.1021/acs.chemmater.7b02847

Al-Nahain A, Lee JE, In I, Lee H, Lee KD, Jeong JH, Park SY (2013) Target delivery and cell imaging using hyaluronic acid-functionalized graphene quantum dots. Mol Pharm 10:3736–3744. https://doi.org/10.1021/mp400219u

Annavaram V, Chen M, Kutsanedzie FYH, Agyekum AA, Zareef M, Ahmad W, Hassan MM, Huanhuan L, Chen Q (2019) Synthesis of highly fluorescent RhDCP as an ideal inner filter effect pair for the NaYF4:Yb,Er upconversion fluorescent nanoparticles to detect trace amount of Hg(II) in water and food samples. J Photochem Photobiol A Chem 382. https://doi.org/10.1016/j.jphotochem.2019.111950

Arvand M, Mirroshandel AA (2019) An efficient fluorescence resonance energy transfer system from quantum dots to graphene oxide nano sheets: application in a photoluminescence aptasensing probe for the sensitive detection of diazinon. Food Chem 280:115–122. https://doi.org/10.1016/j.foodchem.2018.12.069

Bao-Kai M, Mian L, Ling-Zhi C, Xin-Chu W, Cai S, Qing H (2020) Enzyme-MXene nanosheets: fabrication and application in electrochemical detection of H_2O_2. J Inorg Mater 35:132–138. https://doi.org/10.15541/jim20190139

Bonilla JC, Bozkurt F, Ansari S, Sozer N, Kokini JL (2016) Applications of quantum dots in food science and biology. Trends Food Sci Technol 53:75–89. https://doi.org/10.1016/j.tifs.2016.04.006

Bozkurt AG, Buyukgoz GG, Soforoglu M, Tamer U, Suludere Z, Boyaci IH (2018) Alkaline phosphatase labeled SERS active sandwich immunoassay for detection of *Escherichia coli*. Spectrochim Acta Part A Mol Biomol Spectrosc 194:8–13. https://doi.org/10.1016/j.saa.2017.12.057

Caucheteur C, Guo T, Albert J (2015) Review of plasmonic fiber optic biochemical sensors: improving the limit of detection. Anal Bioanal Chem 407:3883–3897. https://doi.org/10.1007/s00216-014-8411-6

Chandra P, Koh WCA, Noh HB, Shim YB (2012) In vitro monitoring of i-NOS concentrations with an immunosensor: The inhibitory effect of endocrine disruptors on i-NOS release. Biosens Bioelectron. https://doi.org/10.1016/j.bios.2011.11.027

Chauhan R, Singh J, Sachdev T, Basu T, Malhotra BD (2016) Recent advances in mycotoxins detection. Biosens Bioelectron 81:532–545. https://doi.org/10.1016/j.bios.2016.03.004

Chen J, Nugen SR (2019) Detection of protease and engineered phage-infected bacteria using peptide-graphene oxide nanosensors. Anal Bioanal Chem 411(12):2487–2492

Chen L, Song J (2017) Tailored graphitic carbon nitride nanostructures: synthesis, modification, and sensing applications. Adv Funct Mater. https://doi.org/10.1002/adfm.201702695

Chen J, Zhao JX (2012) Upconversion nanomaterials: Synthesis, mechanism, and applications in sensing. Sensors 12:2414–2435. https://doi.org/10.3390/s120302414

Chen Z, Li G, Zhang L, Jiang J, Li Z, Peng Z, Deng L (2008) A new method for the detection of ATP using a quantum-dot-tagged aptamer. Anal Bioanal Chem 392:1185–1188. https://doi.org/10.1007/s00216-008-2342-z

Chen H, Shao L, Li Q, Wang J (2013) Gold nanorods and their plasmonic properties. Chem Soc Rev 42:2679–2724. https://doi.org/10.1039/C2CS35367A

Chen Q, Hu W, Sun C, Li H, Ouyang Q (2016a) Synthesis of improved upconversion nanoparticles as ultrasensitive fluorescence probe for mycotoxins. Anal Chim Acta 938:137–145. https://doi.org/10.1016/j.aca.2016.08.003

Chen R, Huang X, Li J, Shan S, Lai W, Xiong Y (2016b) A novel fluorescence immunoassay for the sensitive detection of *Escherichia coli* O157:H7 in milk based on catalase-mediated fluorescence quenching of CdTe quantum dots. Anal Chim Acta 947:50–57. https://doi.org/10.1016/j.aca.2016.10.017

Chen Q, Zhang L, Feng Y, Shi F, Wang Y, Wang P, Liu L (2018) Dual-functional peptide conjugated gold nanorods for the detection and photothermal ablation of pathogenic bacteria. J Mater Chem B 6:7643–7651. https://doi.org/10.1039/C8TB01835A

Chen M, Hassan M, Li H, Chen Q (2020a) Fluorometric determination of lead(II) by using aptamer-functionalized upconversion nanoparticles and magnetite-modified gold nanoparticles. Microchim Acta 187. https://doi.org/10.1007/s00604-019-4030-4

Chen Q, Sheng R, Wang P, Ouyang Q, Wang A, Ali S, Zareef M, Hassan MM (2020b) Ultra-sensitive detection of malathion residues using FRET-based upconversion fluorescence sensor in food. Spectrochim. Acta - Part A Mol. Biomol. Spectrosc. 241:118654. https://doi.org/10.1016/j.saa.2020.118654

Choudhary M, Yadav P, Singh A, Kaur S, Ramirez-Vick J, Chandra P, Arora K, Singh SP (2016) CD 59 Targeted ultrasensitive electrochemical immunosensor for fast and noninvasive diagnosis of oral cancer. Electroanalysis 28:2565–2574. https://doi.org/10.1002/elan.201600238

Danielson E, Dindo M, Porkovich AJ, Kumar P, Wang Z, Jain P, Mete T, Ziadi Z, Kikkeri R, Laurino P, Sowwan M (2020) Non-enzymatic and highly sensitive lactose detection utilizing graphene field-effect transistors. Biosens Bioelectron 165:112419. https://doi.org/10.1016/j.bios.2020.112419

Deka S, Saxena V, Hasan A, Chandra P, Pandey LM (2018) Synthesis, characterization and in vitro analysis of α-Fe_2O_3-$GdFeO_3$ biphasic materials as therapeutic agent for magnetic hyperthermia applications. Mater Sci Eng C 92:932–941. https://doi.org/10.1016/j.msec.2018.07.042

Du J, Singh H, Dong W, Bai Y, Yi T-H (2018) Colorimetric detection of Listeria monocytogenes using one-pot biosynthesized flower-shaped gold nanoparticles. Sensors Actuators B Chem 265: 285–292. https://doi.org/10.1016/j.snb.2018.03.067

Duan N, Shen M, Wu S, Zhao C, Ma X, Wang Z (2017) Graphene oxide wrapped Fe3O4@Au nanostructures as substrates for aptamer-based detection of Vibrio parahaemolyticus by surface-enhanced Raman spectroscopy. Microchim Acta 184:2653–2660. https://doi.org/10.1007/s00604-017-2298-9

Duan H, Li Y, Shao Y, Huang X, Xiong Y (2019) Multicolor quantum dot nanobeads for simultaneous multiplex immunochromatographic detection of mycotoxins in maize. Sensors Actuators B Chem 291:411–417. https://doi.org/10.1016/j.snb.2019.04.101

Duan N, Shen M, Qi S, Wang W, Wu S, Wang Z (2020) A SERS aptasensor for simultaneous multiple pathogens detection using gold decorated PDMS substrate. Spectrochim Acta Part A Mol Biomol Spectrosc 230:118103. https://doi.org/10.1016/j.saa.2020.118103

Erdogan ZO, Akin I, Kucukkolbasi S (2018) A new non-enzymatic sensor based on TiO_2-Ag/polypyrrole for electrochemical detection of tyramine. Synth Met 246:96–100. https://doi.org/10.1016/j.synthmet.2018.10.006

Fu L, Xie K, Wu D, Wang A, Zhang H, Ji Z (2020) Electrochemical determination of vanillin in food samples by using pyrolyzed graphitic carbon nitride. Mater Chem Phys 242:122462. https://doi.org/10.1016/j.matchemphys.2019.122462

Gao N, He C, Ma M, Cai Z, Zhou Y, Chang G, Wang X, He Y (2019) Electrochemical co-deposition synthesis of Au-ZrO2-graphene nanocomposite for a nonenzymatic methyl parathion sensor. Anal Chim Acta 1072:25–34. https://doi.org/10.1016/j.aca.2019.04.043

Govindasamy M, Mani V, Chen S, Chen T (2017) Methyl parathion detection in vegetables and fruits using silver @ graphene nanoribbons nanocomposite modified screen printed electrode. Nat Publ Gr:1–11. https://doi.org/10.1038/srep46471

Guo J, Li Y, Wang L, Xu J, Huang Y, Luo Y, Shen F, Sun C, Meng R (2016) Aptamer-based fluorescent screening assay for acetamiprid via inner filter effect of gold nanoparticles on the fluorescence of CdTe quantum dots. Anal Bioanal Chem 408:557–566. https://doi.org/10.1007/s00216-015-9132-1

Guo P, Yang W, Hu H, Wang Y, Li P (2019) Rapid detection of aflatoxin B 1 by dummy template molecularly imprinted polymer capped CdTe quantum dots. Anal Bioanal Chem 411(12): 2607–2617. https://doi.org/10.1007/s00216-019-01708-2

Gupta BD, Pathak A, Semwal V (2019) Carbon-based nanomaterials for plasmonic sensors: a review. Sensors (Switzerland) 19. https://doi.org/10.3390/s19163536

Haase M, Schäfer H (2011) Upconverting nanoparticles. Angew Chem: Int Ed 50:5808–5829. https://doi.org/10.1002/anie.201005159

Hassan MM, Zareef M, Jiao T, Liu S, Xu Y, Viswadevarayalu A, Li H, Chen Q (2021) Signal optimized rough silver nanoparticle for rapid SERS sensing of pesticide residues in tea. Food Chem 338:127796. https://doi.org/10.1016/j.foodchem.2020.127796

He Q, Liu J, Liu X, Xia Y, Li G, Deng P, Chen D (2018) Novel electrochemical sensors based on cuprous oxide-electrochemically reduced graphene oxide nanocomposites modified electrode toward sensitive detection of sunset yellow. Molecules 23(9):2130. https://doi.org/10.3390/molecules23092130

He D, Wu Z, Cui B, Xu E, Jin Z (2019) Establishment of a dual mode immunochromatographic assay for Campylobacter jejuni detection. Food Chem 289:708–713. https://doi.org/10.1016/j.foodchem.2019.03.106

He D, Wu Z, Cui B, Jin Z, Xu E (2020) A fluorometric method for aptamer-based simultaneous determination of two kinds of the fusarium mycotoxins zearalenone and fumonisin B1 making use of gold nanorods and upconversion nanoparticles. Microchim Acta 187(4):1–8. https://doi.org/10.1007/s00604-020-04236-4

Heo NS, Oh SY, Ryu MY, Baek SH, Park TJ, Choi C, Huh YS, Park JP (2019) Affinity peptide-guided plasmonic biosensor for detection of noroviral protein and human norovirus. Biotechnol Bioprocess Eng 24:318–325. https://doi.org/10.1007/s12257-018-0410-6

Hu G, Sheng W, Zhang Y, Wang J, Wu X, Wang S (2016) Upconversion nanoparticles and monodispersed magnetic polystyrene microsphere based fluorescence immunoassay for the detection of sulfaquinoxaline in animal-derived foods. J Agric Food Chem 64:3908–3915. https://doi.org/10.1021/acs.jafc.6b01497

Hu G, Sheng W, Li J, Zhang Y, Wang J, Wang S (2017) Fluorescent quenching immune chromatographic strips with quantum dots and upconversion nanoparticles as fluorescent donors for visual detection of sulfaquinoxaline in foods of animal origin. Anal Chim Acta 982:185–192. https://doi.org/10.1016/j.aca.2017.06.013

Hu X, Shi J, Shi Y, Zou X, Arslan M, Zhang W, Huang X, Li Z, Xu Y (2019) Use of a smartphone for visual detection of melamine in milk based on Au@Carbon quantum dots nanocomposites. Food Chem 272:58–65. https://doi.org/10.1016/j.foodchem.2018.08.021

Hu G, Gao S, Han X, Yang L (2020a) Comparison of immunochromatographic strips using colloidal gold, quantum dots, and upconversion nanoparticles for visual detection of norfloxacin in milk samples. Food Anal Methods 13:1069–1077. https://doi.org/10.1007/s12161-020-01725-3

Hu Q, Liu LF, Sun H, Han J, Gong X, Liu L, Yang ZQ (2020b) An ultra-selective fluorescence method with enhanced sensitivity for the determination of manganese (VII) in food stuffs using carbon quantum dots as nanoprobe. J Food Compos Anal 88:103447. https://doi.org/10.1016/j.jfca.2020.103447

Hu X, Li Y, Xu Y, Gan Z, Zou X, Shi J, Huang X, Li Z, Li Y (2021) Green one-step synthesis of carbon quantum dots from orange peel for fluorescent detection of *Escherichia coli* in milk. Food Chem 339:127775. https://doi.org/10.1016/j.foodchem.2020.127775

Hua R, Hao N, Lu J, Qian J, Liu Q, Li H, Wang K (2018) A sensitive potentiometric resolved ratiometric photoelectrochemical aptasensor for *Escherichia coli* detection fabricated with non-metallic nanomaterials. Biosens Bioelectron 106:57–63. https://doi.org/10.1016/j.bios.2018.01.053

Jelinek R (2017) Carbon quantum dots. synthesis, properties and applicatons

Jiang M, He J, Gong J, Gao H, Xu Z (2019) Development of a quantum dot-labelled biomimetic fluorescence immunoassay for the simultaneous determination of three organophosphorus pesticide residues in agricultural products. Food Agric Immunol 30:248–261. https://doi.org/10.1080/09540105.2019.1572714

Jin D, Xu Q, Yu L, Mao A, Hu X (2016) A novel sensor for the detection of acetamiprid in vegetables based on its photocatalytic degradation compound. Food Chem 194:959–965. https://doi.org/10.1016/j.foodchem.2015.08.118

Jin B, Wang S, Lin M, Jin Y, Zhang S, Cui X, Gong Y, Li A, Xu F, Lu TJ (2017) Upconversion nanoparticles based FRET aptasensor for rapid and ultrasenstive bacteria detection. Biosens Bioelectron 90:525–533. https://doi.org/10.1016/j.bios.2016.10.029

Jin B, Yang Y, He R, Park YI, Lee A, Bai D, Li F, Lu TJ, Xu F, Lin M (2018) Lateral flow aptamer assay integrated smartphone-based portable device for simultaneous detection of multiple targets using upconversion nanoparticles. Sensors Actuators B Chem 276:48–56. https://doi.org/10.1016/j.snb.2018.08.074

Kahraman M, Mullen ER, Korkmaz A, Wachsmann-Hogiu S (2017) Fundamentals and applications of SERS-based bioanalytical sensing. Nano 6:831–852. https://doi.org/10.1515/nanoph-2016-0174

Karthik R, Kumar JV, Chen SM, Kokulnathan T, Chen TW, Sakthinathan S, Chiu TW, Muthuraj V (2018) Development of novel 3D flower-like praseodymium molybdate decorated reduced graphene oxide: an efficient and selective electrocatalyst for the detection of acetylcholinesterase inhibitor methyl parathion. Sensors Actuators B Chem 270:353–361. https://doi.org/10.1016/j.snb.2018.05.054

Kaushal S, Priyadarshi N, Pinnaka AK, Soni S, Deep A, Singhal NK (2019) Glycoconjugates coated gold nanorods based novel biosensor for optical detection and photothermal ablation of food borne bacteria. Sensors Actuators B Chem 289:207–215. https://doi.org/10.1016/j.snb.2019.03.096

Khan R, Andreescu S (2020) MXenes-based bioanalytical sensors: design, characterization, and applications. Sensors 20:5434. https://doi.org/10.3390/s20185434

Khateb H, Klös G, Meyer RL, Sutherland DS (2020) Development of a Label-Free LSPR-Apta sensor for *Staphylococcus aureus* detection. ACS Appl Bio Mater 3:3066–3077. https://doi.org/10.1021/acsabm.0c00110

Kurt H, Eyüpoğlu AE, Sütlü T, Budak H, Yüce M (2019) Plasmonic selection of ssDNA Aptamers against fibroblast growth factor receptor. ACS Comb Sci 21:578–587. https://doi.org/10.1021/acscombsci.9b00059

Lee S, Jun B-H (2019) Silver nanoparticles: synthesis and application for nanomedicine. Int J Mol Sci 20:865. https://doi.org/10.3390/ijms20040865

Li M, Chen T, Gooding JJ, Liu J (2019) Review of carbon and graphene quantum dots for sensing. ACS Sensors 4:1732–1748. https://doi.org/10.1021/acssensors.9b00514

Li L, Zhang H, Song D, Xu K, Zheng Y, Xiao H, Liu Y, Li J, Song X (2020a) Simultaneous detection of three zoonotic pathogens based on phage display peptide and multicolor quantum dots. Anal Biochem 608:113854. https://doi.org/10.1016/j.ab.2020.113854

Li Y, Lu C, Zhou S, Fauconnier M-L, Gao F, Fan B, Lin J, Wang F, Zheng J (2020b) Sensitive and simultaneous detection of different pathogens by surface-enhanced Raman scattering based on aptamer and Raman reporter co-mediated gold tags. Sensors Actuators B Chem 317:128182. https://doi.org/10.1016/j.snb.2020.128182

Li H, Ahmad W, Rong Y, Chen Q, Zuo M, Ouyang Q, Guo Z (2020c) Designing an aptamer based magnetic and upconversion nanoparticles conjugated fluorescence sensor for screening *Escherichia coli* in food. Food Control 107:106761. https://doi.org/10.1016/j.foodcont.2019.106761

Li H, Huang X, Mehedi Hassan M, Zuo M, Wu X, Chen Y, Chen Q (2020d) Dual-channel biosensor for Hg^{2+} sensing in food using Au@Ag/graphene-upconversion nanohybrids as metal-enhanced fluorescence and SERS indicators. Microchem J 154:104563. https://doi.org/10.1016/j.microc.2019.104563

Lin B, Yu Y, Li R, Cao Y, Guo M (2016) Turn-on sensor for quantification and imaging of acetamiprid residues based on quantum dots functionalized with aptamer. Sensors Actuators B Chem 229:100–109. https://doi.org/10.1016/j.snb.2016.01.114

Liu H, Wu D, Zhou K, Wang J, Sun B (2016a) Development and applications of molecularly imprinted polymers based on hydrophobic CdSe/ZnS quantum dots for optosensing of N-ϵ-carboxymethyllysine in foods. Food Chem 211:34–40. https://doi.org/10.1016/j.foodchem.2016.05.038

Liu J, Chen Y, Wang W, Feng J, Liang M, Ma S, Chen X (2016b) "switch-On" Fluorescent sensing of ascorbic acid in food samples based on carbon quantum dots-MnO_2 Probe. J Agric Food Chem 64:371–380. https://doi.org/10.1021/acs.jafc.5b05726

Liu X, Su L, Zhu L, Gao X, Wang Y, Bai F, Tang Y, Li J (2016c) Hybrid material for enrofloxacin sensing based on aptamer-functionalized magnetic nanoparticle conjugated with upconversion nanoprobes. Sensors Actuators B Chem 233:394–401. https://doi.org/10.1016/j.snb.2016.04.096

Liu Y, Liu X, Guo Z, Hu Z, Xue Z, Lu X (2017a) Horseradish peroxidase supported on porous graphene as a novel sensing platform for detection of hydrogen peroxide in living cells sensitively. Biosens Bioelectron 87:101–107. https://doi.org/10.1016/j.bios.2016.08.015

Liu Y, Zhao C, Fu K, Song X, Xu K, Wang J, Li J (2017b) Selective turn-on fluorescence detection of Vibrio parahaemolyticus in food based on charge-transfer between CdSe/ZnS quantum dots and gold nanoparticles. Food Control 80:380–387. https://doi.org/10.1016/j.foodcont.2017.05.032

Liu J, He H, Xiao D, Yin S, Ji W, Jiang S, Luo D, Wang B, Liu Y (2018a) Recent advances of plasmonic nanoparticles and their applications. Materials (Basel) 11. https://doi.org/10.3390/ma11101833

Liu JM, Hu Y, Yang YK, Liu H, Fang GZ, Lu X, Wang S (2018b) Emerging functional nanomaterials for the detection of food contaminants. Trends Food Sci Technol. https://doi.org/10.1016/j.tifs.2017.11.005

Liu Y, Wang J, Zhao C, Guo X, Song X, Zhao W, Liu S, Xu K, Li J (2019) A multicolorimetric assay for rapid detection of Listeria monocytogenes based on the etching of gold nanorods. Anal Chim Acta 1048:154–160. https://doi.org/10.1016/j.aca.2018.10.020

Liu J, Jalali M, Mahshid S, Wachsmann-Hogiu S (2020a) Are plasmonic optical biosensors ready for use in point-of-need applications? Analyst 145:364–384. https://doi.org/10.1039/C9AN02149C

Liu L, Hu Q, Sun H, Han J, Pan Y, Yang ZQ (2020b) An ultra-sensitive analytical platform based on bluish green emitting carbon quantum dots for the detection of curcumin in dietary foods. J Food Compos Anal 94:103639. https://doi.org/10.1016/j.jfca.2020.103639

Loiseau A, Asila V, Boitel-Aullen G, Lam M, Salmain M, Boujday S (2019a) Silver-based plasmonic nanoparticles for and their use in biosensing. Biosensors 9:78. https://doi.org/10.3390/bios9020078

Loiseau A, Zhang L, Hu D, Salmain M, Mazouzi Y, Flack R, Liedberg B, Boujday S (2019b) Core–shell gold/silver nanoparticles for localized surface plasmon resonance-based naked-eye toxin biosensing. ACS Appl Mater Interfaces 11:46462–46471. https://doi.org/10.1021/acsami.9b14980

Ma X, Xu X, Xia Y, Wang Z (2018) SERS aptasensor for Salmonella typhimurium detection based on spiny gold nanoparticles. Food Control 84:232–237. https://doi.org/10.1016/j.foodcont.2017.07.016

Magesa F, Wu Y, Tian Y, Vianney J-MM, Buza J, He Q, Tan Y (2019) Graphene and graphene like 2D graphitic carbon nitride: electrochemical detection of food colorants and toxic substances in environment. Trends Environ Anal Chem 23:e00064. https://doi.org/10.1016/j.teac.2019.e00064

Mahato K, Kumar S, Srivastava A, Maurya PK, Singh R, Chandra P (2018) Electrochemical immunosensors: fundamentals and applications in clinical diagnostics. In: Handbook of immunoassay technologies. https://doi.org/10.1016/B978-0-12-811762-0.00014-1

Mansoori B, Shotorbani SS, Baradaran B (2014) RNA interference and its role in cancer therapy. Adv Pharm Bull 4:313–321. https://doi.org/10.5681/apb.2014.046

Masdor N, Altintas Z, Tothill I (2017) Surface plasmon resonance immunosensor for the detection of Campylobacter jejuni. Chemosensors 5:16. https://doi.org/10.3390/chemosensors5020016
Matea CT, Mocan T, Tabaran F, Pop T, Mosteanu O, Puia C, Iancu C, Mocan L (2017) Quantum dots in imaging, drug delivery and sensor applications. Int J Nanomedicine 12:5421–5431. https://doi.org/10.2147/IJN.S138624
McElwain TF, Thumbi SM (2017) Animal pathogens and their impact on animal health, the economy, food security, food safety and public health. OIE Rev Sci Tech 36:423–433. https://doi.org/10.20506/rst.36.2.2663
Mishra GK, Barfidokht A, Tehrani F, Mishra RK (2018) Food safety analysis using electrochemical biosensors. Foods. https://doi.org/10.3390/foods7090141
Molaei MJ (2019) Carbon quantum dots and their biomedical and therapeutic applications: a review. RSC Adv 9:6460–6481. https://doi.org/10.1039/c8ra08088g
Na M, Chen Y, Han Y, Ma S, Liu J, Chen X (2019) Determination of potassium ferrocyanide in table salt and salted food using a water-soluble fluorescent silicon quantum dots. Food Chem 288:248–255. https://doi.org/10.1016/j.foodchem.2019.02.111
Naguib M, Kurtoglu M, Presser V, Lu J, Niu J, Heon M, Hultman L, Gogotsi Y, Barsoum MW (2011) Two-dimensional nanocrystals produced by exfoliation of Ti 3AlC 2. Adv Mater 23: 4248–4253. https://doi.org/10.1002/adma.201102306
Nehra M, Dilbaghi N, Hassan AA, Kumar S (2019) Carbon-based nanomaterials for the development of sensitive nanosensor platforms. In: Advances in nanosensors for biological and environmental analysis. Elsevier, pp 1–25. https://doi.org/10.1016/b978-0-12-817456-2.00001-2
Nemati F, Hosseini M, Zare-Dorabei R, Ganjali MR (2018) Sensitive recognition of ethion in food samples using turn-on fluorescence N and S co-doped graphene quantum dots. Anal Methods 10:1760–1766. https://doi.org/10.1039/c7ay02850d
Novoselov KS, Geim AK, Morozov SV, Jiang D, Zhang Y, Dubonos SV, Grigorieva IV, Firsov AA (2004) Electric field effect in atomically thin carbon films. Science (80-) 306:666–669. https://doi.org/10.1126/science.1102896
Nsibande SA, Forbes PBC (2016) Fluorescence detection of pesticides using quantum dot materials – a review. Anal Chim Acta 945:9–22. https://doi.org/10.1016/j.aca.2016.10.002
Oh SY, Heo NS, Shukla S, Cho H-J, Vilian ATE, Kim J, Lee SY, Han Y-K, Yoo SM, Huh YS (2017) Development of gold nanoparticle-aptamer-based LSPR sensing chips for the rapid detection of Salmonella typhimurium in pork meat. Sci Rep 7:10130. https://doi.org/10.1038/s41598-017-10188-2
Pan M, Yin Z, Liu K, Du X, Liu H, Wang S (2019) Carbon-based nanomaterials in sensors for food safety. Nano. https://doi.org/10.3390/nano9091330
Pang B, Zheng Y, Wang J, Liu Y, Song X, Li J, Yao S, Fu K, Xu K, Zhao C, Li J (2019) Colorimetric detection of Staphylococcus aureus using gold nanorods labeled with yolk immunoglobulin and urease, magnetic beads, and a phenolphthalein impregnated test paper. Microchim Acta 186:611. https://doi.org/10.1007/s00604-019-3722-0
Parate K, Pola CC, Rangnekar SV, Mendivelso-Perez DL, Smith EA, Hersam MC, Gomes CL, Claussen JC (2020) Aerosol-jet-printed graphene electrochemical histamine sensors for food safety monitoring. 2D Mater 7(3):034002. https://doi.org/10.1088/2053-1583/ab8919
Park BJ, Hong AR, Park S, Kyung KU, Lee K, Jang HS (2017) Flexible transparent displays based on core/shell upconversion nanophosphor-incorporated polymer waveguides. Sci Rep 7:1–11. https://doi.org/10.1038/srep45659
Pehlivan ZS, Torabfam M, Kurt H, Ow-Yang C, Hildebrandt N, Yüce M (2019) Aptamer and nanomaterial based FRET biosensors: a review on recent advances (2014–2019). Microchim Acta 186:563. https://doi.org/10.1007/s00604-019-3659-3
Quintela IA, de los Reyes BG, Lin C-S, Wu VCH (2019) Simultaneous colorimetric detection of a variety of Salmonella spp. in food and environmental samples by optical biosensing using oligonucleotide-gold nanoparticles. Front Microbiol 10:1–12. https://doi.org/10.3389/fmicb.2019.01138

Ramalingam M, Ponnusamy VK, Sangilimuthu SN (2019) A nanocomposite consisting of porous graphitic carbon nitride nanosheets and oxidized multiwalled carbon nanotubes for simultaneous stripping voltammetric determination of cadmium(II), mercury(II), lead(II) and zinc(II). Microchim Acta 186:69. https://doi.org/10.1007/s00604-018-3178-7

Ren W, Ballou DR, FitzGerald R, Irudayaraj J (2019) Plasmonic enhancement in lateral flow sensors for improved sensing of *E. coli* O157:H7. Biosens Bioelectron 126:324–331. https://doi.org/10.1016/j.bios.2018.10.066

Reshma VG, Mohanan PV (2019) Quantum dots: applications and safety consequences. J Lumin 205:287–298. https://doi.org/10.1016/j.jlumin.2018.09.015

Rodríguez-Lorenzo L, Garrido-Maestu A, Bhunia AK, Espiña B, Prado M, Diéguez L, Abalde-Cela S (2019) Gold nanostars for the detection of foodborne pathogens via surface-enhanced Raman scattering combined with microfluidics. ACS Appl Nano Mater 2:6081–6086. https://doi.org/10.1021/acsanm.9b01223

Rong Y, Li H, Ouyang Q, Ali S, Chen Q (2020) Rapid and sensitive detection of diazinon in food based on the FRET between rare-earth doped upconversion nanoparticles and graphene oxide. Spectrochim Acta: Part A Mol Biomol Spectrosc 239:118500. https://doi.org/10.1016/j.saa.2020.118500

Rouhani M (2019) Fluoro-functionalized graphene as a promising nanosensor in detection of fish spoilage: a theoretical study. Chem Phys Lett 719:91–102. https://doi.org/10.1016/j.cplett.2019.02.001

Rozmysłowska-Wojciechowska A, Karwowska E, Poźniak S, Wojciechowski T, Chlubny L, Olszyna A, Ziemkowska W, Jastrzębska AM (2019) Influence of modification of Ti3C2 MXene with ceramic oxide and noble metal nanoparticles on its antimicrobial properties and ecotoxicity towards selected algae and higher plants. RSC Adv 9:4092–4105. https://doi.org/10.1039/C8RA07633B

Sajwan RK, Lakshmi GBVS, Solanki PR (2021) Fluorescence tuning behavior of carbon quantum dots with gold nanoparticles via novel intercalation effect of aldicarb. Food Chem 340:127835. https://doi.org/10.1016/j.foodchem.2020.127835

Shams S, Bakhshi B, Tohidi Moghadam T, Behmanesh M (2019) A sensitive gold-nanorods-based nanobiosensor for specific detection of Campylobacter jejuni and Campylobacter coli. J Nanobiotechnology 17:43. https://doi.org/10.1186/s12951-019-0476-0

Sharma R, Raghavarao KSMS (2018) Nanoparticle-based aptasensors for food contaminant detection, nanomaterials for food applications. Elsevier Inc. https://doi.org/10.1016/B978-0-12-814130-4.00006-3

Shen Q, Jin R, Xue J, Lu Y, Dai Z (2016) Analysis of trace levels of sulfonamides in fish tissue using micro-scale pipette tip-matrix solid-phase dispersion and fast liquid chromatography tandem mass spectrometry. Food Chem 194:508–515. https://doi.org/10.1016/j.foodchem.2015.08.050

Sheng W, Shi Y, Ma J, Wang L, Zhang B, Chang Q, Duan W, Wang S (2019) Highly sensitive atrazine fluorescence immunoassay by using magnetic separation and upconversion nanoparticles as labels. Microchim Acta:186. https://doi.org/10.1007/s00604-019-3667-3

Sheng W, Huang N, Liu Y, Zhang B, Zhang W, Wang S (2020) An ultrasensitive fluorescence immunoassay based on magnetic separation and upconversion nanoparticles as labels for the detection of chloramphenicol in animal-derived foods. Food Anal Methods 13(11):2039–2049. https://doi.org/10.1007/s12161-020-01820-5

Si F, Zou R, Jiao S, Qiao X, Guo Y, Zhu G (2018) Inner filter effect-based homogeneous immunoassay for rapid detection of imidacloprid residue in environmental and food samples. Ecotoxicol Environ Saf 148:862–868. https://doi.org/10.1016/j.ecoenv.2017.11.062

Song L, Zhang L, Huang Y, Chen L, Zhang G, Shen Z, Zhang J, Xiao Z, Chen T (2017) Amplifying the signal of localized surface plasmon resonance sensing for the sensitive detection of *Escherichia coli* O157:H7. Sci Rep 7:3288. https://doi.org/10.1038/s41598-017-03495-1

Sullivan JJ, Goh KS (2000) Evaluation and validation of a commercial ELISA for diazinon in surface waters. J Agric Food Chem 48:4071–4078. https://doi.org/10.1021/jf000432t

Sun L, Wang T, Sun Y, Li Z, Song H, Zhang B, Zhou G, Zhou H, Hu J (2020) Fluorescence resonance energy transfer between NH2–NaYF4:Yb,Er/NaYF4@SiO_2 upconversion nanoparticles and gold nanoparticles for the detection of glutathione and cadmium ions. Talanta 207:120294. https://doi.org/10.1016/j.talanta.2019.120294

Szuplewska A, Kulpińska D, Dybko A, Chudy M, Jastrzębska AM, Olszyna A, Brzózka Z (2020) Future applications of MXenes in biotechnology, nanomedicine, and sensors. Trends Biotechnol. https://doi.org/10.1016/j.tibtech.2019.09.001

Tabrizi MA, Shamsipur M, Saber R, Sarkar S, Ebrahimi V (2017) A high sensitive visible light-driven photoelectrochemical aptasensor for shrimp allergen tropomyosin detection using graphitic carbon nitride-TiO_2 nanocomposite. Biosens Bioelectron 98:113–118. https://doi.org/10.1016/j.bios.2017.06.040

Takemura K, Lee J, Suzuki T, Hara T, Abe F, Park EY (2019) Ultrasensitive detection of norovirus using a magnetofluoroimmunoassay based on synergic properties of gold/magnetic nanoparticle hybrid nanocomposites and quantum dots. Sensors Actuators B Chem 296:126672. https://doi.org/10.1016/j.snb.2019.126672

Vanegas D, Patiño L, Mendez C, Oliveira D, Torres A, Gomes C, McLamore E (2018) Laser scribed graphene biosensor for detection of biogenic amines in food samples using locally sourced materials. Biosensors 8:42. https://doi.org/10.3390/bios8020042

Verma S, Choudhary J, Singh KP, Chandra P, Singh SP (2019) Uricase grafted nanoconducting matrix based electrochemical biosensor for ultrafast uric acid detection in human serum samples. Int J Biol Macromol. https://doi.org/10.1016/j.ijbiomac.2019.02.121

Wagner AM, Knipe JM, Orive G, Peppas NA (2019) Quantum dots in biomedical applications. Acta Biomater 94:44–63. https://doi.org/10.1016/j.actbio.2019.05.022

Wang B, Park B (2020) Immunoassay biosensing of foodborne pathogens with surface plasmon resonance imaging: a review. J Agric Food Chem. https://doi.org/10.1021/acs.jafc.0c02295

Wang Z, Wu S, Colombi Ciacchi L, Wei G (2018) Graphene-based nanoplatforms for surface-enhanced Raman scattering sensing. Analyst 143:5074–5089. https://doi.org/10.1039/c8an01266k

Wang F, Han Y, Wang S, Ye Z, Wei L, Xiao L (2019) Single-particle LRET Aptasensor for the sensitive detection of Aflatoxin B1 with upconversion nanoparticles. Anal Chem 91:11856–11863. https://doi.org/10.1021/acs.analchem.9b02599

Wang G, Sun J, Yao Y, An X, Zhang H, Chu G, Jiang S, Guo Y, Sun X, Liu Y (2020a) Detection of Inosine Monophosphate (IMP) in meat using double-enzyme sensor. Food Anal Methods 13: 420–432. https://doi.org/10.1007/s12161-019-01652-y

Wang P, Wang A, Hassan MM, Ouyang Q, Li H, Chen Q (2020b) A highly sensitive upconversion nanoparticles-WS2 nanosheet sensing platform for *Escherichia coli* detection. Sensors Actuators B Chem 320:128434. https://doi.org/10.1016/j.snb.2020.128434

Wei C, Li M, Zhao X (2018) Surface-enhanced Raman Scattering (SERS) with silver nano substrates synthesized by microwave for rapid detection of foodborne pathogens. Front Microbiol 9:1–9. https://doi.org/10.3389/fmicb.2018.02857

Wei Q, Liu T, Pu H, Sun DW (2020) Determination of acrylamide in food products based on the fluorescence enhancement induced by distance increase between functionalized carbon quantum dots. Talanta 218:121152. https://doi.org/10.1016/j.talanta.2020.121152

Wei N, Wei MX, Huang BH, Guo XF, Wang H (2021) One-pot facile synthesis of green-emitting fluorescent silicon quantum dots for the highly selective and sensitive detection of nitrite in food samples. Dyes Pigments 184:108848. https://doi.org/10.1016/j.dyepig.2020.108848

Wen S, Zhou J, Zheng K, Bednarkiewicz A, Liu X, Jin D (2018) Advances in highly doped upconversion nanoparticles. Nat Commun 9:2415. https://doi.org/10.1038/s41467-018-04813-5

Wilhelm S (2017) Perspectives for upconverting nanoparticles. ACS Nano 11:10644–10653. https://doi.org/10.1021/acsnano.7b07120

Wu Z (2019) Simultaneous detection of Listeria monocytogenes and Salmonella typhimurium by a SERS-based lateral flow immunochromatographic assay. Food Anal Methods 12:1086–1091. https://doi.org/10.1007/s12161-019-01444-4

Wu Z, Xu E, Chughtai MFJ, Jin Z, Irudayaraj J (2017) Highly sensitive fluorescence sensing of zearalenone using a novel aptasensor based on upconverting nanoparticles. Food Chem 230: 673–680. https://doi.org/10.1016/j.foodchem.2017.03.100

Wu L, Lu X, Dhanjai, Wu ZS, Dong Y, Wang X, Zheng S, Chen J (2018a) 2D transition metal carbide MXene as a robust biosensing platform for enzyme immobilization and ultrasensitive detection of phenol. Biosens Bioelectron 107:69–75. https://doi.org/10.1016/j.bios.2018.02.021

Wu Z, He D, Cui B (2018b) A fluorometric assay for staphylococcal enterotoxin B by making use of platinum coated gold nanorods and of upconversion nanoparticles. Microchim Acta 185:1–8. https://doi.org/10.1007/s00604-018-3058-1

Xie H, Li P, Shao J, Huang H, Chen Y, Jiang Z, Chu PK, Yu XF (2019) Electrostatic self-assembly of Ti3C2Tx MXene and gold nanorods as an efficient surface-enhanced Raman scattering platform for reliable and high-sensitivity determination of organic pollutants. ACS Sensors 4: 2303–2310. https://doi.org/10.1021/acssensors.9b00778

Xiong Y, Pei K, Wu Y, Duan H, Lai W, Xiong Y (2018) Plasmonic ELISA based on enzyme-assisted etching of Au nanorods for the highly sensitive detection of aflatoxin B1 in corn samples. Sensors Actuators B Chem 267:320–327. https://doi.org/10.1016/j.snb.2018.04.027

Xu X, Ma X, Wang H, Wang Z (2018) Aptamer based SERS detection of Salmonella typhimurium using DNA-assembled gold nanodimers. Microchim Acta 185:325. https://doi.org/10.1007/s00604-018-2852-0

Xue L, Zheng L, Zhang H, Jin X, Lin J (2018) An ultrasensitive fluorescent biosensor using high gradient magnetic separation and quantum dots for fast detection of foodborne pathogenic bacteria. Sensors Actuators B Chem 265:318–325. https://doi.org/10.1016/j.snb.2018.03.014

Yadav A, Upadhyay Y, Bera RK, Sahoo SK (2020) Vitamin B6 cofactors guided highly selective fluorescent turn-on sensing of histamine using beta-cyclodextrin stabilized ZnO quantum dots. Food Chem 320:126611. https://doi.org/10.1016/j.foodchem.2020.126611

Yaghubi F, Zeinoddini M, Saeedinia AR, Azizi A, Samimi Nemati A (2020) Design of Localized Surface Plasmon Resonance (LSPR) biosensor for immunodiagnostic of *E. coli* O157:H7 using gold nanoparticles conjugated to the Chicken antibody. Plasmonics 15:1481–1487. https://doi.org/10.1007/s11468-020-01162-2

Yang T, Cai F, Zhang X, Huang Y (2015) Nitrogen and sulfur codoped graphene quantum dots as a new fluorescent probe for Au^{3+} ions in aqueous media. RSC Adv 5:107340–107347. https://doi.org/10.1039/c5ra20060a

Yang P, Zheng J, Xu Y, Zhang Q, Jiang L (2016) Colloidal synthesis and applications of plasmonic metal nanoparticles. Adv Mater 28:10508–10517. https://doi.org/10.1002/adma.201601739

Yang S, Li Y, Wang S, Wang M, Chu M, Xia B (2018) Advances in the use of carbonaceous materials for the electrochemical determination of persistent organic pollutants. A review. Microchim Acta 185:112. https://doi.org/10.1007/s00604-017-2638-9

Yi Y, Zeng W, Zhu G (2021) β-Cyclodextrin functionalized molybdenum disulfide quantum dots as nanoprobe for sensitive fluorescent detection of parathion-methyl. Talanta 222:121703. https://doi.org/10.1016/j.talanta.2020.121703

Yin M, Wu C, Li H, Jia Z, Deng Q, Wang S, Zhang Y (2019) Simultaneous sensing of seven pathogenic bacteria by guanidine-functionalized upconversion fluorescent nanoparticles. ACS Omega 4:8953–8959. https://doi.org/10.1021/acsomega.9b00775

Yola ML, Atar N (2017) Electrochemical detection of Atrazine by platinum nanoparticles/carbon nitride nanotubes with molecularly imprinted polymer. Ind Eng Chem Res 56:7631–7639. https://doi.org/10.1021/acs.iecr.7b01379

You S-M, Luo K, Jung J-Y, Jeong K-B, Lee E-S, Oh M-H, Kim Y-R (2020) Gold nanoparticle-coated starch magnetic beads for the separation, concentration, and SERS-based detection of *E. coli* O157:H7. ACS Appl Mater Interfaces 12:18292–18300. https://doi.org/10.1021/acsami.0c00418

Yüce M, Kurt H (2017) How to make nanobiosensors: surface modification and characterisation of nanomaterials for biosensing applications. RSC Adv 7:49386–49403. https://doi.org/10.1039/c7ra10479k

Yüce M, Kurt H, Hussain B, Ow-Yang CW, Budak H (2018) Exploiting Stokes and anti-Stokes type emission profiles of aptamer-functionalized luminescent nanoprobes for multiplex sensing applications. Chem Select 3:5814–5823. https://doi.org/10.1002/slct.201801008

Zhan S, Fang H, Fu J, Lai W, Leng Y, Huang X, Xiong Y (2019) Gold nanoflower-enhanced dynamic light scattering immunosensor for the ultrasensitive no-wash detection of *Escherichia coli* O157:H7 in milk. J Agric Food Chem 67:9104–9111. https://doi.org/10.1021/acs.jafc.9b03400

Zhan C, Liu B-W, Tian Z-Q, Ren B (2020) Determining the interfacial refractive index via ultrasensitive plasmonic sensors. J Am Chem Soc 142:10905–10909. https://doi.org/10.1021/jacs.0c01907

Zhang B, Li H, Pan W, Chen Q, Ouyang Q, Zhao J (2017) Dual-Color Upconversion Nanoparticles (UCNPs)-based fluorescent immunoassay probes for sensitive sensing foodborne pathogens. Food Anal Methods 10:2036–2045. https://doi.org/10.1007/s12161-016-0758-1

Zhang A, Tao G, Wang J (2018) Assembly of bioconjugated rod-nanotags and multilayer plasmonic nanorod-array for ultrasensitive SERS detection of *S. aureus* bacteria. J Nanopart Res 20:97. https://doi.org/10.1007/s11051-018-4200-z

Zhang D, Liu H, Geng W, Wang Y (2019a) A dual-function molecularly imprinted optopolymer based on quantum dots-grafted covalent-organic frameworks for the sensitive detection of tyramine in fermented meat products. Food Chem 277:639–645. https://doi.org/10.1016/j.foodchem.2018.10.147

Zhang L, Yin S, Hou J, Zhang W, Huang H, Li Y, Yu C (2019b) Detection of choline and hydrogen peroxide in infant formula milk powder with near infrared upconverting luminescent nanoparticles. Food Chem 270:415–419. https://doi.org/10.1016/j.foodchem.2018.07.128

Zhang B, Sheng W, Liu Y, Huang N, Zhang W, Wang S (2020a) Multiplexed fluorescence immunoassay combined with magnetic separation using upconversion nanoparticles as multi-color labels for the simultaneous detection of tyramine and histamine in food samples. Anal Chim Acta 1130:117–125. https://doi.org/10.1016/j.aca.2020.07.043

Zhang S-W, Sun Y-Y, Sun Y-M, Wang H, Li Z-F, Xu Z-L (2020b) Visual upconversion nanoparticle-based immunochromatographic assay for the semi-quantitative detection of sibutramine. Anal Bioanal Chem. https://doi.org/10.1007/s00216-020-02944-7

Zhao L, Wiebe J, Zahoor R, Slavkovic S, Malile B, Johnson PE, Chen JIL (2016) Colorimetric detection of catalase and catalase-positive bacteria (*E. coli*) using silver nanoprisms. Anal Methods 8:6625–6630. https://doi.org/10.1039/C6AY01453D

Zhao F, Yao Y, Jiang C, Shao Y, Barceló D, Ying Y, Ping J (2020) Self-reduction bimetallic nanoparticles on ultrathin MXene nanosheets as functional platform for pesticide sensing. J Hazard Mater 384:121358. https://doi.org/10.1016/j.jhazmat.2019.121358

Zheng L, Cai G, Wang S, Liao M, Li Y, Lin J (2019) A microfluidic colorimetric biosensor for rapid detection of *Escherichia coli* O157:H7 using gold nanoparticle aggregation and smart phone imaging. Biosens Bioelectron 124–125:143–149. https://doi.org/10.1016/j.bios.2018.10.006

Zhou D, Wang M, Dong J, Ai S (2016) A novel electrochemical immunosensor based on mesoporous graphitic carbon nitride for detection of subgroup J of Avian leukosis viruses. Electrochim Acta 205:95–101. https://doi.org/10.1016/j.electacta.2016.04.101

Zhou L, Zhang X, Ma L, Gao J, Jiang Y (2017) Acetylcholinesterase/chitosan-transition metal carbides nanocomposites-based biosensor for the organophosphate pesticides detection. Biochem Eng J 128:243–249. https://doi.org/10.1016/j.bej.2017.10.008

Zhou C, Zou H, Li M, Sun C, Ren D, Li Y (2018) Fiber optic surface plasmon resonance sensor for detection of *E. coli* O157:H7 based on antimicrobial peptides and AgNPs-rGO. Biosens Bioelectron 117:347–353. https://doi.org/10.1016/j.bios.2018.06.005

Zhou J, Ai R, Weng J, Li L, Zhou C, Ma A, Fu L, Wang Y (2020a) A "on-off-on" fluorescence aptasensor using carbon quantum dots and graphene oxide for ultrasensitive detection of the major shellfish allergen Arginine kinase. Microchem J 158:105171. https://doi.org/10.1016/j.microc.2020.105171

Zhou S, Lu C, Li Y, Xue L, Zhao C, Tian G, Bao Y, Tang L, Lin J, Zheng J (2020b) Gold nanobones enhanced ultrasensitive surface-enhanced raman scattering aptasensor for detecting *Escherichia coli* O157:H7. ACS Sensors 5:588–596. https://doi.org/10.1021/acssensors.9b02600

Zhu X, Liu P, Xue T, Ge Y, Ai S, Sheng Y, Wu R, Xu L, Tang K, Wen Y (2020) A novel graphene-like titanium carbide MXene/Au–Ag nanoshuttles bifunctional nanosensor for electrochemical and SERS intelligent analysis of ultra-trace carbendazim coupled with machine learning. Ceram Int. https://doi.org/10.1016/j.ceramint.2020.08.121

Electrochemical Biosensors for Food Safety Control in Food Processing

2

Malvano Francesca, Pilloton Roberto, and Albanese Donatella

Abstract

Foods from animal and plant origin may represent vehicles of different contaminants (chemical and microbiological) which are responsible for many foodborne diseases. Foods can be contaminated during all stages of the food chain by pathogenic bacteria or chemical compounds originated by environmental pollution or uncorrected use of crop protection products. Food safety is therefore a very important issue in the actual context of the intensive development of the food products. Nutrient monitoring and fast screening of contaminants represent some of the key issues in the agri-food field for assessment of food quality and safety. Conventional methods in food safety analysis are laborious, time-consuming, and require skilled technicians. The demand for the development of simple, fast, accurate, low-cost, and portable analytical instruments is growing and biosensors appear to meet these requirements. A biosensor is an analytical device used to quantify the target of interest in a sample. Generally, it comprises a biorecognition element which is specific toward the target. Molecular recognition events between the recognition element and the target compound elicit a physiochemical or biological signal, which is converted into a measurable quantity by the transducer. The choice of biological element and the optimum transducer depends on the properties of the sample of interest and the type of physical magnitude to be measured. The application of biosensors in food safety analysis sheds new light on the efficient and rapid detection of foodborne toxins,

M. Francesca · A. Donatella (✉)
Department of Industrial Engineering, University of Salerno, Fisciano, SA, Italy
e-mail: dalbanese@unisa.it

P. Roberto
Department of Chemistry and Material Technology, Institute of Crystallography of the National Council of Research (CNR), Monterotondo Scalo, Rome, Italy

P. Chandra, P. S. Panesar (eds.), *Nanosensing and Bioanalytical Technologies in Food Quality Control*, https://doi.org/10.1007/978-981-16-7029-9_2

allergens, pathogens, toxic chemicals, heavy metals, and other contaminants. In particular, among the variously reported biosensors, electrochemical biosensors have been very popular and widely used due to their simple and well-understood bio-interaction and detection process. Electrochemical biosensors are based on the measurement of the electrical properties of the sample due to the chemical reaction between immobilized biomolecules and the analyte of interest; they use a transducer where electrochemical signals are generated during biochemical reactions and are monitored using suitable potentiometric, amperometric, conductimetric, impedimetric systems of analysis. Therefore, electrochemical biosensors represent a promising tool for food analysis due to the possibility of satisfying specific demands that the classic methods of analysis do not attain: advantages as high selectivity and specificity, which allows the detection of a broad spectrum of analytes in complex samples with minimum sample pretreatment, relatively low cost of construction, the potential for miniaturization, easier automation, and simple and portable equipment construction. Based on the above, this chapter wants to provide general information about biosensors and to highlight the current situation in the literature on electrochemical biosensors for the detection of some microbiological and chemical hazards in food processing.

Keywords

Food safety · Label-free biosensors · Electrochemical impedance spectroscopy

2.1 Introduction

Illnesses resulting from foodborne diseases have become one of the most widespread public health problems in the world today. Internationally, foodborne diseases associated with microbial pathogens, toxins, and chemical contaminants in food present a serious threat to the health of millions of individuals (Redmond and Griffith 2003). Therefore, the assessment of food safety is one key area for the modern food industry. Food from animal and plant origin may represent vehicles of chemical and microbiological contaminants which are responsible for many foodborne diseases. Foods can be contaminated during all stages of the food chain by pathogenic bacteria or chemical compounds originated by environmental pollution or uncorrected use of crop protection products. Food safety is therefore a very important issue in the actual context of the intensive development of food products. The monitoring and fast screening of contaminants represent some of the key issues in the agri-food field for assessment of food quality and safety. Conventional methods in food safety analysis are expensive, laborious, time-consuming, and require skilled technicians (Campuzano et al. 2017). The demand for the development of simple, fast, accurate, low-cost, and portable analytical instruments able to monitor the presence of food hazards is a primary need in the food industry and the biosensors appear to meet these requirements. International Union of Pure and Applied Chemistry (IUPAC) proposed a very stringent definition of a biosensor: "*A biosensor is a self-confident*

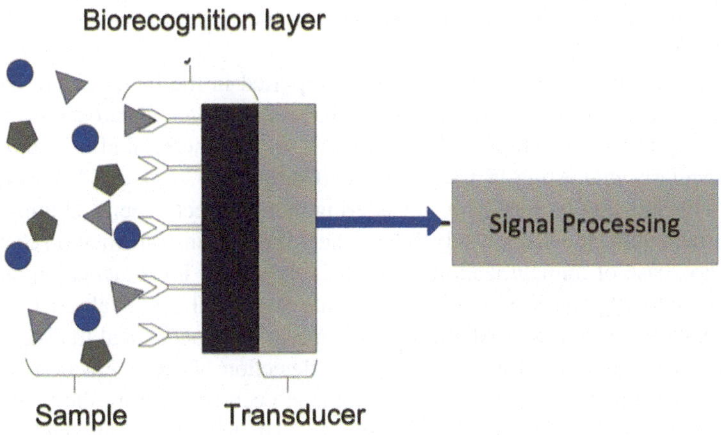

Fig. 2.1 Schematic diagram of a biosensor

integrated device which is capable of providing specific quantitative or semi-quantitative analytical information using a biological recognition element (biochemical receptor) which is in direct spatial contact with a transducer element. A biosensor should be clearly distinguished from a bioanalytical system which requires additional steps, such as reagents addition" (Thevenot et al. 2001). Briefly, a biosensor can be defined as an analytical device characterized by a biological recognition element in close or integrated with a detector to identify the presence of one or more specific analytes and their concentrations in a sample (Fig. 2.1). A biosensor aims to provide rapid, real-time, and reliable information about the biochemical composition of its surrounding environment; ideally, it is a device that is capable of responding continuously, reversibly without perturbing the sample (Chandra et al. 2012; Choudhary et al. 2016; Deka et al. 2018; Mahato et al. 2018; Verma et al. 2019).

Biosensors can be classified in agreement with the type of recognition element or the type of signal transduction. As regards the first classification, biosensors are divided into two main groups: catalytic and affinity biosensors. In the first case, the recognition element can be characterized by enzymes, whole cells (bacteria, fungi, cells, yeast), cell organelles, and plant or animal tissue slices. The catalytic sensors have the longest tradition in the field of biosensors: historically, glucose sensing has dominated the biosensor literature and has delivered huge commercial successes to the field. As concerns the affinity biosensors, the biomolecule can be represented by chemoreceptors, antibodies, nucleic acids; they provide selective interactions with a ligand to form a thermodynamically stable complex. The most developed examples of biosensors using complexing receptors are immunosensors, based on the interaction process between an antigen with its specific antibody.

Related to the classification based on transducers, a wide variety of transduction techniques have been developed in biosensing technology; in particular, the most common are optical, piezoelectric, calorimetric and electrochemical (Thakur and Ragavan 2012).

It is fair to support that most biosensors reported in the literature are based on electrochemical transducers: recent studies have shown that electrochemical-based sensors are the most common and, in particular, electrochemical affinity biosensors are particularly interesting in food analysis (Campuzano et al. 2017; Roariu et al. 2016). This may not be surprising considering that the electrochemical transduction shows, more than others, many advantages including low instrumentation costs, high sensitivity, ease of miniaturization, and relatively simple instrumentation; all these features are highly compatible with portable devices (Malvano et al. 2020).

Furthermore, is worth highlighting that, among electrochemical transducers, the impedimetric ones are optimal for *label-free* detection of bio-interaction, which is based on the direct measurement of phenomena occurring during the biochemical reactions on a transducer surface, concerning a "labeled" detection which relies on the investigation of a specific label (fluorophores, magnetic beads, active enzyme, etc.) (Daniels and Pourmand 2007).

Electrochemical Impedance Spectroscopy is, in fact, a powerful, non-destructive and informative technique, which can be used to study the electrical properties of the sensing device interface and tracing the reaction occurring on it. The application of impedance as a transduction technique, based on the direct monitoring of the interaction between the bioreceptor and its target, enables the production of label-free biosensors for food analysis with significant advantages over labeled ones. By avoiding the laborious and expensive labeling steps, which can cause loss of affinity between the labeled receptor and its target and decrease reproducibility, sensitivity, and selectivity of the biosensor, the use of label-free monitoring reduces biosensor costs and allows analysis in a short time (Rhouati et al. 2016). Thanks to the EIS transduction technique, food biosensor analysis is performed in real-time by studying the change in electrical properties of the electrode surface which depends only on the binding interaction between the analyte and its receptor.

Thus, to respond to the need for food safety control, label-free affinity biosensors can be considered as the most relevant devices for fast measurements of food hazards in food processes, being able to detect a wide range of chemical and microbiological risks through the use of appropriate biomolecules.

In this regard, the last decade has observed phenomenal growth in the field of electrochemical affinity biosensors for analyses of food and beverage, in particular for food safety monitoring.

2.2 Electrochemical Biosensor for Food Safety

This chapter provides an overview of the potential application of electrochemical biosensors for the analysis of chemicals and microorganisms that affect food safety, discussing some examples of the latest advances in this field. A focus on the most commonly responsible for food contaminations, including toxins and mycotoxins, pesticides, pathogenic bacteria will be presented, pointing out the advantages of electrochemical transduction techniques applied on affinity biosensors.

2.2.1 Mycotoxins

Mycotoxins are a varied group of toxic secondary metabolites produced by molds. They are thermally stable and notoriously toxic, teratogenic, mutagenic, and carcinogenic, which can enter into the human food chain causing severe impact on human health. The risk of mycotoxins are well-recognized worldwide and also the incidence of these compounds is a universal problem: they affect a broad range of agricultural products including cereals, cereal-based foods, dried fruits, wine, milk, coffee beans, cocoa, bakery and meat products, which are the basis of the economies of many developing countries (Evtugyn and Hianik 2019).

The most relevant mycotoxins under a toxicological and legislative point of view are the ochratoxins and aflatoxins; their chemical structures are represented in Fig. 2.2.

In the latest years, there has been a significant effort to improve analytical approaches for the effective determination of mycotoxins: common analytical methods like capillary electrophoresis, and chromatography techniques linked to mass spectrometry (LC-MS, GC-MS), reliable but characterized by sophisticated and expensive instruments and not suitable for real-time and on-site application, try to be replaced with innovative biosensor technologies to obtain reliable, fast, and sensitive measurements with high selectivity and reduced cost.

Among mycotoxins, Ochratoxin A (OTA) is one of the most abundant in a wide range of agricultural commodities, ranging from cereals grains to dried fruits to wine and coffee, in a few micrograms per kilogram amounts. In the European Union, the

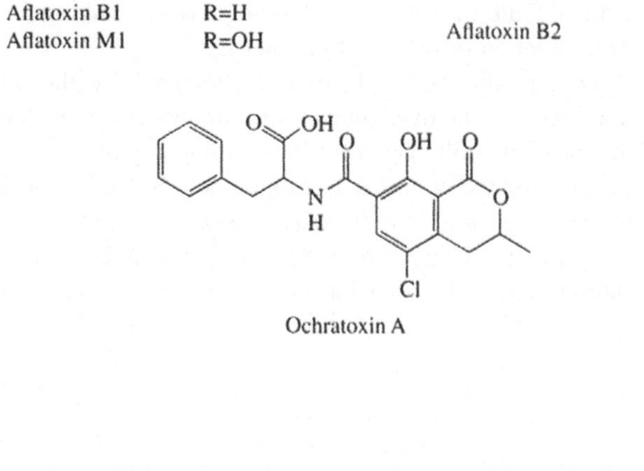

Fig. 2.2 Chemical structures of the main aflatoxins and ochratoxins

maximum limits established for OTA in different food products are fixed in Commission Regulation (EC) N° 1881/2006 and ranged from 10 μg/kg for instant coffee and dried fruits to 0.5 μg/kg for dietary foods intended specifically for infants.

Different electrochemical immunosensors have been reported in the literature for the detection of OTA amount in food matrices at least equal to the acceptable limits of OTA allowed by regulation.

Badea et al. (2016) immobilized monoclonal antibody on screen-printed gold electrodes through bovine serum albumin used as "anchor" for the covalent immobilization of the anti-OTA antibodies: all the steps of the immunosensor construction and also the immunochemical reaction between surface-bound antibody and OTA were analyzed using cyclic voltammetry and electrochemical impedance spectroscopy. The specific interaction between antibody and OTA induces an increase in electron transfer resistance at the interface sensor/solution that is correlated with the concentration of OTA in the sample: the detection of OTA was achieved by EIS in the linear range 2.5–100 ng/mL. The developed immunosensor was also used to detect OTA amounts in licorice extracts samples.

Malvano et al. (2016a) proposed two different antibody immobilization techniques on gold electrodes: oriented and not oriented. The comparison between the two monoclonal anti-OTA immobilization procedures underlined the advantages of oriented one, which showed a more ordered and homogeneous antibody layer that guarantees a higher number of molecules effectively exposed to antigen interaction. The linear range (0.05–25 μg/kg), the very low detection limit (0.05 μg/kg), and high sensitivity (26.45 kΩ mL/ng) showed the potential of the immunosensor as a highly capable analytical device for fast measurement of OTA traces. Tests with cocoa beans were also performed by the authors to study the feasibility of applying the immunosensor for the detection of OTA in food samples.

To exploit the advantages of cheap electrodes, characterized by low-cost fabrication and mass production, Malvano et al. (2016b) proposed a capacitive OTA immunosensor on screen-printed carbon electrode modified with electrodeposited gold nanoparticles. Using the electrochemical impedance spectroscopy it was observed that the capacitance was the best parameter that described the reproducible change in electrical properties of the electrode surface at different OTA concentrations, and it was used to investigate the analytical performances of the developed immunosensor. Under optimized conditions of monoclonal antibody amount, the immunosensor showed a wide linear range between 0.3 and 20 ng/mL with a limit of detection of 0.34 ng/mL, making it suitable for the analytical determination of OTA in food matrices.

Despite the high selectivity guaranteed by the use of antibodies, the main drawback for the development of immunosensors is due to the high cost of specific monoclonal antibodies used for the biorecognition process. Nucleic acid aptamers, obtained by the in vitro selection process SELEX, represent an alternative approach of receptors for affinity biosensors production. The use of aptamers as biomolecular recognition is justified by their low-cost synthesis, high reproducibility, and higher stability due to their nucleic-acid chemical nature. Additionally, they can be easily combined with different chemical labels/groups that provide flexibility for adaptation to different platforms (Miranda-Castro et al. 2016).

In addition to the choice of aptamers as alternative biorecognition molecules, nanostructured platforms based on conductive materials, including conducting polymers, gold nanoparticles (AuNPs), quantum dots (QDs), magnetic beads and carbon nanomaterials, represented, in the latest years, an interesting approach for electrochemical signal enhancement, to improve sensitivity and the stability of biomolecules activity (Campuzano et al. 2017).

To improve the electrical conductivity of the non-homogeneous electrode surfaces, Rivas et al. (2015) developed an impedimetric biosensor using a 3-′-aminated aptamer selective to OTA recognition. The immobilization of the aptamer was carried out, on screen-printed carbon electrodes modified with an electropolymerized film of polythionine and iridium oxide nanoparticles (IrO2 NPs). The aptasensor showed the lowest limits of detection reported so far label-free impedimetric detection of OTA, equal to 5.65 ng/kg.

Mejri-Omrani et al. (2016) covered the surface of a gold electrode with a conductive polypyrrole layer and used fourth-generation polyamide amine dendrimers for the covalent immobilization of an aptamer for OTA detection formed by 36 nucleotides with the sequence NH2-(CH2)6–-5′GATCGGGTGTGGGTGGCGTAAAGGGAGCATCGGACA-3′.

The aptasensor showed a range of up to 5 μg/L and a detection limit of 2 ng/L of OTA, and no matrix effects were observed during the analysis of OTA in red wine.

In more recent years, metallic nanomaterials advantage was exploited for the development of electrochemical label-free aptasensors. Gold nanoparticles combined with carboxylic porous carbon represented an excellent carrier for both the immobilization of DNA-aptamers and the amplification of the impedimetric signal (Wei and Zhang 2017). Under optimized conditions, the change in the charge transfer resistance of the electrode showed a log-linear relationship to OTA concentration in the range 10^{-8}–0.1 ng/mL, with the limit of detection equal to 10^{-8} ng/mL. Recovery studies were performed in soybean samples by spiking 10^{-6} ng/mL and recoveries ranged from 95% to 108%.

A more complex structure based on bimetallic (Cu–Co) Prussian Blue analogs (PBAs) coupled to gold nanoparticles was used to develop an impedimetric aptasensor (Gu et al. 2019). The chemical composition and crystal structure of the bimetallic matrix guaranteed excellent electrochemical conductivity and strong aptamer binding interaction, achieving a very low limit of detection equal to 5.2 fg/mL.

In addition to Ochratoxins, Aflatoxins are a group of mycotoxins characterized by a great carcinogenic power. Coupling the advantages and the effectiveness of monoclonal antibodies with different strategies for signal enhancement, a lot of electrochemical label-free immunosensors were proposed in literature characterized by satisfactory performances.

Li et al. (2017) constructed a label-free impedimetric immunosensor based on gold three-dimensional nanotube ensembles: AFB1 monoclonal antibodies were immobilized on the surface using a staphylococcus protein A layer, obtaining a limit of detection equal to 1 pg/mL. In another example, Costa's group reported an impedimetric immunosensor based on carbon nanotubes and an Au electrode for monitoring AFB1 (Costa et al. 2017): in this immunosensor, the carbon nanotubes exhibited an exceptional surface/volume ratio and excellent electrical properties.

Bhardwaj et al. (2019) showed an immunosensor in which anti-AFB1 was immobilized on the surface of an ITO glass electrode coated with graphene QDs and AuNPs: the electrocatalytic activity of the AuNPs improved the electronic properties of the composite GQDs-AuNPs, reaching a linear range from 0.1 to 3.0 ng/mL. Yagati's group reported an impedimetric immunosensor that selectively detects AFB1 at the lowest level by utilizing polyaniline nanofibers (PANI) coated with gold (Au) nanoparticles composite-based indium tin oxide (ITO) disk electrodes. The Au-PANI acted as an effective sensing platform having high surface area, electrochemical conductivity, and biocompatibility which enabled greater loading deposits of capture antibodies. As a result, the presence of AFB1 has screened in a linear range 0.1–100 ng/mL with a detection limit of 0.05 ng/mL (Yagati et al. 2018).

A platform of Poly(3,4-ethylenedioxythiophene) (PEDOT) and graphene oxide (GO) composite decorated with spherical gold nanoparticles (AuNPs) has been used for the immobilization of anti-aflatoxin B1 covalently immobilized using EDC/NHS coupling. The proposed amperometric immunosensor exhibits a very high sensitivity within two linear range of 0.5–20 ng/mL and 20–60 ng/mL, respectively (Sharma et al. 2018).

A ferrocene-modified gold electrode was proposed by Malvano et al. (2019) as a platform for the immobilization of monoclonal anti-AFB1. In this work, the authors developed a label-free immunosensor, using the impedimetric technique, characterized by linearity in the range 0.01–10 ng/mL and a limit of detection of 0.01 ng/mL.

In more recent years, different electrochemical aptasensors with optimum performances were developed in alternatives to immunosensors. A very novel magnetically assembled aptasensing device has been designed for label-free determination of AFB1 by employing a disposable screen-printed carbon electrode covered with a polydimethylsiloxane (PDMS) film (Wang et al. 2018a, b). The bio-probes were firstly prepared by immobilization of the thiolated aptamers on the Fe_3O_4Au magnetic beads, which were rapidly assembled on the working electrode of SPCE, by using a magnet placed at the opposite side. The developed method allowed the construction of an impedimetric aptasensor with a wide linear range between 20 pg/mL and 50 ng/mL with a low detection limit of 15 pg/mL, opportunely used in peanuts samples.

Aptamer against AFM1 was immobilized on a glassy carbon electrode covered with polymeric neutral red (NR) dye obtained by electropolymerization. In the presence of AFM1, the cathodic peak current related to the NR conversion decreases and an increase of the charge transfer resistance measured by electrochemical impedance spectroscopy was observed. In optimal conditions, this makes it possible to determine AFM1 from 5 to 120 ng/L in standard solutions with a limit of detection of 0.5 ng/L. The aptasensor was validated on the spiked samples of cow and sheep

Table 2.1 Electrochemical affinity biosensors for mycotoxins detection

Analyte	Interface	Transduction technique	Range	LOD	Reference
Patulin	AuE/ZnONRs/ AuNPs/Apt	DPV	0.5–50 ng/mL	0.27 pg/ mL	He and Dong (2018)
Zearalenone	SPCE/BSA/MAb	DPV	0.25–256 ng/ mL	0.25 ng/ mL	Yugender Goud et al. (2017)
Zearalenone	AuE/p-PtNTs/ AuNPs/thionin labeled GO	AMP	0.5 pg/ mL–0.5 μg/ mL	0.17 pg/ mL	He and Yan (2019)
Fumonisin F1	GCE/AuNPs/Apt	EIS	0.1–100 nM	2 pM	Chen et al. (2015)
Zearalenone	GCE/Au-Pt NPs/MAb	DPV	0.005–50 ng/mL	0.5 pg/ mL	Liu et al. (2017)
Zearalenone	GCE/chitosan/ conjugate of zearalenone with BSA	DPV	10 pg/ mL–1000 ng/ mL	4.7 pg/ mL	Xu et al. (2017a, b)
Zearalenone	SPCE/Fe_2O_3/HRP	DPV	1.88–45 ng/mL	0.57 ng/ mL,	Regiart et al. (2018)
DON	GCE/AuNPs/ 4nitrophenylazo	EIS	6–30 ng/ mL	0.3 ng/ mL	Sunday et al. (2015)
DON	SPCE/AuNPs/ Polypyrrole/Ab	DPV	0.05–1 ppm	8.6 ppb	Lu et al. (2016)
Fumonisin B1	SPCE/AuNPs/ Polypyrrole/Ab	DPV	0.2–4.5 ppm	4.2 ppb	Lu et al. (2016)
Fumonisin B1	GCE/chitosan/DON-BSA	DPV	0.01–1000 ng/ mL	5 pg/ mL	Qing et al. (2016)
T-2 Toxin	GCE/chitosan/DON-BSA	DPV	0.01–100 μg/ mL	0.13 μg/ mL,	Wang et al. (2018a, b)

AuE:gold electrode; SPCE: screen-printed carbon electrode; ZnONRs: ZnO nanorods; Apt: aptamer; BSA: bovine serum albumin; MAb: monoclonal antibody; p-PtNTs: porous platinum nanotubes; AuNPs: Gold nanoparticles; GCE: glassy carbon electrode; Au-Pt NPs: gold-platinum nanoparticles; HRP: Horseradish peroxidase

milk, reaching a reliable detection of the 40–160 ng/kg of mycotoxin (Smolko et al. 2018).

Ochratoxins and Aflatoxins are the most common mycotoxins present in the food sample, but there are other substances, less common but not with less harmful effects on human health. The following table (Table 2.1) summarizes some recent example of electrochemical affinity biosensors developed for the detection of different toxins and mycotoxins.

2.2.2 Pathogenic Bacteria

Foodborne illnesses caused by pathogenic bacteria represent an important threat to the health of people. Pathogens are infectious agents that cause disease; they include microorganisms such as fungi, bacteria, and molecular scale infectious agents including viruses and prions. Among these, *Escherichia coli* O157:H7, *Salmonella*, *Listeria monocytogenes*, *Campylobacter*, *Helicobacter*, *Staphylococcus aureus*, and *Bacillus cereus* are the most common and are responsible for approximately 90% of all foodborne diseases (Dye 2014).

Conventional methods for pathogenic bacterial identification involve various culturing techniques and different biochemical tests which are very time-consuming, requiring 2–4 days. Analysis time and sensitivity are the most important limitations related to the usefulness of bacterial testing. An extremely selective detection methodology was also required because low numbers of pathogenic bacteria are often present in a complex biological sample along with many other nonpathogenic bacteria. Tedious and time-consuming detection methods have prompted several groups in recent years to develop other techniques to reduce the detection time like Polymerase Chain Reaction (PCR) and Enzyme-Linked Immunosorbent Assay (ELISA). However, both techniques have limitations that exclude their extensive implementation. These limitations include accurate primer designing, the requirement of specific labeled secondary antibodies, and their failure to distinguish spore viability (Cesewski and Johnson 2020).

Recently, numerous electrochemical biosensors have been developed using impedimetric, potentiometric, and voltammetric techniques for the detection of several bacteria and parasites: a lot of novel approaches of working modification were carried out to develop very sensitive electrochemical biosensors.

E. coli are bacteria that naturally occur in the intestinal tracts of humans and warm-blooded animals to help the body synthesize vitamins. One pathogenic strain, *E. coli* O157:H7, produces toxins that damage the lining of the intestine, causes anemia, stomach cramps and bloody diarrhea, and serious complications called hemolytic uremic syndrome (HUS) and thrombotic thrombocytopenic purpura (TTP). Several electrochemical biosensors have been developed for the detection of this pathogenic bacteria in food products (Doyle 1991).

An electrochemical immunosensor for rapid detection of *E. coli* O157:H7 have been proposed by Xu et al. (2017a, b): the immunosensor was prepared by layer-by-layer assembly involving the formation of 11-amino-1-undecanethiol self-assembled monolayer onto a gold electrode and the immobilization of AuNPs followed by the incorporation of Chitosan-MWCNTs–SiO_2/thionine nanocomposites and AuNPs multilayer films. Finally, anti-E.coli O157:H7 antibodies were covalently bound and electrochemical impedance spectroscopy was used to obtain a calibration curve for heat-killed *E. coli* O157:H7, by measuring the increase in the charge transfer resistance as the antigen concentration increased. The working range was 4.12×10^2–4.12×10^5 CFU/mL.

Gold microelectrodes modified with maleic anhydride/(hycroxyethyl)-methacrylate polymer film were used to immobilize anti-*E.coli* and to develop a capacitive label-free immunosensor able to detect *E.Coli* cells at least equal to 70 CFU/mL (Idil et al. 2017). Graphene electrodes were modified with chitosan/polypyrrole/carbon nanotubes/gold nanoparticles layer (Guner et al. 2017) and CuO/cysteine (Pandey et al. 2017) for the immobilization of monoclonal antibodies to detect *E. Coli* O157:H7 at least equal to 30 CFU/mL and 3.8 CFU/mL, respectively.

Malvano et al. proposed two different impedimetric immunosensors for the sensitive detection of *E. coli* O157:H7: in the first one, monoclonal antibodies were immobilized on a strontium titanate perovskite layer ($SrTiO_3$) synthesized on a platinum electrode. Under optimized conditions, the capacitive immunosensor showed a detection range from 10^1 to 10^7 CFU/mL and an LOD of 10 CFU/mL (Malvano et al. 2018a). A lower limit of detection (3 CFU/mL) was found afterward exploiting the high conductive properties of ferrocene-modified gold electrodes use as a platform for the antibodies immobilization. The immunosensor was used to analyze milk and meat samples obtaining a good agreement with the results of ELISA analysis (Malvano et al. 2018b).

More recently, Jafari et al. (2019) used a TEOS/MTMS sol-gel on gold microelectrodes to immobilize monoclonal antibodies for *E. Coli* O157:H7 detection. Through electrochemical impedance spectroscopy transduction technique, the immunosensor was able to detect the microorganism with a limit of detection equal to 1 CFU/mL. The same limit of detection was reached by Wilson et al. (2019) using an Ag-interdigitated microelectrode array through the immobilization of a peptide as a biorecognition element.

As regards other pathogenic bacteria responsible for foodborne diseases, Sheikhzadeh et al. (2016) reported the combination of poly[pyrrole-co-3-carboxyl-pyrrole] copolymer and aptamer for the development of a label-free electrochemical biosensor suitable for the detection of *S. Typhimurium.* Impedimetric measurements were facilitated by the effect of the aptamer/target interaction on the intrinsic conjugation of the copolymer and subsequently on its electrical properties. The aptasensor detected *S. Typhimurium* in the concentration range 10^2–10^8 CFU/mL with high selectivity and with a limit of quantification of 100 CFU/mL and a limit of detection of 3 CFU/mL. The suitability of the aptasensor for real sample detection was demonstrated via recovery studies performed in spiked apple juice samples.

A label-free impedimetric aptamer-based biosensor for S. *typhimurium* was also fabricated by grafting a diazonium-supporting layer onto SPCEs followed by the immobilization of an aminated-aptamer. This strategy allowed obtaining a dense aptamer layer, which resulted in high sensitivity with a limit of detection of 8 CFU/mL (Bagheryan et al. 2016). Also, a novel outer membrane antigen (OmpD) was used for the first time as a surface biomarker for detecting *S. typhimurium*. Anti-OmpD antibody was used as detector probe in an impedimetric immunosensor using graphene-graphene oxide-modified SPCEs. The developed method was able to selectively detect *S. typhimurium* in spiked water and juice samples with a sensitivity up to 10 CFU/mL (Mutreja et al. 2016).

Izadi et al. (2016) proposed an electrochemical DNA-based biosensor for *Bacillus cereus* in milk and infant formula. They explored AuNPs to prepare a modified pencil graphite electrode that could detect Bacillus cereus as low as 100 CFU/mL.

Gold-interdigitated electrode arrays were realized for the detection of *L. monocytogenes*, using polyclonal antibodies: the devices were able to detect until 160 CFU/mL (Chen et al. 2016) and 39 CFU/mL of bacteria (Wang et al. 2017a, b), using different antibodies immobilization techniques.

Other electrochemical biosensing platforms have also been reported for the determination of *S.aureus*. CNT-coated Au-tungsten microwire electrodes (Yamada et al. 2016) and PEI/CNT composite on Au microwire electrode (Lee and Jun 2016) were used as a platform for the immobilization of polyclonal antibodies. Both the biosensors were able to show the same LOD of 100 CFU/mL. Higher performance in the detection limit was reached by Primiceri et al. (2016) who proposed a biochip based on an interdigitated microelectrode array able to quantitatively detect two of the most common food-associated pathogens, *Listeria monocytogenes* and *Staphylococcus aureus*, with a detection limit as low as 5.00 CFU/mL for *L. monocytogenes* and 1.26 CFU/mL for *S. aureus*.

2.2.3 Pesticides

According to the US Environmental Protection Agency (EPA), pesticides are defined as any substance or mixture of substances intended for repelling, destroying, or controlling any pest. Due to their high insecticidal activity, they are widely used in agriculture to protect crops and seeds by destroying insects, bacteria, and rodents and other weed animals (World Health Organization 2016).

However, the presence of pesticide residue in food, water, and soil has become a very critical problem in environmental chemistry.

Pesticides are classified in several ways, according to their toxicity (dangerous, highly dangerous, moderately dangerous, and slightly dangerous) and their lifetime (permanent, persistent, moderately persistent, and not persistent). Often, they are classified according to the use as insecticides, miticides, herbicides, nematicides, fungicides, molluscicides, and rodenticides. Referring to the chemical structure, the commonly reported main classes are organochlorines, organophosphates, carbamates, and pyrethroids. In addition to these common classes of pesticides, there are other chemical classes employed as herbicides, hormonal, amides, nitro compounds, benzimidazoles, bipyridyl compounds, ethylene dibromide, sulfur-containing compounds, copper, or mercury, among others. (Garcia et al. 2012).

The monitoring and the fast quantification of pesticides and their residues have become extremely important to ensure compliance with legal limits. The analysis of these compounds is an important issue due to their potential bioaccumulation, high toxicity, and their long-term damage risk, also for the use at low concentration. Food safety assurance requires fast and easy analytical tools to work alongside confirmatory methods such as chromatography coupled to mass spectrometry that require very expensive equipment, long analysis times, high reagent sample volumes, and qualified personnel (Kumar et al. 2015). Due to these limits, alternative methodologies for pesticide detection have been recommended in the last few years: the most relevant ones are those based on electrochemical methods. Table 2.2 summarizes the strategies and features of the electrochemical immunosensors developed for the quantification of different kinds of pesticides in food products.

The most used approach for the electrochemical label-free biosensors based on non-competitive pesticides detection was Electrochemical Impedance Spectroscopy but also voltammetry technologies were adopted.

In 2017, a very innovative enzyme inhibition-based biosensor, immobilizing AChE enzyme on cysteamine-modified electrode, was proposed to sensitively detect carbamate and organophosphate compounds with an extremely fast response. The working principle of the biosensor is based on the high-affinity interaction between the investigated pesticides (Carbaryl, Paraquat, Kresomix–Methyl, Dichlorvos, Chlorpyrifos–Methyl Pestanal, Phosmet) and the active site of the enzyme. The capability of CBs and OPs compound to form a very stable complex with the enzyme causes an impedimetric change, allowing to go up very fast to the presence of the toxic compounds in food matrices. The proposed biosensor showed linearity between 5 and 170 ppb for carbamates and 2.5–170 ppb for organophosphate compounds (Malvano et al. 2017).

As highlighted above, also for pesticide detection nucleic acid aptamers have represented an alternative approach in the biosensor field. Novel aptasensors based on the impedimetric and voltammetric transduction techniques were developed in the last years; strategies and features are summarized in Table 2.3.

Detection limits at the picomolar level are reported for a lot of the developed assays and the proposed sensors show that the combination of novel transduction materials and strategies with improved recognition elements can push toward lower and lower achievable detection limits.

2.3 Future Perspectives

Ensuring food safety is the main interest both for the food industry and for consumers. The guarantee of food safety requires fast and specific controls for all contaminants, chemicals, and bacteria, which are harmful to human health.

Despite common analytical techniques that are time-consuming, require highly trained personnel, are expensive and require steps of sample pretreatment, increasing the time of analysis, among food and beverage industries exists a growing demand in biosensing technologies as simple, rapid, cheap, low-cost, and portable analytical devices for the monitoring of chemical and microbiological contaminants (toxins, mycotoxins, pathogenic bacteria, pesticides, and allergens) that endanger the food safety. In particular, the electrochemical biosensors systems have been demonstrated to have advantages like portability, shortened analysis time, ease of operation, novice-friendly, and direct analysis with no sample preparation procedures. Thus, the electrochemical sensing arrays have been acknowledged as reliable tools for

Table 2.2 Electrochemical immunosensors for pesticides detection in food

Analyte	Interface	Transduction technique	Range [M]	LOD [M]	Reference
Carbofuran	gelatin/Ab/GA/L-Cys/Au electrode	EIS	4.52×10^{-10}–4.52×10^{-6}	4.52×10^{-10}	Liu et al. (2015)
Chlorpyrifos	Ab/protein A/AuNPs/PDDA/gold IDAMs	EIS	1.43×10^{-9}–1.43×10^{-6}	1.43×10^{-9}	Jia et al. (2015)
Chlorpyrifos	BSA/Ab/protein A/gold IDAMs	EIS	2.85×10^{-9}–2.85×10^{-4}	3.99×10^{-11}	Cao et al. (2015)
2,4-D	Ab/AuNPs-PANABAMWCNTs/SPE	EIS	4.52×10^{-9}–4.52×10^{-7}	1.36×10^{-9}	Fusco et al. (2017)
Parathion	Ab/fG/SPE	EIS	3.43×10^{-13}–3.43×10^{-9}	1.79×10^{-13}	Mehta et al. (2016)
Atrazina	BSA/Ab/AuNPs/Au	DPV	$2.32{*}10^{-10}$–2.32×10^{-9}	7.42×10^{-11}	Liu et al. (2014a, b)
Chlorpyrifos	BSA/Ab/GS-MB/AuNPs/GCE	CV	2.85×10^{-9}–1.43×10^{-6}	1.60×10^{-10}	Qiao et al. (2014)
Atrazine	ATR-BSA/PAMAM/AET/Au/GCE	CV	4.64×10^{-11}–4.64×10^{-6}	5.56×10^{-9}	Giannetto et al. (2014)
Chlorpyrifos	BSA/Antigen/Co3O4/PANI/ITO	CV	0–2.85×10^{-5}	2.85×10^{-8}	Wang et al. (2017a, b)
Paraquat	GEC	SWV	1.20×10^{-8}–2.63×10^{-7}	3.11×10^{-11}	Liu et al. (2014a, b)

GA: glutaraldehyde; L-cys: L-cysteine; PDDA: poly (diallydimethylammonium chloride); IDAMs: interdigitated array microelectrodes; BSA: bovine serum albumin; PANABA: polymer poly-(aniline-co-3-aminobenzoic acid); MWCNTs: multi-walled carbon nanotubes; SPE: screen-printed electrode; fG: graphene sheets functionalized; GS-MB: graphene sheets-methylen blue; GS-PEI: graphene sheets-ethyleneimine polymer; ATR: atrazine; PAMAM: poliamidoaminic dendrimers; AET: 2-aminoethanethiol; ITO: indium tin oxide; SWNTs; GEC: graphite composite electrode

Table 2.3 Electrochemical aptasensors for pesticides detection in food

Analyte	Interface	Transduction technique	Range [M]	LOD [M]	Reference
Carbendazim	MCH/aptamer/Au electrode	EIS	5.23×10^{-11}–5.23×10^{-8}	4.29×10^{-11}	Eissa and Zourob (2017)
Acetamiprid and atrazine	MCH/aptamer/GOPTS/PtNPs microwires modified Au IDEs	EIS	1.00×10^{-11}–1.00×10^{-7} (acetamiprid) 1.00×10^{-10}–1.00×10^{-6} (atrazine)	1.00×10^{-12} (acetamiprid) 1.00×10^{-11} (atrazine)	Madianos et al. (2018)
Acetamiprid	MCH/aptamer/Ag-NG/GCE	EIS	1.00×10^{-13}–5.00×10^{-9}	3.30×10^{-14}	Jiang et al. (2015)
Acetamiprid	MCH/aptamer (oligo 1)/AuNPs/PANI/GSPEs	DPV	2.50×10^{-7}–2.00×10^{-6}	8.60×10^{-8}	Rapini et al. (2016)
Malathion	Aptamer/SA/CHIT-IO/FTO	DPV	3.03×10^{-12}–3.03×10^{-8}	3.03×10^{-12}	Prabhakar et al. (2016)
Chlorpyrifos	Aptamer/AMP/CuONFs-SWCNTs/Nafion/GCE	DPV	2.85×10^{-10}–4.28×10^{-7}	2.00×10^{-10}	Xu et al. (2018)
Chlorpyrifos	BSA/aptamer/Fc/MWCNTs/OMC/GCE	CV	2.85×10^{-9}–2.85×10^{-4}	9.41×10^{-10}	Jiao et al. (2016)
Chlorpyrifos	BSA/aptamer/GO/Fe_3O_4/CB-CS/GCE	CV	2.85×10^{-10}–2.85×10^{-4}	9.41×10^{-11}	Jiao et al. (2017)

MCH: 6-Mercap-1-hexanol; GOPTS: (3-glycidyloxypropyl)trimethoxysilane; PtNPs: platinum nanoparticles; IDEs: interdigitated electrodes; GSPEs: graphite screen-printed electrodes; SA: streptavidin; CHIT-IO: chitosan-iron oxide nanocomposite; FTO: fluorine tin oxide; AMP: amino-modified capture probe; CuONFs: copper oxide nanoflowers; SWCNTs: single-walled carbon nanotubes; Fc: ferrocene; OMC: mesoporous carbon; GO: graphene oxide; CB: carbon black automated on-site analysis of mycotoxins in food processing and manufacturing industries.

Moreover, research results achieved in recent years confirm that nanomaterial usage had been rapidly growing in the development of electrochemical biosensors. Analytical performances of the biosensors.

systems increased enormously with the incorporation of nanomaterials: low detection limits up to sub/picomolar and sub/femtomolar levels and wide linear analytical ranges were achieved with nanomaterials and nanocomposites of synergetic combinations.

Therefore, the development of new materials and the application of nanostructures to biosensor systems could lead to the development of highly sophisticated analytical systems.

The speed of analysis and the low cost of the transduction instrumentation makes the electrochemical biosensors the most promising devices, for routine applications by common users, ensuring high analytical performance in terms of sensitivity and low detection limits.

References

Badea M, Floroian L, Restani P, Codruta S, Cobzac A, Moga M (2016) Ochratoxin A detection on antibody immobilized on BSA-functionalized gold electrodes. PLoS ONE (7):11, e0160021

Bagheryan Z, Raoof J-B, Golabi M, Turner APF, Beni V (2016) Diazonium-based impedimetric aptasensor for the rapid label-free detection of Salmonella typhimurium in food sample. Biosens Bioelectron 80:566–573

Bhardwaj H, Pandey MK, Rajesh, and Sumana, G. (2019) Electrochemical Aflatoxin B1 immunosensor based on the use of graphene quantum dots and gold nanoparticles. Microchim Acta 186:592

Campuzano S, Yanez-Sedeno P, Pingarron JM (2017) Electrochemical affinity biosensors in food safety. Chemosensors 5(1):8

Cao Y, Sun X, Guo Y, Zhao W, Wang X (2015) An electrochemical immunosensor based on interdigitated array microelectrode for the detection of chlorpyrifos. Bioprocess Biosyst Eng 38: 307–313

Cesewski E, Johnson BN (2020) Electrochemical biosensors for pathogen detection. Biosens Bioelectron 159:112214

Chandra P, Koh WCA, Noh HB, Shim YB (2012) In vitro monitoring of i-NOS concentrations with an immunosensor: the inhibitory effect of endocrine disruptors on i-NOS release. Biosens Bioelectron

Chen X, Huang Y, Ma X, Jia F, Guo X, Wang Z (2015) Impedimetric aptamer-based determination of the mold toxin fumonisin B1. Microchim Acta 182:1709–1714

Chen Q, Wang D, Cai G, Xiong Y, Li Y, Wang M, Huo H, Lin J (2016) Fast and sensitive detection of foodborne pathogen using electrochemical impedance analysis, urease catalysis and microfluidics. Biosens Bioelectron 86:770–776

Choudhary M, Yadav P, Singh A, Kaur S, Ramirez-Vick J, Chandra P, Arora K, Singh SP (2016) CD 59 Targeted ultrasensitive electrochemical immunosensor for fast and noninvasive diagnosis of oral cancer. Electroanalysis 28(10):2565–2574

Costa MP, Frías IAM, Andrade CAS, Oliveira MDL (2017) Impedimetric immunoassay for aflatoxin B1 using a cysteine modified gold electrode with covalently immobilized carbon nanotubes. Microchim Acta 184:3205–3213

Daniels JS, Pourmand N (2007) Label-free impedance biosensors: opportunities and challenges. Electroanalysis 19:1239–1257

Deka S, Saxena V, Hasan A, Chandra P, Pandey LM (2018) Synthesis, characterization and in vitro analysis of α-Fe_2O_3-$GdFeO_3$ biphasic materials as therapeutic agent for magnetic hyperthermia applications. Mater Sci Eng C 92:932–941

Doyle MP (1991) *Escherichia coli* O157:H7 and its significance in foods. Int J Food Microbiol 12: 289–301

Dye C (2014) After 2015: infectious diseases in a new era of health and development. Philos Trans Royal Soc Lond Ser B Biol Sci 369:20130426–20130426

Eissa S, Zourob M (2017) Selection and characterization of dna aptamers for electrochemical biosensing of carbendazim. Anal Chem 89:3138–3145

Evtugyn G, Hianik T (2019) Electrochemical immuno- and aptasensors for mycotoxin determination. Chemosensors 7:10

Fusco G, Gallo F, Tortolini C, Bollella P, Ietto F, De Mico A, D'Annibale A, Antiochia R, Favero G, Mazzei F (2017) AuNPS-functionalized panaba-mwcnts nanocomposite-based impedimetric immunosensor for 2,4-dichlorophenoxy acetic acid detection. Biosens Bioelectron 93:52–56

Garcia FP, Ascencio SYC, Oyarzun JCG, Hernandez AC, Alavarado PV (2012) Pesticides: Classification,uses and toxicity. Measures of exposure and genotoxic risks. Journal of Environmental Science and Toxicology 1:279–293

Giannetto M, Umiltà E, Careri M (2014) New competitive dendrimer-based and highly selective immunosensor for determination of atrazine in environmental, feed and food samples: the importance of antibody selectivity for discrimination among related triazinic metabolites. Anal Chim Acta 806:197–203

Goud KY, Hayat A, Satyanarayana M, Sunil Kumar V, Catanante G, Gobi KV, Marty JL (2017) Aptamer-based zearalenone assay based on the use of a fluorescein label and a unctional graphene oxide as a quencher. Microchim Acta 184:4401–4408

Gu C, Yang L, Wang M, Zhou N, He L, Zhang Z, Du M (2019) A bimetallic (cu-co) Prussian blue analogue loaded with gold nanoparticles for impedimetric aptasensing of ochratoxin A. Microchim Acta 186:343

Guner A, Cevik E, Senel M, Alpsoy L (2017) An electrochemical immunosensor for sensitive detection of Escherichia coli O157:H7 by using chitosan, MWCNT, polypyrrole with gold nanoparticles hybrid sensing platform. Food Chem 229:358–365

He B, Dong X (2018) Aptamer based voltammetric patulin assay based on the use of ZnO nanorods. Microchim Acta 185:462

He B, Yan X (2019) An amperometric zearalenone aptasensor based on signal amplification by using a composite prepared from platinum nanotubes, gold nanoparticles and thioninelabelled graphene oxide. Microchim Acta 186:383

Idil N, Hedstrom M, Denizli A, Mattiasson B (2017) Whole cell based microcontact imprinted capacitive biosensor for the detection of *Escherichia coli*. Biosens Bioelectron 87:807–815

Izadi Z, Sheikh-Zeinoddin M, Ensafi AA, Soleimanian-Zad S (2016) Fabrication of an electrochemical DNA-based biosensor for Bacillus cereus detection in milk and infant formula. Biosensors Bioelectron 80:582–589

Jafari H, Amiri M, Abdi E, Navid SL, Bouckaert J, Jijie R, Boukherroub R, Szunerits S (2019) Entrapment of uropathogenic E. coli cells into ultra-thin sol-gel matrices on gold thin films: a low cost alternative for impedimetric bacteria sensing. Biosens Bioelectron 124–125:161–166

Jia H, Guo Y, Sun X, Wang X (2015) An electrochemical immunosensor based on microfluidic chip for detection of chlorpyrifos. Int J Electrochem Sci 10:8750–8758

Jiang D, Du X, Liu Q, Zhou L, Dai L, Qian J, Wang K (2015) Silver nanoparticles anchored on nitrogen-doped graphene as a novel electrochemical biosensing platform with enhanced sensitivity for aptamer-based pesticide assay. Analyst 140:6404–6411

Jiao Y, Jia H, Guo Y, Zhang H, Wang Z, Sun X, Zhao J (2016) An ultrasensitive aptasensor for chlorpyrifos based on ordered mesoporous carbon/ferrocene hybrid multiwalled carbon nanotubes. RSC Adv 6:58541–58548

Jiao Y, Hou W, Fu J, Guo Y, Sun X, Wang X, Zhao J (2017) A nanostructured electrochemical aptasensor for highly sensitive detection of chlorpyrifos. Sensors Actuators B 243:1164–1170

Kumar P, Kim K-H, Deep A (2015) Recent advancements in sensing techniques based on functional materials for organophosphate pesticides. Biosens Bioelectron 70:469–481

Lee I, Jun S (2016) Simultaneous detection of *E. coli* K12 and *S. aureus* using a continuous flow multijunction biosensor. J Food Sci 81(6):1530–1536

Li X, Cao L, Zhang Y, Yan P, Kirk DW (2017) Fabrication and modeling of an ultrasensitive label free impedimetric immunosensor for Aflatoxin B1 based on Protein A self-assembly modified gold 3D nanotube electrode ensembles. Electrochim Acta 247:1052–1059

Liu X, Li WJ, Li L, Yang Y, Mao L-G, Peng Z (2014a) A label-free electrochemical immunosensor based on gold nanoparticles for direct detection of atrazine. Sensors Actuators B 191:408–414

Liu G, Guo W, Song D (2014b) A multianalyte electrochemical immunosensor based on patterned carbon nanotubes modified substrates for detection of pesticides. Biosensors Bioelectron 52: 360–366

Liu L, Xu D, Hu Y, Liu S, Wei H, Zheng J, Wang G, Hu X, Wang C (2015) Construction of an impedimetric immunosensor for label-free detecting carbofuran residual in agricultural and environmental samples. Food Control 53:72–80

Liu N, Nie D, Tan Y, Zhao Z, Liao Y, Wang H, Sun C, Wu A (2017) An ultrasensitive amperometric immunosensor for zearalenones based on oriented antibody immobilization on a glassy carbon electrode modified with MWCNTs and AuPt nanoparticles. Microchim Acta 184:147–153

Lu L, Seenivasan R, Wang Y-C, Yu J-H, Gunasekaran S (2016) An electrochemical immunosensor for rapid and sensitive detection of mycotoxins fumonisin B1 and deoxynivalenol. Electrochim Acta 213:89–97

Madianos L, Tsekenis G, Skotadis E, Patsiouras L, Tsoukalas D (2018) A highly sensitive impedimetric aptasensor for the selective detection of acetamiprid and atrazine based on microwires formed by platinum nanoparticles. Biosens Bioelectron 101:268–274

Mahato K, Kumar S, Srivastava A, Maurya PK, Singh R, Chandra P (2018) Electrochemical immunosensors: fundamentals and applications in clinical diagnostics. In: Handbook of immunoassay technologies

Malvano F, Albanese D, Pilloton R, Di Matteo M (2016a) A highly sensitive impedimetric label free immunosensor for Ochratoxin measurement in cocoa beans. Food Chem 212:688–694

Malvano F, Albanese D, Crescitelli A, Pilloton R, Esposito E (2016b) Impedimetric label-free immunosensor on disposable modified screen-printed electrodes for ochratoxin A. Biosensors 6(3):33

Malvano F, Albanese D, Pilloton R, Di Matteo M, Crescitelli A (2017) A new label-free impedimetric affinity sensor based on cholinesterases for detection of organophosphorous and carbamic pesticides in food samples: impedimetric versus amperometric detection. Food Bioprocess Technol 10:1834–1843

Malvano F, Pilloton R, Albanese D (2020) Label-free impedimetric biosensors for the control of food safety—a review. Int J Environ Anal Chem 100(4):468–491

Malvano F, Albanese D, Pilloton R (2019) Label-free impedimetric immunosensors for sensitive detection of aflatoxin B1 in food. Chem Eng Trans 74:1567–1572

Malvano F, Maritato L, Carapella G, Orgiani P, Pilloton R, Di Matteo M, Albanese D (2018a) Fabrication of $SrTiO_3$ layer on Pt electrode for label-free capacitive biosensors. Biosensors 8:26

Malvano F, Pilloton R, Albanese D (2018b) Sensitive detection of *Escherichia* coli O157:H7 in food products by impedimetric immunosensors. Sensors 18:2168

Mehta J, Vinayak P, Tuteja SK, Chhabra VA, Bhardwaj N, Paul A, Kim K-H, Deep A (2016) Graphene modified screen printed immunosensor for highly sensitive detection of parathion. Biosensors Bioelectron 83:339–346

Mejri-Omrani N, Miodek A, Zribi B, Marrakchi M, Hamdi M, Marty JL (2016) Direct detection of OTA by impedimetric aptasensor based in modified polypyrrole-dendrimers. Anal Chim Acta 920:37–46

Miranda-Castro R, de-los-Santos-Alvarez N, Miranda-Ordieres AJ, Lobo-Castanon MJ (2016) Harnessing aptamers to overcome challenges in gluten detection. Biosensors 6(2):16

Mutreja R, Jariyal M, Pathania P, Sharma A, Sahoo DK, Raman Suri C (2016) Novel surface antigen based impedimetric immunosensor for detection of Salmonella typhimurium in water and juice samples. Biosens Bioelectron 85:707–713

Pandey CM, Tiwari I, Singh VN, Sood KN, Sumana G, Malhotra BD (2017) Highly sensitive electrochemical immunosensor based on graphene-wrapped copper oxide-cysteine hierarchical structure for detection of pathogenic bacteria. Sensors Actuators B Chem 238:1060–1069

Prabhakar N, Thakur H, Bharti A, Kaur N (2016) Chitosan-iron oxide nanocomposite based electrochemical aptasensor for determination of malathion. Anal Chim Acta 939:108–116

Primiceri E, Chiriaco MS, de Feo F, Santovito E, Fusco V, Maruccio G (2016) A multipurpose biochip for food pathogen detection. Anal Methods 8(15):3055–3060

Qiao L, Wang X, Sun X (2014) A novel label-free amperometric immunosensor based on graphene sheets-methylene blue nanocomposite/gold nanoparticles. International Journal of Electrochemistry Science 9:1399–1414

Qing Y, Li CR, Yang XX, Zhou XP, Xue J, Luo M, Xu X, Chen S, Qiu JF (2016) Electrochemical immunosensor using single-walled carbon nanotubes/chitosan for ultrasensitive detection of deoxynivalenol in food samples. J Appl Electrochem 46:1049–1057

Rapini R, Cincinelli A, Marrazza G (2016) Acetamiprid multidetection by disposable electrochemical dna aptasensor. Talanta 161:15–21

Redmond EC, Griffith CJ (2003) Consumer food handling in the home: a review of food safety studies. J Food Prot 66:130–161

Regiart M, Fernández O, Vicario A, Villarroel-Rocha J, Sapag K, Messina GA, Raba J, Bertolino FA (2018) Mesoporous immunosensor applied to zearalenone determination in Amaranthus cruentus seeds. Microchem J 141:388–394

Rhouati A, Catanante G, Nunes G, Hayat A, Marty JL (2016) Label-free Aptasensors for the detection of mycotoxins. Sensors 16(12):2178

Rivas L, Mayorga-Martinez CC, Quesada-Gonzalez D, Zamora-Galvez A, de la Escosura-Muniz A, Markoci A (2015) Label-free impedimetric Aptasensor for Ochratoxi-A detection using iridium oxide nanoparticles. Anal Chem 87:5167–5172

Roariu L, Lagarde F, Jaffrezic-Renault N, Bala C (2016) Electrochemical biosensors for fast detection of food contaminantys – trendas and perspective. Trends Anal Chem 79:80–87

Sharma A, Kumar A, Khan R (2018) A highly sensitive amperometric immunosensor probe based on gold nanoparticle functionalized poly (3, 4-ethylenedioxythiophene) doped with graphene oxide for efficient detection of aflatoxin B1. Synth Met 235:136–144

Sheikhzadeh E, Chamsaz M, Turner APF, Jager EWH, Beni V (2016) Label-free impedimetric biosensor for Salmonella Typhimurium detection based on poly [pyrrole-co-3-carboxyl-pyrrole] copolymer supported aptamer. Biosens Bioelectron 80:194

Smolko V, Shurpik D, Porfireva A, Evtugyn G, Stoikov I, Hianik T (2018) Electrochemical aptasensor based on poly(Neutral red) and carboxylated pillar[5]arene for sensitive determination of aflatoxin M1. Electroanalysis 30:486–496

Sunday CE, Masikini M, Wilson L, Rassie C, Waryo T, Baker PCL, Iwuoha EI (2015) Application on gold nanoparticles-dotted 4-nitrophenylazo graphene in a label-free impedimetric deoxynivalenol immunosensor. Sensors 15:3854–3871

Thakur MS, Ragavan KV (2012) Biosensor in food processing. J Food Sci Technol 50:625–641

Thevenot DR, Toth K, Durst RA, Wilson GS (2001) Electrochemical biosensors: recommended defintions and classification. Biosens Bioelectron 16:121–131

Verma S, Choudhary J, Singh KP, Chandra P, Singh SP (2019) Uricase grafted nanoconducting matrix based electrochemical biosensor for ultrafast uric acid detection in human serum samples. Int J Biol Macromol 130:333–341

Wang C, Qian J, An K, Ren C, Lu X, Hao N, Liu Q, Li H, Huang X, Wang K (2018a) Fabrication of magnetically assembled aptasensing device for label-free determination of aflatoxin B1 based on EIS. Biosens Bioelectron 108:69–75

Wang Y, Zhang L, Peng D, Xie S, Chen D, Pan Y, Tao Y, Yuan Z (2018b) Construction of electrochemical immunosensor based on gold-nanoparticles/carbon nanotubes/chitosan for sensitive determination of T-2 toxin in feed and swine meat. Int J Mol Sci 19:3895

Wang D, Chen Q, Huo HL, Bai SS, Cai GZ, Lai WH, Lin JH (2017a) Efficient separation and quantitative detection of Listeria monocytogenes based on screen-printed interdigitated electrode, urease and magnetic nanoparticles. Food Control 73:555–561

Wang W, Han Z, Liang P, Guo D, Xiang Y, Tian M, Song Z, Zhao H (2017b) Co_3O_4/pan magnetic nanoparticle-modified electrochemical immunosensor for chlorpyrifos. Digest J Nanomater Biostruct 12:1–9

Wei M, Zhang W (2017) A novel impedimetric aptasensor based on AuNPs–carboxylic porous carbon for the ultrasensitive detection of ochratoxin A. Royal Soc Chem 7:28655–28660

Wilson D, Materon EM, Ibanez-Redin G, Faria RC, Correa DS, Oliveira ON Jr (2019) Electrical detection of pathogenic bacteria in food samples using information visualization methods with a sensor based on magnetic nanoparticles functionalized with antimicrobial peptides. Talant 194: 611–618

World Health Organization (2016) Manual on development and use of FAO and WHO specifications for pesticide; 3rd Revisions. World Health Organization, Geneva, Switzerland

Yagati AK, Chavan SG, Baek C, Lee M-H, Min J (2018) Label-free impedance sensing of aflatoxin B1 with polyaniline nanofibers/Au nanoparticle electrode array. Sensors 18:1320

Yamada K, Choi W, Lee I, Cho BK, Jun S (2016) Rapid detection of multiple foodborne pathogens using a nanoparticle-functionalized multi-junction biosensor. Biosens Bioelectron 77:137–143

Xu G, Huo D, Hou C, Zhao Y, Bao J, Yang M, Fa H (2018) A regenerative and selective electrochemical aptasensor based on copper oxide nanoflowers-single walled carbon nanotubes nanocomposite for chlorpyrifos detection. Talanta 178:1046–1052

Xu W, Qing Y, Chen S, Chen J, Qin Z, Qiu JF, Li CR (2017a) Electrochemical indirect competitive immunoassay for ultrasensitive detection of zearalenone based on a glassy carbon electrode modified with carboxylated multi-walled carbon nanotubes and chitosan. Microchim Acta 184: 3339–3347

Xu M, Wang R, Li Y (2017b) Electrochemical biosensors for rapid detection of *Escherichia coli* O157:H7. Talanta 162:511–522

Aptamer-Based Sensors for Drug Analysis

3

Zahra Khoshbin, Elham Sameiyan, Khalil Abnous, and Seyed Mohammad Taghdisi

Abstract

The uncontrolled use of drugs endangers human health and causes drastic economic losses and irreparable consequences. Today, there is a great demand to introduce accurate, potent, real-time, and rapid methodologies for sensitive detection and quantification of drugs. To overcome this challenging difficulty, biosensors have been introduced as valuable tools. Among the diverse kinds of biosensors, aptamer-based biosensors (aptasensors) have evolved as novel candidates for the sensitive evaluation of different groups of drugs, owing to their superior specificity, sensitivity, and selectivity. This chapter encompasses the recent progress in the development of aptasensors to quantitatively monitor various types of drugs. Besides, sensing mechanisms associated with the aptasensors are given that provide ideas to develop the novel aptasensing platforms as the portable test kits for the on-site detection of drugs.

Z. Khoshbin · K. Abnous (✉)
Pharmaceutical Research Center, Pharmaceutical Technology Institute, Mashhad University of Medical Sciences, Mashhad, Iran

Department of Medicinal Chemistry, School of Pharmacy, Mashhad University of Medical Sciences, Mashhad, Iran
e-mail: abnouskh@mums.ac.ir

E. Sameiyan
Department of Pharmaceutical Biotechnology, School of Pharmacy, Mashhad University of Medical Sciences, Mashhad, Iran

S. M. Taghdisi
Targeted Drug Delivery Research Center, Pharmaceutical Technology Institute, Mashhad University of Medical Sciences, Mashhad, Iran

P. Chandra, P. S. Panesar (eds.), *Nanosensing and Bioanalytical Technologies in Food Quality Control*, https://doi.org/10.1007/978-981-16-7029-9_3

Keywords

Aptasensor · Drug · Antibiotics · Chemotherapeutic drug · Respiratory drug · Nanoparticles

3.1 Introduction

The rapid, sensitive, and on-site detection of chemicals, toxins, pollutants, pesticides, disease markers, and drugs is a challenging issue. There are diverse conventional analytical methods for detection and quantitative measurements of these substances, such as liquid chromatography-mass spectrometry (LC-MS), mass spectroscopy (MS), liquid chromatography (LC), gas chromatography (GC), and high-performance capillary electrophoresis (HPCE). Despite their high accuracy, the methods are intricate and time-consuming under the restrictions like high cost, cumbersome sample preparation, operational complexity, and poor anti-inference that make them impractical for the rapid and on-site analyte detection (Cheng et al. 2020; Sharma et al. 2020; Yan et al. 2018). Therefore, new strategies for the facile, sensitive, fast, and selective detection of the analytes are intensively desired. To achieve the requirement, biosensors are introduced as the efficient additions to analytical sensing assays with high potential for the quantitative high-throughput analyte monitoring (Chandra et al. 2012; Choudhary et al. 2016; Deka et al. 2018; Mahato et al. 2018; Verma et al. 2019). The enzyme-linked immunosorbent assay (ELISA) is applied as the common biochemistry test strips for the rapid target detection. Unfortunately, the deficiencies of limited lifetime, instability, affecting the sample matrix, storage difficulty, and denaturing of enzyme induced by severe experiential conditions restrict their application (Cao et al. 2014; Bahadır and Sezgintürk 2015; Han et al. 2013; Syshchyk et al. 2015; Othman et al. 2016). Hence, aptamers have been introduced as the efficient biorecognition elements for designing aptamer-based biosensors (aptasensors). Aptamers are short single-stranded oligonucleotide sequences, obtained by systematic evolution of ligands exponential enrichment (SELEX) as an artificial screening method. They capture the specific targets with high specificity and binding affinity through conformational changes. Aptamers possess supreme features, such as cost-effective synthesis, high thermal and chemical stability, and adaptive modification, that convert them to the promising segments for biosensing assays (Ha et al. 2017; Dirkzwager et al. 2016; Jin et al. 2017).

Despite the usage of aptasensors for the highly sensitive detection of targets, there is still an urgent requirement for facile, portable, user-friendly, and disposable point-of-care (POC) diagnostic assays. Hence, microfluidic biosensing devices have received a great attention as the equipment-free lab-on-chip approaches for the robust consumer diagnostics, particularly in the regions lacking laboratory analytical tools. The supreme characteristics of microfluidic assays, such as low-reagent consumption, high-throughput, and rapidity accompanied with significant features of aptasensors provides the opportunity to design microfluidic aptasensor platforms for the highly sensitive on-site diagnostics (Jin et al. 2017; Ma et al. 2018; Lin et al. 2016). This chapter provides a comprehensive study on the recent advances in the aptasensor technology to detect diverse drug groups. This study can provide an advanced perspective for designing the promising portable aptasensing test strips for drug detection.

3.2 Antibiotic Drugs

3.2.1 Aminoglycosides

Aminoglycoside antibiotics are effective clinical drugs derived from *Streptomyces* species or generated synthetically that include the diverse sub-classes, such as kanamycin, tobramycin, gentamicin, neomycin, and so on (Hu et al. 2018a; Edson and Terrell 1999). Aminoglycosides are extensively utilized to prevent or treat bacterial infections through binding to the prokaryotic ribosomal sites that results in the mRNA mistranslation, message readout imperfection, and finally, bacterial cell death (Hermann 2007; Vicens and Westhof 2003; Auerbach et al. 2004; Tor 2006; Purohit and Stern 1994). The marvelous inhibition effect to bacteria makes aminoglycosides advantageous in therapy, pharmaceutical industry, fishery, etc. However, their overuse arouses some serious problems for environmental safety and human health. The accumulation of aminoglycosides in human body causes the inevitable threats, such as renal toxicity, hearing loss, respiratory failure, and allergic reactions (Zhou et al. 2020a; Azadbakht and Abbasi 2019). Hence, their monitoring is very substantial for human safety (Derbyshire et al. 2012; McGlinchey et al. 2008).

3.2.1.1 Kanamycin

Wang et al. (2020) developed a label-free fluorescent aptasensor for the kanamycin detection by using gold nanoparticles (AuNPs) for quenching the carbon dots (CDs) fluorescence. In the absence of target, the specific aptamer was adsorbed on the AuNPs surface that induced a dispersion state of AuNPs. Subsequently, the effective quenching of CDs was achieved based on the inner filter effect (IFE) strategy. Upon adding kanamycin, the specific aptamer interacted with the target, resulting in the AuNPs aggregation in the high salt concentration. So, the CDs fluorescence appeared as a sign of the kanamycin presence. The aptasensor showed a linearity to kanamycin in the range of 0.04–0.24 μM with the limit of detection (LOD) of 18 nM.

Jiang et al. (2019) introduced an ultrasensitive aptasensor for the monitoring of kanamycin by using AuNPs-coated Ag shell layer and surface-enhanced Raman spectroscopy (SERS). In the absence of kanamycin, the DNA probe embedded on the Au@Ag core-shell was hybridized with the Cy3-labeled aptamer. So, the close proximity of the Cy-3 molecule to the Ag shell triggered the signal enhancement. With adding kanamycin, the release of the specific aptamer from the Au@Ag core-shell was induced that weakened the Raman signals of Cy3 molecule. The designed aptasensor is advantageous for the fast and accurate sensing of the antibiotic. The aptasensor could detect kanamycin in the concentration range of 0 μg mL^{-1} to 100 ng mL^{-1} with the LOD of 0.90 pg mL^{-1}.

In addition, our team recently designed a fluorescent aptasensor to determine kanamycin, which used Rhodamine B with the role of fluorescent probe (Dehghani et al. 2018). The mesoporous silica nanoparticles (MSNs) were applied whose pores were filled with the fluorescent probe, and then, capped with the aptamer/complementary strand (dsDNA). Upon the addition of kanamycin, the aptamer-target complex was formed. Thus, the aptamer was separated from the complementary strand, leading to uncapping of the MSNs gates, the release of Rhodamine B, and increasing the fluorescence response. The lowest detectable concentration of kanamycin with this aptasensor was reported to be 7.5 nM with a linear ranges of 24.75–137.15 nM.

3.2.1.2 Tobramycin

Shang et al. (2019) constructed a sensitive photoelectrochemical aptasensing platform to monitor tobramycin as illustrated in Fig. 3.1. An amorphous MoS_x (a-MoS_x) @ZnO core-shell nanorod (NR) array was grown on the indium tin oxide (ITO). Then, the platform was coated with chitosan to obtain more stable sensing platform and provide the anchors for attaching the aptamer. The specific aptamer was immobilized on the ITO/a-MoS_x@ZnO NR array modified electrode through the phosphoramidate bond between the phosphate segment of the aptamer and amino group of chitosan. With adding tobramycin, the aptamer-tobramycin was obtained that reduced the photocurrent. The LOD was obtained as low as 5.7 pg mL^{-1}. The designed aptasensor provided a label-free, facile, and cost-effective strategy to sensitively monitor tobramycin.

Our team developed a fluorescent aptasensor for tobramycin detection, which used PicoGreen (PG) as a fluorescence light-up probe (Khajavian et al. 2021). The aptasensor was engineered in such a way that a triple structure, called three-way junction (3WJ) structure, was formed with participation of a complementary strand. By the formation of the structure, more double-stranded regions were created for the PG intercalation, resulting in an improved fluorescent intensity. Conversely, in the presence of tobramycin, the aptamers interacted with the target. Hence, the complementary strand could no longer bind to its matching sequences. So, fewer double-stranded regions were available for PG, leading to drastically reduced fluorescence intensity. The proposed sensor was tested to measure the different amounts of the spiked concentrations of tobramycin in the real human serum sample with the LOD of 321.2 nM and two linear zones of 5–20 μM and 20–300 μM.

3.2.1.3 Streptomycin

Yin et al. (2017) designed a novel electrochemical aptasensing platform for the highly sensitive detection of streptomycin by using carbon nanorods and Fe_3O_4-AuNPs-doped graphene (CNR-Fe_3O_4-AuNPs-GR) nanocomposites as the signal amplifier agent through providing the high surface area and large pore volume. First, the GCE electrode was modified by the CNR-Fe_3O_4-AuNPs-GR

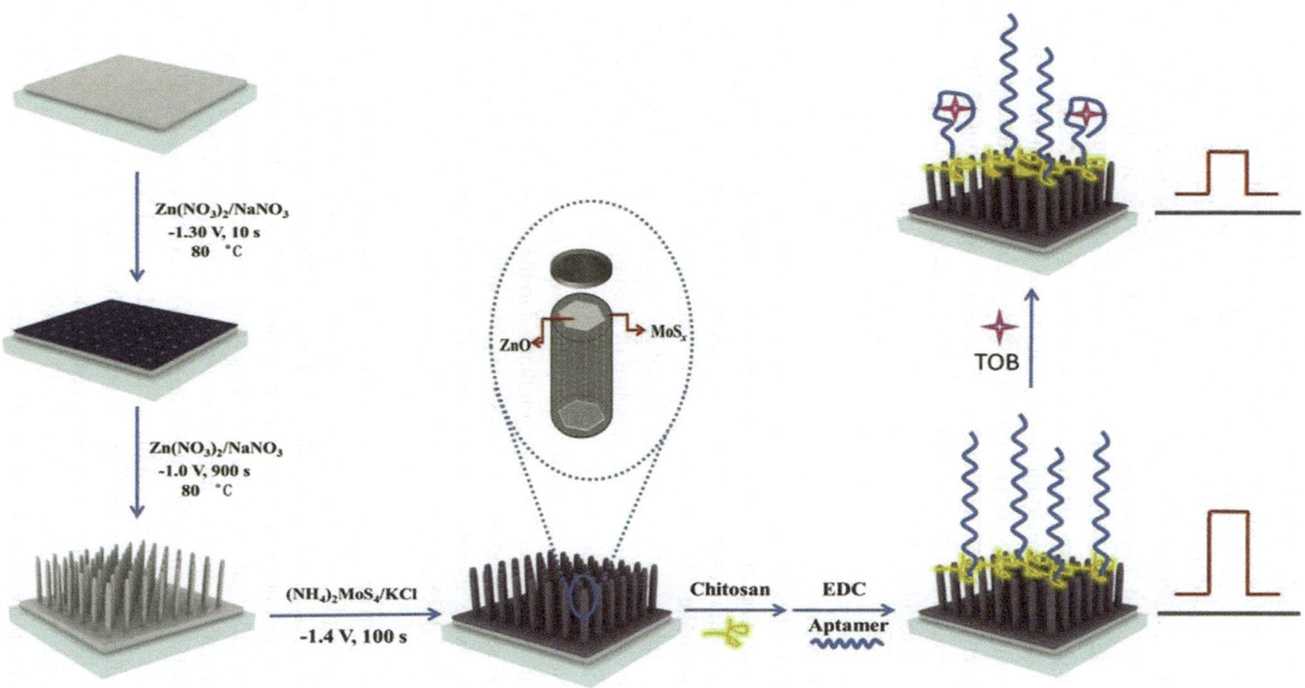

Fig. 3.1 Schematic illustration of the photoelectrochemical aptasensor to detect tobramycin based on the MoSx@ZnO core-shell nanorod array. Reprinted with permission from reference Derbyshire et al. (2012)

nanocomposite that supplied a great number of binding sites for the aptamer immobilizations on the electrode through the strong interaction between AuNPs and the thiol group of the aptamer. In the presence of streptomycin, the immobilized specific aptamer on the CNR-GR-modified electrode interacted with the target and changed the electron transfer. The aptasensor exhibited a wide linearity to streptomycin in the range of 0.05–200 ng mL^{-1} with the LOD of 0.028 ng mL^{-1}.

Taghdisi et al. (2016) introduced a fluorescence aptasensor by utilization of SYBR Gold dye and Exonuclease III (EXO III) enzyme for the streptomycin detection. In the absence of the target, by adding EXO III into the system, the double-stranded DNA was degraded, resulting in a low fluorescence response. However, the aptamer/target conjugate was formed with the addition of streptomycin, and the complementary strand separated from the aptamer. Subsequently, it led to more aptamer protection against Exo III enzyme. After adding SYBR Gold, a strong fluorescence intensity was achieved. The low detection limit of streptomycin was obtained up to 54.5 nM.

3.2.1.4 Neomycin

Khavani et al. (2019) theoretically designed new RNA aptamers for Neomycin B based on the mutation of the SELEX-introduced aptamer. The binding affinity and sensing ability of the designed aptamers toward the target were examined from the theoretical viewpoint. The theoretically designed aptamers with better binding affinity were then applied to experimentally develop a simple colorimetric aptasensor by using AuNPs. In the absence of the target, a red color solution was observed by the stabilized AuNPs with aptamers after addition of NaCl. In the presence of Neomycin B, the interacted aptamer with the target could not protect AuNPs from the salt-induced aggregation that led to the change of the color solution to blue. As the advantageous of the approach, Neomycin B could be detected more sensitively by using the theoretically designed aptamer in comparison with the SELEX-introduced RNA strand. The LOD was obtained 27 nM for the Neomycin B detection.

Ling et al. (2016) designed a sensitive fluorescent aptasensor for the Neomycin B detection based on the aptamer self-assembly on AuNPs. The Neomycin B-specific aptamer was first cleaved between the rA14 and rG15 ribonucleotides at the end of pentaloop zone to obtain two shorter segments (NEO1 and NEO2). The NEO1 strand was modified through the attachment of the polythymidine spacer and polyadenosine tail to have an upright structure for ligand binding and adopt the anchoring block for the preferential attachment to the AuNPs surface. The presence of Neomycin B induced the assembly of the FAM-labeled NEO2 on the surface of AuNPs through forming the binding pockets with the NEO1 strand. So, fluorescence quenching of the aptasensor was achieved as the result of the proximity of the FAM-labeled NEO2 to AuNPs. A low LOD of 0.01 μM with a linear concentration range of 0.1–10 μM was achieved by the aptasensor.

3.2.2 β-Lactams

β-Lactam antibiotics are a broad class of the antibacterial agents. They include penicillins, monobactams, cephalosporins, and carbapenems. β-Lactams are efficiently applied to treat the bacterial infection, and their presence in the dairy products have great impact on human health. However, the imprudent utilization of β-Lactam antibiotics may damage human health by antibiotic resistance and allergic reactions. So, it is necessary to develop sensing techniques for their sensitive detection (He et al. 2020; Zhao et al. 2016; Pandey and Cascella 2020).

Yan et al. (2020) developed a photoelectrochemical sensing platform to monitor ampicillin by using $AgBiS_2$ dual-sensitized Zn/Co bimetallic oxide ($Zn_xCo_{3-x}O_4$) dodecahedron and nitrogen-doped graphene quantum dots (N-GQDs) (Fig. 3.2). The $Zn_xCo_{3-x}O_4$ dodecahedron with the unique porous structure and large surface area provided the impressive active sites that shortened the photoelectrons transmission path. The N-GQDs and eco-friendly $AgBiS_2$ with the up-conversion effect and optimal band gap, respectively, were applied for strengthening the aptasensor

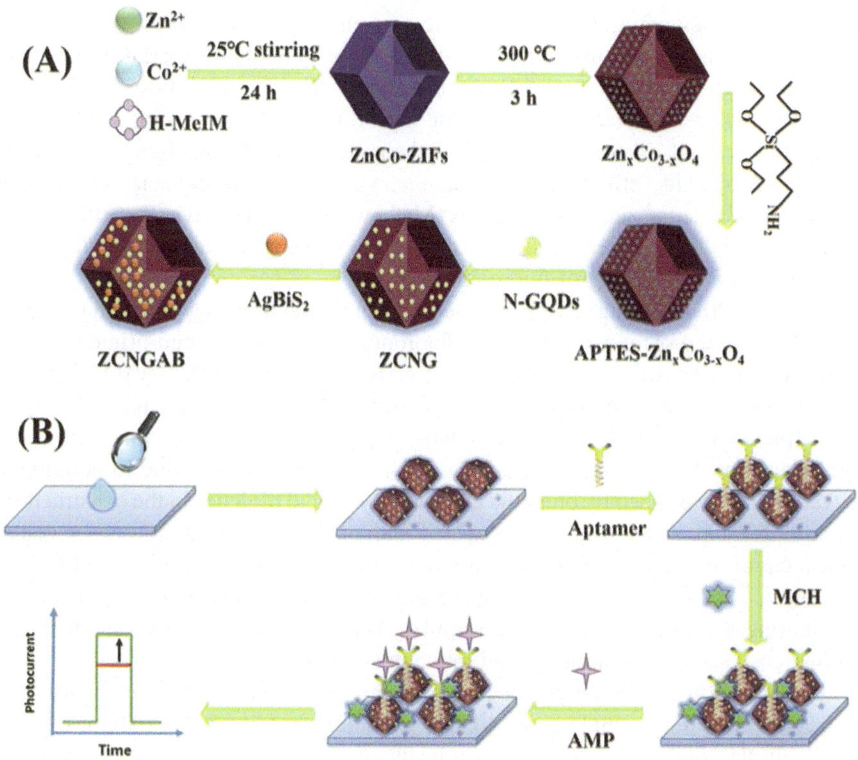

Fig. 3.2 Schematic illustration of the synthesis of $Zn_xCo_{3-x}O_4$/N-GQDs/$AgBiS_2$ composite (**a**), and the construction process of the aptasensor for detection of ampicillin (**b**). Reprinted with permission from reference Yan et al. (2020)

response. To construct the aptasensor, the specific aptamer was fixed on the surface of the modified electrode by the $Zn_xCo_{3-x}O_4$/N-GQDs/$AgBiS_2$. After adding ampicillin, the ampicillin-aptamer complex was formed on the modified electrode that improved the photocurrent. The aptasensor exhibited a linearity to ampicillin in the range of 0.5 pM–10 nM with the LOD of 0.25 pM.

Recently, our team introduced a fluorescent aptasensor for the ampicillin monitoring, by using 3,4,9,10-perylenetetracarboxylic acid diimide (PTCDI) as a low-cost and accessible fluorophore (Esmaelpourfarkhani et al. 2020). The ampicillin-specific aptamer was immobilized on the AuNPs surface. Under the addition of ampicillin, the aptamer-target complex was formed; hence, the complementary strand (CS) was separated from the aptamer and exposed to PTCDI after the centrifugation, leading to a reduction in the fluorescence intensity, because of the PTCDI aggregation of by the CS. The LOD of the proposed aptasensor was reported 29.2 pM with a broad linear range from 100 pM to 1000 pM.

3.2.3 Fluoroquinolones

Fluoroquinolones are a class of antibiotics, commonly utilized to prevent or treat certain bacterial infections, such as sinusitis, pneumonia, bacterial bronchitis, typhoid fever, anthrax, and urinary infections. These antibiotics contain the sub-groups of gemifloxacin, delafloxacin, ciprofloxacin, levofloxacin, norfloxacin, moxifloxacin, and ofloxacin. In spite of their great application, fluoroquinolones include serious side effects, such as nausea, vomiting, diarrhea, seizures, tendon rupture, angioedema, and hallucinations (Mathews et al. 2019; LiverTox 2020; Grill and Maganti 2011). Hence, it is required to develop ultrasensitive detection approaches for monitoring of fluoroquinolones.

Hu et al. (2018b) designed a novel electrochemical aptasensing platform to detect ciprofloxacin as a second generation of fluoroquinolones. The screen-printed carbon electrode (SPCE) was functionalized by carbon nanotube (CNT)-V_2O_5-chitosan (CS) nanocomposites that provided a large surface for the aptamer immobilization. In the presence of ciprofloxacin, the aptamer-ciprofloxacin complex was formed on the surface of the electrode that decreased the electron transfer, due to additional negative charges caused by phenolic and carboxylic acid moieties in the construction of ciprofloxacin, repelling ($[Fe(CN)_6]^{4-/3-}$ anions. The designed aptasensor could detect ciprofloxacin in the concentration range of 0.5–8.0 ng mL^{-1} with the LOD of 0.5 ng mL^{-1}. The applicability of the aptasensor was verified through the successful monitoring of ciprofloxacin in milk samples. The aptasensor possesses the features of low-cost, stability, simplicity, and miniaturized device.

Our team constructed an electrochemical sensing approach for the ciprofloxacin detection by using a single-stranded DNA-binding protein (SSB) (Abnous et al. 2017). In the target absence, SSB as a negative charge barrier interacted with the immobilized single-stranded DNA aptamer with high affinity. So, the access of $[Fe(CN)_6]^{3-/4-}$ redox probe to the surface of the gold electrode strongly reduced, resulting in a weak current signal. When the ciprofloxacin was introduced into the system, the SSB could no longer bind to the aptamer because of the formation of the target-aptamer complex, eventually leading to an enhancement in the obtained signal. The aptasensor was tested in the different media, such as water, serum, and milk samples with the LODs of 261, 336, and 351 pM, respectively.

3.2.4 Sulfonamides

Sulfonamides are synthetic bacteriostatic antibiotics that competitively inhibit the enzyme dihydropteroate synthase (DHPS) as an effective enzyme in folate synthesis. Sulfonamides include mafenide, sulfacetamide, sulfadiazine, sulfanilamide, sulfasalazine, sulfadimethoxine, sulfamethizole, sulfamethoxazole, sulfisoxazole, and sulfonamidochrysoidine. They are applied to treat cough and allergies, as well as antifungal functions. However, they are potent to cause some untoward reactions, e.g., porphyria, hemopoietic and urinary tract disorders, and hypersensitivity reactions (Carta et al. 2012; Liu et al. 2019; Tian et al. 2020).

Yan et al. (2017) developed a simple colorimetric aptasensor to quantify sulfadimethoxine based on the peroxidase-like function of AuNPs. In the absence of sulfadimethoxine, the aptamer covered the surface of AuNPs and disabled their peroxidase-like activity. With adding sulfadimethoxine, the specific aptamer desorbed from the surface of AuNPs through binding to its target. Subsequently, the peroxidase-like activity of AuNPs was recovered and catalyzed the oxidation of 3,3′,5′,5′-tetramethylbenzidine (TMB) by H_2O_2, leading to a change of the color solution from red to blue. Sulfadimethoxine could be detected by the designed aptasensor in the concentration range of 0.01–1000 μg mL^{-1} with the LOD of 10 ng mL^{-1}. The potential applicability of the aptasensor was confirmed by monitoring sulfadimethoxine in milk samples. The aptasensing method is simple and low cost with a response time less than 15 min that makes it well suited for rapid target screening in foodstuffs.

Chen et al. introduced a fluorescent aptasensor for quantitative monitoring of sulfadimethoxine (Chen et al. 2020). In the absence of sulfadimethoxine, the specific aptamer interacted with AuNPs; and after adding the complementary strand of aptamer and SYBR Green I dye to the supernatant, no fluorescence response was obtained. SYBR Green I is a dye which can selectively interact with double-stranded DNA (dsDNA) and make sharp fluorescence emission. While, the aptamer specifically interacted with the target in the target presence. So, a significant fluorescence emission was observed after adding the complementary strand and SYBR Green I to the supernatant because of the formation of dsDNA. The linear concentration range of 2–300 ng mL^{-1} with the LOD of 3.41 ng mL^{-1} was achieved for detecting sulfadimethoxine by the aptasensor. The properties of simple, label-free, and fast detection make the aptasensor appropriate for applying it in the real sample analysis.

Bai et al. constructed a simple electrochemical aptasensor for the sulfadimethoxine determination based on the cleavage function of nuclease P1 (Bai et al. 2019). The complementary strand was fixed on the electrode surface. The specific aptamer was also fixed on the electrode surface through its interaction with the complementary strand and formed dsDNA on it. Owing to dsDNA formation, the activity of Nuclease P1 as a single-stranded-specific endonuclease was efficiently blocked and a great number of the anti-dsDNA antibody was placed on the electrode surface, obtaining a weak electrochemical signal. In the target presence, the sulfadimethoxine-aptamer complex was formed that released the aptamer from the electrode and facilitated digestion of the complementary strand by nuclease P1. So, the anti-dsDNA antibody could not be immobilized on the electrode surface, enhancing the response signal. Sulfadimethoxine could be determined with the linear range of 0.1–500 nM and the LOD of 0.038 nM. Although the aptasensing method possesses highly specificity and reproducibility, being time-consuming restricts its widespread application.

3.2.5 Chloramphenicol

Chloramphenicol antibiotic is greatly used in the treatment of many infectious diseases, such as plague, cholera, meningitis, and typhoid fever, due to the properties of effectiveness and cheapness (Lofrano et al. 2016; Wu et al. 2015). However, its residues found in animal-derived foods such as meat, milk, and honey are inherently toxic and possess side effects on the human, which can lead to some diseases, e.g., anemia, bone marrow, kidney damage, and leucopenia (Hanekamp and Bast 2015; Roushani et al. 2020; Bangemann 1994). Hence, the development of the rapid, sensitive, and facile sensing methods for the chloramphenicol identification is very important.

Wu et al. introduced a novel colorimetric aptasensor for the rapid monitoring of chloramphenicol (Wu et al. 2019a). In the target absence, the specific aptamer was adsorbed on the surface of AuNPs through the interaction between Au and the aptamer bases. Following the addition of La^{3+} ion, it strongly interacted with the aptamer and altered AuNPs form dispersed to the aggregated state, accompanying with altering the color from red to blue. With adding the target, the aptamer combined with chloramphenicol and did not bind to AuNPs. Then, the aptamer/chloramphenicol complex interacted with La^{3+} ion. Consequently, the dispersed AuNPs induced a red color solution. The linear detection range from 0 to 450 nM with the LOD of 5.88 nM were obtained for chloramphenicol detection.

Tu et al. (2020) developed a sensitive aptasensing strategy based on triple-helix molecular switch (THMS) aptamer. The signal transduction probe (STP) strand modified with the FAM fluorophore and BHQ1 quencher showed a weak fluorescence, due to the proximity of the BHQ1 and FAM segments. After adding the specific aptamer, the STP strand hybridized with aptamer and formed THMS that significantly enhanced the fluorescence response through elongation of the distance between the FAM and BHQ1 segments. With introducing the target, the aptamer interacted with it and released the STP strand, resulting in a quenched fluorescence response. The linear range of 5 nM to 0.1 μM with the LOD of 1.2 nM were achieved for detecting the target by the aptasensor. It was successfully utilized to quantify chloramphenicol in honey samples. Besides, the aptasensing method possesses the features of low cost, simple operation, and short response time.

Wang et al. (2018) described a photoelectrochemical aptasensor to monitor chloramphenicol by using TiO_2 nanorod array (NRA) and Eu(III)-doped CdS quantum dots (QDs). The modification of the fluorine-doped tin oxide (FTO) electrode was done by NRA and QDs that provided a great number of the binding sites for the aptamer attachment. Chloramphenicol interacted with the specific aptamer on the electrode, reducing the photocurrent response as a sign for the target presence. The aptasensor could detect chloramphenicol in the linear range of 1.0 pM–3.0 nM with the LOD of 0.36 pM. As a consequence, utilizing QDs in designing the aptasensors can be novel to improve their efficiency.

3.2.6 Tetracyclines

Tetracyclines as a group of important broad-spectrum antibiotics are extensively applied in medicine and foodstuff to prevent diseases and promote animal growth. They are divided into some sub-groups, including tetracycline, chlortetracycline, oxytetracycline, and doxycycline. Unfortunately, the abuse of tetracyclines causes their accumulation in animal-derived foods that results in serious threats to human health, such as chronic toxicity, liver damage, allergic reactions, and gastrointestinal disturbance (Sun et al. 2018; Rad and Azadbakht 2019; Jalalian et al. 2018).

Ramezani et al. (2015) designed a simple colorimetric aptasensor to quantitatively monitor tetracycline. A THMS aptamer structure was formed by interacting the STP strand and specific aptamer. In the absence of the target, the THMS structure remained stable and was not able to protect AuNPs from the salt-induced aggregation, inducing in a change of the solution color from red to blue. With adding tetracycline, the specific aptamer interacted with the target and released the STP strand. So, the STP strand was adsorbed on the AuNPs surface that led to the well-dispersed AuNPs with a red color. Tetracycline could be detected by the aptasensor in the linear range of 0.3–10 nM with the LOD of 266 pM.

Rouhbakhsh et al. (2020) constructed a novel liquid crystal-based aptasensor for the ultrasensitive tetracycline detection. Based on Fig. 3.3, the aptasensor was assembled by the aptamer immobilization onto the silane-modified glass that induced the homeotropic alignment of liquid crystals. In the presence of tetracycline, the aptamer-target complex was formed that disturbed the orientation of liquid crystals, leading to the change of the optical response from a dark to bright. The label-free aptasensor especially monitored the trace level of tetracycline as low as 0.5 pM. As a consequence, utilization of liquid crystals in designing aptasensors provides the advantage of detecting targets in ultra-low levels.

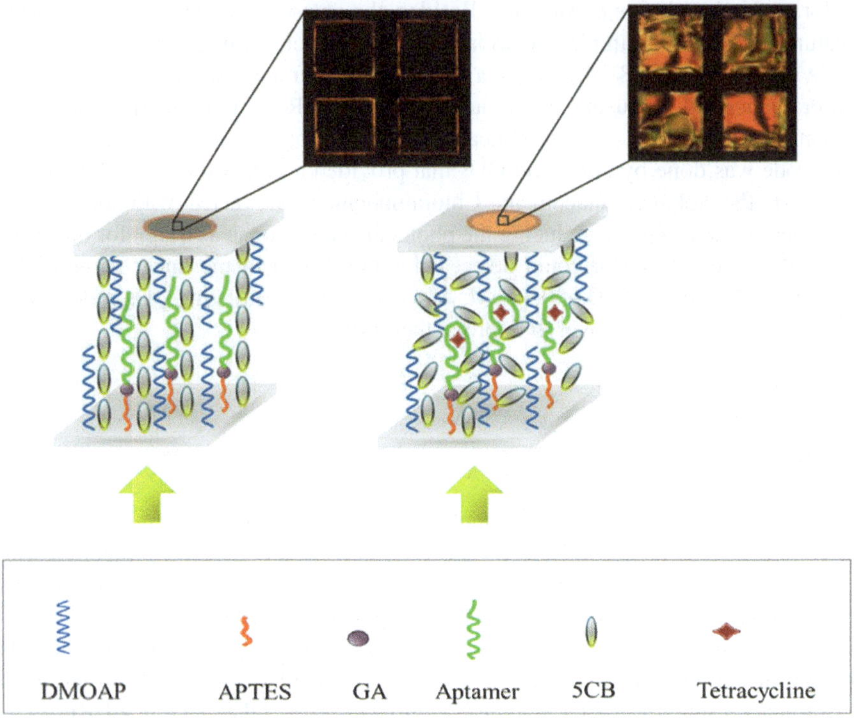

Fig. 3.3 Schematic illustration of the liquid-based aptasensor for the highly sensitive detection of tetracycline. Reprinted with permission from reference Rouhbakhsh et al. (2020)

3.3 Chemotherapeutic Drugs

Chemotherapy has been introduced to the medical field more than 50 years ago as an advantageous way to treat cancer. This type of cancer treatment applies one or more anticancer drugs, called chemotherapeutic agents, which serves a variety of purposes, including intention of treatment, prolonging lifespan, and reducing symptoms (Alfarouk et al. 2015; Johnstone et al. 2002). To quantitatively monitor the level of chemotherapeutic drugs, diminish their toxicity, and attain the desired therapeutic purpose, it is crucial to develop reliable and sensitive sensing methods. Over the past few decades, aptamer-based sensors have demonstrated the benefits of high sensitivity, fast response, and low cost as the promising option for rapid and real-time detection of chemotherapeutic agents during the cancer treatment (Sun et al. 2019; Bahner et al. 2018; Alkahtani 2020; Blum et al. 1973; Akiyama et al. 2008; Kato et al. 2018).

3.3.1 Bleomycin

Bleomycin (BLM) is an antitumor antibiotic obtained from *Streptomyces verticillus*, with chemotherapeutic application for clinical treatment of some special cancers, such as Hodgkin's disease, malignant pleural effusions, non-Hodgkin lymphoma, and testicular cancer (Blum et al. 1973). The BLM antitumor activity is thought to result from its ability through a metal-dependent oxidative cleavage of DNA in the presence of oxygen (Akiyama et al. 2008). However, sometimes BLM can cause some serious side effects, such as particularly pulmonary fibrosis and renal toxicity (Kato et al. 2018). Thus, it is important to evaluate BLM concentration in clinical and pharmaceutical samples.

Zhou et al. (2018) designed a novel aptasensor to detect trace amounts of BLM in the different aqueous environments. Two types of aptasensors were synthesized based on bimetallic core-shell Prussian blue analog (PBA), including CuFe@FeFe and AgNCs/Apt@CuFe@FeFe nanospheres. The Fe(II)·BLM complex was formed through the potent coordination interaction between BLM and plentiful Fe(II) ions involved in CuFe@FeFe PBA nanospheres. The Fe(II)·BLM complexes were able to cause an irreversible cleavage in the aptamer strands, leading to the significant changes in the electrochemical signals (Fig. 3.4). The coupling of AgNCs/Apt to the CuFe@FeFe-based aptasensor could both improve its sensing performance and shorten its fabrication process. The target was detected with the LOD of 0.49 fg mL^{-1} in the concentration range of 1.0 fg mL^{-1} to 2.0 ng mL^{-1}. The AgNCs/Apt@CuFe@FeFe-based aptasensor could monitor BLM in the concentration range of 0.01 fg mL^{-1} to 0.1 pg mL^{-1} with an extremely low LOD of 0.0082 fg mL^{-1}. Both the CuFe@FeFe- and AgNCs/Apt@CuFe@FeFe-based aptasensors displayed good stability, great selectivity, and excellent functionality in water, milk, and human serum samples.

Sun et al. (2019) described a dual-mode aptasensor based on the CdS-In_2S_3 composite for the BLM detection. The dual-mode response of the aptasensor was based on the photofuel cell (PFC) and photoelectrochemical (PEC) performance. In the PEC aptasensor, CdS-In2S3 plays the role of the photoactive substantial. After BLM interaction with the immobilized binding aptamer on the surface of CdS-In2S3, the photocurrent response reduced as a sign for the BLM presence. In the PFC aptasensor, CdS-In_2S_3 plays the role of photoanode, which caused the oxidation of water to oxygen under the visible light illumination. The produced oxygen reduced on the Pt cathode, resulting in the generation of electricity. Hence, the detection signal could be provided without the presence of an external electrical power source. The PEC and PFC aptasensors could detect BLM with the LODs of 0.85 nM and 1.0 nM in the concentration linear range of 5.0–250 nM and 10–250 nM, respectively.

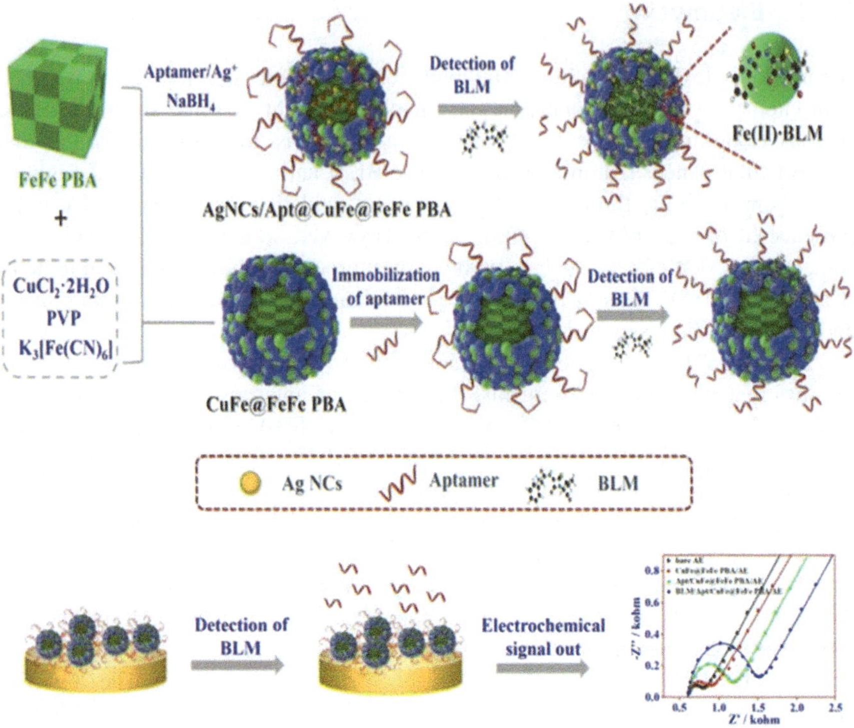

Fig. 3.4 Schematic representation of (**a**) the construction of the BLM aptasensors based on AgNCs/Apt@CuFe@FeFe PBA and CuFe@FeFe PBA; and (**b**) Detection of BLM and electrochemical signal out. Reprinted with permission from reference Zhou et al. (2018)

3.3.2 Daunorubicin

Daunorubicin (DNR) is an anticancer drug that can intercalate into the DNA structure in cell nuclei and subsequently prevent the proliferation of cancer cells. Lin et al. (2014) designed an aptasensor for the DNR diagnosis based on $CuInS_2$ quantum dot ($CuInS_2$ QDs) conjugate with the MUC1 aptamer that could detect it by altering the photoluminescence response of $CuInS_2$ QDs. The fluorescence intensity of the MUC1-QDs conjugate clearly diminished with the increase of the DNR concentration. The LOD of 19 nM with the linear detection range of 33–88 nM were obtained by the aptasensor.

Chandra et al. (2011) developed an effective aptasensor for the electrochemical detection of DNR, in which phosphatidylserine (PS) and DNR aptamer were used as the bioreceptors. PS and DNR-binding aptamer were fixed on AuNPs, modified by a conducting polymer [2,2′:5′,2″-terthiophene-3′-(*p*-benzoic acid)] (polyTTBA). The DNR-aptamer and DNR-PS interactions on the surface of AuNPs induced the current response. The LOD of 52.3 ± 2.1 pM with the dynamic range of 0.1–60.0 nM were obtained for the DNR analysis.

3.3.3 Doxorubicin

The diagnosis of doxorubicin (Dox) as a chemotherapeutic agent is very important because of its toxic effect on proliferating cells and strong side effects, especially cardiac toxicity (Weiss 1992).

Bahner et al. (2018) developed an aptasensor to electrochemically detect Dox with a three-electrode system in which the modified aptamer was placed on the gold electrodes through a self-assembly process. The interaction between Dox and immobilized aptamer prevented the access of $[Fe(CN)_6]^{3-/4}$ to the gold electrode that led to an enhancement in the impedance spectra. The LOD of 28 nM with a linear range of 31–125 nM were obtained for the Dox monitoring.

3.3.4 Irinotecan

Idili et al. (2019) developed a special and sensitive electrochemical aptasensor which could measure irinotecan as a chemotherapeutic agent in the living body directly. The modified aptamer was deposited onto a gold electrode through the thiol group at its 5′-end while a methylene blue redox reporter was attached to its 3′-end. Through the irinotecan interaction with the aptamer, a conformational change induced the formation of the G-quadruplex structure that brought the methylene blue molecule closer to the electrode surface that induced a change in the electron transfer.

3.3.5 Amifostine

Alkahtani et al. (2020) developed an efficient aptasensor for electrochemical detection of amifostine (AMF) as a DNA-binding chemotherapeutic drug. A glassy carbon electrode was modified by a nanocomposite containing silver nanoparticles@MnFe Prussian blue nanospheres (AgNPs@MnFePBA NS). The AMF aptamer was placed onto the nanocomposite (aptamer/AgNPs@MnFePBA NS/GCE). In the presence of AMF, a successful interaction was occurred between the aptamer and its target, leading to an increment in the charge transfer resistance. The LOD and linear detection range in the human blood plasma samples were obtained 0.11 nM and 0.34–45 nM, respectively.

3.4 Cardiovascular Drugs

Digoxin is one of the oldest antiarrhythmic drugs, used in the treatment of the different heart diseases, such as atrial fibrillation and congestive heart failure with the general effect of strengthening muscles with heart failure. However, due to the very narrow therapeutic index of digoxin (0.5–2.0 ng mL^{-1}), its concentration above 2 ng mL^{-1} in plasma is extremely toxic (Juillière and Selton-Suty 2010). Therefore, the exact monitoring of digoxin levels in the bloodstream is essential.

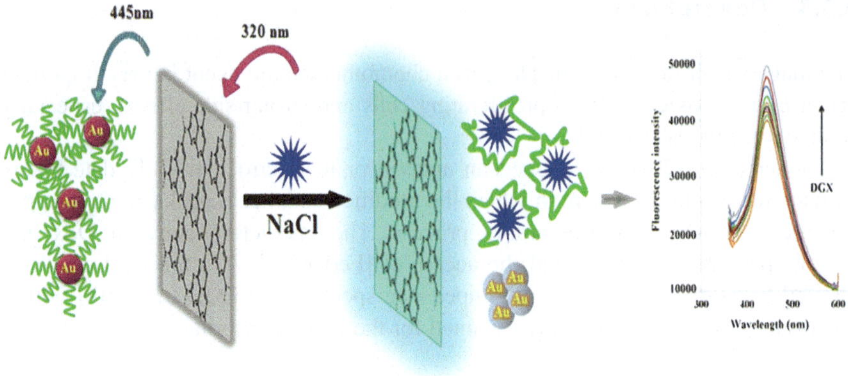

Fig. 3.5 Schematic illustration of the designed aptasensor by using graphitic carbon nitride nanosheet as the fluorescent probe and aptamer/AuNPs conjugate to detect digoxin. Reprinted with permission from reference Shirani et al. (2020)

For this purpose, some aptasensors have been introduced analytical tools for detection of digoxin (Shirani et al. 2020; Mashhadizadeh et al. 2017; Emrani et al. 2015). Zhang et al. (2020) developed a unique self-powered and dual-photoelectrode aptasensor on the basis of a photofuel cell (PFC), consisted of black TiO_2 (B-TiO_2) and CuBr as a photoanode and photocathode, respectively. The selective aptamer was immobilized on photoanode. After introducing digoxin molecule into the system, the complex of aptamer-target was formed and subsequently was oxidized by the photo-induced holes. So, an enhancement in the power output was then obtained as the criterion for the target presence. The LOD was 0.33 pM in fetal calf serum samples with a broad linear detection range of 1 pM–10 μM.

Shirani et al. (2020) constructed a label-free fluorescent aptasensor for digoxin detection by using graphitic carbon nitride nanosheet (g-C_3N_4NS) as the fluorescent probe. The interaction of aptamer/AuNPs conjugate with the fluorescent probe reduced the fluorescence intensity. Although the researchers mistakenly cited the mechanism of this interaction as FRET, while the term surface energy transfer (SET) was often used in the case of energy transfer for metal surfaces (such as AuNPs) (Saini et al. 2007; Ray et al. 2014). In the presence of digoxin and in the salt-containing medium, the aptamers were separated from the surface of AuNPs and bound to their target. This is followed by the aggregation of AuNPs that obtained the restoration of the fluorescence intensity (Fig. 3.5). The designed aptasensor was tested in the plasma samples and the LOD was reported to be 3.2 ng L^{-1} with the linear range of 10–500 ng L^{-1}.

3.5 Antidiabetic Drugs

Diabetes is a metabolic disease that occurs if there is not sufficient insulin in the body for a long time, leading to an increase in blood glucose levels in the circulatory system (Ling et al. 2016; Alberti and Zimmet 1998). Insulin is produced by the pancreatic islet β cells as the main anabolic hormone of body, which plays a significant role in the glycophysiological metabolism (Voet and Voet 2011). Since modern models measure the insulin dose to control blood glucose levels on the basis of insulin-related glucose uptake estimations (Zhao et al. 2019; Barnes 2013), it is substantial to determine insulin in biomedicine and clinical treatment of diabetes. Hence, aptasensors have been introduced to provide sensitive, accurate, highly selective, simple, and low-cost insulin detection (Zhou et al. 2020b; Shang et al. 2020).

Taib et al. (2020) designed an optical aptasensor for the insulin monitoring, in which Nickel-salphen type complex [Ni(II)-SP] was utilized as an optical label. Besides, aminated porous silica microparticles (PSiMPs) were applied to prepare large area for the immobilization of the insulin-binding aptamer (IGA3). Due to the π-π stacking interaction between planar aromatic groups of Ni(II)-SP and aromatic rings of the IGA3 aptamer, the complex of IGA3-Ni(II)-SP was formed, which resulted in a yellow color. With adding insulin, the G-quadruplex aptamer was bound to it that led to a decrease of the optical reflectance response, accompanying with a yellow to brownish orange color change. The insulin levels were measured in the healthy human serums. The LOD of 3.71 μIU mL^{-1} with a linear range of 10–50 μIU mL^{-1} were obtained for the insulin detection.

Zhao et al. (2019) fabricated a dual-signaling aptasensor to monitor insulin sensitively, as shown in Fig. 3.6. The insulin-binding aptamer (IBA) was modified by methylene blue (MB), used as a "signal-off" probe. The AuNPs were functionalized by DNA2/Ferrocene (Fc) [DNA2Fc@GNPs], applied as a "signal-on" probe. Two probes were integrated through the linker mDNA that eventually formed the DNA2Fc@GNPs/mDNA/MB-IBA-modified electrode as the sensing interface. With incubation of the target, the IBA aptamer was separated from the complex, resulting in proximity of the DNA2Fc@GNPs to the electrode surface. So, an increase in the response of Fc and a reduction in the response of MB were achieved simultaneously. The LOD and linear concentration range were obtained as 0.1 pM and 10 pM–10 nM, respectively, for the insulin detection in the serum samples.

3.6 Respiratory Drugs

Theophylline (TP) is applied as a bronchodilator drug to cure many respiratory failures, such as chronic obstructive pulmonary disease, neonatal apnea, and bronchial asthma. (Barnes 2013) The most important obstacle in the clinical usage of TP is its narrow therapeutic range (20–100 μM), which can be toxic and lethal at higher concentrations and cause permanent nerve damage (Dawson and Whyte 1999).

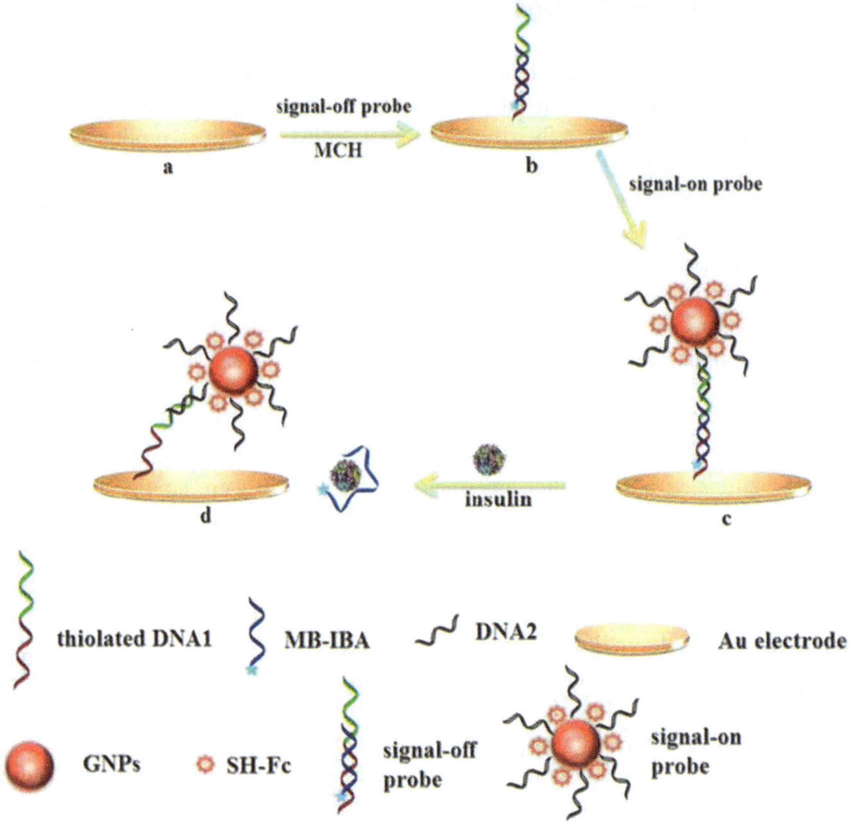

Fig. 3.6 Schematic illustration of the dual-signaling aptasensor for insulin detection. Reprinted with permission from reference Zhao et al. (2019)

Therefore, monitoring of the TP serum levels is very crucial to avoid the serious problems. Aptasensors can be a grate suggestion with low cost, great sensitivity, and high speed for analyzing TP.

Wu et al. (2019b) designed a novel aptasensor for the fluorescent detection of TP by using a RNA aptamer (RNA1), functionalized by QDs as a fluorescent label, and also graphene oxide (GO) as a fluorescence quencher. In the absence of TP, the RNA1 was adsorbed on the GO platform through the stacking interactions. So, the quenching process was occurred from QD to GO, leading to a decline in the QD fluorescence. When RNA2 and TP were introduced into the system, a conformational change was happened and the dsRNA-TP complex was formed. Subsequently, the fluorescent intensity was recovered by moving the QDs away from the GO platform. The LOD of the aptasensing strategy was 4 nM with a linear range of 10–300 nM.

Katiyar et al. (2013) fabricated a colorimetric sensing method for the TP detection based on the RNA aptamer attached to AuNPs via the electrostatic interaction that prevented AuNPs from the salt-induced aggregation. With the addition of TP, AuNPs were aggregated in the presence of salt, due to the aptamer binding to the target that resulted in a red to purple color change. The LOD of the introduced colorimetric aptasensor was obtained as low as 50 ng mL^{-1}.

3.7 Conclusions

Abuse of drugs may increase the risks and harms to human and animal health. Hence, the low cost, rapid, sensitive, accurate, and on-site detection of drugs is an enormous challenge. As one of the most efficient types of biosensing arrays, aptasensors are novel diagnostic tools for this purpose, in which aptamers are applied as the bioreceptor elements. Aptamers not only have supreme properties over antibodies in generation, modification, and stability, but also possess the unique benefits of nucleic acid nature and adaptive binding that facilitates the design of ultrasensitive aptasensors. Optical and electrochemical aptasensing platforms, owing to their aforementioned advantages, suggest excellent strategies to monitor diverse drugs by employing a bioreceptor section against certain targets. With their capability in naked-eye detection, the colorimetric aptasensing methods are successfully designed for drug detection mainly involving nanoparticles. The colorimetric-based aptasensors are appropriate for the on-site monitoring of drugs because of no requirement for the fluorescent labels and intercalating dyes. The significant advances in the electrochemical-based aptasensors represents the superior ability to quantify drugs with high sensitivity. Generally, the electrochemical-based aptasensors provide less detection limit, and hence, greater functionality to detect drugs in comparison with the optical types. In spite of all developments in aptasensors, a few number of commercial products are available for the simultaneous determination of drugs in the matrix of the biological contents, such as blood, saliva, urine, and so on. Therefore, future scientific efforts should be centralized on the fabrication of commercial portable aptasensors with the capability of simultaneous detection of drugs in the complex biological samples.

References

Abnous K, Danesh NM, Alibolandi M, Ramezani M, Taghdisi SM, Emrani ASJS et al (2017) A novel electrochemical aptasensor for ultrasensitive detection of fluoroquinolones based on single-stranded DNA-binding protein. Sensors Actuators B: Chem 240:100–106

Akiyama Y, Ma Q, Edgar E, Laikhter A, Hecht SM (2008) Identification of strong DNA binding motifs for bleomycin. J Am Chem Soc 130(30):9650–9651

Alberti KGMM, Zimmet PZ (1998) Definition diagnosis and classification of diabetes mellitus and its complications. Part 1: diagnosis and classification of diabetes mellitus. Provisional report of a WHO consultation. Diabet Med 15(7):539–553

Alfarouk KO, Stock C-M, Taylor S, Walsh M, Muddathir AK, Verduzco D et al (2015) Resistance to cancer chemotherapy: failure in drug response from ADME to P-gp. Cancer Cell Int 15(1):71

Alkahtani SA (2020) Silver nanoparticles conjugated MnFe-based Prussian blue analogue for voltammetric and impedimetric bioaptasensing of amifostine (ethyol). Microchim Acta 187(10):1–8

Auerbach T, Bashan A, Yonath A (2004) Ribosomal antibiotics: structural basis for resistance, synergism and selectivity. Trends Biotechnol 22(11):570–576

Azadbakht A, Abbasi AR (2019) Impedimetric aptasensor for kanamycin by using carbon nanotubes modified with MoSe 2 nanoflowers and gold nanoparticles as signal amplifiers. Microchim Acta 186(1):23

Bahadır EB, Sezgintürk MK (2015) Applications of commercial biosensors in clinical, food, environmental, and biothreat/biowarfare analyses. Anal Biochem 478:107–120

Bahner N, Reich P, Frense D, Menger M, Schieke K, Beckmann D (2018) An aptamer-based biosensor for detection of doxorubicin by electrochemical impedance spectroscopy. Anal Bioanal Chem 410(5):1453–1462

Bai Z, Chen Y, Li F, Zhou Y, Yin H, Ai S (2019) Electrochemical aptasensor for sulfadimethoxine detection based on the triggered cleavage activity of nuclease P1 by aptamer-target complex. Talanta 204:409–414

Bangemann M (1994) Commission Regulation (EC) No 1430/94. Official Journal of the European Communities, Brussels, Belgium

Barnes PJ (2013) Theophylline. Am J Respir Crit Care Med 188(8):901–906

Blum RH, Carter SK, Agre K (1973) A clinical review of bleomycin—a new antineoplastic agent. Cancer 31(4):903–914

Cao J, Sun T, Grattan KT (2014) Gold nanorod-based localized surface plasmon resonance biosensors: a review. Sensors Actuators B Chem 195:332–351

Carta F, Scozzafava A, Supuran CT (2012) Sulfonamides: a patent review (2008–2012). Expert Opin Ther Pat 22(7):747–758

Chandra P, Noh H-B, Won M-S, Shim Y-B (2011) Detection of daunomycin using phosphatidylserine and aptamer co-immobilized on Au nanoparticles deposited conducting polymer. Biosens Bioelectron 26(11):4442–4449

Chandra P, Koh WCA, Noh HB, Shim YB (2012) In vitro monitoring of i-NOS concentrations with an immunosensor: the inhibitory effect of endocrine disruptors on i-NOS release. Biosens Bioelectron

Chen X-X, Lin Z-Z, Hong C-Y, Yao Q-H, Huang Z-Y (2020) A dichromatic label-free aptasensor for sulfadimethoxine detection in fish and water based on AuNPs color and fluorescent dyeing of double-stranded DNA with SYBR Green I. Food Chem 309:125712

Cheng S, Liu H, Zhang H, Chu G, Guo Y, Sun X (2020) Ultrasensitive electrochemiluminescence aptasensor for kanamycin detection based on silver nanoparticle-catalyzed chemiluminescent reaction between luminol and hydrogen peroxide. Sensors Actuators B Chem 304:127367

Choudhary M, Yadav P, Singh A, Kaur S, Ramirez-Vick J, Chandra P, Arora K, Singh SP (2016) CD 59 targeted ultrasensitive electrochemical immunosensor for fast and noninvasive diagnosis of oral cancer. Electroanalysis 28:2565–2574

Dawson AH, Whyte IM (1999) Therapeutic drug monitoring in drug overdose. Br J Clin Pharmacol 48(3):278–283

Dehghani S, Danesh NM, Ramezani M, Alibolandi M, Lavaee P, Nejabat M et al (2018) A label-free fluorescent aptasensor for detection of kanamycin based on dsDNA-capped mesoporous silica nanoparticles and Rhodamine B. Anal Chim Acta 1030:142–147

Deka S, Saxena V, Hasan A, Chandra P, Pandey LM (2018) Synthesis, characterization and in vitro analysis of α-Fe2O3-GdFeO3 biphasic materials as therapeutic agent for magnetic hyperthermia applications. Mater Sci Eng C 92:932–941

Derbyshire N, White SJ, Bunka DH, Song L, Stead S, Tarbin J et al (2012) Toggled RNA aptamers against aminoglycosides allowing facile detection of antibiotics using gold nanoparticle assays. Anal Chem 84(15):6595–6602

Dirkzwager RM, Liang S, Tanner JA (2016) Development of aptamer-based point-of-care diagnostic devices for malaria using three-dimensional printing rapid prototyping. ACS Sensors 1(4): 420–426

Edson RS, Terrell CL, editors. The aminoglycosides. In: Mayo clinic proceedings; 1999, Elsevier

Emrani AS, Taghdisi SM, Danesh NM, Jalalian SH, Ramezani M, Abnous K (2015) A novel fluorescent aptasensor for selective and sensitive detection of digoxin based on silica nanoparticles. Anal Methods 7(9):3814–3818

Esmaelpourfarkhani M, Abnous K, Taghdisi SM, Chamsaz MJB (2020) A novel turn-off fluorescent aptasensor for ampicillin detection based on perylenetetracarboxylic acid diimide and gold nanoparticles. Biosensors Bioelectron 164:112329

Grill MF, Maganti RK (2011) Neurotoxic effects associated with antibiotic use: management considerations. Br J Clin Pharmacol 72(3):381–393

Ha N-R, Jung I-P, Kim S-H, Kim A-R, Yoon M-Y (2017) Paper chip-based colorimetric sensing assay for ultra-sensitive detection of residual kanamycin. Process Biochem 62:161–168

Han Y, Zheng J, Dong S (2013) A novel nonenzymatic hydrogen peroxide sensor based on Ag–MnO2–MWCNTs nanocomposites. Electrochim Acta 90:35–43

Hanekamp JC, Bast A (2015) Antibiotics exposure and health risks: chloramphenicol. Environ Toxicol Pharmacol 39(1):213–220

He H, Wang S-Q, Han Z-Y, Tian X-H, Zhang W-W, Li C-P et al (2020) Construction of electrochemical aptasensors with Ag (I) metal– organic frameworks toward high-efficient detection of ultra-trace penicillin. Appl Surf Sci 531:147342

Hermann T (2007) Aminoglycoside antibiotics: old drugs and new therapeutic approaches. J Cell Mol Life Sci 64(14):1841–1852

Hu S-W, Qiao S, Pan J-B, Kang B, Xu J-J, Chen H-Y (2018a) A paper-based SERS test strip for quantitative detection of Mucin-1 in whole blood. Talanta 179:9–14

Hu X, Goud KY, Kumar VS, Catanante G, Li Z, Zhu Z et al (2018b) Disposable electrochemical aptasensor based on carbon nanotubes-V2O5-chitosan nanocomposite for detection of ciprofloxacin. Sensors Actuators B Chem 268:278–286

Idili A, Arroyo-Currás N, Ploense KL, Csordas AT, Kuwahara M, Kippin TE et al (2019) Seconds-resolved pharmacokinetic measurements of the chemotherapeutic irinotecan in situ in the living body. Chem Sci 10(35):8164–8170

Jalalian SH, Karimabadi N, Ramezani M, Abnous K, Taghdisi SM (2018) Electrochemical and optical aptamer-based sensors for detection of tetracyclines. Trends Food Sci Technol 73:45–57

Jiang Y, Sun D-W, Pu H, Wei Q (2019) Ultrasensitive analysis of kanamycin residue in milk by SERS-based aptasensor. Talanta 197:151–158

Jin B, Wang S, Lin M, Jin Y, Zhang S, Cui X et al (2017) Upconversion nanoparticles based FRET aptasensor for rapid and ultrasenstive bacteria detection. Biosens Bioelectron 90:525–533

Johnstone RW, Ruefli AA, Lowe SW (2002) Apoptosis: a link between cancer genetics and chemotherapy. Cell 108(2):153–164

Juillière Y, Selton-Suty C (2010) Digoxin therapy: a persisting interest despite contrary winds. Arch Cardiovasc Dis 103(5):281–284

Katiyar N, Selvakumar LS, Patra S, Thakur MS (2013) Gold nanoparticles based colorimetric aptasensor for theophylline. Anal Methods 5(3):653–659

Kato S, Inui N, Hakamata A, Suzuki Y, Enomoto N, Fujisawa T et al (2018) Changes in pulmonary endothelial cell properties during bleomycin-induced pulmonary fibrosis. Respir Res 19(1):127

Khajavian Z, Esmaelpourfarkhani M, Ramezani M, Alibolandi M, Abnous K, Taghdisi SM (2021) A highly sensitive, simple and label-free fluorescent aptasensor for tobramycin sensing based on PicoGreen intercalation into DNA duplex regions of three-way junction origami. Microchem J 160:105657

Khavani M, Izadyar M, Housaindokht MR (2019) Theoretical design and experimental study on the gold nanoparticles based colorimetric aptasensors for detection of neomycin B. Sensors Actuators B Chem 300:126947

Lin Z, Ma Q, Fei X, Zhang H, Su X (2014) A novel aptamer functionalized CuInS2 quantum dots probe for daunorubicin sensing and near infrared imaging of prostate cancer cells. Anal Chim Acta 818:54–60

Lin X, Sua X, Luo S, Liu B, Yang C (2016) Development of DNA-based signal amplification and microfluidic technology for protein assay: a review. TrAC Trends Anal Chem 80:132–148

Ling K, Jiang H, Zhang L, Li Y, Yang L, Qiu C et al (2016) A self-assembling RNA aptamer-based nanoparticle sensor for fluorometric detection of Neomycin B in milk. Anal Bioanal Chem 408(13):3593–3600

Liu Z, Xiao H, Zhang B, Shen H, Zhu L, Li C (2019) Copper-catalyzed remote C (sp^3)−H trifluoromethylation of carboxamides and sulfonamides. Angew Chem Int Ed 58(8):2510–2513

LiverTox N. Clinical and research information on drug-induced liver injury. National Institute of Diabetes and Digestive and Kidney Diseases, Bethesda (MD); 2020

Lofrano G, Libralato G, Adinolfi R, Siciliano A, Iannece P, Guida M et al (2016) Photocatalytic degradation of the antibiotic chloramphenicol and effluent toxicity effects. Ecotoxicol Environ Safety 123:65–71

Ma C, Liu H, Zhang L, Li H, Yan M, Song X et al (2018) Multiplexed aptasensor for simultaneous detection of carcinoembryonic antigen and mucin-1 based on metal ion electrochemical labels and Ru (NH3) 63+ electronic wires. Biosens Bioelectron 99:8–13

Mahato K, Kumar S, Srivastava A, Maurya PK, Singh R, Chandra P (2018) Electrochemical immunosensors: fundamentals and applications in clinical diagnostics. In: Handbook of immunoassay technologies

Mashhadizadeh MH, Azhdeh A, Naseri N (2017) 3-Mercapto propionic acid self-assembled on gold nano-particles applied for modification of screen-printed electrode as a new digoxin electrochemical aptasensor using graphene oxide-based signal-on strategy. J Electroanal Chem 787:132–138

Mathews B, Thalody AA, Miraj SS, Kunhikatta V, Rao M, Saravu K (2019) Adverse effects of fluoroquinolones: a retrospective cohort study in a South Indian tertiary healthcare facility. Antibiotics 8(3):104

McGlinchey TA, Rafter PA, Regan F, McMahon GP (2008) A review of analytical methods for the determination of aminoglycoside and macrolide residues in food matrices. Anal Chim Acta 624(1):1–15

Othman A, Karimi A, Andreescu S (2016) Functional nanostructures for enzyme based biosensors: Properties, fabrication and applications. J Mater Chem B 4(45):7178–7203

Pandey N, Cascella M (2020) Beta lactam antibiotics. StatPearls [Internet]

Purohit P, Stern S (1994) Interactions of a small RNA with antibiotic and RNA ligands of the 30S subunit. Nature 370(6491):659–662

Rad AO, Azadbakht A (2019) An aptamer embedded in a molecularly imprinted polymer for impedimetric determination of tetracycline. Microchim Acta 186(2):56

Ramezani M, Danesh NM, Lavaee P, Abnous K, Taghdisi SM (2015) A novel colorimetric triple-helix molecular switch aptasensor for ultrasensitive detection of tetracycline. Biosens Bioelectron 70:181–187

Ray PC, Fan Z, Crouch RA, Sinha SS, Pramanik AJCSR (2014) Nanoscopic optical rulers beyond the FRET distance limit: fundamentals and applications. Chem Soc Rev 43(17):6370–6404

Rouhbakhsh Z, Verdian A, Rajabzadeh G (2020) Design of a liquid crystal-based aptasensing platform for ultrasensitive detection of tetracycline. Talanta 206:120246

Roushani M, Rahmati Z, Farokhi S, Hoseini SJ, Fath RH (2020) The development of an electrochemical nanoaptasensor to sensing chloramphenicol using a nanocomposite consisting of graphene oxide functionalized with (3-Aminopropyl) triethoxysilane and silver nanoparticles. Mater Sci Eng C 108:110388

Saini S, Bhowmick S, Shenoy VB, BJJoP B, Chemistry PA (2007) Rate of excitation energy transfer between fluorescent dyes and nanoparticles. J Photochem Photobiol A: Chem 190(2–3): 335–341

Shang M, Zhang J, Qi H, Gao Y, Yan J, Song W (2019) All-electrodeposited amorphous MoSx@ZnO core-shell nanorod arrays for self-powered visible-light-activated photoelectrochemical tobramycin aptasensing. Biosens Bioelectron 136:53–59

Shang M, Gao Y, Zhang J, Yan J, Song W (2020) Signal-on cathodic photoelectrochemical aptasensing of insulin: plasmonic Au activated amorphous MoSx photocathode coupled with target-induced sensitization effect. Biosens Bioelectron 165:112359

Sharma N, Selvam SP, Yun K (2020) Electrochemical detection of amikacin sulphate using reduced graphene oxide and silver nanoparticles nanocomposite. Appl Surf Sci 512:145742

Shirani M, Kalantari H, Khodayar MJ, Kouchak M, Rahbar N (2020) A novel strategy for detection of small molecules based on aptamer/gold nanoparticles/graphitic carbon nitride nanosheets as fluorescent biosensor. Talanta 219, 121235

Sun C, Su R, Bie J, Sun H, Qiao S, Ma X et al (2018) Label-free fluorescent sensor based on aptamer and thiazole orange for the detection of tetracycline. Dyes Pigments 149:867–875

Sun M, Zhu Y, Yan K, Zhang J (2019) Dual-mode visible light-induced aptasensing platforms for bleomycin detection based on CdS–In2S3 heterojunction. Biosens Bioelectron 145:111712

Syshchyk O, Skryshevsky VA, Soldatkin OO, Soldatkin AP (2015) Enzyme biosensor systems based on porous silicon photoluminescence for detection of glucose, urea and heavy metals. Biosens Bioelectron 66:89–94

Taghdisi SM, Danesh NM, Nameghi MA, Ramezani M, Abnous K (2016) A label-free fluorescent aptasensor for selective and sensitive detection of streptomycin in milk and blood serum. Food Chem 203:145–149

Taib M, Tan LL, Abd Karim NH, Ta GC, Heng LY, Khalid B (2020) Reflectance aptasensor based on metal salphen label for rapid and facile determination of insulin. Talanta 207:120321

Tian S, Zhang C, Huang D, Wang R, Zeng G, Yan M et al (2020) Recent progress in sustainable technologies for adsorptive and reactive removal of sulfonamides. Chem Eng J 389:123423

Tor Y (2006) The ribosomal A-site as an inspiration for the design of RNA binders. Biochimie 88(8):1045–1051

Tu C, Dai Y, Zhang Y, Wang W, Wu LJSAPAM, Spectroscopy B (2020) A simple fluorescent strategy based on triple-helix molecular switch for sensitive detection of chloramphenicol. Spectrochim Acta Part A: Mol Biomol Spectrosc 224:117415

Verma S, Choudhary J, Singh KP, Chandra P, Singh SP (2019) Uricase grafted nanoconducting matrix based electrochemical biosensor for ultrafast uric acid detection in human serum samples. Int J Biol Macromol

Vicens Q, Westhof E (2003) Molecular recognition of aminoglycoside antibiotics by ribosomal RNA and resistance enzymes: an analysis of x-ray crystal structures. Biopolym: Original Res Biomol 70(1):42–57

Voet D, Voet JG (2011) Biochemistry, 4th edn. John Wiley& Sons Inc, New York, p 492

Wang Y, Bian F, Qin X, Wang Q (2018) Visible light photoelectrochemical aptasensor for chloramphenicol by using a TiO_2 nanorod array sensitized with Eu (III)-doped CdS quantum dots. Microchim Acta 185(3):161

Wang J, Lu T, Hu Y, Wang X, Wu Y (2020) A label-free and carbon dots based fluorescent aptasensor for the detection of kanamycin in milk. Spectrochim Acta Part A: Mol Biomol Spectrosc 226:117651

Weiss RB (1992) The anthracyclines: will we ever find a better doxorubicin? Semin Oncol 19(6)

Wu S, Zhang H, Shi Z, Duan N, Fang C, Dai S et al (2015) Aptamer-based fluorescence biosensor for chloramphenicol determination using upconversion nanoparticles. Food Control 50:597–604

Wu Y-y, Liu B-w, Huang P, Wu F-Y (2019a) A novel colorimetric aptasensor for detection of chloramphenicol based on lanthanum ion–assisted gold nanoparticle aggregation and smartphone imaging. Anal Bioanal Chem 411(28):7511–7518

Wu J-F, Gao X, Ge L, Zhao G-C, Wang G-F (2019b) A fluorescence sensing platform of theophylline based on the interaction of RNA aptamer with graphene oxide. RSC Adv 9(34): 19813–19818

Yan J, Huang Y, Zhang C, Fang Z, Bai W, Yan M et al (2017) Aptamer based photometric assay for the antibiotic sulfadimethoxine based on the inhibition and reactivation of the peroxidase-like activity of gold nanoparticles. Microchim Acta 184(1):59–63

Yan S, Lai X, Du G, Xiang Y (2018) Identification of aminoglycoside antibiotics in milk matrix with a colorimetric sensor array and pattern recognition methods. Anal Chim Acta 1034:153–160

Yan T, Zhang X, Ren X, Lu Y, Li J, Sun M et al (2020) Fabrication of N-GQDs and AgBiS2 dual-sensitized ZIFs-derived hollow ZnxCo3-xO4 dodecahedron for sensitive photoelectrochemical aptasensing of ampicillin. Sensors Actuators B: Chem 320:128387

Yin J, Guo W, Qin X, Zhao J, Pei M, Ding F (2017) A sensitive electrochemical aptasensor for highly specific detection of streptomycin based on the porous carbon nanorods and multifunctional graphene nanocomposites for signal amplification. Sensors Actuators B Chem 241:151–159

Zhang M, Zhang Z, Xu Y, Wen Z, Ding C, Guo Y et al (2020) A novel self-powered aptasensor for digoxin monitoring based on the dual-photoelectrode membrane/mediator-free photofuel cell. Biosens Bioelectron 156:112135

Zhao J, Guo W, Pei M, Ding F (2016) GR–Fe3O4 NPs and PEDOT–AuNPs composite based electrochemical aptasensor for the sensitive detection of penicillin. Anal Methods 8(22): 4391–4397

Zhao Y, Xu Y, Zhang M, Xiang J, Deng C, Wu H (2019) An electrochemical dual-signaling aptasensor for the ultrasensitive detection of insulin. Anal Biochem 573:30–36

Zhou N, Yang L, Hu B, Song Y, He L, Chen W et al (2018) Core-shell heterostructured CuFe@FeFe Prussian blue analogue coupling with silver nanoclusters via a one-step bioinspired approach: efficiently nonlabeled aptasensor for detection of bleomycin in various aqueous environments. Anal Chem 90(22):13624–13631

Zhou Y, Zuo L, Wei Y, Dong C (2020a) Development of fluorescent aptasensing system for ultrasensitive analysis of kanamycin. J Lumin 222:117124

Zhou X, Zhang W, Wang Z, Han J, Xie G, Chen S (2020b) Ultrasensitive aptasensing of insulin based on hollow porous C3N4/S2O82−/AuPtAg ECL ternary system and DNA walker amplification. Biosens Bioelectron 148:111795

Paper-Based Devices for the Detection of Food-Related Analyte

4

Maryam Mousavizadegan, Amirreza Roshani, and Morteza Hosseini

Abstract

This chapter mainly focuses on recently developed paper-based analytical devices for the detection of specific biomarkers for health threatening pathogens and contaminants in food which persist as a major problem for public safety and a financial burden on agricultural and food industries. Current detection methods are often time-consuming and expensive and require professional instrumentations. Because of their simplicity, portability, and low-cost, paper-based devices have proven to be efficient for the rapid detection of many biomarkers including those related to food safety. In this chapter, we will first present an introduction and a brief description of the fabrication methods used for paper-based devices. Thus, recent developments regarding food-related analyte detection using paper-based devices based on various readout approaches including colorimetric, fluorescent, chemiluminescence, surface-enhanced Raman scattering and electrochemical will be provided. In the end, current limitations and future perspectives for the commercialization of paper-based sensors will be discussed.

Keywords

Paper device · Food biomarkers · Detection · Microfluidics · Bioanalysis

M. Mousavizadegan · A. Roshani · M. Hosseini (✉)
Department of Life Science Engineering, Faculty of New Sciences and Technologies, University of Tehran, Tehran, Iran
e-mail: smhosseini@khayam.ut.ac.ir

P. Chandra, P. S. Panesar (eds.), *Nanosensing and Bioanalytical Technologies in Food Quality Control*, https://doi.org/10.1007/978-981-16-7029-9_4

4.1 Introduction

One of the main concerns of all nations is the continuous supply of safe, healthy food for all citizens. Food safety involves the identification and control of contaminants, both chemical and biological, during various stages starting from food preparation and processing to storage, transportation, and consumption. Pertaining to this, various identification methods including high-performance liquid chromatography (HPLC), gas chromatography (GC), and mass spectrometry for chemical contaminants and cell culture methods, biochemical identification, and polymerase chain reaction (PCR) for pathogen analysis are being used for food safety monitoring. However, most of these approaches require high-tech instrumentation and professional technicians. They are also expensive and time-consuming, and require laborious preparation steps. Furthermore, many of these methods are not feasibly accessible in remote, developing areas, which can lead to the accumulation of food safety hazards in those places. To overcome these problems, rapid, simple, and cost-effective sensing methods are required for food contaminant detection and control (Ragavan and Neethirajan 2019; Shams et al. 2020).

Sensor, derived from the Latin word *sentire* meaning recognition, is defined as a system or device that can respond to a chemical, physical, or biological analyte by producing measurable output signals which correspond to the amount of analyte (Choudhary et al. 2016; Chandra et al. 2012; Mahato et al. 2018; Deka et al. 2018; Verma et al. 2019). Sensors are generally composed of three segments including a recognition element, a transducing element, and a detector. Biosensors are a subclass of sensors in which biological elements such as enzymes, antibodies, nucleic acids, aptamers, and whole cells are generally used as the recognition unit (Patel 2002). Rapid, sensitive sensors and biosensors are widely used in numerous fields including biomedicine, environmental monitoring, agriculture, and food safety. They can overcome the challenges of traditional detection methods as they are simple, inexpensive, and portable. Nanotechnology, which refers to the science and technology of designing, and manipulating materials and systems at nanoscale have impacted a myriad of applications in science and engineering, including the design of efficient sensing platforms. Nanomaterials possess distinctive physicochemical properties such as outstanding electrical conductivity, unique optical features, and catalytic behavior. Furthermore, increased surface-to-volume ratios, they provide a larger surface for biofunctionalization (Pérez-López and Merkoçi 2011). Thus, implementing nanomaterials with detection platforms can significantly enhance the sensing parameters.

Paper technology was first developed in the second century AD in China. Since then it has revolutionized the human civilization in numerous areas. Besides being abundant and easily accessible, paper is flexible, lightweight, and easily manufactured and modified, thus making it an excellent candidate for the development of point-of-care sensing platforms. First introduced in the seventeenth century for the detection of uric using, paper-based analytical devices (PADs) have rapidly become one of the forerunners in the field of sensing (Schiff 1866). The hydrophobicity and porous nature of paper provide a platform for fluid flow through capillary actions. Also, paper can easily be functionalized with biomolecules such as antibodies, enzymes, and aptamers or modified with nanomaterials, which can enhance the specificity and sensitivity of the detection. All these attributes make paper-based sensors an ideal platform for food monitoring and point-of-care testing (Busa et al. 2016).

In this chapter, we will focus on paper-based sensing strategies for the detection of food-related analytes. First, an overview on the various types of food contaminants will be provided. We will then elaborate on the various fabrication methods commonly used for the development of PADs, followed by the different formats of PADs. The different signal readout techniques used for paper-based sensors including colorimetric, fluorescence, chemiluminescence, surface-enhanced Raman scattering (SERS), and electrochemical methods will then be discussed. Finally, the limitations and future perspectives of PADs in food monitoring will be mentioned.

4.2 Food Safety Analytes

The ability to attain nutritious food has ample significance in maintaining a healthy lifestyle. In the past, there was much neglect towards the importance of food safety and its impact on human health, which resulted from lack of knowledge regarding the various contaminants that can compromise the wholesome nutritious value of our food. In 2015, the World Health Organization (WHO) published an article estimating the significance of food safety and the global burden of food-borne illnesses. According to this article, 42,000 deaths each year from 32 food-related diseases occur. Developing countries carry a heavier burden pertaining to food-related diseases and deaths (Griesche and Baeumner 2020). These contaminants also pose a great economic threat to the agriculture and food industry in all countries. Thus, the identification of these contaminants in early stages is highly important.

To be able to develop rapid, sensitive detection approaches for food safety monitoring, a thorough understanding of the various food contaminants is necessary. Examples of these contaminants include heavy metals, pesticides, pathogens, toxins, etc. (Marklinder et al. 2020). In Table 4.1, an overview of these contaminants with certain examples has been provided.

4.3 Fabrication and Development of Paper-Based Sensors

Paper technology has become one of the most significant technologies since its development during the second century AD in China. In the field of sensing, paper has numerous advantageous properties such as being environmentally friendly, lightweight, flexible, available in a range of thicknesses, and easily accessible. Also, paper provides a porous platform in which fluid flow occurs through capillary actions, which reduces reagent waste. Since their first introduction in the seventeenth century, paper-based analytical devices (PADs) have become the focus of attention

Table 4.1 Examples of common food contaminating agents

Common food contaminants		
Pathogens	Bacteria	1. *Escherichia coli* 2. *Listeria monocytogenes* 3. *Salmonella typhimurium* 4. *Brucella* spp. 5. *Campylobacter* spp. 6. *Clostridium botulinum* 7. *Mycobacterium bovis* 8. *Shigella* spp. 9. *Mycobacterium tuberculosis* 10. *Staphylococcus aureus* 11. *Streptococcus* spp. 12. *Vibrio cholerae*
	Viruses	1. Astrovirus 2. Hepatitis A virus 3. Norovirus 4. Rotavirus
	Parasites	1. *Cryptosporidium* spp. 2. *Cyclospora cayetanensis* 3. *Giardia intestinalis* 4. *Toxoplasma gondii*
Toxins	**Bacterial toxins**	Toxins produced by: 1. *Staphylococcus aureus* 2. *Bacillus cereus* 3. *Clostridium perfringens* 4. *Clostridium botulinum*
	Mycotoxins	1. Aflatoxins: produced by *Aspergillus flavus* and *Aspergillus parasiticus* 2. Ochratoxins: produced by *Aspergillus* and *Penicillium* species 3. Deoxynivalenol: produced by *Fusarium* fungi 4. Zearalenone: produced by *Fusarium* fungi 5. Ergot alkaloids: produced by *Claviceps* species
	Phytotoxins	1. Cyanogenic glycosides: produced in several plant species including cassava, sorghum, stone fruits, bamboo roots, and almonds 2. Furocoumarins: present in many plants such as parsnips, celery roots, citrus plants, and some medicinal plants 3. Lectins: present in various beans, especially kidney beans 4. Solanine and chaconine: present in tomatoes, potatoes, and eggplants 5. Pyrrolizidine alkaloids: produced in many plants and have been detected in herbal teas, honey, herbs, spices, and cereals
	Aquatic toxins	1. Algal toxins: formed by algae 2. Ciguatoxins: produced by dinoflagellates
Heavy metals	Metals with an atomic weight between 63.5 and 200.6 $gr.mol^{-1}$ and a specific gravity greater than 5 $gr.cm^{-1}$ such as arsenic (As), beryllium (Be), cadmium (Cd), chromium (Cr), lead (Pb), and mercury (Hg)	
Pesticides	Insecticides	Organophosphates, pyrethroids, and carbamates
	Fungicides	Captan, sulfur, and mancozeb
	Herbicides	Diclofop, dinoseb, diquat, and paraquat

(continued)

Table 4.1 (continued)

Common food contaminants		
	Rodenticides	Bromadiolone, chlorophacinone, difethialone, brodifacoum, and warfarin
	Bactericides	Streptomycin sulfate and oxytetracycline
Veterinary drugs	Veterinary drugs are used in animal husbandry for many reasons, including for the prevention and curing of diseases in herd and flock, to improve meat quality, and to promote growth. The chemical classes of drugs used are broad, but major classes include antibiotics, antiparasitics, and hormones.	
Illegal additives	A variety of additives including certain preservatives, artificial colors, flavoring agents, etc. have been banned. The list of illegal additives varies in each country and region.	

in the field of sensing in medical diagnosis, pharmaceuticals, and food and environment monitoring.

Cellulose, being the main substrate in paper fabrication is the most abundantly available biomolecule in nature. For the design of PADs, several derivatives of cellulose have been used including Whatman papers which are the most commonly used. Whatman papers are composed of cotton cellulose and come in a range of thicknesses and porosities. Nitrocellulose membranes are another substrate commonly used for PAD fabrication with very uniform pores of 0–0.45 μm. Other substrates include polyester–cellulose blended-paper and glass microfiber filters (Manisha et al. 2018).

In this section, we will first discuss the common fabrication routes for PAD development, followed by a brief description of the various formats of paper-based sensors.

4.3.1 Fabrication Methods of Paper-Based Sensors

In general, a paper-based sensor is composed of hydrophilic channels for reagent flow, which are separated by hydrophobic regions for liquid confinement. As paper is an extremely flexible platform, thus it enables the opting of various fabrication techniques. The overall process of paper fabrication can be divided into two stages; first, the hydrophobic regions must be patterned on the substrate thus creating the hydrophilic channels, and then the device must be assembled. An established approach for the first stage is the blocking of the paper pores with hydrophobic materials. Different water impervious materials including PDMS, polystyrene, ethyl cellulose, alkenyl ketene dimer, silicones, rosin, paraffin, printer varnish, cellulose esters, hydrophobic gels, SU-8, and other photoresist materials have been used to implement the hydrophobic barriers (Fakhri et al. 2018; Dixit et al. 2016). Numerous methods have been opted for the implementation of such materials, the most common of which will be briefly discussed.

I. *Photolithography:* Photolithography is the non-contact process of coating a photoresist polymer such as SU-8 on paper using UV light. In this process, paper is first soaked with the photoresist polymer and baked to remove unwanted substances. The modified paper is then exposed to UV light, which leads to the formation of crosslinks between the polymers, thus creating the hydrophobic regions. The hydrophilic channels can be functionalized with enzymes, antibodies, or oligonucleotides for targeted detection (Martinez et al. 2007). The advantages of this approach is that it leads to distinct hydrophobic borders which facilitate the flow of fluids through capillary actions and is suitable for large-scale production, but its requirement of expensive reagents, organic solvents, and complex preparations can be a downfall. Furthermore, papers fabricated via this approach tend to be quite fragile.
II. *Inkjet printing:* This non-contact method is among the earlier approaches used for PAD fabrication. In this method, using a nozzle, the hydrophobic ink such as polydimethylsiloxane (PDMS) is directly transferred on the paper surface following the required pattern. Inkjet printing can be carried out in two modes: continuous mode, in which the ink is continuously pumped in evenly spaced micrometer droplets; or drop-by-drop mode (DOD), in which the ink is pumped on paper as a result of a piezo-electrically or thermally activated acoustic pulse (Soleimani-Gorgani 2016). Through inkjet printing, rapid fabrication can be achieved, although the requirement of a customized inkjet printer and an extra heating step for curing purposes can be limiting.
III. *Screen Printing:* Among the most commonly used PAD fabrication approaches, screen printing involves the layer-by-layer transfer of liquid material onto the substrate using a screen. This can be done either manually or automatically, in which the pressure applied on the substrate and the amount of ink is regulated. The paper is left to dry after the application of the liquid material and can then be subjected to further treatment and modifications (Dungchai et al. 2011). The major advantages of this approach are simplicity, rapid fabrication, low costs, flexibility to various materials, and compatibility with mass production. This technique is also widely applied to generate electrodes on paper platforms for electrochemical detections (Timur et al. 2004). Low resolution and the requirement of different screens for the generation of various patterns can be mentioned as the downfalls of this approach.
IV. *Wax printing:* As wax dipping has the fewest number of steps, it is considered among the simplest fabrication methods. Three different methods of wax patterning have so far been introduced including painting with a wax pen, printing with a normal inkjet printer followed by tracing the patterns with a wax pen, and direct printing by a wax printer. In all methods, melted wax is transferred on paper to create the hydrophobic barriers (Carrilho et al. 2009). The many merits of this approach include being environmentally friendly, rapid, cost-effective, and simple. It is also quite suitable for mass production. Although the requirement for a patterned mesh and several heating steps can be a limiting factor.

V. *CO_2 laser cutting:* In this approach, a CO_2 laser is employed to cut and make patterns on the paper surface. The hydrophilic patterns are then remodeled to create the hydrophobic barriers for the hydrophilic channels. This approach is widely used in the fabrication of microfluidic devices. It is simple and creates sharp patterns, but the requirement for expensive equipment and its susceptibility to contamination can be limiting factors (Spicar-Mihalic et al. 2013).

VI. *Laser printing:* This technique enables the fabrication of microfluidic paper devices. Through this method, toner layers are deposited on the paper surface using a laser printer, creating white regions which represent the channels, which are then sealed via a laminating process (Ng and Hashimoto 2020). This is an inexpensive and simple method and can create sharp barriers although it requires special equipment such as laser printers and laminators.

VII. *Wax dipping:* In this method, melted wax is used to generate the hydrophobic regions. Iron molds are placed on the hydrophilic channels using a magnetic field for protection. The paper attached to the iron molds is then dipped in molten wax, thus creating the hydrophobic barriers. This method is very simple and cheap, and it requires easily accessible solid wax. But, because of variations in dipping, batch variations can be seen, making this method unsuitable for mass production (Songjaroen et al. 2011).

VIII. *Plasma treatment:* In plasma treatment, the filter paper is first dipped in a hydrophobic solution and placed in an oven to create the hydrophobic paper. This paper is then placed between two metal masks containing the desired patterns and submitted to plasma treatment to create the hydrophilic regions for fluid flow. The main advantage of this method over other methods which focus on barrier design is the possibility to build simple functional elements such as filters, switches, and separators in the paper device (Li et al. 2008). This method is also quite inexpensive and renders flexible paper devices, but it requires special masks for each desired pattern.

IX. *Flexographic printing:* In this method, flexible relief plates are used. Polystyrene is printed on the raised parts of the plates which forms the hydrophobic barriers. Regions without polystyrene are the hydrophilic channels for fluid flow. The advantage of this method over lithography is its ability to use a wider range of inks including water-based inks, and it can also be used to print on a variety of materials. This method also enables rapid commercial production of paper devices (Gonzalez-Macia et al. 2010). Although its requirement of complex steps, various reagents and advanced instruments can be limiting. Furthermore, papers generated through this method are highly prone to contamination.

4.3.2 Formats of Paper-Based Sensors

Since the development of paper-based analytical devices, various formats have been proposed to control fluid flow and biochemical analysis. Dipstick assays were among the first approaches proposed, subsequently followed by the more advanced lateral flow assays. Today, microfluidic PADs are dominating this field due to their enhanced fluid flow capabilities. In this section, we will briefly elaborate on these different formats.

4.3.2.1 Dipstick Assays

Making the simplest paper devices, dipsticks are generally known for pH test strips and the detection of glucose and other biochemicals in urine (Mori et al. 2010). In this method, the specific indicators are predisposed on the filter paper usually through soaking the paper in solutions containing the indicators. When the analyte reacts with the indicator, a signal is generated. Dipsticks are mainly coupled with colorimetric readout methods. Although these assays are simple and convenient, they have limited fluid flow capacities, thus cannot be used for multistep detection procedures.

4.3.2.2 Lateral Flow Assays

Lateral flow assays (LFAs) are generally composed of a nitrocellulose membrane, which can provide a platform both for reaction and detection. The platform is divided into four zones: (i) the sample pad for sample filtration and buffer storage; (ii) conjugation pad which is adjacent to the sample pad and contains dried reagent; (iii) detection pad, where the reagents are captured on the nitrocellulose and detection signal is developed; and (iv) absorbent pad which provides a driving force through the loose mass of nitrocellulose fibers. The adjacent sections overlap to ensure a coordinated fluid flow (Fenton et al. 2009). To perform the assay, the sample is first applied on the pretreated sample pad. The mixture flows to the conjugation pad which may contain nanomaterials which provide unique optical, electronic, catalytic properties (Wong and Tse 2008).

Generally, two formats of LFAs have been reported, sandwich and competitive formats. In the sandwich format, the conjugated particles and the analyte for particle-analyte complexes, which are captured at the detection pad by special capturing molecules. In the competitive format, both analytes and particles have affinity towards the capture molecules. The analytes have more affinity compared to the conjugated particles, thus in the presence of the analytes the attach to the capture molecules, which results in the non-aggregation of the particles (Nery and Kubota 2013). Both these formats have been used for food monitoring. Because of the nitrocellulose membrane, LFAs have higher fluid handling capabilities. But precise quantification of analytes still remains a challenge in this platform.

4.3.2.3 Microfluidic Devices

Microfluidic paper-based analytical devices (μPADs) have gained increasing attention since their first introduction in 2007 (Morbioli et al. 2017). These devices combine paper-based detection with microfluidic technology, which refers to the control of fluids in the range of micro to picoliters. μPADs can be divided into two-dimensional and three-dimensional platforms. The cellulose paper used in the fabrication of μPADs generally has an average fiber diameter of 1–100 μm and average pore size of 1–10 μm. Capillary-driven flow of fluids occurs between the porous medium. Sample loss caused by undesired soaking is minimalized in μPADs, which leads to more accurate quantifications of samples.

4.4 Detection methods

Numerous optical, spectroscopy and electrochemical readout methods have been coupled with PADs for qualitative and quantitative analysis of food-related markers. In this section, we will discuss paper-based food analysis based on colorimetric, fluorescence, chemiluminescence, surface-enhanced Raman scattering (SERS), and electrochemical approaches. A brief explanation of each method, followed by example paper assays in food monitoring will be presented.

4.4.1 Colorimetric

Colorimetric detection techniques are among the most prevalently used methods in PADs since they are simple, inexpensive, and offer the possibility for qualitative and semi-quantitative signal analysis by the naked eye. In these sensors, the presence and concentration of the analyte can be assessed with the help of a reagent able to undergo color change with altering conditions. As paper substrates provide a bright, white background, color change signals can be easily discerned (Nery and Kubota 2013). The incorporation of various chemicals, enzymes, nanoparticles, targeting molecules such as antibodies and aptamers have led to the development of advanced colorimetric sensors with the ability for qualitative, semi-quantitative, and fully quantitative assessment of food-related analytes (Morbioli et al. 2017; Sharma et al. 2018; Dehghani et al. 2019a; Abarghoei et al. 2019). The Beer–Lambert Law, which states that the intensity of the color signal is directly proportional to the concentration of the analyte, is generally used for the quantitative analysis in colorimetric PADs. The widespread use of digital cameras, smartphone cameras, and scanners, coupled with image processing software such as ImageJ has further eased signal interpretation and color intensity assessment in colorimetric PADs (Dehghani et al. 2018).

One of the most common colorimetric readout methods relies on the enzymatic transformation of chromogenic substrates like 3,3′,5,5′-tetramethylbenzidine (TMB) to colored products. This reaction is mostly catalyzed by horseradish peroxidase (HRP) (Busa et al. 2016), numerous enzyme-mimicking nanostructures have also been reported with the ability to oxidize TMB. Numerous nanomaterials including Fe_3O_4 nanoparticles, noble metal nanostructures, and transition metal nanostructures have been proven to show catalytic activity (Dehghani et al. 2018; Hosseini et al. 2017; Dehghani et al. 2019b; Kermani et al. 2018). Enzyme-mimicking nanomaterials have comparable catalytic behavior but are more stable in various sensing conditions compared to conventional enzymes. Exploiting the intrinsic peroxidase activity of $ZnFe_2O_4$ magnetic nanoparticles, a microfluidic paper-based colorimetric assay was proposed for the detection of bisphenol A (BPA), an

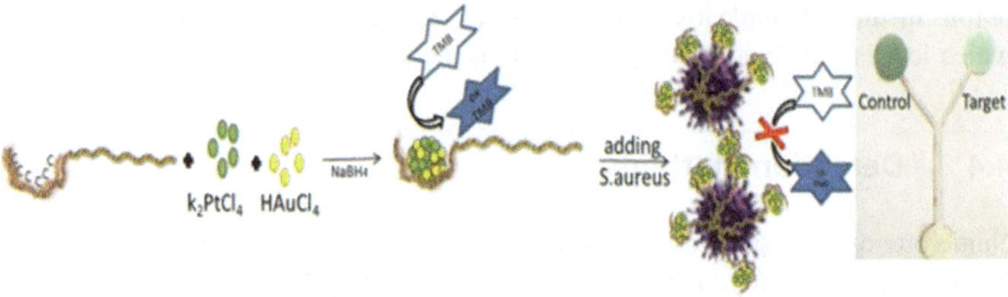

Fig. 4.1 Schematic representation of a paper-based colorimetric aptasensor for the detection of *S. aureus* using the intrinsic catalytic activity of Au/Pt bimetallic nanoclusters (Hosseini 2020)

endocrine disrupting chemical widely used in packaging products. Molecularly imprinted polymer (MIP) membranes were used for the absorption of BPA. The absorbance of BPA inhibited the oxidation of TMB and the formation of a blue color signal (Kong et al. 2017).

Metallic nanoclusters, which are composed of a few metal atoms and have discrete energy levels, have displayed favorable properties including enzyme-mimicking behavior. In an interesting approach for the detection of *Staphylococcus aureus*, DNA sequences were used both as recognition units (aptamers) and templates for the synthesis of Au/Pt bimetallic nanoclusters to fabricate a microfluidic paper-based aptasensor (Hosseini 2020). As shown in Fig. 4.1, in the absence of the target bacteria, the nanoclusters oxidized TMB to create a blue signal.

Another commonly used technique for colorimetric detection is based on the color change resulting from the aggregation and disassociation of AuNPs. When aggregated, AuNPs appear purple whereas upon dispersion a color change to red can be seen. Based on this phenomenon, a microfluidic colorimetric paper-based aptasensor was developed for the identification of aflatoxin B1. The presence of aflatoxin B1 leads to the aggregation of AuNPs, generating a grayish-purple color (Kasoju et al. 2020).

4.4.2 Fluorescence

Fluorescence is the process in which the excited electrons of a molecule emit light upon their return to the ground state. The emitted light is of a longer wavelength and lower energy levels compared to the absorbed light. Fluorescence has many advantages, including simplicity, rapidity, outstanding sensitivity, and convenient operation which make it a prominent readout technique in sensors (Ahmad et al. 2017). Fluorescence paper-based assay has had some challenges resulting from some innate features of paper for instance its opacity and fibrous natures which cause backscattering noise leading to reduced sensitivity (Ulep et al. 2020). Self-fluorescence and auto-bleaching are also major problems that can arise in PADs because of some additives used during paper synthesis (Yang et al. 2017).

Nevertheless, many fluorescence paper-based approaches have been proposed for the identification of food-based analytes.

To date many molecules have been proven to display intrinsic fluorescence including aromatic amino acids, NADH, flavins, etc. Fluorescein isothiocyanate (FITC) and carboxyfluorescein (FAM), both derivatives of fluorescein, are among the most widely used fluorophores for biological applications. Fluorescent proteins GFP (green), YFP (yellow), and RFP (red) can be fused with other proteins for labeling. Since the discovery of fluorescent nanomaterials including quantum dots (QDs), fluorescent nanoparticles and nanoclusters, conventional dyes have been replaced to a great extent as these nanomaterials display higher stability towards photo bleaching and superior sensitivity (Wang et al. 2016; Nemati et al. 2018a, b; Borghei et al. 2017).

Quantum dots (QDs), defined as nanoscale semiconductor materials, have dominated the field of fluorescence in the recent year because of their tunable emission wavelength, stability, wide excitation and narrow emission spectra, and high quantum yield (Bera et al. 2010). In one study, QDs were used to develop a microfluidic sensor for the detection of Cu^{2+} and Hg^{2+} ions. This approach was based on the quenching effect of ion imprinted polymers on the QDs in the absence of the respective ions (Qi et al. 2017). In another study, two different QDs including CdTe ($\lambda_{em} \approx 635$ nm) and ZnCdSe ($\lambda_{em} \approx 480$ nm) were used for the paper-based detection of organophosphorus pesticides (OPPs). Nanoporphyrins, which have a quenching effect on QDs, were used to capture OPPs. The fluorescence of both QDs was recovered through the interaction of OPPs with nanoporphyrins (Huang et al. 2019). The same methodology was implemented for the detection of carbamate pesticides using just CdTe QDs. The presence of the target restored the fluorescence of the QDs (Chen et al. 2020).

Graphene oxide (GO) is a two-dimensional nanomaterial proven to have fluorescence quenching properties for different dyes through resonance energy transfer (Chang et al. 2010). Exploiting this characteristic, a PAD was developed for the multiplex detection of heavy metals including Hg^{2+} and Ag^{+} and aminoglycoside antibiotics residues in food using cyanine 5-labeled DNA sequences (Zhang et al. 2015). The quenching effect of GO was also used in another study for the development of a fluorescence paper-based aptasensor for the identification of food allergens and food toxins using QD-functionalized aptamers. As seen in Fig. 4.2, the presence of the target released the aptamers from GO-modified paper, leading to emission recovery in QDs (Weng and Neethirajan 2018).

4.4.3 Chemiluminescence

Since its introduction in the 1970s (Deo and Roda 2011), and as a result of numerous advantages such as high sensitivity, simplicity, and its need of inexpensive reagents, chemiluminescence (CL) has been applied to numerous sensing applications. In CL, a chemical reaction which generates unstable intermediate species, which emit light upon their return to the ground state. As an external light source is omitted in CL,

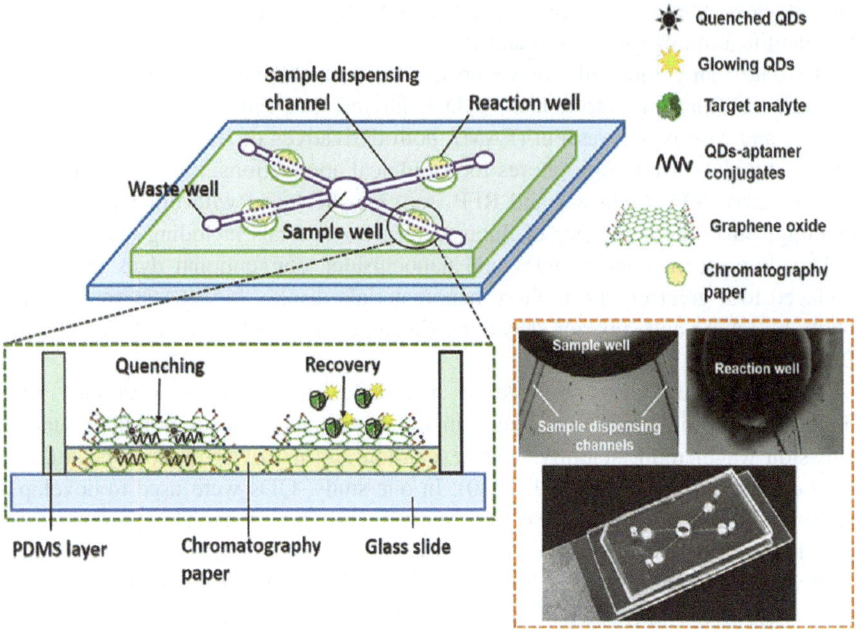

Fig. 4.2 Schematic representation of a fluorescence PAD for the detection of food allergens and food toxins based on QD-labeled aptamers and GO-modified paper (Weng and Neethirajan 2018)

background noise is noticeably reduced leading to heightened sensitivity. Controlled emission, simplicity, and rapidity are among the other merits of CL (Xu et al. 2016; Beigi et al. 2019; Mesgari et al. 2020).

Luminol is the most common CL reagent which can be oxidized by hydrogen peroxide, permanganate, periodate, etc. in an alkaline medium, thus forming the excited 3-aminophthalate anion, which emits light upon its return to the ground state (Liu et al. 2010). Various enzymes such as horseradish peroxidase (HRP) or alkaline phosphatase (ALP), and also enzyme-mimicking nanomaterials, such as AuNPs, can be used in CL systems. In one study, a CL PAD was developed for the detection of deltamethrin (DM), a commonly used pesticide. This sensor was based on the CL signal generated from the oxidation of graphene quantum dots (GQDs) by $KMnO_4$ in an acidic environment. The excited GQDs emit light at 490 nm upon their relaxation. Polyphosphate (PP) was used for signal enhancement. As shown in Fig. 4.3, the presence of DM led to reduced CL intensities as a result of the consumption of oxidant by (DM) and interaction with GQDs (Al Yahyai et al. 2021).

4.4.4 Surface-Enhanced Raman Scattering

Raman scattering is among the well-established vibrational spectroscopy methods for detection, in which the sample is irradiated with monochrome light within the

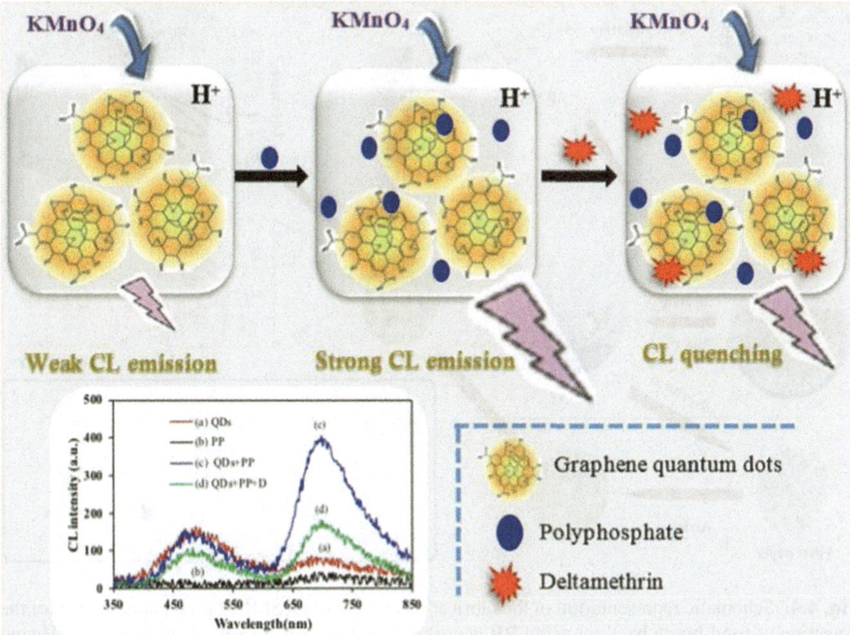

Fig. 4.3 Schematic representation of a CL PAD for the detection of deltamethrin based on the interaction between graphene quantum dots and $KMnO_4$ (Al Yahyai et al. 2021)

visible or near-infrared (IR) region resulting in excited vibrational states. Raman scattering is when relaxation occurs with photon emission with lower energy levels. This method generally has low intensity as only a small number of molecules under Raman scattering (Thygesen et al. 2003). Surface-enhanced Raman scattering aims to enhance this intensity through the absorption of the target within a distance of 10–100 nm on roughened noble metal surfaces or metal nanostructures (Doering et al. 2007). Noble metals such as gold, silver, and copper are generally used for this purpose.

Various forms of gold and silver nanostructures are commonly used for SERS detection. In a recent study, a spraying process was employed to modify the paper substrate with AuNPs and AgNPs. The prepared Au/AgNP-modified paper was used to detect residual fishery drugs in aquatic samples (Yang et al. 2020). Silver nanostructures possess better enhancing ability whereas gold nanostructures tend to show more stability. For this, many researchers aim to use nanostructures containing both noble metals. In one study, a core–shell Au@Ag nanorod monolayer was coated on a paper platform for the SERS-based detection thiram residues both on fruit surfaces and in juice. In this approach, hydrophobic polydimethylsiloxane (PDMS), a commonly used support material for SERS substrates for its high optical transparency and stability, was used to support the paper, resulting in higher sensitivities and nanomaterial protection (Lin et al. 2020).

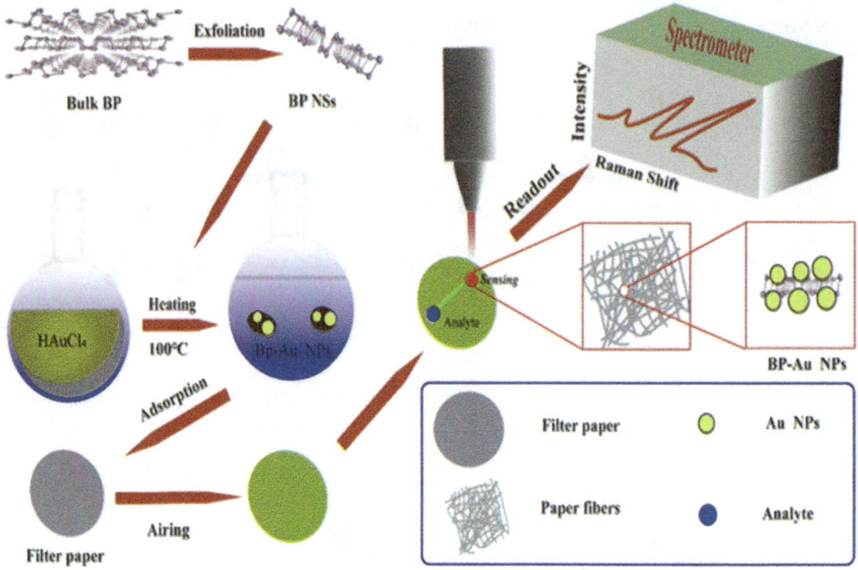

Fig. 4.4 Schematic representation of the fabrication process of an SERS-based paper device for the detection of food-borne bacteria using BP-nanosheets and AuNPs for signal enhancement (Huang et al. 2019)

The SERS effect can arise from two mechanisms; charge transfer between the substrate and analyte known as the chemical mechanism and enhancement resulting from plasmonic nanostructures known as the electromagnetic mechanism (Sharma et al. 2012). Signal enhancement resulting from noble metal nanomaterials is provided through the electromagnetic mechanism. In one study, a black phosphorous (BP) nanosheets, an inorganic two-dimensional nanomaterial, was combined with AuNPs to develop a paper-based sensor for the detection of food-borne bacteria (Fig. 4.4). This approach combined both chemical enhancement and electromagnetic enhancement for heightened sensitivity (Dungchai et al. 2009).

4.4.5 Electrochemical

Electrochemical paper-based analytical devices (ePADs) have rapidly turned into a prominent sensing platform since their first introduction in 2015 (Dungchai et al. 2009). In addition to their compact size and potable nature, they also display outstanding selectivity and sensitivity and have rapid response time (Mettakoonpitak et al. 2016). Besides the mentioned merits, ePADs are also among the best analytical tools for multiplexed sensing (Huang et al. 2019).

The efficiency of the electrodes has a great impact on the performance of ePADs. Thus, ample research has been devoted to the assessment of a variety of materials and fabrication methods for the design of electrodes on paper platforms. Some of the

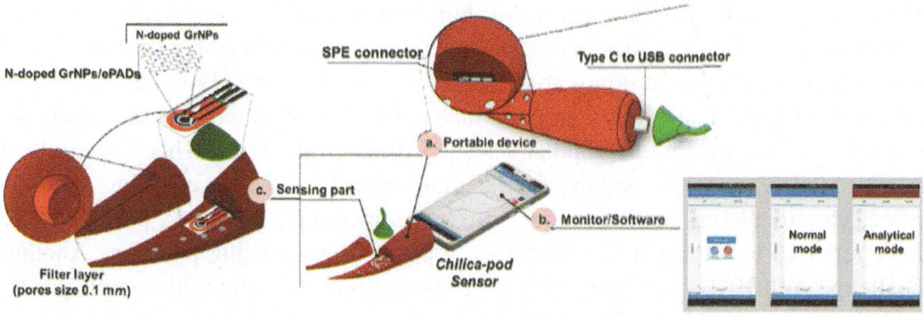

Fig. 4.5 Schematic representation of the portable chili-shaped ePAD based on N-doped graphene nanoplates for the detection of capsaicin in chili peppers (Soleh et al. 2020)

commonly used approaches for electrode fabrication are screen printing, pencil-drawing and inkjet printing, sputtering deposition, metallic wire tape, and nanoparticle growth (Mettakoonpitak et al. 2016). The most commonly employed signal transduction method in ePADs is voltammetry, amperometry, and potentiometry (Adkins et al. 2015).

Various nanomaterials, including carbon nanotubes, graphene, Pt nanomaterials, etc. have been incorporated with the electrodes in electrochemical systems in order to increase the detection system. Because of its high conductivity and outstanding electronic features, graphene, a two-dimensional carbon-based material, is among the most commonly used materials in ePADs. Studies have shown that the presence of heteroatoms such as N, S, O, or B in the graphene lattice results in more electrochemically active sites, thus further enhancing the sensitivity of the detection method (Akyazi et al. 2018). N-doped graphene nanoplates (GrNPs) were used to develop a portable electrochemical sensor for the detection of capsaicin in chili peppers. As seen in Fig. 4.5, this portable sensor was designed in the shape of a chili pepper and included an interface to connect to smartphones for point-of-testing (Soleh et al. 2020).

Carbon nanotubes (CNTs) display unique electrochemical properties and are commonly used in ePADs. Single-walled CNTs (SWCNTs) were functionalized with antibodies and used to modify carbon paste electrodes for the development of an ePAD for the detection of *Staphylococcus aureus* in one study (Bhardwaj et al. 2017). In another study, multi-walled CNTs (MWCNTs) and chitosan were used to modify the electrodes in an ePAD for the detection of aflatoxin B1 using specific antibodies (Migliorini et al. 2020). In both studies, it was proven that the use of CNTs significantly enhanced the sensitivity of the ePAD.

4.5 Current Limitations

Paper-based analytical devices provide an ideal sensing platform for rapid, portable, and inexpensive detection of food contaminants. However, many challenges need to be addressed before these devices can be commercially available. First, the cost and fabrication of PADs need improvement. Furthermore, the functionalization of paper substrates with biomolecules persists as a challenge as nonspecific attachment and loss of functionality commonly arise. Moreover, varying sensitivity and selectivity have been reported for PADs, in particular those manufactured in large scales for commercial use, leading to imprecise results (Pike et al. 2013). This is a result of batch-to-batch variations and different combinations of substrate coupling. Furthermore, various testing conditions in terms of temperature, humidity, pH, etc. have been reported to impact the efficiency of a paper-based sensing platform. Another challenge with PADs for food controlling arises from difficulty with sample preparation and extraction from different food samples. Overall, problems with large-scale manufacturing, biofunctionalization, and sample preparation should be addressed before the full potential of PADs for point-of-test food monitoring to be exploited.

4.6 Conclusion and Future Perspectives

Food safety monitoring has significant importance as despite the numerous efforts for contamination control, countries are still plagued by the health and economic burdens of chemically and biologically contaminated food samples. Paper-based approaches have great potential for the development of portable and cost-efficient detection platforms. The emergence of microfluidics and its implementation with paper sensors has made the precise quantification of analyte with PADs possible. Furthermore, nanotechnology and the use of nanomaterials with outstanding optical and electrical characteristics has helped to further enhance the sensitivity of PADs. Moreover, integrating smart phone technologies with PADs has eased the progress of point-of-test food monitoring in remote, underdeveloped areas.

In this chapter, we have provided an overview on the applications of paper-based sensing for food analyte analysis. After a brief introduction into the various food contaminants, we discussed the various formats and fabrication methods commonly used for the design of paper devices. Thus, various detection technologies including colorimetric, fluorescence, chemiluminescence, surface-enhances Raman scattering, and electrochemical methods coupled with PADs employed for food analysis were discussed.

References

Abarghoei S, Fakhri N, Borghei YS, Hosseini M, Ganjali MR (2019) A colorimetric paper sensor for citrate as biomarker for early stage detection of prostate cancer based on peroxidase-like activity of cysteine-capped gold nanoclusters. Spectrochim Acta A Mol Biomol Spectrosc 210: 251–259

Adkins J, Boehle K, Henry C (2015) Electrochemical paper-based microfluidic devices. Electrophoresis 36(16):1811–1824

Ahmad MH, Sahar A, Hitzmann B (2017) Fluorescence spectroscopy for the monitoring of food processes. In: Measurement, modeling and automation in advanced food processing. Springer, pp 121–151

Akyazi T, Basabe-Desmonts L, Benito-Lopez F (2018) Review on microfluidic paper-based analytical devices towards commercialisation. Anal Chim Acta 1001:1–17

Al Yahyai I, Hassanzadeh J, Al-Lawati HA (2021) A novel and selective multi-emission chemiluminescence system for the quantification of deltamethrin in food samples. Sens Actuators B 327:128927

Beigi SM, Mesgari F, Hosseini M, Aghazadeh M, Ganjali MR (2019) An enhancement of luminol chemiluminescence by cobalt hydroxide decorated porous graphene and its application in glucose analysis. Anal Methods 11(10):1346–1352

Bera D, Qian L, Tseng T-K, Holloway PH (2010) Quantum dots and their multimodal applications: a review. Materials 3(4):2260–2345

Bhardwaj J, Devarakonda S, Kumar S, Jang J (2017) Development of a paper-based electrochemical immunosensor using an antibody-single walled carbon nanotubes bio-conjugate modified electrode for label-free detection of foodborne pathogens. Sens Actuators B 253:115–123

Borghei Y-S, Hosseini M, Ganjali MR (2017) Fluorescence based turn-on strategy for determination of microRNA-155 using DNA-templated copper nanoclusters. Microchim Acta 184(8): 2671–2677

Busa LSA, Mohammadi S, Maeki M, Ishida A, Tani H, Tokeshi M (2016) Advances in microfluidic paper-based analytical devices for food and water analysis. Micromachines 7(5):86

Carrilho E, Martinez AW, Whitesides GM (2009) Understanding wax printing: a simple micropatterning process for paper-based microfluidics. Anal Chem 81(16):7091–7095

Chandra P, Koh WCA, Noh HB, Shim YB (2012) In vitro monitoring of i-NOS concentrations with an immunosensor: The inhibitory effect of endocrine disruptors on i-NOS release. Biosens Bioelectron

Chang H, Tang L, Wang Y, Jiang J, Li J (2010) Graphene fluorescence resonance energy transfer aptasensor for the thrombin detection. Anal Chem 82(6):2341–2346

Chen H, Hu O, Fan Y, Xu L, Zhang L, Lan W, Hu Y, Xie X, Ma L, She Y (2020) Fluorescence paper-based sensor for visual detection of carbamate pesticides in food based on CdTe quantum dot and nano ZnTPyP. Food Chem 327:127075

Choudhary M et al (2016) CD 59 Targeted ultrasensitive electrochemical immunosensor for fast and noninvasive diagnosis of oral cancer. Electroanalysis 28(10):2565–2574. https://doi.org/10.1002/elan.201600238

Dehghani Z, Hosseini M, Mohammadnejad J, Bakhshi B, Rezayan AH (2018) Colorimetric aptasensor for Campylobacter jejuni cells by exploiting the peroxidase like activity of Au@Pd nanoparticles. Microchim Acta 185(10):1–9

Dehghani Z, Mohammadnejad J, Hosseini M (2019a) A new colorimetric assay for amylase based on starch-supported Cu/Au nanocluster peroxidase-like activity. Anal Bioanal Chem 411(16): 3621–3629

Dehghani Z, Hosseini M, Mohammadnejad J, Ganjali MR (2019b) New Colorimetric DNA Sensor for Detection of Campylobacter jejuni in Milk Sample Based on Peroxidase-Like Activity of Gold/Platinium Nanocluster. Chem Select 4(40):11687–11692

Deka S, Saxena V, Hasan A, Chandra P, Pandey LM (2018) Synthesis, characterization and in vitro analysis of α-Fe_2O_3-$GdFeO_3$ biphasic materials as therapeutic agent for magnetic hyperthermia applications. Mater Sci Eng C 92:932–941

Deo SK, Roda A (2011) Chemiluminescence and bioluminescence: past, present and future. Anal Bioanal Chem 401(5):1457

Dixit CK, Kaushik AK, Kaushik A (2016) Microfluidics for biologists. Springer

Doering WE, Piotti ME, Natan MJ, Freeman RG (2007) SERS as a foundation for nanoscale, optically detected biological labels. Adv Mater 19(20):3100–3108

Dungchai W, Chailapakul O, Henry CS (2009) Electrochemical detection for paper-based microfluidics. Anal Chem 81(14):5821–5826

Dungchai W, Chailapakul O, Henry CS (2011) A low-cost, simple, and rapid fabrication method for paper-based microfluidics using wax screen-printing. Analyst 136(1):77–82

Fakhri N, Hosseini M, Tavakoli O (2018) Aptamer-based colorimetric determination of Pb^{2+} using a paper-based microfluidic platform. Anal Methods 10(36):4438–4444

Fenton EM, Mascarenas MR, López GP, Sibbett SS (2009) Multiplex lateral-flow test strips fabricated by two-dimensional shaping. ACS Appl Mater Interfaces 1(1):124–129

Gonzalez-Macia L, Morrin A, Smyth MR, Killard AJ (2010) Advanced printing and deposition methodologies for the fabrication of biosensors and biodevices. Analyst 135(5):845–867

Griesche C, Baeumner AJ (2020) Biosensors to support sustainable agriculture and food safety. TrAC Trends Anal Chem:115906

Hosseini M (2020) Fast and selective whole cell detection of Staphylococcus aureus bacteria in food samples by paper based colorimetric nanobiosensor using peroxidase-like catalytic activity of DNA-Au/Pt bimetallic nanoclusters. Microchem J 159:105475

Hosseini M, Aghazadeh M, Ganjali MR (2017) A facile one-pot synthesis of cobalt-doped magnetite/graphene nanocomposite as peroxidase mimetics in dopamine detection. New J Chem 41(21):12678–12684

Huang D, Zhuang Z, Wang Z, Li S, Zhong H, Liu Z, Guo Z, Zhang W (2019) Black phosphorus-Au filter paper-based three-dimensional SERS substrate for rapid detection of foodborne bacteria. Appl Surf Sci 497:143825

Kasoju A, Shrikrishna NS, Shahdeo D, Khan AA, Alanazi AM, Gandhi S (2020) Microfluidic paper device for rapid detection of aflatoxin B1 using an aptamer based colorimetric assay. RSC Adv 10(20):11843–11850

Kermani HA, Hosseini M, Miti A, Dadmehr M, Zuccheri G, Hosseinkhani S, Ganjali MR (2018) A colorimetric assay of DNA methyltransferase activity based on peroxidase mimicking of DNA template Ag/Pt bimetallic nanoclusters. Anal Bioanal Chem 410(20):4943–4952

Kong Q, Wang Y, Zhang L, Ge S, Yu J (2017) A novel microfluidic paper-based colorimetric sensor based on molecularly imprinted polymer membranes for highly selective and sensitive detection of bisphenol A. Sens Actuators B 243:130–136

Li X, Tian J, Nguyen T, Shen W (2008) based microfluidic devices by plasma treatment. Anal Chem 80(23):9131–9134

Lin S, Hasi W, Han S, Lin X, Wang L (2020) A dual-functional PDMS-assisted paper-based SERS platform for the reliable detection of thiram residue both on fruit surfaces and in juice. Anal Methods 12(20):2571–2579

Liu M, Lin Z, Lin J-M (2010) A review on applications of chemiluminescence detection in food analysis. Anal Chim Acta 670(1–2):1–10

Mahato K, Kumar S, Srivastava A, Maurya PK, Singh R, Chandra P (2018) Electrochemical immunosensors: fundamentals and applications in clinical diagnostics. In: Handbook of immunoassay technologies

Manisha H, Shwetha PP, Prasad K (2018) Low-cost paper analytical devices for environmental and biomedical sensing applications. In: Environmental, chemical and medical sensors. Springer, pp 315–341

Marklinder I, Ahlgren R, Blücher A, Börjesson S-ME, Hellkvist F, Moazzami M, Schelin J, Zetterström E, Eskhult G, Danielsson-Tham M-L (2020) Food safety knowledge, sources thereof and self-reported behaviour among university students in Sweden. Food Control 113: 107130

Martinez AW, Phillips ST, Butte MJ, Whitesides GM (2007) Patterned paper as a platform for inexpensive, low-volume, portable bioassays. Angew Chem 119(8):1340–1342

Mesgari F, Beigi SM, Fakhri N, Hosseini M, Aghazadeh M, Ganjali MR (2020) based chemiluminescence and colorimetric detection of cytochrome c by cobalt hydroxide decorated mesoporous carbon. Microchem J 157:104991

Mettakoonpitak J, Boehle K, Nantaphol S, Teengam P, Adkins JA, Srisa-Art M, Henry CS (2016) Electrochemistry on paper-based analytical devices: a review. Electroanalysis 28(7):1420–1436

Migliorini FL, Santos DMD, Soares AC, Mattoso LH, Oliveira ON, Correa DS (2020) Design of A Low-Cost and Disposable Paper-Based Immunosensor for the Rapid and Sensitive Detection of Aflatoxin B1. Chemosensors 8(3):87

Morbioli GG, Mazzu-Nascimento T, Stockton AM, Carrilho E (2017) Technical aspects and challenges of colorimetric detection with microfluidic paper-based analytical devices (μPADs)-A review. Anal Chim Acta 970:1–22
Mori R, Yonemoto N, Fitzgerald A, Tullus K, Verrier-Jones K, Lakhanpaul M (2010) Diagnostic performance of urine dipstick testing in children with suspected UTI: a systematic review of relationship with age and comparison with microscopy. Acta Paediatr 99(4):581–584
Nemati F, Zare-Dorabei R, Hosseini M, Ganjali MR (2018a) Fluorescence turn-on sensing of thiamine based on Arginine–functionalized graphene quantum dots (Arg-GQDs): central composite design for process optimization. Sens Actuators B 255:2078–2085
Nemati F, Hosseini M, Zare-Dorabei R, Salehnia F, Ganjali MR (2018b) Fluorescent turn on sensing of Caffeine in food sample based on sulfur-doped carbon quantum dots and optimization of process parameters through response surface methodology. Sens Actuators B 273:25–34
Nery EW, Kubota LT (2013) Sensing approaches on paper-based devices: a review. Anal Bioanal Chem 405(24):7573–7595
Ng JS, Hashimoto M (2020) Fabrication of paper microfluidic devices using a toner laser printer. RSC Adv 10(50):29797–29807
Patel P (2002) (Bio) sensors for measurement of analytes implicated in food safety: a review. TrAC Trends Anal Chem 21(2):96–115
Pérez-López B, Merkoçi A (2011) Nanomaterials based biosensors for food analysis applications. Trends Food Sci Technol 22(11):625–639
Pike J, Godbert S, Johnson S (2013) Comparison of volunteers' experience of using, and accuracy of reading, different types of home pregnancy test formats. Expert Opin Med Diagn 7(5): 435–441
Qi J, Li B, Wang X, Zhang Z, Wang Z, Han J, Chen L (2017) Three-dimensional paper-based microfluidic chip device for multiplexed fluorescence detection of Cu^{2+} and Hg^{2+} ions based on ion imprinting technology. Sens Actuators B 251:224–233
Ragavan K, Neethirajan S (2019) Nanoparticles as biosensors for food quality and safety assessment. In: Nanomaterials for food applications. Elsevier, pp 147–202
Schiff H (1866) Eine neue reihe organischer diamine. Justus Liebigs Ann Chem 140(1):92–137
Shams R, Singh J, Ashraf S, Manzoor M, Dar A (2020) Application of biosensors in food quality control. J Postharvest Technol 8(1):53–74
Sharma B, Frontiera RR, Henry A-I, Ringe E, Van Duyne RP (2012) SERS: Materials, applications, and the future. Mater Today 15(1–2):16–25
Sharma N, Barstis T, Giri B (2018) Advances in paper-analytical methods for pharmaceutical analysis. Eur J Pharm Sci 111:46–56
Soleh A, Saisahas K, Promsuwan K, Thavarungkul P, Kanatharana P, Limbut W (2020) N-Doped graphene nanoplatelets for direct capsaicin detection in chili pepper samples. ACS Appl Nano Mater 3(10):10094–10104
Soleimani-Gorgani A., 14.1 Fundamentals of Inkjet Printing Technology 231 14.1. 1 CIJ Printing Systems 232 14.1. 2 Drop on Demand Inkjet Printing Systems 232. 2016
Songjaroen T, Dungchai W, Chailapakul O, Laiwattanapaisal W (2011) Novel, simple and low-cost alternative method for fabrication of paper-based microfluidics by wax dipping. Talanta 85(5): 2587–2593
Spicar-Mihalic P, Toley B, Houghtaling J, Liang T, Yager P, Fu E (2013) CO_2 laser cutting and ablative etching for the fabrication of paper-based devices. J Micromech Microeng 23(6): 067003
Thygesen LG, Løkke MM, Micklander E, Engelsen SB (2003) Vibrational microspectroscopy of food. Raman vs. FT-IR. Trends Food Sci Technol 14(1–2):50–57
Timur S, Della Seta L, Pazarlioğlu N, Pilloton R, Telefoncu A (2004) Screen printed graphite biosensors based on bacterial cells. Process Biochem 39(11):1325–1329
Ulep T-H, Zenhausern R, Gonzales A, Knoff DS, Diaz PAL, Castro JE, Yoon J-Y (2020) Smartphone based on-chip fluorescence imaging and capillary flow velocity measurement for

detecting ROR1+ cancer cells from buffy coat blood samples on dual-layer paper microfluidic chip. Biosens Bioelectron 153:112042

Verma S, Choudhary J, Singh KP, Chandra P, Singh SP (2019) Uricase grafted nanoconducting matrix based electrochemical biosensor for ultrafast uric acid detection in human serum samples. Int J Biol Macromol

Wang Y, Zhang C, Chen X, Yang B, Yang L, Jiang C, Zhang Z (2016) Ratiometric fluorescent paper sensor utilizing hybrid carbon dots–quantum dots for the visual determination of copper ions. Nanoscale 8(11):5977–5984

Weng X, Neethirajan S (2018) Paper-based microfluidic aptasensor for food safety. J Food Saf 38(1):e12412

Wong R, Tse H (2008) Lateral flow immunoassay. Springer Science & Business Media

Xu Y, Liu M, Kong N, Liu J (2016) Lab-on-paper micro-and nano-analytical devices: fabrication, modification, detection and emerging applications. Microchim Acta 183(5):1521–1542

Yang Y, Noviana E, Nguyen MP, Geiss BJ, Dandy DS, Henry CS (2017) based microfluidic devices: emerging themes and applications. Anal Chem 89(1):71–91

Yang G, Fang X, Jia Q, Gu H, Li Y, Han C, Qu L-L (2020) Fabrication of paper-based SERS substrates by spraying silver and gold nanoparticles for SERS determination of malachite green, methylene blue, and crystal violet in fish. Microchim Acta 187(5):1–10

Zhang Y, Zuo P, Ye B-C (2015) A low-cost and simple paper-based microfluidic device for simultaneous multiplex determination of different types of chemical contaminants in food. Biosens Bioelectron 68:14–19

Optical Detection of Targets for Food Quality Assessment

5

Maryam Mousavizadegan, Aida Alaei, and Morteza Hosseini

Abstract

Current technics used in food quality assessment require complex equipment and professional personnel, which has financially burdened the food industry. Optical devices are a powerful candidate which can provide simple, cost-effective, and rapid detection approaches with the possibility to be easily manufactured in a large scale to cover the vast need of this industry. This chapter will first present an overview regarding food safety and common analytes. Various optical sensing technics, including colorimetric, fluorescent, chemiluminescent, surface-enhanced Raman scattering (SERS), and surface plasmon resonance (SPR) methods will then be explored with examples of novel sensing platforms developed for food monitoring.

Keywords

Optical detection · Food quality · Biomarkers · Biological receptors · Analytical performance

5.1 Introduction

Continuous supply of healthy and quality food for citizens has always been one of the main concerns in all countries. Over the past three decades, hazard analysis, control of critical points, food production techniques, and standard health performance methods have aimed to meet this need. The identification of biological and

M. Mousavizadegan · A. Alaei · M. Hosseini (✉)
Department of Life Science Engineering, Faculty of New Sciences and Technologies, University of Tehran, Tehran, Iran
e-mail: smhosseini@khayam.ut.ac.ir

P. Chandra, P. S. Panesar (eds.), *Nanosensing and Bioanalytical Technologies in Food Quality Control*, https://doi.org/10.1007/978-981-16-7029-9_5

chemical contaminants is of critical importance in order to ensure the healthy nutrition of consumers. In that regard, different analysis techniques for food safety analysis, preservation, and elevation of food quality have been developed which vary depending on the type of food. Despite the many efforts for preservation and quality control, countries are still plagued by the economical and health burdens of food fraud, the entry of substandard food into the market, and the resulting foodborne illness. To overcome these problems, food safety control and quality assessment are of vital significance which persist as a continuous discussion point for nations (Ragavan and Neethirajan 2019; Shams et al. 2020).

Food safety is the process of ensuring zero contamination, be it chemical or biological, during the preparation and storage of food. So far, suitable analytical techniques such as high-performance liquid chromatography (HPLC), gas chromatography (GC), and mass spectrometry for chemical contaminants and cell culture methods, biochemical identification, and polymerase chain reaction (PCR) for pathogen analysis have been developed to analyze food contaminants as standard techniques. However, these approaches have many downfalls including being laborious and costly, requiring complex sample preparation steps, and needing the assistance of a professional technician and high-tech instrumentation (Narsaiah et al. 2012). Hence, developing rapid, simple detection methods is a dire requirement, especially in developing areas which do not have advanced laboratories for conventional detection. Biosensors offer selectivity and compact size, relatively high sensitivity, low cost, rapid response times, and are user friendly to operate (Chandra et al. 2012; Choudhary et al. 2016; Deka et al. 2018; Mahato et al. 2018; Verma et al. 2019).

Optical techniques, as a result of their simplicity, selectivity, and stability have dominated the world of sensing. Owing to their merits, they have become a popular approach for the identification of food analytes. Optical detection assays are techniques in which the identification of the target analyte is transduced to visible, ultraviolet or infrared (IR) radiations which can be detected by the naked eye or using spectroscopy methods (Damborský et al. 2016).

In this chapter, optical methods for food safety analysis will be discussed. We will first introduce the various kinds of food contamination including both chemical and biological contaminants. A scope of the developed optical sensing assays, classified in colorimetric, fluorescence, chemiluminescence, SERS, and SPR-based methods for the detection of these analytes will then be presented.

5.2 Food Safety Analytes

Acquiring nutritious food is very important to maintain a vigorous lifestyle and promote health. In the past, the significance of food contamination was underestimated due to the lack of reports related to food safety and the difficulty in analyzing the relationship between contaminated food and various diseases. There were also many shortcomings internationally in the knowledge, attitude, and behavior of food consumers suffering from serious diseases (Marklinder et al. 2020).

Finally, the significance of food safety in health was realized and in 2015, the World Health Organization (WHO) published an article estimating the global burden of foodborne diseases. This article showed that only 31 types of food contaminants lead to 32 diseases that cause 42,000 deaths per year (Griesche and Baeumner 2020). Compared to industrial countries, developing nations are harder afflicted by the repercussions of food contamination. Recently, food safety has attracted much attention due to the understanding of its heightened importance and also, the accumulation of knowledge regarding the presence of various contaminants in food samples and how they impact the health of the consumer. Some examples of such contaminants are heavy metal in the environment and how they pollute the water resources, improper use of pesticides and antibiotics, and other pollutants and toxins that accumulate in various plants and animals. Also, numerous pathogens including viruses, bacteria, and parasites can also contaminate food samples at various stages of agricultural plantation, production, processing, storage, and delivery (Lu et al. 2020).

In order to effectively develop sensing platforms for food monitoring, an understanding of common contaminants is necessary. Furthermore, the development of new receptors, including antibodies and aptamers, with a strong affinity toward specific molecules has been instrumental in developing sensitive and rapid diagnostic methods for hazardous substances in food samples (Caglayan et al. 2020). In this section, a brief description of the various food contaminants has been presented.

5.2.1 Pathogens

Microorganisms have always been present in food, drinking water, the sea and rivers, soil, and human intestines. Many of these microorganisms are quite beneficial for the environment and also our overall health like probiotic bacteria, but some are pathogenic and can cause many major or sometimes quite fatal problems. Pathogenic microorganisms include bacteria, viruses, and parasites.

Bacteria are among the most common of foodborne pathogens and exist in a variety of types, many of which are responsible for deadly food-related illnesses. Numerous pathogenic bacteria have been identified in food samples so far, with *Escherichia coli*, *Listeria monocytogenes*, and *Salmonella typhimurium* being three of the most important and fatal pathogens (O'Kennedy et al. 2005; Jokerst et al. 2012). Also some bacterial pathogens generate spores with high resistance to heat treatments such as *Clostridium botulinum, Clostridium perfringens, Bacillus subtilis*, and *Bacillus cereus*, and some are able to secrete heat-resistant toxins such as *Staphylococcus aureus* and *Clostridium botulinum*. Although most bacteria are mesophilic and their growth is inactivated in temperatures below 20 °C, some species such as *Listeria monocytogenes*, and *Yersinia enterocolitica* can propagate in the refrigerator and contaminate food samples in colder temperatures (Bacon et al. 2003).

Viruses are another group of foodborne pathogens. To date, over 100 different enteric viruses have been reported to be the source of foodborne illnesses, among which Hepatitis A and noroviruses are the most common. Most viruses are transmitted via bivalve molluscs such as clams, cockles, mussels, and oysters (Gosling 2008). Waters are also increasingly prone to virus contamination through sewage discharge and waste disposal from infected shellfish harvesters. These viruses accumulate for days in the shellfish digestive tract. Health problems arise as most shellfish are eaten raw, with their digestive tracts in place. The shellfish themselves also act as a protective barrier for the viruses against thermal inactivation (DiGirolamo et al. 1970).

Another important group of foodborne pathogens are parasites, which are known as single-celled microorganisms with an organized nucleus. Similar to viruses, the propagation of parasites is dependent on the host, thus they do not multiply in food samples. Cysts are the transmissible form of parasites which can be transmitted from animals or other humans to humans. These organisms live and procreate usually in the digestive tract of the infected host and are excreted in the feces, thus fecal transmission is quite common. A parasite infection can lead to severe malnourishment in the host. *Cyclospora cayetanensis*, *Toxoplasma gondii*, and *Trichinella spiralis* are the most common foodborne parasites (Bintsis 2017).

5.2.2 Toxins

Toxins are toxic chemicals produced by a living organism such as microorganisms, plants, and animals, which can pose serious health risks. Toxins are naturally present in food and can contaminate food during various stages, from production to processing and packaging, and even delivery. Although many foodborne illnesses are caused by bacteria, such as salmonellosis, campylobacteriosis, toxins secreted by bacteria during growth can also lead to severe health problems. *Staphylococcus aureus* and *Clostridium botulinum* are two bacterial species with the ability to produce heat-resistant toxins (Ligler et al. 2003; Hodnik and Anderluh 2009).

Mycotoxins, defined as toxins produced by various moulds (fungi), can contaminate a wide range of food samples from fruits and vegetables, to nuts and dry cereals. Aflatoxins are among the most poisonous kind which are generally produced by *Aspergillus flavus* and *Aspergillus parasiticus* on cereals, tree nuts, oilseeds, and even some spices. This toxin is also found in the milk of animals fed with contaminated feed. Aflatoxins are especially hazardous as they are known to damage the DNA which can in turn lead to cancer. Other mycotoxins include ochratoxins (produces by *Aspergillus* and *Penicillium* species), deoxynivalenol and zearalenone (produced by *Fusarium* fungi), and ergot alkaloids (produced by *Claviceps* species) (Abrunhosa et al. 2016).

Plants, as a defense mechanism against predators, insects, or microorganisms, are also known to create toxins (Yamane et al. 2010). Microscopic algae and planktons in oceans and in lakes create chemical compounds which are non-toxic to fish and shellfish, but can have an adverse effect on humans who eat seafood containing these toxins (Campàs et al. 2007).

5.2.3 Heavy Metals

With the development of countries, heavy metal industries ensued which led to their entrance in the human life. Heavy metals and their respective ions are metals with an atomic weight between 63.5 and 200.6 g mol^{-1} and a specific gravity greater than 5 g cm^{-1}. Organisms need a minimum amount of specific heavy metals such as cobalt (Co), molybdenum (Mo), vanadium (V), zinc (Zn), and strontium (Sr). However, other highly toxic metals such as arsenic (As), beryllium (Be), cadmium (Cd), chromium (Cr), lead (Pb), and mercury (Hg) have irreversible effects on the human health and ecosystems, even in very small amounts. Heavy metals are present in the earth's crust. Therefore, they become incorporated with food and water resources through agriculture and industrial processes, and as a result become consumed by humans and other organisms in various ways.

The toxicity of heavy metals derives from their interference in the body's biochemical and metabolic reactions, such as digestion. If heavy metal ions in drinking water are higher than the permitted amount, severe and dangerous diseases such as cancer, cardiovascular disease, brain damage, kidney failure, and nervous system complications will occur. Traditional heavy metal diagnostic methods such as atomic absorption spectroscopy have an acceptable detection limit but, have limitations such as their requirement of complex instrumentation and being costly (Wang et al. 2020a).

5.2.4 Pesticide Residues

Pesticides are widely used in agriculture to destroy or control various pests that can blemish the quality of crop. Pesticides are used as insecticides, fungicides, herbicides, and other types which help ensure the maximum quality of food. Based on their chemical structures and functionality, synthetic pesticides are classified into five classes: organochlorine, organophosphate, carbamate, neonicotinoid, and pyrethroid. Most pesticides are designed to attack pests, but they also adversely endanger humans, the environment, and wildlife and are considered among the most hazardous pollutants (Wang and Zhou 2014; Trojanowicz and Hitchman 1996). Furthermore, the more they perpetuate in the natural cycle, their concentration, and toxic intensity accumulates. Therefore, the study of different ways to diagnose pesticides is of crucial importance.

5.2.5 Veterinary Drug

Different types of veterinary drugs, including growth factors and antibiotics, are extensively used in livestock and poultry these days to prevent disease and help fight infections and also to induce growth. These drugs can accumulate in animal tissues and cause problems in humans in a variety of ways. 6051 ton of various active ingredients are used as veterinary medicines. Some of these drugs get excreted as urine and feces from the animals' body and can then be used as fertilizers thus entering the environment. The continuous entry of these substances into the environment and long-term accumulations in the human body can cause persistent side effects such as allergies and bacterial resistance (Wang et al. 2020b).

5.2.6 Illegal Additives

In the food industry, certain substances known as preservatives are used to prevent microbial growth and make food seem pristine. For example, sodium and potassium salts of nitrite and nitrate are added to meat, fish, sausages, and cheese to preserve the products. They can also cause a better appearance by impacting the color and change the taste of these foods. However, excessive use of nitrite leads to immediate toxic reactions such as abdominal pain, vomiting, and decreased blood oxygenation. WHO has set a high daily intake of 3.7 mg kg^{-1} of nitrate and 0.07 nitrite mg kg^{-1} of body weight (Thongkam and Hemavibool 2020).

Melamine, which is another type of food additive, has become the focus of the food industry in the world. Melamine is added to milk and dairy products to increase their protein efficiency. Maximum residue levels (MRLs) are allowed to be added to milk powder in China and the United States at 1 and 2.5 mg kg^{-1}, respectively. Given the importance of dietary supplements and the adverse effect they can have when exceeding the permitted level, they should be identified and quantified by rapid and sensitive laboratory techniques (Boutillier et al. 2020).

5.3 Detection Based on Optical Sensors

The word sensor is derived from the Latin word *sentire*, meaning recognition. These devices are made of three sections including a sensing element, a transducing element, and a detector. The sensing element (antibody, enzyme, nucleic acid, cell, aptamer, bacteriophage, and microorganism) receives a physiological response and transfers it to the transducer (optical, electrochemical, mass-based), and finally, the signal is detected by the detector. Depending on the transducer, sensors can be divided into different groups including optical, electrochemical, acoustic wave, and piezoelectric sensors (Bahadır and Sezgintürk 2017). Biosensors are used for a large number of applications within the field of biotechnology, including medical diagnosis, pharmaceutical analysis, and food. One of the best types of sensors for the detection of contaminating microorganisms and substances in food samples are optical and electrochemical sensors (Huet et al. 2010).

Due to numerous advantages such as the possibility of remote control in hazardous environments, acceptable sensitivity and high stability, optical sensors are extensively used for the rapid detection of food contaminants. Furthermore, they enable the detection at different wavelengths and consequently provide a platform to simultaneously detect various parameters (Ohk and Bhunia 2013). The various optical geometries used in sensing assays include optical fibers, planar wave guides, surface plasmon resonance, and microarrays. Optical biosensors measure changes in the phase, frequency, and amplitude of light (Narsaiah et al. 2012). They are divided into different types according to the transmission mechanism, including colorimetric, fluorescence, chemiluminescence, surface-enhanced Raman scattering (SERS), and surface plasmon resonance (SPR) (Silva et al. 2018).

Nanotechnology has impacted various fields of science and technology including the research for the development of efficient detection assays. NMs, with a high surface-to-volume ratio, demonstrate interesting optical and electrical properties which have led to their immense employment in the development of optical sensors. Compared to other nanomaterials, gold nanoparticles (AuNPs) have been tremendously used in many optical sensors due to their unique properties such as catalytic behavior, SPR, unique optical properties which results in the color change as they aggregate or dissociate, fluorescence emission (as seen in gold nanocrystals), fluorescence quenching, etc. To this day, and by taking advantage of the many merits of AuNPs, numerous sensing approaches have been proposed based on these nanomaterials for a multitude of reasons. The first reason is related to the compatibility of gold nanoparticles with various organic and metallic molecules, which leads to their natural reaction with molecules. Second, when gold nanoparticles are synthesized by the citrate method, due to their charge properties, the probability of their interaction with other molecules increases. The third is the high surface-to-volume ratio, which increases analyte detection. The last and fourth reason is associated with their ability to adjust to different conditions through creating a change in morphology, size, and synthetic environment (Naderi et al. 2018; Upadhyayula 2012).

In this section, we will examine various types of optical sensors for food analysis and present novel proposed sensing assays based on colorimetric, fluorescence, chemiluminescence, SERS and SPR methods for food safety analysis.

5.3.1 Colorimetric

Colorimetric sensors have been developed as an emerging and suitable sensing platform for food analysis and chemical screening. The colorimetric method has advantages over other approaches such as cost-effectiveness, facile portability and on-site application, repeatability, no need for a specialized operator, and the possibility to detect the results by the naked eye. Currently, colorimetric sensors play an important role in improving the safety and quality of food. In addition, it is possible to strengthen them through the development of nanotechnology, sample preparation and handling, and incorporation of more efficient reagents (Mesgari et al. 2020; Dehghani et al. 2019).

Colorimetric sensors determine sensitivity, response time, specificity, and signal-to-noise ratio by two essential factors:

1. Cognition elements that cause proper interaction between the analyte and the sensor targeting component.

2. Conductive materials or NMs with suitable optical properties that provide the appropriate color change response in the visible region (390–700 nm) after analyte recognition.

Colorimetric sensing approaches depending on the targeting technique can generally be classified in immunosensors, in which antibodies act as the target detection unit, and aptasensors, in which aptamers are used to capture the target; although less common targeting molecules are also used. Numerous approaches have been reported to generate the color readout signal in these sensors, the most predominant of which are NM-based sensors especially with Au or Ag nanoparticles, enzyme-based sensors such as horseradish peroxidase (HRP), DNAzyme-based sensors, and sensors using enzyme mimics. (Maduraiveeran and Jin 2017).

Many targeting elements can be used for specific detection on colorimetric assays, including antibodies, enzymes, nucleic acids and aptamers, receptor ligands, etc. Aptamers, defined as single-stranded oligonucleotides with a specific secondary structure which can bind to and detect various molecules, are gaining a lot of attention due to being more cost-effective and stable compared to antibodies (Hosseini et al. 2015).

AuNPs are used in colorimetric sensors due to their molecular interactions and visible color change which arises from the aggregation or dissociation of the NPs. In general, when AuNPs are dispersed, they appear red, but when they accumulate the solution color changes to purple. Based on this phenomenon, several colorimetric sensors have been designed to detect food pathogens. Acetamiprid in fruits and vegetables was detected in a simple and sensitive colorimetric method based on gold nanoparticles. Acetamiprid has a cyano group that induces accumulation of gold nanoparticles and changes the solution color from red to purple. The linear range of acetamiprid detection in this approach based on gold nanoparticles with a diameter of 15 nm and 22 nm is 6.6–66 μM and 0.66–6.6 μM, respectively (Yang et al. 2017).

Due to their high light absorption properties, silver nanoparticles also have many applications in colorimetric biosensors and, like with gold nanoparticles, the amount of light absorption by AgNPs can be adjusted by controlling their size, shape, and environmental conditions. Silver-based nanomaterials have unique electrical and catalytic properties and optical properties such as photoluminescence and SPR (Abou El-Nour et al. 2010). A rapid, sensitive, selective, and simple colorimetric assay based on AgNPs was reported to detect melamine in raw milk. Melamine can induce the aggregation of AgNPs via three amine groups in its structure and leading to solution color change from yellow to red. The proposed method can be used to detect melamine in raw milk, with a detection limit of 0.01 mM (Alam et al. 2017).

Numerous nanostructures have proven to possess peroxidase-mimicking activity; thus, they can be used to replace HRP in colorimetric sensors (Dehghani et al. 2019; Alam et al. 2017). Exploiting this phenomenon, a colorimetric biosensor was developed for the detection of *Campylobacter jejuni* with a detection limit of 100 CFU mL^{-1}. In this method, specific aptamers and Au@Pd nanoparticles were used. Interaction of free aptamers with the surface of the nanoparticle inhibits their catalytic activity and decreases 3,3′,5,5′-tetramethylbenzidine (TMB) oxidation;

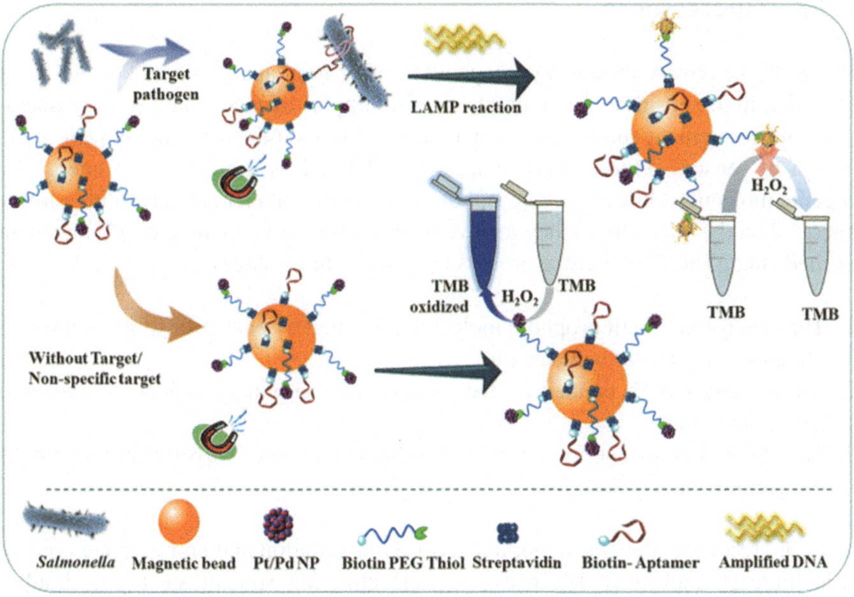

Fig. 5.1 Schematic representation of a colorimetric aptasensor used for the detection of *Salmonella* based on loop-mediated isothermal amplification (LAMP) and Pt/Pd nanoparticles with peroxidase-mimicking activity (Dehghani et al. 2021)

whereas in the presence of the target bacteria, the aptamers attach to it, and TMB can be oxidized generating a blue-colored solution (Dehghani et al. 2018). In another study, magnetic beads modified with Pt/Pd nanoparticles and aptamers were used for the specific detection of *S. typhimurium* based on loop-mediated isothermal amplification (LAMP) (Dehghani et al. 2021). In the presence of the target bacteria, LAMP reaction based on *Salmonella*-specific primers led to the generation of amplified DNA sequences which, when absorbed on Pt/Pd NPs, inhibited their peroxidase-mimicking activity. In the absence of the target bacteria, Pt/Pd NPs created oxidized TMB, which led to a blue-colored signal (Fig. 5.1).

Another colorimetric method was developed to detect ochratoxin A (OTA) in wine based on a DNAzyme-aptamer sensing element. The DNAzyme part mimics the peroxidase properties of HRP enzymes and is connected to the OTA aptamer sequence through a hairpin loop. As shown in Fig. 5.1, when the OTA is in the environment, it interacts with the aptamer forming a complex. Thus, the hairpin opens, and the DNAzyme is free to create a blue color by oxidation of TMB to TMBox based on its peroxidase-mimicking property. The DNAzyme-based sensor is linearly correlated with the OTA concentration to 10 nM, showing a limit of detection of 2.5 nM (Yang et al. 2012).

5.3.2 Fluorescence

Fluorescence is very promising compared to other food analysis methods due to its low cost, high sensitivity, and simplicity, rapid hybridization kinetics, easy operation, and convenient automation, which have led to its immense employment in the design of sensors. Fluorescence emission is defined as the release of energy by specific molecules, known as fluorophores, at longer wavelengths and with a lower energy than the wavelength they receive upon excitation from the ground state to an excited state. The fluorescence process consists of three stages:

1. The electron of the fluorophore molecule is excited by energy absorption through photons at a specific wavelength.
2. An unstable state is created for the excited electron with a tendency to return to the ground state.
3. The electron emits energy in the form of light at a longer wavelength to return to the ground state.

In fluorescence, the two important spectra of excitation and emission are considered for each molecule. The excitation and emission spectra are taken simultaneously by determining a constant distance between the excitation and the emission wavelength (Ahmad et al. 2017).

The principles of fluorescence analysis depend on the analyte interaction with the sensing elements, which can alter the fluorescence properties. There are five major principles for fluorescence signal output as follows:

1. Fluorescence quenching.
2. Increased fluorescence intensity.
3. Shifts in the emission wavelength.
4. Fluorescence lifetime variations.
5. Fluorescence resonance energy transfer (FRET).

Many molecules display natural fluorescent, some of which may display fluorescence in one state and but be non-fluorescent in another. For example, the NADH molecule is fluorescent whereas NAD^+ lacks this property. Therefore, through inducing the change from NADH to NAD^+ and creating a change in the fluorescent behavior in the sensors, the target molecule can be detected.

Most analytes are non-fluorescent, thus labels must be used to detect them by fluorescence methods. Labels are materials that can be attached to the analyte by covalent bonding with the help of hydroxyl, amine, carboxyl, or sulfhydryl groups. Probes are another solution to create fluorescence for this purpose, but probes have a high response to environmental variations such as ions, pH, and oxygen. Organic dyes and nanomaterials like AuNPs, AgNPs, and carbon nanotube (CNTs) can be used as labels. Required criteria for selecting fluorescent labels include high excitability, the ability to generate a strong and recognizable signal, having a specific functional group for connection, high fluorescence quantum yield, and high molar absorption coefficient. Fluorescent organic dyes such as rhodamine, fluorescein, cyanine, and coumarin also have some limitations for applications in sensing due to their low solubility, poor bioavailability, toxicity, and narrow excitation. Fluorescent nanomaterials such as metal nanoparticles and nanoclusters, quantum dots, and carbon and graphene dot have many merits over organic dyes but their wide applications for food analysis is also limited due to their complex synthesis, high toxicity, and low biocompatibility (Kermani et al. 2017; Borghei et al. 2017; Pebdeni et al. 2020; Dehghani et al. 2020).

In this sensing platform, as with others, a target molecule can increase the sensitivity and selectivity of the sensor by a multitude. In fluorescent biosensors, aptamers are an excellent option for targeting elements. Unlike antibodies, peptides and enzymes, aptamers have high flexibility for easy functionalization and chemical modification, and are more stable to environmental change. Therefore, fluorescent aptasensors to detect a wide range of analytes (Sharma et al. 2018). Aptabeacons, a new class of molecular beacons, have attracted a lot of attention in the development of fluorescent sensors. Aptabeacons consist of a hairpin structure attached to a fluorophore and a quencher. When the target is in the environment, it connects to the aptamer and activates the fluorescence signal by disrupting the energy transfer between the quencher and the fluorophore since the hairpin structure opens and the distance between the quencher and the fluorophore increases (Yamamoto and Kumar 2000).

Another type of aptamer-based fluorescence sensor can be designed based on an aptamer switch probe. An aptamer switch probe consists of an aptamer, a small DNA molecule that complements the end and a fluorophore is attached to the other end of the aptamer. When the analyte is in the environment, it binds to the aptamer leading to intramolecular displacement and converting the aptamer to a fluorescence probe (Tang et al. 2008). This technique was used in the detection of aflatoxin M1 in milk samples as depicted in Fig. 5.2 (Sharma et al. 2016).

Another approach to design fluorescent sensors is based on pyrene. Pyrene is a dye with little fluorescence, but if two monomer units come together to form an eximer, the fluorescence lifetime increases, which can be used to analyze the analyte. This approach has been used for the detection of Hg^{2+} in samples (Wu et al. 2019).

Recently, most fluorescent sensing methods are based on FRET. FRET is a process in which an energy donor gives its energy (electrons) to the nearest acceptor (about 10 nm). The emission spectrum of the donor has to overlap with the excitation spectrum of the acceptor for FRET to take place. In a study, in an attempt to detect Aflatoxin B1(AFB) in rice and peanuts, a FRET biosensor was designed. In this method, specific aptamers, attached to quantum dot (QDs), are absorbed on AuNPs. When the target is not in the environment, emission of QDs and the absorption of AuNPs overlap, leading to fluorescence quenching by FRET. By attaching to the target, the aptamer is released from the gold nanoparticles, and the energy transfer is interrupted due to the increase in the distance between the QDs and AuNPs. Thus, the quantum dot's fluorescence emission is activated. The linear range of this method is 10–400 nM (Sabet et al. 2017).

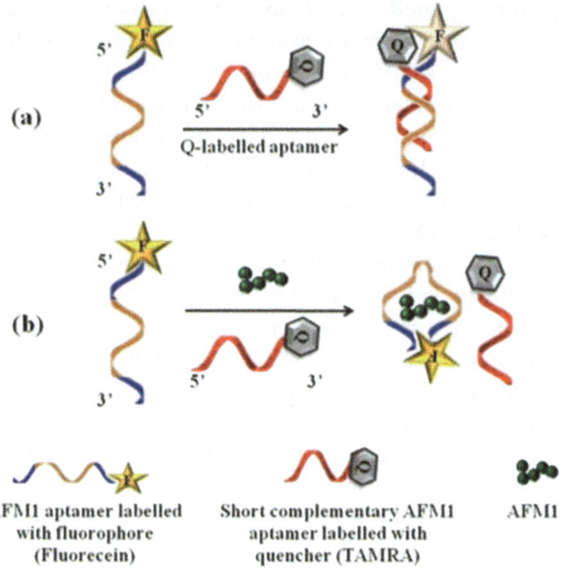

Fig. 5.2 Schematic representation of structure switching signaling aptasensing platform for the detection of aflatoxin M1 (Sharma et al. 2016)

5.3.3 Chemiluminescence

Chemiluminescence (CL), defined as the process of light generation through a chemical reaction, has grown into a well-established optical detection technique for the analysis of various liquid phase samples. A typical chemiluminescent reaction is as follows:

$$[A] + [B] \rightarrow [\Diamond] \rightarrow [\text{Products}] + \text{light}$$

in which an interaction between A and B leads to the production of excited intermediate species (◊), which emit light as they return to their grounded state. Compared to other optical techniques, CL has superior sensitivity as a result of reduced background noise due to the elimination of an external light source. Other advantages of this technique include controlled emission rate, rapidity, stability, and safety (Vacher et al. 2018).

The most commonly used CL reagent in luminol, which can be oxidized by hydrogen peroxide, permanganate, periodate, etc. in an alkaline medium to generate the excited 3-aminophthalate anion. The latter emits light upon its return to the ground state. Besides luminol, other reagents such as peroxyoxalate derivatives and tris(2,2-bipyridine) ruthenium(II) (Ru(bpy)3 2+) have also been used in the development of numerous sensors (Liu et al. 2010).

In order to gain higher efficiencies and quantum yields, various catalysts such as horseradish peroxidase (HRP) or alkaline phosphatase (ALP) are used in CL systems. In one study, HRP was used in a CL aptasensor for the detection of aflatoxin B1 (AFB1). In this approach, capture probes, which hybridize to selective

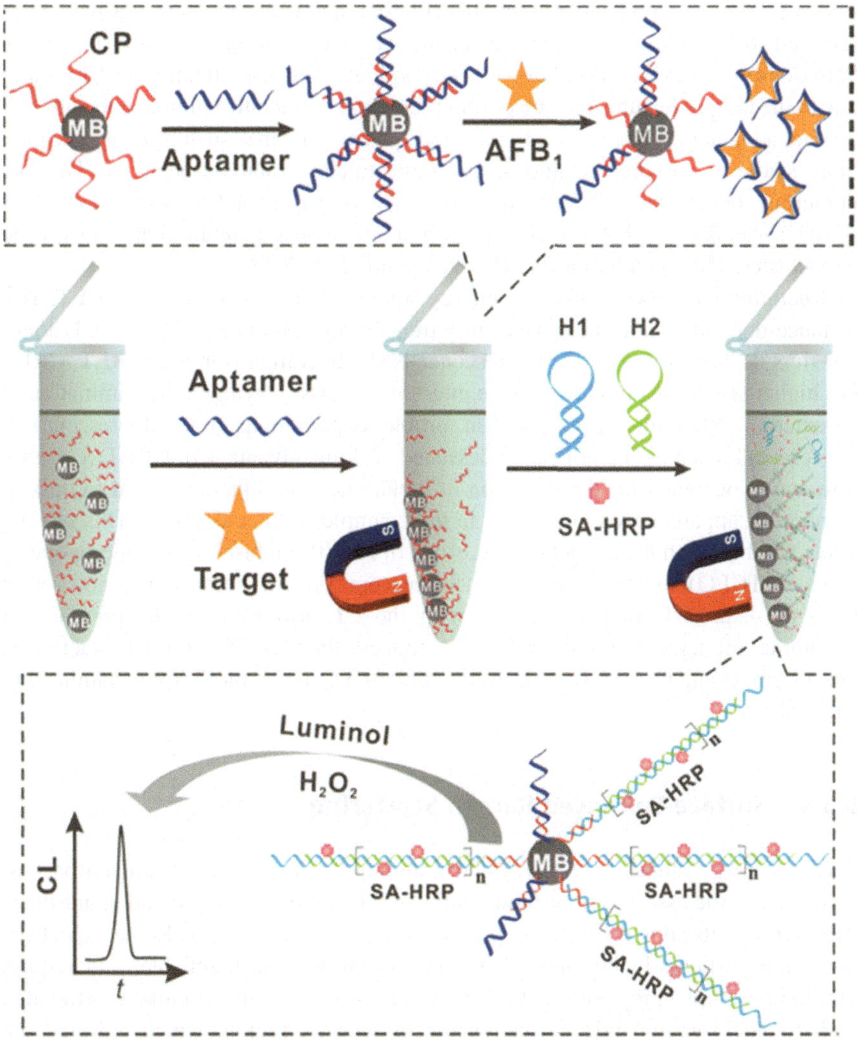

Fig. 5.3 Schematic representation of a CL aptasensor for the detection of AFB1 based on HCR for signal enhancement (Xie et al. 2019)

aptamers for AFB1, were immobilized on magnetic beads in order to enable magnetic separation to reduce the background signal. Hybridization chain reaction (HCR) was also used for further signal enhancement. In Fig. 5.3, a schematic of the proposed sensor is presented (Xie et al. 2019).

As previously mentioned, nanomaterials possess unique optical and often catalytic features which has to an increased interest in employing them for signal enhancement in CL-assays. It has been proven that incorporating metal NPs or semiconductor crystals can significantly enhance the luminescence signal in CL systems. AuNPs, AgNPs, platinum NPs (PtNPs), and CdTe nanocrystals have been reported to be able to enhance the sensitivity of CL sensors (Chen et al. 2011). Peroxidase-mimicking NMs have recently gained immense attention in CL-assays due to the unique features that arise from their large surface-to-volume ratio and their distinct advantages over conventional catalysts and enzymes such as being cheaper, more stable in varying conditions, and more tunable catalytic behavior. To date, numerous peroxidizing NMs have been discovered including iron oxide NPs (Fe_3O_4), AuNPs, graphene oxide and carbon dots, some quantum dots, and several nanoclusters (NCs) including PtNCs (Dehghani et al. 2018).

Chemiluminescence resonance energy transfer (CRET), similar to FRET, is a distance-dependent non-radiative energy transfer phenomenon in which a CL donor transfers its energy to a dye or fluorescence NMs. In comparison with FRET, CRET has higher sensitivity and lower noise interference. Also, owing to the elimination of an external light source for excitation, problems pertaining to autofluorescence or fluorescent bleaching are noticeably lessened. Taking advantage of CRET, a turn-on sensor was designed for the detection of melamine, an adulterant added to milk to increase its apparent protein content, in milk samples (Du et al. 2015). This sensing assay was based on the quenching CL effect of AuNPs on bis(2,4,6-trichlorophenyl) oxalate (TCPO)–hydrogen peroxide–fluorescein system because the absorption band of dispersed AuNPs overlaps with the CL spectrum. In the presence of melamine, the aggregation of AuNPs is induced, thus restoring the CL reaction in the system. Using this sensor, a detection limit of 3×10^{-13} mol/L for melamine was attained.

5.3.4 Surface-Enhanced Raman Scattering

Spectroscopic methods have always been among the favored detection approaches, especially in the assessment of food quality as a result of their rapid and nondestructive nature. Vibrational spectroscopy, including infrared (IR) and Raman spectroscopy, are particularly highlighted as they are simple, fast, reliable, and require minimum sample preparation. In Raman spectroscopy, the sample is irradiated with monochromatic light in the visible or near-IR region, which elevated the vibrational energy levels to higher, short-lived states. Molecular relaxation can occur either by photon emission with the same wavelength (Rayleigh scattering), or by photon emission with a lower frequency (Raman scattering). Since only a small percentage of molecules undergo Raman scattering, the intensity of this technique is quite low (Thygesen et al. 2003). Surface-enhanced Raman scattering is an approach that aims to enhance Raman scattering of molecules either absorbed on metal surfaces or in vicinity of metal particles. Noble metals, such as Ag, Au, and Cu, due to their special plasmon resonance properties, are typically used for signal enhancement in SERS. A selection of SERS substrates, based on noble metals, are available which can be divided into colloidal and solid substrates. Samples are either dropped on a solid substrate, or put in a solution with colloidal substrates to be analyzed through SERS.

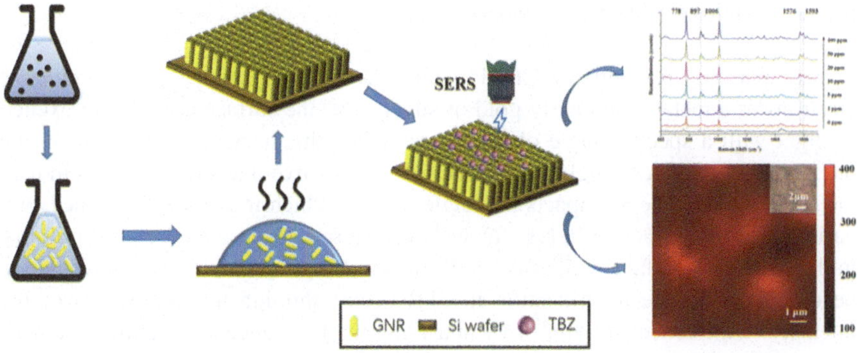

Fig. 5.4 Schematic representation of a gold nanorod (GNR)-based SERS sensor for the specific identification of thiabendazole (Fu et al. 2019)

Nanotechnology has had a big impact on SERS, as noble metal NMs display higher plasmon resonance properties and can thus further enhance Raman scattering. Many strategies have been opted to improve the SERS signals through the optimization of the structure, shape, size, and components of noble metal NMs (Sharma et al. 2013). As Raman enhancement occurs at nanoscale gaps in between the NPs, decreasing metal particle size influences the particle density and increases SERS hotspots, thus generally increasing SERS signal (He et al. 2017). AgNPs are among the strongest signal enhancing substances used in SERS assays. Pertinent to these findings, Tang et al. hypothesized that a nonplanar SERS substrate can be more effective in signal enhancement. Using AgNP-coated glass beads, they designed a rapid sensor for the identification of two pesticides, chlorpyrifos and imidacloprid, in apple extracts, attaining a detection limit of 10 ng/mL and 50 ng/mL, respectively (Tang et al. 2019).

Various nanostructures containing gold have also been extensively used in SERS sensors for food analysis from simple AuNPs (Luo et al. 2018), to more complex structures such as densely arranged AuNPs templated from mesoporous silica film (MSF) as used for the detection of pesticides in food samples via SERS (Xu et al. 2020); and also Fe_3O_4@Au core-shell gold nanostructures as used in the SERS-based assessment of synthetic food colors such as acid orange II and brilliant blue (Xie et al. 2019). Au@AgNPs core-shell nanostructures combine the excellent enhancing properties of silver with the high stability of gold. This nanostructure was used for the simultaneous detection of thiram and dicyandiamide in liquid milk (Hussain et al. 2020). Gold nanorods (AuNRs) have also recently been under the spotlight for Raman enhancement as they display unique surface plasmon resonance and tunable aspect ratio, which can easily be adjusted for signal enhancement. As seen in Fig. 5.4, AuNRs were employed in an SERS-based sensing platform for the selective detection of thiabendazole, a pesticide, in apples with a detection limit of 0.037 mg/L (Fu et al. 2019).

5.3.5 Surface Plasmon Resonance

Surface plasmon resonance (SPR) is defined as the oscillation of the conductive electrons of a metal (commonly gold or silver) near the surface upon being excited by light with a specific angle of incidence. When this happens, a decrease in the intensity of the reflected light can be seen. The refractive index of the medium near the surface of the metal impacts the angle of light which triggers SPR. Hence, any variation in the refractive index, for instance caused because of the absorption of molecules to the surface, will affect SPR, thus making it possible to the molecule to be detected. Analyte identification in SPR occurs through the assessment of the intensity of the reflected light, or by analyzing the resonance angle shifts in a real-time manner (Zhu and Gao 2019). As SPR assays provide a label-free and real-time detection route and require minimum reagents, they have extensively been applied for the analysis of numerous analytes including food-related analytes.

In a typical SPR assay, targeting molecules such as antibodies or aptamers are absorbed on a gold or silver film. The presence of a target and its interaction with the respective target molecule lead to a mass change near the surface, which thus impacts the refractive index and the reflective light beam. This general approach was used to develop an SPR-based immunoassay for the specific detection of amantadine (AM), an antiviral drug, in animal-derived food samples with a detection limit of 4.0 ng mL^{-1} (Pan et al. 2019). Multiplexed detection based on the SPR technique has also been extensively explored, as done in one study for the simultaneous detection of aflatoxin B1, ochratoxin A, zearalenone, and deoxynivalenol in cereal samples (Wei et al. 2019).

Localized surface plasmon resonance (LSPR) occurs in the interface of nanotechnology and SPR. In this approach, instead of a metallic film, metallic nanostructures, which have enhanced plasmon resonance and provide a bigger surface for functionalization, are employed. LSPR has been used in several novel sensing approaches for food monitoring. In one study, an LSPR-aptasensor was used for the recognition of *Staphylococcus aureus* in milk samples with a detection limit of 10^3 CFU/mL (Khateb et al. 2020). In another study, gold nanorods were employed to develop an aptasensor for the in situ detection of ochratoxin A with a limit of detection of 12.0 pM (Lee et al. 2018). In a different approach, Au@Pt nanozymes were used in an LSPR-based sensor for the identification of silver ions Ag^+ with a detection limit of 500 nM. The presence of Ag^+ disrupts the peroxidizing effect of the nanozymes on H_2O_2. In this situation, and in the presence of a weak acid, the residual H_2O_2 leads to the reduction of the silver ions and a blue shift in the SPR spectrum (Fig. 5.5) (Tian et al. 2020).

5.4 Conclusion and Future Perspectives

Recent advances in optical biosensing assays have revolutionized the attempts for rapid analyte detection and can evidently provide many benefits in food safety control. Optical biosensors possess many profound advantages such as being able

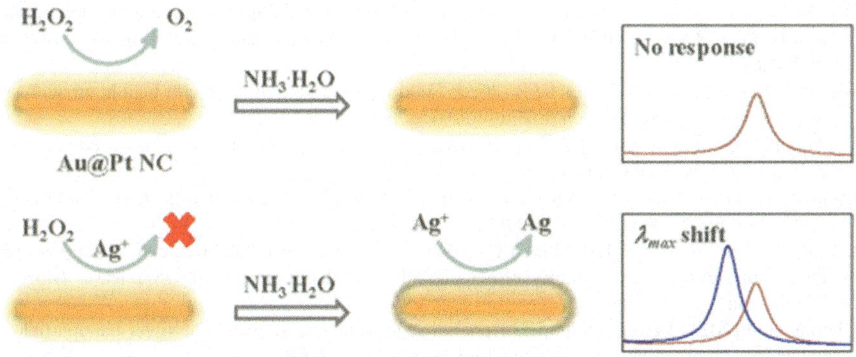

Fig. 5.5 Schematic representation of an LSPR sensor based on the peroxidase-mimicking of Au@Pt nanozymes for the detection of silver ions (Tian et al. 2020)

to provide rapid, selective, and sensitive recognition methods and being able to provide more compact and portable sensors for in situ food analysis. The incorporation of functional nanomaterials, such as enzymes, DNAzymes, antibodies, and aptamers, can enhance the sensing performance through effective targeting or signal enhancing. Novel nanomaterials, due to their unique optical properties which arise from their increased surface-to-volume ratio, have also had a great impact on the improvement of optical strategies. Also, wireless-communication technologies and the advancement of smartphones can further help with the development of portable, in situ sensing platforms. Although the broad practical application of optical assays is still limited because of the numerous challenges mostly relating to sample preparation and creating compact systems for portability, optical sensors have proven to be among the most efficient detection routes for food safety monitoring.

References

Abou El-Nour KM et al (2010) Synthesis and applications of silver nanoparticles. Arab J Chem 3(3):135–140

Abrunhosa L et al (2016) A review of mycotoxins in food and feed products in Portugal and estimation of probable daily intakes. Crit Rev Food Sci Nutr 56(2):249–265

Ahmad MH, Sahar A, Hitzmann B (2017) Fluorescence spectroscopy for the monitoring of food processes. In: Measurement, modeling and automation in advanced food processing. Springer, pp 121–151

Alam MF et al (2017) Colorimetric method for the detection of melamine using in-situ formed silver nanoparticles via tannic acid. Spectrochim Acta A Mol Biomol Spectrosc 183:17–22

Bacon R, et al. (2003) Food safety handbook

Bahadır EB, Sezgintürk MK (2017) Biosensor technologies for analyses of food contaminants. In: Nanobiosensors. Elsevier, pp 289–337

Bintsis T (2017) Foodborne pathogens. AIMS Microbiol 3(3):529

Borghei Y-S, Hosseini M, Ganjali MR (2017) Fluorescence based turn-on strategy for determination of microRNA-155 using DNA-templated copper nanoclusters. Microchim Acta 184(8): 2671–2677

Boutillier S, Fourmentin S, Laperche B (2020) Food additives and the future of health: an analysis of the ongoing controversy on titanium dioxide. Futures 122:102598

Caglayan MO, Şahin S, Üstündağ Z (2020) Detection strategies of Zearalenone for food safety: a review. Crit Rev Anal Chem:1–20

Campàs M, Prieto-Simón B, Marty J-L (2007) Biosensors to detect marine toxins: assessing seafood safety. Talanta 72(3):884–895

Chandra P, Koh WCA, Noh HB, Shim YB (2012) In vitro monitoring of i-NOS concentrations with an immunosensor: the inhibitory effect of endocrine disruptors on i-NOS release. Biosens Bioelectron

Chen X et al (2011) Determination of bisphenol A in water via inhibition of silver nanoparticles-enhanced chemiluminescence. Anal Chim Acta 689(1):92–96

Choudhary M, Yadav P, Singh A, Kaur S, Ramirez-Vick J, Chandra P, Arora K, Singh SP (2016) CD 59 Targeted ultrasensitive electrochemical immunosensor for fast and noninvasive diagnosis of oral cancer. Electroanalysis 28(10):2565–2574

Damborský P, Švitel J, Katrlík J (2016) Optical biosensors. Essays Biochem 60(1):91–100

Dehghani Z et al (2018) Colorimetric aptasensor for Campylobacter jejuni cells by exploiting the peroxidase like activity of Au@ Pd nanoparticles. Microchim Acta 185(10):448

Dehghani Z et al (2019) New colorimetric DNA sensor for detection of Campylobacter jejuni in milk sample based on peroxidase-like activity of gold/platinium nanocluster. Chem Sel 4(40): 11687–11692

Dehghani Z et al (2020) Whole cell FRET immunosensor based on graphene oxide and graphene dot for Campylobacter jejuni detection. Food Chem 309:125690

Dehghani Z et al (2021) Magnetic beads modified with Pt/Pd nanoparticle and aptamer as a catalytic nano-bioprobe in combination with loop mediated isothermal amplification for the on-site detection of Salmonella Typhimurium in food and fecal samples. Food Control 121:107664

Deka S, Saxena V, Hasan A, Chandra P, Pandey LM (2018) Synthesis, characterization and in vitro analysis of α-Fe_2O_3-$GdFeO_3$ biphasic materials as therapeutic agent for magnetic hyperthermia applications. Mater Sci Eng C 92:932–941

DiGirolamo R, Liston J, Matches J (1970) Survival of virus in chilled, frozen, and processed oysters. Appl Microbiol 20(1):58–63

Du J, Wang Y, Zhang W (2015) Gold nanoparticles-based chemiluminescence resonance energy transfer for ultrasensitive detection of melamine. Spectrochim Acta A Mol Biomol Spectrosc 149:698–702

Fu G et al (2019) Fabrication of gold nanorods for SERS detection of thiabendazole in apple. Talanta 195:841–849

Gosling E (2008) Bivalve molluscs: biology, ecology and culture. John Wiley & Sons

Griesche C, Baeumner AJ (2020) Biosensors to support sustainable agriculture and food safety. TrAC Trends Anal Chem:115906

He RX et al (2017) Effect of the size of silver nanoparticles on SERS signal enhancement. J Nanopart Res 19(8):267

Hodnik V, Anderluh G (2009) Toxin detection by surface plasmon resonance. Sensors 9(3): 1339–1354

Hosseini M et al (2015) Aptamer-based colorimetric and chemiluminescence detection of aflatoxin B1 in foods samples. Acta Chim Slov 62(3):721–728

Huet A-C et al (2010) Advances in biosensor-based analysis for antimicrobial residues in foods. TrAC Trends Anal Chem 29(11):1281–1294

Hussain A, Sun D-W, Pu H (2020) Bimetallic core shelled nanoparticles (Au@ AgNPs) for rapid detection of thiram and dicyandiamide contaminants in liquid milk using SERS. Food Chem 317:126429

Jokerst JC et al (2012) Development of a paper-based analytical device for colorimetric detection of select foodborne pathogens. Anal Chem 84(6):2900–2907

Kermani HA et al (2017) DNA methyltransferase activity detection based on graphene quantum dots using fluorescence and fluorescence anisotropy. Sensors Actuators B Chem 241:217–223

Khateb H et al (2020) Development of a label-free LSPR-Apta sensor for *Staphylococcus aureus* detection. ACS Appl Bio Mater 3(5):3066–3077

Lee B et al (2018) An optical fiber-based LSPR aptasensor for simple and rapid in-situ detection of ochratoxin A. Biosens Bioelectron 102:504–509

Ligler FS et al (2003) Array biosensor for detection of toxins. Anal Bioanal Chem 377(3):469–477

Liu M, Lin Z, Lin J-M (2010) A review on applications of chemiluminescence detection in food analysis. Anal Chim Acta 670(1–2):1–10

Lu Y, Yang Q, Wu J (2020) Recent advances in biosensor-integrated enrichment methods for preconcentrating and detecting the low-abundant analytes in agriculture and food samples. TrAC Trends Anal Chem:115914

Luo H et al (2018) Rapid and sensitive surface-enhanced Raman spectroscopy (SERS) method combined with gold nanoparticles for determination of paraquat in apple juice. J Sci Food Agric 98(10):3892–3898

Maduraiveeran G, Jin W (2017) Nanomaterials based electrochemical sensor and biosensor platforms for environmental applications. Trends Environ Anal Chem 13:10–23

Mahato K, Kumar S, Srivastava A, Maurya PK, Singh R, Chandra P (2018) Electrochemical immunosensors: fundamentals and applications in clinical diagnostics. In: Handbook of immunoassay technologies

Marklinder I et al (2020) Food safety knowledge, sources thereof and self-reported behaviour among university students in Sweden. Food Control 113:107130

Mesgari F et al (2020) based chemiluminescence and colorimetric detection of Cytochrome C by cobalt hydroxide decorated mesoporous carbon. Microchem J:104991

Naderi M, Hosseini M, Ganjali MR (2018) Naked-eye detection of potassium ions in a novel gold nanoparticle aggregation-based Aptasensor. Spectrochim Acta A Mol Biomol Spectrosc 195: 75–83

Narsaiah K et al (2012) Optical biosensors for food quality and safety assurance—a review. J Food Sci Technol 49(4):383–406

O'Kennedy R et al (2005) In: van Amerongen A, Barug D, Lauwaars M (eds) Advances in biosensors for detection of pathogens in food and water. Rapid methods for biological and chemical contaminants in food and feed. Wagningen Academic Publishers, Wagningen, The Netherlands, pp 85–104

Ohk SH, Bhunia AK (2013) Multiplex fiber optic biosensor for detection of *Listeria monocytogenes*, *Escherichia coli* O157: H7 and *Salmonella enterica* from ready-to-eat meat samples. Food Microbiol 33(2):166–171

Pan M et al (2019) A reproducible surface plasmon resonance immunochip for the label-free detection of amantadine in animal-derived foods. Food Anal Methods 12(4):1007–1016

Pebdeni AB, Hosseini M, Ganjali MR (2020) Fluorescent turn-on Aptasensor of *Staphylococcus aureus* based on the FRET between green carbon quantum dot and gold nanoparticle. Food Anal Methods 13(11):2070–2079

Ragavan K, Neethirajan S (2019) Nanoparticles as biosensors for food quality and safety assessment. In: Nanomaterials for food applications. Elsevier, pp 147–202

Sabet FS et al (2017) FRET-based aptamer biosensor for selective and sensitive detection of aflatoxin B1 in peanut and rice. Food Chem 220:527–532

Shams R et al (2020) Application of biosensors in food quality control. J Postharvest Technol 8(1): 53–74

Sharma B et al (2013) High-performance SERS substrates: advances and challenges. MRS Bull 38(8):615

Sharma A et al (2016) Development of structure switching aptamer assay for detection of aflatoxin M1 in milk sample. Talanta 158:35–41

Sharma A et al (2018) Designed strategies for fluorescence-based biosensors for the detection of mycotoxins. Toxins 10(5):197

Silva NF et al (2018) Electrochemical biosensors for Salmonella: state of the art and challenges in food safety assessment. Biosens Bioelectron 99:667–682

Tang Z et al (2008) Aptamer switch probe based on intramolecular displacement. J Am Chem Soc 130(34):11268–11269

Tang J, Chen W, Ju H (2019) Rapid detection of pesticide residues using a silver nanoparticles coated glass bead as nonplanar substrate for SERS sensing. Sensors Actuators B Chem 287: 576–583

Thongkam T, Hemavibool K (2020) An environmentally friendly microfluidic paper-based analytical device for simultaneous colorimetric detection of nitrite and nitrate in food products. Microchem J 159:105412

Thygesen LG et al (2003) Vibrational microspectroscopy of food. Raman vs. FT-IR. Trends Food Sci Technol 14(1–2):50–57

Tian Y et al (2020) Peroxidase-like Au@ Pt nanozyme as an integrated nanosensor for Ag+ detection by LSPR spectroscopy. Talanta 221:121627

Trojanowicz M, Hitchman ML (1996) Determination of pesticides using electrochemical biosensors. TrAC Trends Anal Chem 15(1):38–45

Upadhyayula VK (2012) Functionalized gold nanoparticle supported sensory mechanisms applied in detection of chemical and biological threat agents: a review. Anal Chim Acta 715:1–18

Vacher M et al (2018) Chemi-and bioluminescence of cyclic peroxides. Chem Rev 118(15): 6927–6974

Verma S, Choudhary J, Singh KP, Chandra P, Singh SP (2019) Uricase grafted nanoconducting matrix based electrochemical biosensor for ultrafast uric acid detection in human serum samples. Int J Biol Macromol

Wang J, Zhou HS (2014) Colorimetric biosensor for food chemical hazards detection. In: Food chemical hazard detection: development and application of new technologies. John Wiley & Sons, Inc., pp 291–313

Wang L et al (2020a) Recent advances in the development of electrochemical aptasensors for detection of heavy metals in food. Biosens Bioelectron 147:111777

Wang B et al (2020b) Broad spectrum detection of veterinary drugs with a highly stable metal-organic framework. J Hazard Mater 382:121018

Wei T et al (2019) Simultaneous detection of aflatoxin B1, ochratoxin A, zearalenone and deoxynivalenol in corn and wheat using surface plasmon resonance. Food Chem 300:125176

Wu Y, Wen X, Fan Z (2019) An AIE active pyrene based fluorescent probe for selective sensing Hg^{2+} and imaging in live cells. Spectrochim Acta A Mol Biomol Spectrosc 223:117315

Xie Y et al (2019) Rapid SERS detection of acid orange II and brilliant blue in food by using Fe_3O_4@ Au core–shell substrate. Food Chem 270:173–180

Xu Y et al (2020) Mesoporous silica supported orderly-spaced gold nanoparticles SERS-based sensor for pesticides detection in food. Food Chem 315:126300

Yamamoto R, Kumar PK (2000) Molecular beacon aptamer fluoresces in the presence of Tat protein of HIV-1. Genes Cells 5(5):389–396

Yamane, H., et al. (2010) Chemical defence and toxins of plants

Yang C et al (2012) Aptamer-DNAzyme hairpins for biosensing of Ochratoxin A. Biosens Bioelectron 32(1):208–212

Yang W et al (2017) Ultrasensitive and selective colorimetric detection of acetamiprid pesticide based on the enhanced peroxidase-like activity of gold nanoparticles. Anal Methods 9(37): 5484–5493

Zhu X, Gao T (2019) Spectrometry. In: Nano-inspired biosensors for protein assay with clinical applications. Elsevier, pp 237–264

Nanotechnology in Food Security and Quality

6

Mehdi Dadmehr, Behnaz Korouzhdehi, and Morteza Hosseini

Abstract

Growth of world population demands increasing food production and subsequently requires more food quality control. Low sensitivity and undesirable detection limits affect some of the present quality control tests which require development of novel method to address above challenges. Nanotechnology as a developing field is regarded almost in all aspects of science nowadays and attracts great attention and wide application. Main application of nanotechnology includes using of nanoparticles in food packaging processes, pathogen and toxin detection through nanobiosensors, and modification of food-related surfaces in nanoscale dimension to prevent pathogen attachment and contamination. In this chapter, we describe the recent improvement in application of nanotechnology in mentioned aspects of food industry. Due to distinguished nature of nanomaterials, most of the introduced nano-based methods showed higher sensitivity and improvement compared to conventional methods.

Keywords

Nanotechnology · Food · Food coverage · Nanobiosensors · Disinfection

M. Dadmehr
Department of Biology, Payame Noor University, Tehran, Iran

B. Korouzhdehi
Department of Biotechnology, University of Tehran, Tehran, Iran

M. Hosseini (✉)
Department of Life Science Engineering, Faculty of New Sciences & Technologies, University of Tehran, Tehran, Iran
e-mail: smhosseini@khayam.ut.ac.ir

P. Chandra, P. S. Panesar (eds.), *Nanosensing and Bioanalytical Technologies in Food Quality Control*, https://doi.org/10.1007/978-981-16-7029-9_6

6.1 Introduction

Food quality and security is regarded as the most important factor to guarantee the health of food supply chain. Conventional methods toward determination of food quality have shown some inefficiencies and drawbacks, and there is a need to develop novel methods that would be most efficient and address some inefficiencies in current methods. Nanotechnology as the developing and high demand technology currently showed its superiorities in food industry sector and its related branches. The involvement of nanomaterials has been demonstrated in packaging, biosensors for detection of food contaminations, and their application as antimicrobial agents. A biosensor is an analytical device used for the detection of a chemical substance that combines a biological component with a physicochemical detector (Chandra et al. 2012; Choudhary et al. 2016; Deka et al. 2018; Mahato et al. 2018; Verma et al. 2019). Meanwhile, there are some concerns for consumers about nanomaterial safety and their effect on human health and environment. Increasing demand for production of food products results in some challenges such as modification in food production chain and also following intrinsic risks in climate condition and water shortage. Using less preservations and chemicals during food process is also another concern in this industry. So, being sure about the safety of foods alongside with their production at the high scale should be considered as much as possible.

6.2 Conventional Methods for Food Safety

The food safety includes the control processes which started after harvesting of primary product from farm to its edible form in consumer plate. The aim of these strategies is to minimize all possible risks without affecting food quality. In order to remove any microbial contamination in food products, many physical and chemical approaches are applied (Parish et al. 2003). The physical methods are categorized into thermal approaches which are subdivided into heat and freezing treatments, ultrasonication (Sala et al. 1995), radiation which involves ionization (gamma) and non-ionization (UV) radiation (Fedorova 1964), drying and filtration methods. Chemical methods are exploiting chemical disinfectant agents such as chlorine solutions, hydrogen peroxide, and ozone (Poult Sci 2017). These are current approaches which commonly used in microbial control protocols. The applied methods show some failures in their food application such as their high cost, high rates of instability and degradability, environmental side effects, and human health problems. Regarding these problems, there is a global demand for using green methods instead of chemical agents. Nanotechnology as an emerging field can address all the challenges present in food industry safety and quality.

6.3 Application of Nanotechnology in Food Industry

After widespread application of nanotechnology in industries, it was also used in food industry. Contribution of nanomaterials in food sector includes their bactericide role, nano encapsulation of food ingredients, nutrient nano additives, and food packaging process (Peters et al. 2016). Nanoparticles such as Ag, ZnO, and TiO_2 show their intrinsic properties as effective antimicrobial particles to exploit on the surfaces of foods and food covers (Noimark et al. 2015). So, surface disinfection would be possible through nanoparticle application and improve the food safety and quality.

6.3.1 Application of Nanomaterials in Surface Disinfection

Deposition of specific nanoparticles with antimicrobial activity can help to provide the coating with hygiene property which could be kept during the food preparation process from production to the next steps in food chain. This layer could help us to prevent any contamination through inactivation of bacteria and inhibit the biofilm formation on the surface of food product covers. This strategy is also available for fabrication of the antimicrobial kitchenware. The applied nanomaterials include nanoparticles of metal oxides such as the particles with photocatalytic activity such as TiO_2 and ZnO and the metal oxide with intrinsic antimicrobial property toward bacterial disinfection such as Ag. Modification in nanoscale dimension in surfaces topography is another nano-based strategy which create the anti-fouling traits could also help to antimicrobial activity.

Nanoparticles with photocatalytic activity such as TiO_2 and ZnO have the ability to generate the reactive oxygen species (ROS) upon exposure by UV light and acquire the oxidization and degradation function on the organic matter substrates. Efficiency of these nanoparticles on bacteria inactivation also investigated (Gamage et al. 2010). UV irradiation is considered as the main limitation challenge, and the novel approaches have been developed for activation of these nanoparticles by visible light (Banerjee et al. 2014). The applied strategies are involving the dye sensitization method (Aponiene and Luksiene 2015) and also incorporating of different elements such as Cu (Yadav et al. 2014). As shown in Fig. 6.1, core-shell structure of TiO_2-Cu nanoparticle with photocatalytic activity showed efficient antibacterial activity under visible light illumination.

A range of nanomaterials have been applied yet to disinfect the food surfaces and coating. Among them silver nanoparticles are the most frequently used metal nanoparticles. These nanoparticles can be used in coating surface of food products to inactivate the bacteria. Direct application of silver nanoparticles has been done in preparation process of some food products such as banana powder (Orsuwan et al. 2016) and gelatin (Kanmani and Rhim 2014). It has been demonstrated that the presence of silver nanoparticles on the graphene oxide sheet inhibits 100% of bacteria attachment on applied surfaces (De Faria et al. 2014). Other surfaces such as glass can also be functionalized by deposition of silver nanoparticles. Silicone

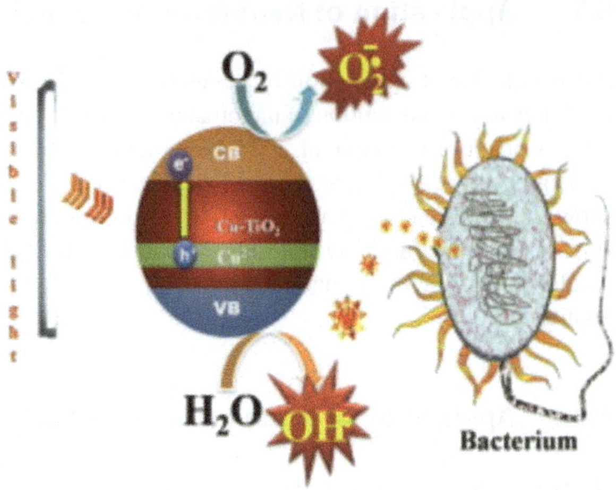

Fig. 6.1 schematic representation of synthesized Cu-coated TiO_2 effect on bacteria (Yadav et al. 2014)

functionalization with silver nanoparticles and crystal violet could also help for better efficiency of crystal violet dye for inactivation of microbial source under the visible light. Photocatalytic activity of nanoparticles such as TiO_2, ZnO, and CeO_2 has been regarded recently due to its antimicrobial activities. This process requires UV light illumination and results in creation of reactive oxygen species (ROS) which oxidize any organic substance such as bacteria. UV radiation may result in unpredicted damage to operators. So recently, alternative approaches have been developed for activation of nanoparticles with visible light instead of UV radiation.

Moreover, some food surfaces have been functionalized with the green natural extracts instead of chemical compounds that could exhibit the antimicrobial activity. These structures show their efficiencies in nano capsulation and nano emulsion forms in cinnamaldehyde and soybean polysaccharide, respectively (Noimark et al. 2014; Donsì et al. 2015). Above-mentioned studies showed antimicrobial activity against *E. coli* bacteria. Biofilms are resulted from formation of bacteria aggregation on surfaces which also could include the pathogenic bacteria. These biofilms show resistance to conventional disinfection methods and are considered as one of the main important sources of food products contamination. Modification of food surfaces morphology in nanoscale dimension and generation of specific nanoscale topography is an emerging development toward inhibition of microbial surface attachment. Some examples of these modifications are deposition of nanoparticles on surfaces and nanolithography.

6.3.2 Nanotechnology-Based Application in Food Chain

Agri/food/feed is the novel phrase which indicates the important role of nanotechnology in agriculture, food, and feed industries. Nano pesticides and nano fertilizers which include the chemical compounds capsulated in nanocarriers are among common commercial products in agriculture section. Other examples of these

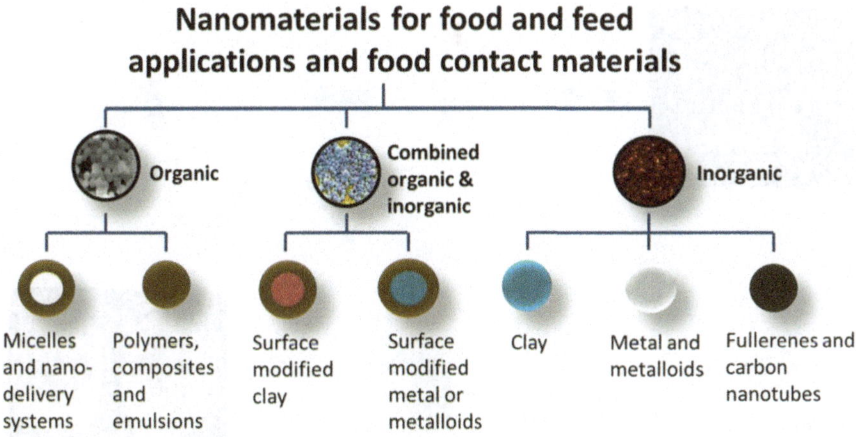

Fig. 6.2 Different types of applied nanomaterials for food coverage (Peters et al. 2014)

products in food industry include food additives and food supplements in nano size form and antimicrobial agents in food packaging process.

According to EFSA inventory report, about 276 agri/food/feed nanotechnology-based applications have been developed and present in market and many other applications are under development process (Peters et al. 2014). The applied nanoparticles for construction of food coverage include organic, inorganic, and combination of organic and inorganic materials (Fig. 6.2). Most popular organic originated nanoparticles include micelles and polymers and also inorganic nanoparticles include metallic nanoparticles, clay, and carbon-based materials. The combined nanoparticles have core/shell structure which in most of the organic part is used as the shell of nanoparticles.

Many of these introduced applications are related to microbiological food safety which boost the food quality and regarded as alternative future goals. Engineering of food surface coating with engineered nanoparticles such as Ag, ZnO, and TiO_2 are the most and frequent approaches toward illustrated applications. Many nanobiosensors have been developed to detect the trace amounts of pathogens in food products (Rashidi and Khosravi-Darani 2011). Invention of food package with antibacterial and also biodegradable properties makes the new generation of intelligent packages.

6.3.3 Nanotechnology in Food Packaging

Inhibition of microbial contamination requires two main approaches, the first one is microbial inactivation which helps to provide the hygiene environment for food during food preparation and storage processes, and the next approach is prevention of microbial attachment which finally led to biofilm formation. Using clay in nanoscale form and in combination with biopolymer-based plastic resulted to

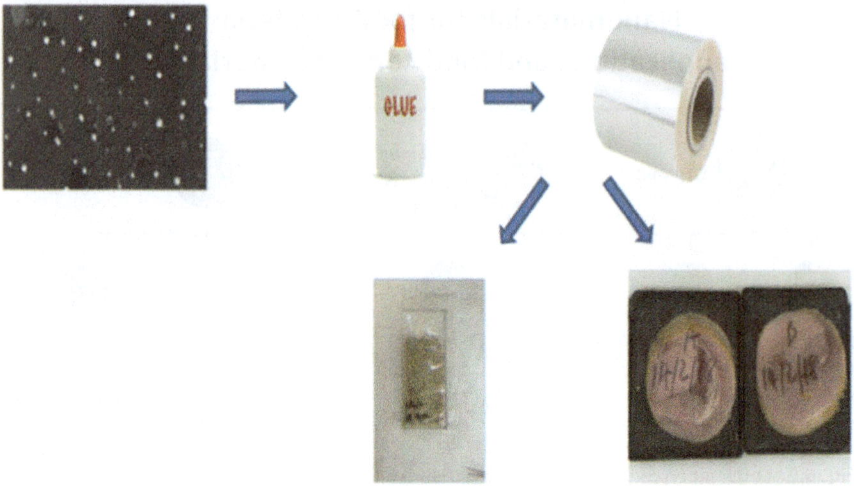

Fig. 6.3 Multilayer coverage incorporated with selenium nanoparticles for antioxidative protection in food samples (Vera et al. 2016)

construction of biodegradable and ecofriendly food package (Ghanbarzadeh et al. 2013). Titanium nitride nanoparticles also has been used for development of resistant packaging to CO_2 leakage in carbonated drinks (EFSA Publication and Frandsen 2012).

Integration of nanoparticles in bioactive compounds prevents food oxidation and spoilage. Some examples that have been used for this purpose include cellulose and selenium integrated in food packages which inhibit the ROS to decrease the food quality (Vera et al. 2016). As shown in Fig. 6.3, a flexible multilayer incorporated with selenium nanoparticles developed and showed its antioxidant efficiency in hazelnuts, walnuts, and potato chips samples.

Oxidative damage is the common spoilage process, especially in fatty foods. So, similar strategy had been reported for fatty food protection using aromatic groups such as phenol (Liu et al. 2017). In another study, encapsulated tea polyphenol nanoparticles incorporated in gelatin films and the obtained results showed oxidation protective of presented coverage due to release of tea polyphenols in the contacting surface (Fig. 6.4).

6.3.4 Nanobiosensors in Food Industry

One of the main important factors in food quality is sensitive detection of pathogens which cause spoilage in food products. Nanobiosensors are recently developed recognition tools toward sensitive detection of microorganisms which their higher performance originated from their nanoscale dimensions property. This feature makes them more sensitive due to their higher surface to volume ratio and also provides the efficient platform with rapid response for detection of many pathogens.

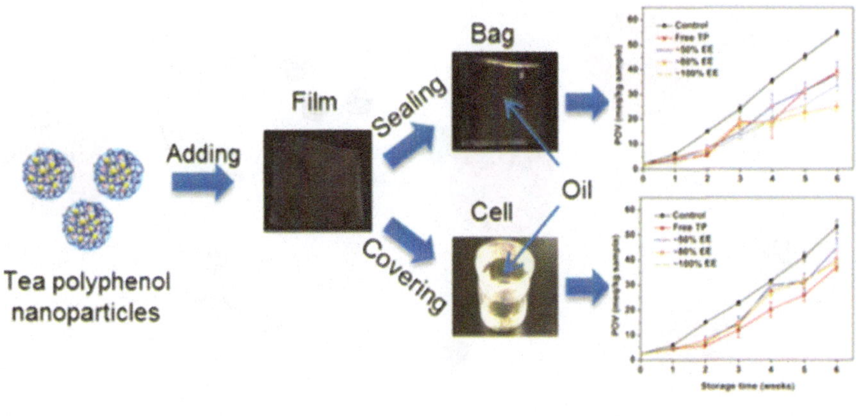

Fig. 6.4 Schematic presentation of tea polyphenol nanoparticles incorporated in gelatin films and their antioxidant property (Liu et al. 2017)

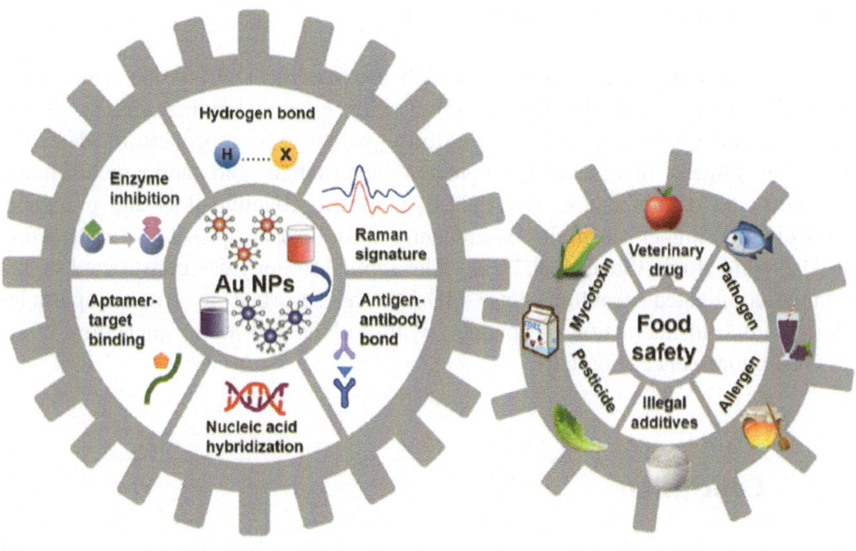

Fig. 6.5 Different strategies for detection of varieties food contamination (Hua et al. 2021)

The most frequently used nanoparticles for fabrication of these sensors are gold and silver. The gold-based approaches provide colorimetric methods for visual detection of food pathogens (Wang and Alocilja 2015). Synthesis of gold nanoparticles with different size and modification not only improves their pathogen detection range but also improves the sensitivity of detection methods (Hua et al. 2021) As shown in Fig. 6.5, different gold-based biosensors and strategies have been designed to detect a wide range of food contaminants. Silver nanoparticles have also been used for detection of some pathogens such as *E. coli* and salmonella bacteria through chemiluminescent assays (He et al. 2015).

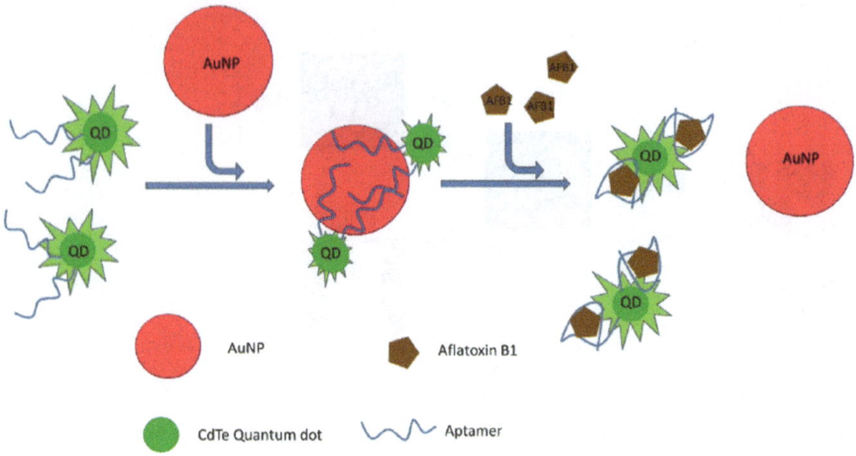

Fig. 6.6 Schematic representation of applied strategy based on FRET mechanism between quantum dots and gold nanoparticles (Sabet et al. 2017)

Aptasensors are other biosensors which use specific DNA probe with distinguished sequence toward toxin detection with specific binding affinity. These biosensors have gained great attention during recent years especially for detection of food pathogens. Involvement of nanoparticles, such as gold nanoparticles, quantum dots and carbon nanotubes (CNTs), graphene quantum dots and graphene, results in their higher sensitivity. We recently introduced an aptasensor for detection of B1 food aflatoxin (Sabet et al. 2017). The applied strategy was constructed so that aptamer-conjugated quantum dots (QDs) were adsorbed to Au nanoparticles (AuNPs) due to interaction of aptamers with AuNPs and led to quenching effect on QDs fluorescence (Fig. 6.6).

6.4 Conclusion

The benefit and widespread applications of nanotechnology in food industries offer many significant and efficient results toward increasing the microbial food safety and subsequently improving the quality of food products. Construction of food coverage with antimicrobial property through different strategies results in prevention of microorganism attachment and their inactivation in order to prevent the biofilm formation on surfaces. Another application of nanotechnology in food industry is construction of nanobiosensors for detection of pathogens which has benefits such as higher sensitivity and rapidness. However, the application of nanotechnology in determination of food contamination is relatively limited compared to other industrial applications. Due to great application of nanomaterials in food industry, many researches are still conducted and most of them are in development stage to provide more safety in food products to address the required and necessary standards and obligation. It should be noted that any application of nanomaterials must be under global regulations which finally provide environmental sustainability and global economic growth.

References

Aponiene K, Luksiene Z (2015) Effective combination of LED-based visible light, photosensitizer and photocatalyst to combat Gram (−) bacteria. J Photochem Photobiol B Biol 142:257–263

Banerjee S, Pillai SC, Falaras P, O'Shea KE, Byrne JA, Dionysiou DD (2014) New insights into the mechanism of visible light photocatalysis. J Phys Chem Lett 5:2543–2554

Chandra P, Koh WCA, Noh HB, Shim YB (2012) In vitro monitoring of i-NOS concentrations with an immunosensor: The inhibitory effect of endocrine disruptors on i-NOS release. Biosens Bioelectron. https://doi.org/10.1016/j.bios.2011.11.027

Choudhary M, Yadav P, Singh A, Kaur S, Ramirez-Vick J, Chandra P, Arora K, Singh SP (2016) CD 59 Targeted ultrasensitive electrochemical immunosensor for fast and noninvasive diagnosis of oral cancer. Electroanalysis 28:2565–2574. https://doi.org/10.1002/elan.201600238

De Faria AF, Martinez DST, Meira SMM, de Moraes ACM, Brandelli A, Filho AGS, Alves OL (2014) Anti-adhesion and antibacterial activity of silver nanoparticles supported on graphene oxide sheets. Colloids Surf B Biointerfaces 113:115–124

Deka S, Saxena V, Hasan A, Chandra P, Pandey LM (2018) Synthesis, characterization and in vitro analysis of α-Fe_2O_3-$GdFeO_3$ biphasic materials as therapeutic agent for magnetic hyperthermia applications. Mater Sci Eng C 92:932–941. https://doi.org/10.1016/j.msec.2018.07.042

Donsì F, Marchese E, Maresca P, Pataro G, Vu KD, Salmieri S, Lacroix M, Ferrari G (2015) Green beans preservation by combination of a modified chitosan based-coating containing nanoemulsion of mandarin essential oil with high pressure or pulsed light processing. Postharvest Biol Technol 106:21–32

EFSA Publication, Frandsen HL (2012) EFSA Panel on food contact materials, enzymes, flavourings and processing aids (CEF): Scientific Opinion on the safety evaluation of the substance, titanium nitride, nanoparticles, for use in food contact materials. EFSA J 10:2641–2649

Fedorova RI (1964) Effect of ultraviolet radiation on microorganisms as a principal extremal factor of space environment. Life Sci Space Res 2:305–310

Gamage J, Zhang Z, Gamage J, Zhang Z (2010) Applications of photocatalytic disinfection. Int J Photoenergy 2010:1–11

Ghanbarzadeh B, Oleyaei SA, Almasi H (2013) Nanostructured materials utilized in biopolymer-based plastics for food packaging applications. Crit Rev Food Sci Nutr 55:1699–1723

He Y, Xu B, Li W, Yu H (2015) Silver nanoparticle-based chemiluminescent sensor array for pesticide discrimination. J Agric Food Chem 63:2930–2934

Hua Z, Yu T, Liu D, Xianyu Y (2021) Recent advances in gold nanoparticles-based biosensors for food safety detection. Biosens Bioelectron 179:113076

Kanmani P, Rhim J-W (2014) Physicochemical properties of gelatin/silver nanoparticle antimicrobial composite films. Food Chem 148:162–169

Liu F, Avena-Bustillos RJ, Chiou B-S, Li Y, Ma Y, Williams TG, Wood DF, McHugh TH, Zhong F (2017) Controlled-release of tea polyphenol from gelatin films incorporated with different ratios of free/nanoencapsulated tea polyphenols into fatty food simulants. Food Hydrocoll 62:212–221

Mahato K, Kumar S, Srivastava A, Maurya PK, Singh R, Chandra P (2018) Electrochemical immunosensors: fundamentals and applications in clinical diagnostics. In: Handbook of immunoassay technologies

Noimark S, Allan E, Parkin IP (2014) Light-activated antimicrobial surfaces with enhanced efficacy induced by a dark-activated mechanism. Chem Sci 5:2216–2223

Noimark S, Weiner J, Noor N, Allan E, Williams CK, Shaffer MSP, Parkin IP (2015) Antimicrobial surfaces: dual-mechanism antimicrobial polymer–ZnO nanoparticle and crystal violetencapsulated silicone. Adv Funct Mater 25:1366

Orsuwan A, Shankar S, Wang L-F, Sothornvit R, Rhim J-W (2016) Preparation of antimicrobial agar/banana powder blend films reinforced with silver nanoparticles. Food Hydrocoll 60:476–485

Parish ME, Beuchat LR, Suslow TV, Harris LJ, Garrett EH, Farber JN, Busta FF (2003) Methods to reduce/eliminate pathogens from fresh and fresh-cut produce. Comprehensive Rev Food Sci Food Safety 2:161–173

Peters R, Brandhoff P, Weigel S, Marvin H, Bouwmeester H, Aschberger K, Rauscher H, Amenta V, Arena M, Botelho Moniz F et al (2014) Inventory of nanotechnology applications in the agricultural, feed and food sector. EFSA Support Publ 11(7):621E

Peters RJB, Bouwmeester H, Gottardo S, Amenta V, Arena M, Brandhoff P, Marvin HJP, Mech A, Moniz FB, Pesudo LQ et al (2016) Nanomaterials for products and application in agriculture, feed and food. Trends Food Sci Tech 54:155–164

(2017) Effects of chlorine and hydrogen peroxide sanitation in low bacterial content water on biofilm formation model of poultry brooding house waterlines. Poult Sci 96(7:2145–2150

Rashidi L, Khosravi-Darani K (2011) The applications of nanotechnology in food industry. Crit Rev Food Sci Nutr 51:723–730

Sabet FS, Hosseini M, Khabbaz H, Dadmehr M, Ganjali MR (2017) FRET-based aptamer biosensor for selective and sensitive detection of aflatoxin B1 in peanut and rice. Food Chem 220:527–532

Sala FJ, Burgos J, Condón S, Lopez P, Raso J (1995) Effect of heat and ultrasound on microorganisms and enzymes. In: Gould GW (ed) New methods of food preservation. Springer, Boston, MA. https://doi.org/10.1007/978-1-4615-2105-1_9

Vera P, Echegoyen Y, Canellas E, Nerín C, Palomo M, Madrid Y, Cámara C (2016) Nano selenium as antioxidant agent in a multilayer food packaging material. Anal Bioanal Chem. https://doi.org/10.1007/s00216-016-9780-9

Verma S, Choudhary J, Singh KP, Chandra P, Singh SP (2019) Uricase grafted nanoconducting matrix based electrochemical biosensor for ultrafast uric acid detection in human serum samples. Int J Biol Macromol. https://doi.org/10.1016/j.ijbiomac.2019.02.121

Wang Y, Alocilja EC (2015) Gold nanoparticle-labeled biosensor for rapid and sensitive detection of bacterial pathogens. J Biol Eng 9:1

Yadav HM, Otari SV, Koli VB, Mali SS, Hong CK, Pawar SH, Delekar SD (2014) Preparation and characterization of copper-doped anatase TiO_2 nanoparticles with visible light photocatalytic antibacterial activity. J Photochem Photobiol A Chem 280:32–38

7 Electrochemical Sensors for the Detection of Food Adulterants in Miniaturized Settings

Ananthappan Periyasamy, Seena Selvam, Arunbalaji Chellakannu, Vasantha Vairathevar Sivasamy, and Jayalakshmi Mariakuttikan

Abstract

Intentional and unintentional adulteration of food deceives the people and can cause risk to their health. This book chapter provides an overview about the classification of adulterants and types of food adulteration in different food samples such as milk, meat, fish, vegetables, water, and fruit juices and also the conventional methods employed to detect adulteration. The main focus of this book chapter is to provide a detailed review of the different kinds of miniaturized electrochemical sensors, applied for the detection of food adulterants like melamine, urea, hydrogen peroxide, formaldehyde, inappropriate animal meat, synthetic dyes, pesticides, fungicides, veterinary drugs, microbial toxins, and heavy metal ions. Finally, the overall observation, challenges, and future recommendation about the electrochemical sensors were discussed towards the transformation of the research findings into sensitive and affordable miniaturized devices.

Keywords

Electrochemical sensors · Food adulterants · Miniaturization · Nanocomposites · Food stuffs

A. Periyasamy · S. Selvam · A. Chellakannu · V. V. Sivasamy (✉)
Department of Natural Products Chemistry, School of Chemistry, Madurai Kamaraj University, Madurai, Tamil Nadu, India
e-mail: vasantha.chem@mkuniversity.org

J. Mariakuttikan
Department of Immunology, School of Biological Sciences, Madurai Kamaraj University, Madurai, Tamil Nadu, India

P. Chandra, P. S. Panesar (eds.), *Nanosensing and Bioanalytical Technologies in Food Quality Control*, https://doi.org/10.1007/978-981-16-7029-9_7

7.1 Introduction

Food consumption in India is closely connected with climate, religion, and culture. Food is the main source for survival and providing food without adulterant to the public is a challenging task (Banerjee et al. 2017). Food adulteration mainly affects the quality of the food; which may be due to the addition of intentional or accidental second-grade ingredients or by the removal of important components (Bansal et al. 2017). It is a major abuse to the public health (Syed Imran et al. 2020; Pal and Meenu 2020; Campuzano et al. 2020) and causes more than 200 diseases including Delhi belly to cancer. Among the world population, 4,20,000 people were dead, including 1,25,000 children within the age group of 5 (Syed Imran et al. 2020).

7.1.1 Types of Food Adulteration

Food adulteration can be classified as

(a) **Intentional Adulteration**
The intentional addition of similar inferior ingredients to food is called intentional adulteration, i.e., addition of melamine, urea, synthetic dyes, stones, marbles, sand, mud, filth, chalk powder, and contaminated water (Zhang et al. 2020a, b). Melamine and urea are added to increase the protein level in milk which can cause kidney failure, urinary tract obstruction, gastrointestinal bleeding, liver failure, cachexia, and nephritic disorder (Naik et al. 2020). Addition of synthetic dyes to soft drinks, chocolates, cakes, etc. causes cancer, asthma, contact anaphylaxis, immunosuppression, angioneurotic edema, and contact urticaria.

(b) **Unintentional Adulteration**
Mixing of unwanted ingredients to the food samples due to carelessness, ignorance, and poor hygiene is said to be accidental adulteration. In addition, veterinary drugs are given to animals for immunization stay in animal's body and these drug residues persist in the exported meat. Similarly, pesticides sprayed on fruits and vegetables remain in agricultural products, for example, organophosphorus, organochlorines, carbamates, and pyrethroids are causing carcinogenic diseases. Apart from these unintentional adulterants, the accidental adulteration is also due to microbial pathogens and heavy metals in food stuffs and water. The flow chart for classification of adulteration is given in Fig. 7.1.

All the country has developed Food Quality Assurance Systems to determine the quality of food nationally and globally (Campuzano et al. 2016) through various conventional techniques. Although conventional techniques are accurate and sensitive in detection of majority of the illegal food additives, they are time-consuming, labor-intensive procedures and involve expensive instrumentation. Thus, cost-effective, simple, efficient, and rapid identification of hazardous additives in food is a crucial task in analytical research across the globe (Mansouri et al. 2020a, b). A

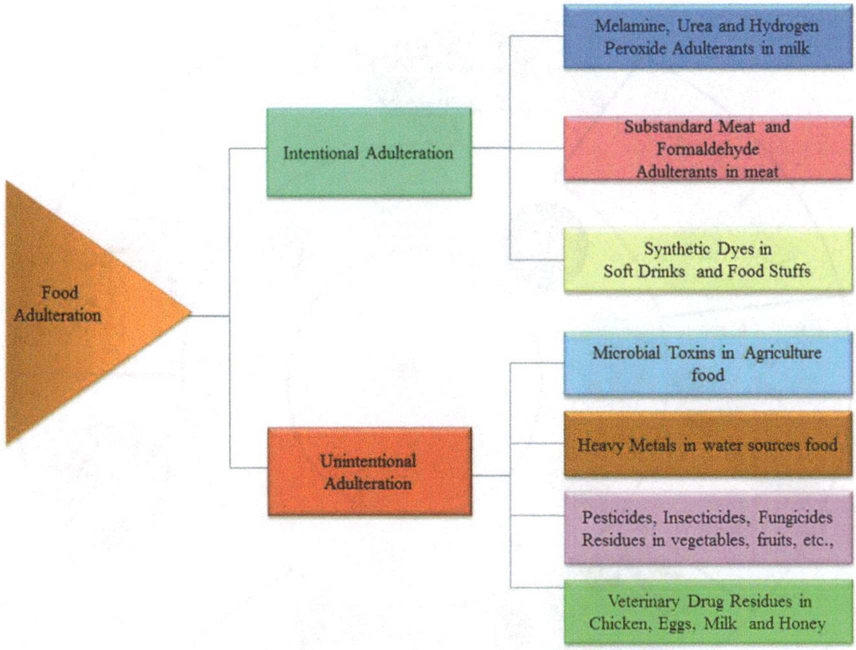

Fig. 7.1 Classification of food adulterations

biosensor is a device that measures biological or chemical reactions by generating signals proportional to the concentration of an analyte in the reaction (Chandra et al., 2012; Choudhary et al., 2016; Deka et al., 2018; Mahato et al., 2018; Verma et al., 2019).

Electrochemical sensors offer numerous advantages in the detection of food adulterants. Different families of electrochemical sensors are classified based on the nature of electrical signals used in the transduction during analysis: potentiometric (change of membrane potential); conductometric (change of conductance); impedimetric (change of impedance); and voltammetric or amperometric (change of current for an electrochemical reaction with the applied voltage ramp in the first case, or with time at a fixed applied potential in the latter) (Fig. 7.2).

To facilitate miniaturization and large-scale production of electrochemical sensors at low cost, screen-printing methodology has been adopted. In recent decades, the screen-printed electrodes (SPE) satisfy the growing demands for cost-efficient, stable, disposable, and portable electrochemical sensors which are crucial in food industries.

One of the most successful and profitable biosensors till date is the disposable glucose biosensor. The miniaturization can allow for the development of "lab-on-a-chip" and bedside sensor types. Point-of-Care (POC) diagnostic devices should satisfy the following requirements to be used compared to sophisticated devices:

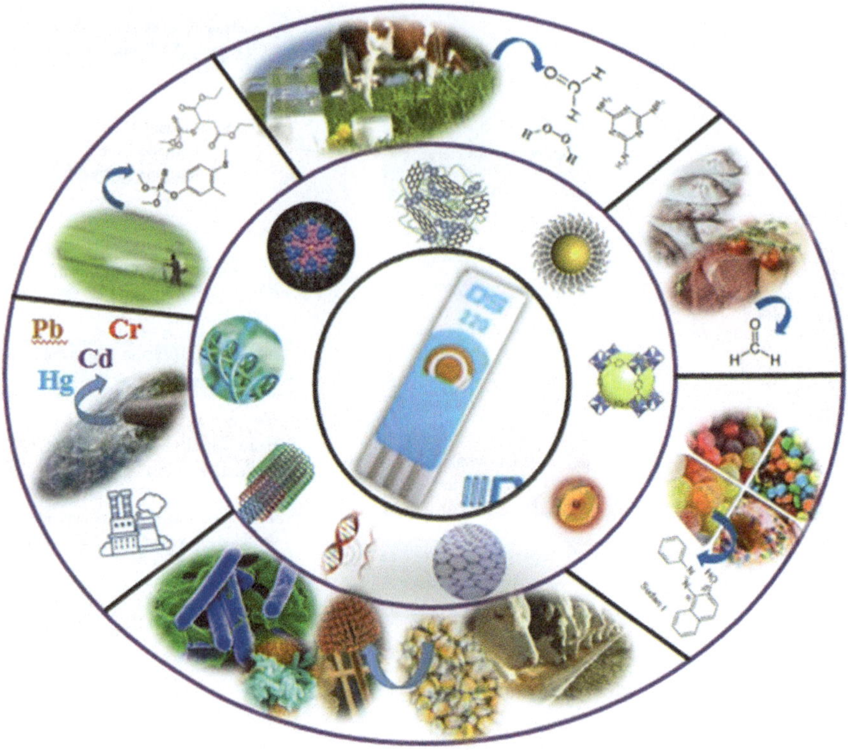

Fig. 7.2 Schematic representation of electrochemical sensors for food adulterants

low Limit of Detection (LOD), high sensitivity and selectivity, low sample volume, low cost, rapidity, and user-friendly format (Marzuki et al. 2012).

7.2 Electrochemical-Miniaturized Sensors for Intentional Adulteration

Food trading is one of the most important businesses in the world. There are many standard and sensitive methods like HPLC, LC-MS, UPLC, LFIA, GC, SPR, Raman Spectroscopy, Fluorescence Spectroscopy (Wang et al. 2019; Khalil et al. 2020) etc. are available to detect the various food adulterants. Nevertheless, they are time-consuming, expensive, and need trained personnel to detect adulterants. But electrochemical methods provide distinctive opportunity to develop portable, rapid, accurate, and sensitive sensors for the detection of food adulterants (Nascimento et al. 2017; Mansouri et al. 2020a, b) by overcoming the restrictions and limitations of the standard methods. The electrochemical sensors are playing a vital role in analyzing diagnosing diseases, agro-foods, and ecological inspection (Zeng et al. 2018).

7.2.1 Electrochemical Sensors for Milk Adulteration

Milk is one of the most important and essential nutritional drinks to humans as well as animals which contains 32 g/L of protein content. Mixing of water with milk reduces the protein and fat content of the milk. In order to adjust the milk properties, a variety of chemicals are mixed, for example, melamine, urea, hydrogen peroxide, olive oil, soap, caustic soda, benzoic acid, boric acid, and salicylic acid, (Azad and Ahmed 2016) and also animal fat (Rebechi et al. 2016). Consecutive consumption of such adulterated milk can lead to harmful effects in human body (Yu et al. 2015), and they can be detected by changes in their bio-physical properties like pH, conductivity, and impedance during adulteration (Tripathy et al. 2017).

7.2.1.1 Electrochemical Detection of Melamine in Milk

In 2008, WHO/FAO reported that in China, melamine (20.298 M) was added with milk by a food industry which was 1706 times higher than the WHO-recommended limit. Due to this melamine addition incident, 2,94,000 infants, 51,900 children, and 13 adults were hospitalized and suffered from kidney problems. After this report, the limit dosage of melamine was set by the US Food and Drug Administration (FDA) as 1 ppm for infant milk and 2.5 ppm for other dairy products (Jun-shi, 2009).

Shang et al. (2014) have reported, an electrochemical sensor based on porous graphene/PdCu alloy nanoparticles framework and hydrazine was used as signaling element for the detection of melamine in raw milk and milk powder. Molecularly imprinted polymers (MIP) using copolymer poly(acrylic acid-co-(7-(4-vinylbenzyloxy)-4-methyl coumarin)-co-ethylhexyl acrylate) (poly(AA-co-VMc-co-EHA)/MWCNTs) (Xu et al. 2018) and *in situ* co-electropolymerized aniline/acrylic acid by employing melamine as template molecules.

Melamine spiked in infant and raw milk were analyzed by the electrochemical sensors (Regasa et al. 2020). A robust and reliable electrochemical sensor based on CuNFs/L-arginine (L-Arg)/rGO-Cu nanoflower composite/GCE was developed in the presence of ascorbic acid by Daizy et al. (2019) for the determination of melamine in commercial infant milk samples (Fig. 7.3).

7.2.1.2 Electrochemical Detection of Urea in Milk

Excessive consumption of urea-adulterated milk can lead to indigestion, gastrointestinal bleeding, urinary tract obstruction, renal failure, and cancer (Migliorini et al. 2018). The Food Safety and Standards Authority of India (FSSAI) Act 2006 and Prevention of Food Adulteration (PFA) Rule 1995 recommended that the amount of urea should be less than 70 mg/100 mL of milk (Azad and Ahmed 2016). Hence, a simple, portable, and reliable biosensor needs to be developed to detect urea in milk.

Electrochemical sensors for detection of urea in milk as follows; electrochemically synthesized Fe_3O_4/MWCNT-polyaniline nanocomposite was used for the fabrication of electrochemical biosensor for the detection of urea. The developed sensor has exhibited stability up to 60 days along with good performance (Singh et al. 2020). AgNPs/glucose strip was used to fabricate a portable electrochemical biosensor for urea detection (Liu et al. 2020). Ezhilan et al. (2017) developed a

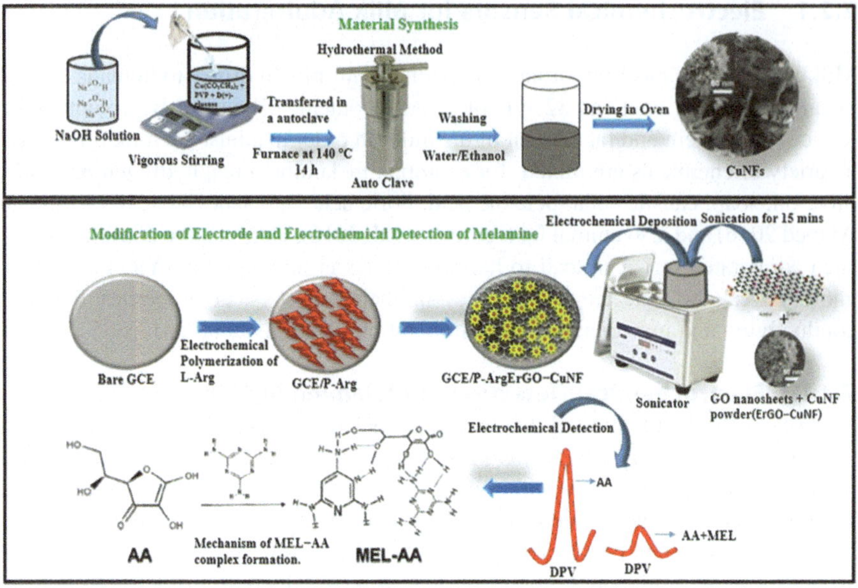

Fig. 7.3 Schematic illustration of electrochemical detection of melamine using molecular imprinting electrode. (Adopted with permission from (Shang et al. (2014), ACS Appl. Mater. Interfaces, 6, (21), 18721–18727). Copyright (2014) American Chemical Society)

highly sensitive electrochemical biosensor based on Pt/ZnO/AChE/Chitosan for the simultaneous determination of melamine and urea. The biosensor has detected urea and melamine with LOD of 1 and 3 pM, respectively. *Arthrobacter creatinolyticus* urease immobilized polyaniline membrane-based potentiometric biosensor was fabricated for the detection of urea (Ramesh et al. 2015).

7.2.1.3 Electrochemical Detection of Hydrogen Peroxide in Milk

Hydrogen peroxide (H_2O_2) is one of the most important and widely used antibacterial agents in food-packing industries. Higher concentration of H_2O_2 leads to different disorders like atherosclerosis, renal disease, Parkinson's disease, Alzheimer's disease, and cancer (Pramanik and Dey 2011).

A hybrid polypyrrole-dodecylbenzene sulfate-cerium oxide was applied for the detection of H_2O_2 in the development of an enzyme-free electrochemiluminescence (ECL) sensor. The CeO_2 in polymer matrix enhanced the ECL signal with the LOD of 0.12 fM (Karimi et al. 2018). Prussian blue nanoparticles-doped PEDOT composite was synthesized electrochemically on the glassy carbon electrode and applied for fabrication of enzyme-free voltammetric hydrogen peroxide sensor (Wang et al. 2017a, b).

7.2.2 Electrochemical Sensors for Adulteration in Meat

Adulteration of meat is the mixing of one animal meat with another animal meat. Meat is also adulterated with preservative chemicals during preservation process. Therefore, there is a need to detect adulteration in meat and prevent it (Mansouri et al. 2020a)

7.2.2.1 Electrochemical Detection of Meat Adulteration

Pork adulteration was detected in raw meat by using label-free 4-carcoxyphenyl layer/carbon-nanofiber SPE immunoassay for the detection of porcine serum albumin (Lim and Ahmed 2016). Montiel et al. (2017) have developed a disposable nucleic acid-based amperometric biosensor to detect horse meat mixed with meat. The 40-mer RNA probe labeled magnetic micro-carriers specific to mitochondrial DNA of horse was used as the sensing probe. This biosensor is highly sensitive, selective, rapid, and simple for detection of horse meat adulterant (Fig. 7.4).

Similarly, AuNPs-DNA bioconjugate was used as a platform to detect the pork meat using pork mitochondrial DNA in raw and processed meat by electrochemical method (Hartati et al. 2019). Zhang et al. (2020a, b) also detected the pork adulteration in meat by using electro-entrapment of IgG in polymer-modified graphite paste electrode. Mandli et al. (2018) have detected the cattle meat in the meat mixer containing cattle (target), sheep, pig, horse, donkey, dog, fox, rabbit, mouse, rat, chicken, duck, and goose by using ELISA/immunosensor techniques. An inexpensive DNA-based electrochemical genosensor was developed and applied for detection of donkey adulteration in cooked sausages (Mansouri et al. 2020a, b).

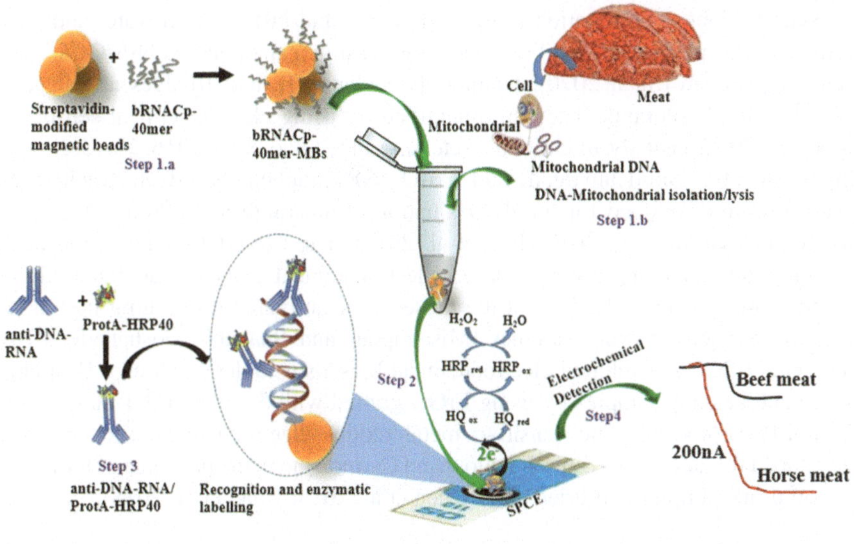

Fig. 7.4 Schematic display of steps involved in the development of electrochemical immunoassay for horse meat adulterant

7.2.2.2 Electrochemical Detection of Formaldehyde in Fish

Formalin (40% formaldehyde in water) is commonly used to keep the sea foods fresh. In 2004, the International Agency for Research on Cancer classified formaldehyde as a human carcinogen. Formalin causes cancer in mouth, nose, throat, digestive tract, etc.

A miniaturized Capillary Electrophoresis with Electrochemical Detection method has been developed for fast determination of formaldehyde and acetaldehyde in several food products without pre-concentration (Zhang et al. 2011). Enzymatic biosensor (formaldehyde dehydrogenase) was developed for the detection of formaldehyde in Indian Mackeral (Rastrelliger kanagurta) fish (Marzuki et al. 2012) and in fruit juices (Kundu et al. 2019) A non-enzymatic electrochemical biosensor was designed using CdS NPs for the detection of formalin in fish sample solutions (Baabu et al. 2020). Au NPs/polypyrrole composite was synthesized and applied for sensing of formalin by Differential Pulse Voltametry (DPV) method (Xi et al. 2020).

7.2.3 Electrochemical Detection of Synthetic Food Dyes in Soft Drinks

Synthetic dyes are extensively added into different types of food items and beverages in order to attract the consumers and promote the business. The main synthetic food dyes are triphenylmethane, xanthene, indigotine, quinolone, and azo dyes. The synthetic dyes are toxic and carcinogenic which cause asthma, contact anaphylaxis, immunosuppression, angioneurotic edema, and contact urticaria.

Simultaneous determination of metanil yellow and fast green in water and juice samples was performed by DPV technique using calixarene/AuNPs/GCE as a sensing platform (Shah 2020). Commonly Sudan I, II, and III dyes are illegally added as food colorants. They are quantified by using electrochemical sensors in food samples especially in chili and ketchup sauce (Yao et al. 2016; Heydari et al. 2019). A silver solid amalgam and Co_3O_4 NPs/graphene-based electrochemical sensors were developed for the determination of amaranth and allura red, respectively (Tvorynska et al. 2019; Jing et al. 2017). The CdO/rGO nanocomposite/1, 3-dipropylimidazolium bromide-based electrochemical sensor was reported for determination of Food Red 17 in the presence of tartrazine in washing liquid, ice creams, and jelly powder samples (Misaghpour and Shabani-Nooshabadi 2018). Yun et al. (2015) developed electrochemical biosensor to detect Orange II in chili sauce and ketchup samples by using ErGO grafted with 5-amino-1,3,4-thiadiazole-2-thiol-Pt composite. The sensitive metal oxides electrochemical sensors were developed for analysis of Sunset Yellow and Carmoisine in the presence of tartrazine in soft drink samples and dried fruit (Chen et al. 2013; Bijad et al. 2018).

7.3 Electrochemical Sensors for Unintentional Adulteration

7.3.1 Electrochemical Sensing of Pesticides, Insecticides, and Fungicides

In agriculture, as mentioned in the introduction many pesticides are used to protect crops from pests and insects. The sprayed pesticides stay in the tissue area of fruits, vegetables, and also mix with the soil and groundwater. These get bio-accumulated in the living organisms leading to induce oxidative stress and damage the central nervous system.

The nanomaterials play an important role in the development of electrochemical sensor for the analysis of pesticides in food products. For example, GOS@$CuFeS_2$ nanocomposite film SPE electrochemical sensor detected selectively methyl paraoxon in vegetable samples (Rajaji et al. 2019). Similarly, silver-doped ZnO nanorods and nanostructured diatom-ZrO_2 composite were applied as sensing material in the development of electrochemical sensors for carbamoate and methyl parathion, respectively, in foodstuffs and environmental samples (Wang et al. 2020; Gannavarapu et al. 2019). Further, NiO and MnO_2 NPs-based electrochemical sensors were developed for the analysis of carbofuran and organophosphates, respectively, in vegetables and beverages (Baksh et al. 2020;Ravi et al. 2020; Thangarasu et al. 2019). Carbendazim and methyl parathion were simultaneously detected by nanoporous gold-driven electrochemical sensor in water samples (Gao et al. 2019).

Electrochemically synthesized porous polymer-based materials were also applied for development of electrochemical sensors for the detection of para-nitrophenol and fenuron (Arulraj et al. 2015; Abraham and Vasantha 2020). 3D SWCNT-BODIPY hybrid material was used for ultrasensitive detection of eserine (Şenocak et al. 2019). Highly porous nanomaterials CNT and metal-organic frameworks have been utilized in the fabrication of electrochemical sensors for determination of different pesticides like malathion and total pesticides in vegetables (Al'Abri et al. 2019;Bakytkarim et al. 2019; Bhardwaj et al. 2020). Brilliant blue dye was used to modify the working electrode and applied for the determination of Triazophos in vegetables (Shalini Devi et al. 2018). MIP-based electrochemical sensor was reported for malathion in olive fruits and oils (Aghoutane et al. 2020). Yola and Atar (2017) designed an electrochemical sensor using ATZ-polyphenol MIP/Pt NPs/C_3N_4 NTs film, for the detection of atrazine and its LOD is calculated as 15 pg/L (Fig. 7.5).

7.3.2 Electrochemical Detection of Veterinary Drug Residues

Cows, goats, pigs, chickens, etc. are reared on farms for milk, meat, and eggs. Such animals are given large doses of veterinary drugs, either through food or injection, to protect them from disease or to increase body weight. Eating the meat of these animals allows veterinary drugs to accidentally enter the human body and create gray baby syndrome, leukemia, aplastic anemia, etc. (Zhaoling et al. 2014).

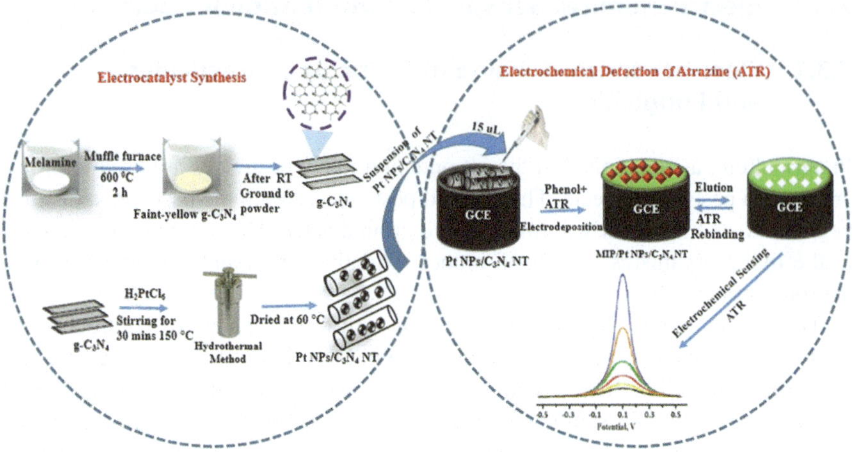

Fig. 7.5 Schematic illustration of electrochemical detection of atrazine using molecular imprinting electrode

MIP materials play a crucial role in the development of electrochemical sensor for veterinary drugs. For example, toltrazuril, streptomycin, chloramphenicol, neomycin, and olaquindox were detected electrochemically by using MIP/nanomaterial composite in chicken muscle, eggs, milk, and honey (Huang et al. 2019; Liu et al. 2013; Yang and Zhao 2015; Lian et al. 2013; Bai et al. 2020). Kanamycin is an aminoglycoside antibiotic drug which can cause a variety of side effects such as ototoxicity, nephrotoxicity, antibiotic resistance, and allergic reactions during intake of other drugs. Li and Wang et al. (2017) have developed the aptasensor with dsDNA for rapid and highly sensitive detection of kanamycin in milk. Another research group has developed a carbon nitride nanotubes/ionic liquid nanohybrid-based electrochemical sensor for sensing of ractopamine in the presence of other β-agonists (Mert et al. 2018).

7.3.3 Electrochemical Sensing of Pathogenic Bacteria and Fungus in Food Samples

Human beings are vulnerable to microbial toxins which are carcinogenic, neurotoxic, and hepatotoxic (Hoffmann and Anekwe 2013). To overcome the demerits of conventional methods, researchers urge to develop rapid, selective, and miniaturized sensors.

Aflatoxins (AFB_1) causes vomiting, abdominal convulsion, and teratogenic effect to animal and humans. Tang et al. (2009) developed a multifunctional and reusable immunoassay using magnetic composites of $SiO_{2(shell)}$-$CoFe_2O_{4(core)}$/ITO to detect AFB_1 with LOD of 6.0 pg/mL in red paprika, and the doped Prussian blue nanoparticles was mediator. Since, AFB_1 is accidently adulterated in milk and

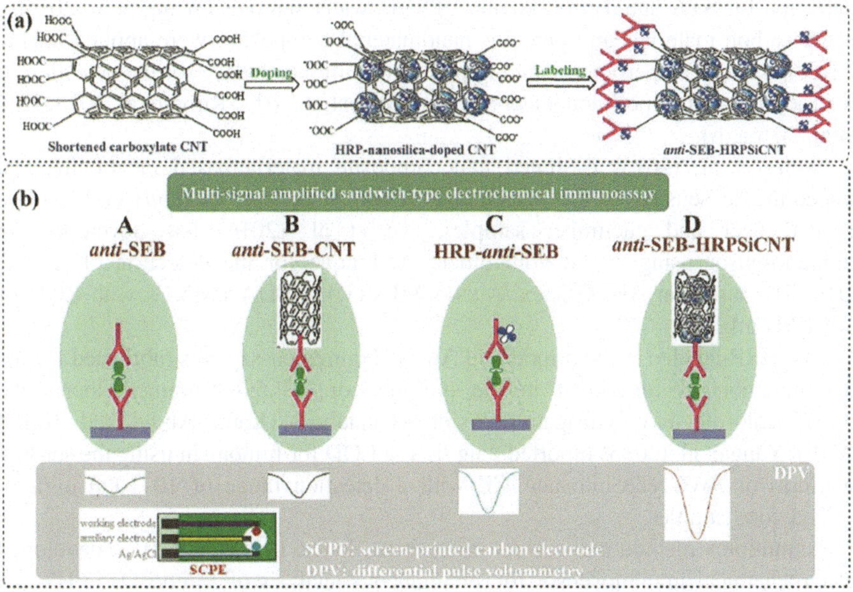

Fig. 7.6 (**a**) Preparation and functionalization of HRPSiCNTs and (**b**) Multisignal amplification and comparison of the sandwich-type electrochemical immunoassay with various immunoassay protocols (Reprinted with permission from (Tang et al. (2010), J. Agric. Food Chem. 58(20);10824-10830. Copyright (2010) American Chemical Society)

meat samples frequently, many research groups focused and developed a impedimetric-based miniaturized biosensor using nanomaterials composite and applied them for the analysis of AFB_1 (Ma et al. 2016). Further, DPV was used for sensing of AFB_1 in fg and proved to be better than ELISA (Wang et al. 2015a, b).

Botulism is considered as neurotic bioweapon produced by *Clostridium botulinum*. Therefore, some research groups studied electrochemical sensing of botulinum using nanocomposite as sensing material in milk and orange juice (Narayanan et al. 2015). Afkhami et al. (2017) developed a highly sensitive impedimetric sensor using GCE/COOH/Au-Gr-Cs materials and detected the botulinum neurotoxin serotype A.

Staphylococcal enterotoxin B (SEB) exhibits its adverse effect with vomiting, abdominal convulsion, and massacre within 2–6 h. Tang et al. (2010) has developed a HRP-labeled pulse technique for the detection of SEB, and its LOD was calculated as 0.1 ng/L in milk and juices of watermelon, apple, and soy milk (Fig. 7.6), whereas Arun et al. (2016) has reported ascorbic acid 2-phosphate-labeled miniaturized pulse voltammetry immunoassay for the detection of SEB.

An enzyme label-free electrochemical aptamer-based sensor was fabricated by Xiaoyan et al. (2019) for the determination of saxitoxin (STX) in shell fish. Bratakou et al. (2016) reported N-methyl-pyrrolidone-labeled polymer-stabilized lipid film-based miniaturized potentiometric biosensor for STX toxins in lake water. A label-free impedimetric immunoassay has also been designed for real-time simultaneous detection for STX and Tetrodotoxin (TTX) in water samples (Wang et al. 2015b). Nano carbon materials and precious nanomaterial composites were applied for the development of amperometric and impedimetric immunoassays for the detection of microcystein-LR in polluted water samples (Wei et al. 2011; Ruiyi et al. 2013; Hou et al. 2016a, b).

Wang et al. (2013) designed nanocomposite material-based electrochemical impedimetric sensor for the detection of *Escherichia coli* (*E. coli*) O157:H7 in ground beef and cucumber samples. Xu et al. (2016) has developed an immunosensor using cyclic voltammetry technique for the detection of *E. coli* O157:H7 based on ABs/GOxext/AuNPs/MBs-GOx@PDA PMNCs with LOD of 10^2 CFU/mL.

The HRP-labeled AuNPs-modified Au-SPE immunoassay was fabricated for the determination of fumonisin in the presence of 3,3′,5,5′-Tetramethylbenzidine dihydrochloride and hydrogen peroxide as catalysts (Kadir Abdul and Tothill 2010). Yang et al. (2015) reported 2 ng /L as a LOD for fumonisin using the sensing platform of SWCNT/Chitosan/GCE with a detection range of 10–1000 μg/L in spiked corn samples.

A multiplex method reduces detection time. Hence, Dou et al. (2013) developed electrochemical immunosensor for simultaneous analysis of *E. coli* O157:H7 and *Enterobacter sakazakii* detection using 4-Channel SPE. Balaji Viswanath et al. (2018) designed an enzyme-free multiplex electrochemical sensor to simultaneous detection for the *P. aeruginosa (Ps)* and *A. hydrophila (Ah)*. Thionine-labeled-*Ah* and Ferrocene-labeled-*Ps* are used as signaling molecules on the ZIF-8/AuNPs-modified platform.

7.3.4 Detection of Heavy Metal Ions Using Electrochemical-Miniaturized Sensors

Heavy metal ions (HMI) are classified based on their density (5×10^3 kg/m^3) and toxicity, for example, As, Cd, Pb, Cr, and Hg. HMI are released from mines, industries, and refineries wastes. They are causing environmental and biological dysfunction due to their polluting and toxic nature (Bansod et al. 2017). The WHO (2011) provides guideline values for metal ions as 10 ppb, 10 ppm, 50 ppm, 3 ppm, and 6 ppm for As^{3+}, Pb^{2+}, Cr^{6+}, Cd^{2+}, and Hg^{2+} ions, respectively.

HMI interacts with functional groups of proteins, DNA, enzymes, etc. and causes irreversible damage for enzymes and proteins (Zhou et al. 2016; Munir et al. 2018). The unethical amassment of HMI in the telluric and aquatic ecosystem led to dreadful events such as Minamata and Itai-Itai diseases during 1940s in Japan (Bansod et al. 2017; Li et al. 2018). The As^{3+} ions have been identified as a serious threat in the drinking water in various countries such as Bangladesh, India, China, Taiwan, and Canada. It causes cancer in the kidney, urinary bladder, liver, skin, and lungs (WHO Guidelines 2011; www.who.int).

Inductive coupled plasma-mass spectrometry, inductively coupled plasma optical emission spectrometry, atomic absorption spectroscopy, and electrochemical-inductively coupled plasma-mass spectrometry methods have been used to detect the HMI (Aline and Clésia 2013; Yukirda et al. 2018). These methods are all laborious and complicated. To overcome these problems, electrochemical sensors with unique characteristics such as rapidity, affordability, user-friendly and portability should be developed (Wei et al. 2012).

In earlier times, HMI was detected using a hanging mercury drop electrode (HMDE). This provided a 5- to 10-fold signal increase when compared to other electrochemical techniques due to the amalgamation with analytes (Rodrigues et al. 2011). However, this HMDE was avoided due to its toxicity and tedious handling (Xia et al. 2018). Hence, the researchers developed modified electrodes with environment-friendly materials for the detection of more than one HMI simultaneously. For example, the Au electrodes embedded in miniaturized settings are used to detect As^{3+} ions in food sources (Walsh et al. 2010). The AuNPs decorated on Fe_3O_4 nanosphere materials was used to detect As^{3+} ions electrochemically in the presence of room temperature ionic liquid. The Fe_3O_4 nanosphere attributed a high loading of AuNPs (3–9 nm) and also enhanced the sensitivity of 458.66 μA ppb^{-1} cm^{-2} (Wei et al. 2016). Mani et al. (2020) reported sensors for Roxarsone using $SrWO_4$ NPs/GrO based nanocomposite as sensing materials with high sensitivity. Yukirda et al. (2018) synthesized ZnO@Graphene composite with ZnO intercalation to detect both Cd^{2+} and Pb^{2+} ions. The anodic peak current response was 1.5 and 4 folds higher than those of graphene-modified electrodes. A research group reported GO-MWCNTs hybrid nanocomposites catalyst based on simultaneous detection of Pb^{2+} and Cd^{2+} ions (Huang et al. 2014). Luong et al. (2014) stated an overview of the impact of As^{3+} ions and its various detection methods. Cui et al. (2016), developed a label-free signal-on aptasensor to detect the As^{3+} by impedance analysis.

Metal-organic framework has unique properties such as high thermal stability and tuneable with high surface area. Hence, UiO-66-NH_2@PANI-modified electrochemical sensor was developed for the detection of Cd^{2+} ions in the lake and tap water (Wang et al. 2017a, b). A simple and inexpensive electrochemically reduced p-nitrobenzoic acid film-modified differential pulse anodic stripping voltammetry sensor was developed for the simultaneous detection of Zn^{2+}, Cd^{2+} , Pb^{2+}, and Hg^{2+} ions in tap water with sensitivity towards Hg^{2+} ions up to 240 pM (Arulraj et al. 2014). A femtomolar detection of Hg^{2+} ions could be achieved by Arulraj et al. (2016) using PPy/Pct/GR as sensing materials which is better than G/PANI/PS nanoporous fiber/SPCE and PA/PPy/GO/GCE for the detection of Hg^{2+} ions in river and tap water (Promphet et al. 2015; Dai et al. 2016). Rattanarat et al. (2014) developed a miniaturized paper-based sensor which simultaneously detected Cd^{2+} and Pb^{2+} ions using anodic stripping square wave voltammetry and Fe^{2+}, Ni^{2+}, Cr^{6+}, and Cu^{2+} ions using colorimetry (Fig. 7.7).

Xuan et al. (2018) proposed a Bi/TRGO/Au-modified micro-patterned device with the fully integrated electrochemical sensor in a miniaturized and portable manner to detect the Pb^{2+} and Cd^{2+} ions. Wang et al. (2011) fabricated amine-functionalized graphene oxide (AGO) electrochemical sensor for selective analysis of Pb^{2+} ions in water samples.

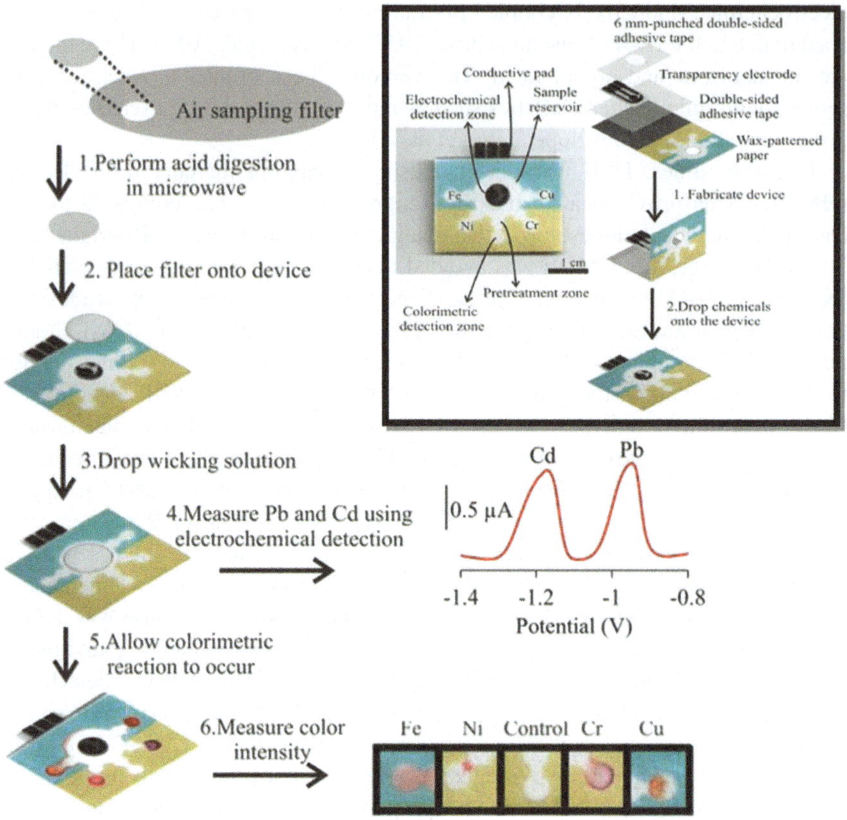

Fig. 7.7 Analytical procedure for metal assays using the PADs. The assay requires three steps: (i) microwave-assisted acid digestion, (ii) anodic stripping voltammetry detection of Cd^{2+} and Pb^{2} ions, (iii) colorimetric detection of Fe^{2+}, Ni^{2+}, Cr^{6+}, and Cu^{2+} ions. Protocols (Reprinted with permission from (Rattanarat et al. (2014), Anal. Chem. 86;3555-3562). Copyright (2014) American Chemical Society)

This book chapter deals about adulterant and their classification and also about electrochemical sensors for the detection of intentional and unintentional adulterants. Based on the overview of the above content, the overall observations, challenges, and future recommendations are given as follows:

Observations

- The molecularly imprinted polymer methodology is exploited for the sensing of melamine in milk samples. Here, MIP-MWCNTs nanohybrids/GCE-based electrochemical sensor showed lowest LOD among earlier reports.
- Polymer-stabilized metal oxide nanoparticle composites are mostly used for the detection of urea in milk without enzyme. The report on urease-based sensor is limited.

- Conducting polymer/nano metal oxide composites have been utilized for the detection of H_2O_2.
- Labeled DNA-based immunoassays have been designed and applied for the detection of meat adulteration in meat products and pathogens in food samples.
- Conducting polymer-stabilized metal and metal sulfide-based nanocomposites are showing good performance towards formalin.
- Metal or metal oxide nanoparticles/carbon nanomaterial composites are widely used for the detection of synthetic dye adulterants.
- ZnO or ZrO_2 or MnO_2 or NiO stabilized with conducting polymers-MIP technology have been used for the detection of veterinary drugs, heavy metals, and pesticides in vegetables, fruits, meat, and milk.

Challenges to be Addressed

- No single catalytic material is used as sensing element in the development of electrochemical sensors for the above adulterants.
- Electrode modification needs vast experience in catalyst selection and modification methodology.
- Since, nanomaterials are used for the fabrication of the sensors; retaining stability of the sensors is crucial step.
- The aptamers or DNA or antibodies are utilized in the development of electrochemical immunoassays. Therefore, they are not cost-effective.
- Almost all the electrochemical sensor is applied to demonstrate above adulterants in spiked samples not tested in real condition.

Future Recommendation

- Generally, the adulterants create dreadful diseases like cancers, dysfunction of kidney, immunosuppression, asthma, etc. Determination and prevention of these adulterants in food stuffs are essential to maintain a healthy life.
- Portable, affordable, and reliable analytical tools for testing of food adulterants are very much essential in day-to-day life.
- The research findings should not be stopped in the publication level. They should be transformed into miniaturized and affordable devices to ensure the quality of food stuffs.
- In order to take all the above scientific findings to common people, collaboration with electronic engineers and medical doctors becomes indispensable (Table 7.1).

Table 7.1 Miniaturized electrochemical sensors for the detection of food adulterants

Adulterants	Food samples	Sensing platform	LOD	Linear range	Reference
Melamine	Raw milk and milk powder	MIP-PGN-pPdCu/GCE	2 nM	0.01–1 μM	Shang et al. (2014)
	Milk	MIP-MWCNTs Nanohybrids/GCE	0.56 pM/L	1.0 pM–1 μM	Xu et al. (2018)
	Infant milk	MIP-PANI-PAA/GCE	0.0573 and 0.0172 nM	0.1–180 nM	Regasa et al. (2020)
	Infant milk	P-Arg/ErGO–CuNFs/GCE	5.0 nM	10–90 nM	Daizy et al. (2019)
Melamine and urea	Milk	ZnO/AChE/Chitosan/Pt	3 pM and 1 pM	1–20 nM	Ezhilan et al. (2017)
Urea	Milk	Urease/Fe_3O_4/MWCNT/PANI-NF/GCE	67 μM	1–25 mM	Singh et al. (2020)
	Milk	AgNPs coated/Glucose strip	0.14 mM	1–8 mM	Liu et al. (2020)
	Milk	Urease/PAN Membrane electrode	0.3 mM	1–100 mM	Ramesh et al. (2015)
H_2O_2	Milk	PPy/DBS/CeO_2/Pt	0.12 fM	0.05 mM–0.5 fM	Karimi et al. (2018)
	Milk	PEDOT/PB/GCE	0.16 μM	0.5–839 μM	Wang et al. (2017a, b)
Pork meat	Fresh meat	4-carboxyphenyl diazonium/CNF-SPE	0.9 pg/mL	0.5–500 pg/mL	Lim and Ahmed (2016)
	Raw and processed meat	AuNPs-DNA probe/SPCE-Au surface	0.58 μg/mL	0.1–5 μg/mL	Hartati et al. (2019)
	Raw and cooked meat	PPy/IgG film/GPE	0.01 %	1–100%	Mandli et al. (2018)

Horse meat	Beef meat	Anti-DNA-RNA/ProtA-HRP40/SPE	0.12 pM	–	Montiel et al. (2017)
Cattle meat	Various animals and birds meat	Methylene blue-P-labeled species-specific DNA sequences/ITO	8.2 fM	0.01–100 pM	Zhang et al. (2020a, b)
Formaldehyde	Fish	FDH/NF/Au electrode	0.016 ppm	0.1–10 ppm	Marzuki et al. (2012)
	Orange juice	CNT-Fe_3O_4	0.05 mg/L	0.05–0.5 mg/L	Kundu et al. (2019)
	Fish	CdS/Chitosan/Pt	5 mg/L	5–50 mg/L	Baabu et al. (2020)
	Water and milk	Au@PPY composite/GCE	20 µM and 0.5 mM	–	Xi et al. (2020)
Metanil yellow and fast green	Water and juice	Calix8/AuNPs/GCE	9.8 & 19.7 nM	0.05–45 µM	Shah et al. (2020)
Sudan I	Ketchup and chili powder	Au-CuNP/rGO/GCE	0.4 nM	1 nM–10 µM	Yao et al. (2016)
Sudan II and III	Ketchup and chili powder	ZnO NPs/CPE	1.87 and 2.62 nM	0.05–20 and 0.05–15 µM	Heydari et al. (2019)
Amaranth and allura red	Soft drinks	Silver solid amalgam electrode	2.1 and 3.4 nM	4–8.9 and 8–6 nM	Tyorynska et al. (2019)
Amaranth	Soft drinks	PSS-GN/CO_3O_4/GCE	4 nM	0.01–6 nM	Jing et al. (2017)
Food red 17	Ice cream and jelly	CdS NP/rGO/1,3DPZBr/CPE	0.1 nM	4 nM–8 µM	Misaghpour et al. (2018)
Orange II	Chili sauce and ketchup	ERGO-ATDT-Pt/GCE	3.4 pM	0.01 µM–0.6 µM	Yun et al. (2015)
Sunset Yellow	Soft Drinks	Alumina microfibre/CPE	0.16 nM	0.5–100 nM	Chen et al. (2013)

(continued)

Table 7.1 (continued)

Adulterants	Food samples	Sensing platform	LOD	Linear range	Reference
Carmoisine and tartrazine	Dried fruit and soft drink	CNTs/NiO/1-M-3BIBr /CPE	20 nM and 0.06 μM	70–650 and 0.1–750 μM	Bijad et al. (2018)
Methyl paraoxon	Vegetables	GOS@$CuFeS_2$ NC film/SPCE	4.5 nM	0.073–801.5 μM	Rajaji et al. (2019)
Carbamate	Food	Ag-doped ZnO/GO/GCE	0.34 nM	–	Wang et al. (2020)
Methyl parathion	Water	Nanoporous Si-ZrO_2 composite/GCE	54.3 pM	3.4 nM–64 μM	Gannavarapu et al. (2019)
Carbofuran	Vegetables	NiO NPs/GCE	0.5 μM	5.0–305 μM	Baksh et al. (2020)
4-Nitrophenyl phosphate disodium salt hexahydrate	Vegetables and water	4-NPP/MnO_2/Pt	10 nM	100 and 900 nM	Ravi et al. (2020)
Methyl parathion	Water	MnO_2/PANI/rGO/GCE	7.4 nM	0.5–50 μM	Thangarasu et al. (2019)
Carbendazim and methyl parathion	Water	NPG/GCE	0.24 and 0.02 mM	3–120 and 0.5–150 mM	Gao et al. (2019)
Fenuron	–	Hollow PPy/AAS	5 nM	10 nM–110 μM	Abraham et al. (2020)
Para-nitrophenol	Water	ENPPy/SDS film/GCE	0.1 nM	0.1 nM –100 mM	Arulraj et al. (2015)
Malathion	Olive fruits and oil	Polyacrylamide-malathion/Au-SPE	0.06 pg/mL	1–1000 pg/mL	Aghoutane et al. (2020)
Eserine	Orange juice	3D SWCNT-BODIPY hybrid/GCE	160 and 528 nM	0.25–2.25 μM	Şenocak et al. (2019)
Malathion	Vegetables	BTCA-P-Cu-CP/CPE	0.17 nM	0.6–24 nM	Al'Abri et al. (2019)

Parathion	Vegetables	CS-MWCNTs@SiC NPs/PVC	20 ng/mL	–	Bakytkarim et al. (2019)
Various pesticides	Vegetables	BSA/Chi-AuNP-rIgG-BSA/MOF/ITO	4 ng/L	1–100 ng/L	Bhardwaj et al. (2020)
Triazophos	Vegetables	CoomBB@nano CB/GCE	119 pg/20 μL	16–632 ng/20 μL	Shalini Devi et al. (2018)
Toltrazuril	Chicken muscle and eggs	TZR-CM-β-CD/TiO_2/rGO/Pt	0.21 μg/L	0.43–42.54 μg/L	Huang et al. (2019)
Streptomycin	Milk and honey	MB /PVP/Au-STR-OPD/MIP	10 pg/mL	0.05–20 ng/ mL	Liu et al. (2013)
Chloramphenicol	Milk and Honey	MWCNTs@MIP/CKM-3/P-r-GO/GCE	0.1 nM/L	5.0 nM–0.5 μM/L	Yang and Zhao (2015)
Neomycin	Milk	PPy-Neomycin/GR-MWCNTs/CS-SNP/Au	7.63 nM/L	9 nM/L–7 μM/L	Lian et al. (2013)
Olaquindox	Fish and feedstuffs	OLA-PPy/DGr/GCE	7.5 nM/L	50 nM/L–500 nM/L	Bai et al. (2020)
Kanamycin	Milk	dsDNA/CdS NPs/AuNPs/GE	2.85 nM	10.0–450.0 nM	Li and Wang (2017)
Ractopamine	–	C_3N_4NTs/ILs/CPE in the presence of β-agonists	0.1 pM	1 nM–1 pM	Mert et al. (2018)
Aflatoxin B_1	Red paprika	$SiO_{2(shell)}$-$CoFe_2O_{4(core)}$)/ITO	6.0 pg/mL	0.05–12 ng/mL	Tang et al. (2009)
	Bee pollen	Ab/SGIL/GCE	0.01 ng/mL	0.1–10 ng/mL	Zaijun et al. (2010)
	Pea nut	pAb-MSC–Thi-Nf-GCE	0.3 ng/mL	0.01–20 ng/mL	Lin et al. (2015)
	Corn	PPy/PPa/rGO	10 fg/mL	10 pg/mL–10 fg/mL	Wang et al. (2015a)
	Maize	CHI-AuNPs	0.06 ng/mL	1–30 ng/mL	Ma et al. (2016)

(continued)

Table 7.1 (continued)

Adulterants	Food samples	Sensing platform	LOD	Linear range	Reference
Botulinum	Milk	Covalent AuNPs/GCE	1 pg/mL	4–35 pg/mL	1Liu et al. (2014)
	Orange juice	GNS/Ph–Modified-Au $PhNH_2$/GCE	5 pg/mL	0 pg/mL–10 ng/mL	Narayanan et al. (2015)
	Milk	GCE/COOH/Au-Gr-Cs/MPA/Ab/BSA	0.11 pg/mL	0.27–268 pg/mL	Afkhami et al. (2017)
E. coli	Food	PB-modified SP-IDME	190 CFU/g	10^3–10^6 CFU/g	Xu et al. (2016)
	Water	Au/ssDNA+MCH/cDNA	0.16 CFU/mL	5×10^5 CFU/ml	Yang et al. (2016)
	Beef and cucumber	Rgo/AuNPs	1.5×10^4 and 1.5×10^3 CFU/mL	1.5×10^2–1.5×10^7 CFU/mL	Wang et al. (2013)
	Milk protein	MWCNT/SA/CMC/SPCE	4.57×10^3 CFU/mL	10^4–10^{10} CFU/mL	Dou et al. (2013)
Ps and *Ah*	Fish	Au-Fc-anti Ps /Ps/BSA/antiPs/AuNPs/ZIF8/GCE and Au-Thi-anti-Ah Ah/BSA/antiAh/AuNPs/ZIF-8/GCE	5.60 and 8.053 CFU/mL	10^1–10^3 and 10^1–10 CFU/mL	Balaji Viswanath et al. (2018)
Microcystein-LR	Drinking water	SWCNT-PSS	10–40 nmol/L	0.6 nmol/L	Wang et al. (2009)
	Polluted water	Ab_2PtRu-Ag/Ab_1/GS/GCE	9.63 pg /mL	0.01–28 ng/mL	Wei et al. (2011)
	Water	AuNP-polyDPB-G-AuNP/GCE	0.037 pM	0.01 × pM–0.08 pM	Ruiyi et al. (2013)
	Fortified water	AuNPs/GCE	0.004 μg/L.	10–1000 μg/L	Hou et al. (2016a, b)

Saxitoxin and tetrotoxin	Water	Au Electrode	0.087 ng/mL	0.6–1.3 2.3–4.2 ng/mL	Wang et al. (2015a)
Saxitoxin	Shell fish and lake water	Lipid-polymer-based graphene electrode	1 nM	1 nM–1 μM	Bratakou et al. (2016)
	Shell fish	Au/ODT/MWCNT/MB/aptamer	0.38 nM	0.9–30 nM	Hou et al. (2016a, b)
	Sea water	Au-SPE	0.92 nM	1–400 nM	Xiaoyan et al. (2019)
SEB	Milk and Juices from watermelon, soymilk apple and pork.	Anti-SEB-HRPSiCNTs/SPE	10 pg/mL	0.05–15 ng/mL	Tang et al. (2010)
	Skim Milk Milk and Rice Powder	Au electrode	0.17 ng/mL	0.5–500 ng/mL	Xiong et al. (2018)
	Water	AA-2P-AntiSEB/SEB/Anti-SEB/SPE	100 ng/mL	100–100,000 ng/mL	Arun et al. (2020)
Hg^{2+}	Tap water	ERpNBA/GCE	240 pM	5–500 nM 0.5–15.5 μM	Arulraj et al. (2016)
Pb^{2+} and Cd^{2+}	Water	Bi/K_3 $[Fe(CN_6)]$/SPE	0.25 ng 0.25 ng	5–150 μg/L 5–150 μg/L	Rattanarat et al. (2014)
Cd^{2+}, Pb^{2+} and Cu^{2+}	Water	GNS/GCE	10 pM 10 pM 100 nM	NA	Bin et al. (2011)
Pb^{2+}	Water	AGOs/GCE	0.1 pM	0.5–50 μM	Wang et al. (2011)
Cd^{2+}	Lake and Tap water	UiO-66-NH_2@PANI/GCE	0.3 μg L^{-1}	0.5–600 μg/L	Wang et al. (2017a, b)
	Water	AuNPs@Fe_3O_4/SPCE	0.0022 ppb	0.1–1 ppb	Wei et al. (2016)
Cd^{2+} and Pb^{2+}	Laoshan honey and Wuxi rice	L-cys/GR–CS/GCE	0.45 & 0.12 μg/L	0.56–67.2 1.04–62.1 μg/L	Zhou et al. (2016)

(continued)

Table 7.1 (continued)

Adulterants	Food samples	Sensing platform	LOD	Linear range	Reference
As^{3+}	Waters from India and UK	Au- Micro wire electrode	0.5 nM	1 nM–100 μM	Walsh et al. (2010)
	Lake and Tap water	Ars-2/AuNPs/SPCE	0.15 nM	0.2–100 nM	Cui et al. (2016)
	Chicken	$SrWO_4$ NPs/GrO/GCE	22.5 nM	35 nM–1816.5 μM	Mani et al. (2020)

References

Abraham DA, Vasantha VS (2020) Hollow Polypyrrole composite synthesis for detection of trace-level toxic herbicide. ACS Omega 5(34):21458–21467

Afkhami A, Hashemi P, Bagheri H, Salimian J, Ahmadi A, Madrakian T (2017) Impedimetric immunosensor for the label-free and direct detection of botulinum neurotoxin serotype A using Au nanoparticles/graphene chitosan composite. Biosens Bioelectron 93:124–131

Aghoutane Y, Diouf A, Österlund BB, El Bari N (2020) Development of a molecularly imprinted polymer electrochemical sensor and its application for sensitive detection and determination of malathion in olive fruits and oils. Bioelectrochemistry 132:107404

Al'Abri AM, Abdul Halim SN, Abu Bakar NK, Saharin SM, Sherino B, Rashidi Nodeh H, Mohamad S (2019) Highly sensitive and selective determination of malathion in vegetable extracts by an electrochemical sensor based on Cu-metal organic framework. J Environ Sci Health B 54(12):930–941

Arulraj AD, Vijayan M, Samseya J, Vasantha VS (2014) A simple and highly sensitive electrochemically reduced p-nitrobenzoic acid film modified sensor for determination of mercury. Electroanalysis 26:2773–2782

Arulraj AD, Vijayan M, Vasantha VS (2015) Highly selective and sensitive simple sensor based on electrochemically treated nanopolypyrrole-sodium dodecyl sulphate film for the detection of para-nitrophenol. Anal Chim Acta 899:66–74

Arulraj AD, Devasenathipathy R, Chen SM, Vasantha VS, Wang SF (2016) Femtomolar detection of mercuric ions using polypyrrole, pectin and graphene nanocomposites modified electrode. J Colloid Interface Sci 483:268–274

Arun S, Rao VK, Kamboj DV, Gaur R, Shaik M, Shrivastava AR (2016) Enzyme free detection of staphylococcal enterotoxin B (SEB) using ferrocene carboxylic acid labeled monoclonal antibodies: an electrochemical approach. New J Chem 40:8334–8341

Arun S, Vepa KR, Dev VK (2020) Electrochemical immunosensor for the detection of Staphylococcal Enterotoxin B using screen-printed electrodes. Indian J Chem 59A:174–180

Azad T, Ahmed S (2016) Common milk adulteration and their detection techniques. Int J Food Contam 3(1):1–9

Baabu PRS, Srinivasan P, Kulandaisamy AJ, Robinson J, Geevaretnam J, Rayappan JB (2020) A non-enzymatic electrochemical biosensor for the detection of formalin levels in fishes: realization of a novel comparator effect based on electrolyte. Anal Chim Acta 1139:50–58

Bai X, Zhang B, Liu M, Hu X, Fang G, Wang S (2020) Molecularly imprinted electrochemical sensor based on polypyrrole/dopamine@ graphene incorporated with surface molecularly imprinted polymers thin film for recognition of olaquindox. Bioelectrochemistry 132:107398

Baksh H, Buledi JA, Khand NH, Solangi AR, Mallah A, Sherazi ST, Abro MI (2020) Ultra-selective determination of carbofuran by electrochemical sensor based on nickel oxide nanoparticles stabilized by ionic liquid. MonatsheftefürChemie-Chem Monthly 151(11): 1689–1696

Bakytkarim Y, Tursynbolat S, Zeng Q, Huang J, Wang L (2019) Nanomaterial ink for on-site painted sensor on studies of the electrochemical detection of organophosphorus pesticide residuals of supermarket vegetables. J Electroanal Chem 841:45–50

Balaji Viswanath K, Krithiga N, Jayachitra A, Sheik Mideen AK, Amali AJ, Vasantha VS (2018) Enzyme-free multiplex detection of pseudomonas aeruginosa and aeromonashydrophila with ferrocene and thionine-labeled antibodies using ZIF-8/Au NPs as a platform. ACS Omega 12: 17010–17022

Banerjee et al (2017) Recent advances in detection of food adulteration. Academic Press, Food Safety in the 21st Century, Public Health Perspective. pp. 145-160

Bansal S, Singh A, Mangal M, Mangal AK, Kumar S (2017) Food adulteration sources health risks and detection methods. Crit Rev Food Sci Nutr 57(6):1174–1189

Bansod B, Kumar T, Thakur R, Rana S, Singh I (2017) A review on various electrochemical techniques for heavy metal ions detection with different sensing platforms. Biosens Bioelectron 94:443–455

Bhardwaj R, Rao RP, Mukherjee I, Agrawal PK, Basu T, Bharadwaj LM (2020) Layered construction of nanoimmuno-hybrid embedded MOF as an electrochemical sensor for rapid quantification of total pesticides load in vegetable extract. J Electroanal Chem 873:114386

Bijad M, Karimi-Maleh H, Farsi M, Shahidi SA (2018) An electrochemical-amplified-platform based on the nanostructure voltammetric sensor for the determination of carmoisine in the presence of tartrazine in dried fruit and soft drink samples. J Food Meas Charact 12(1):634–640

Bin WA, Chang YH, Zhi LJ (2011) High yield production of graphene and its improved property in detecting heavy metal ions. New Carbon Mater 26:31–35

Bratakou S, Nikoleli GP, Siontorou CG, Nikolelis DP, Karapetis S, Tzamtzis N (2016) Development of an electrochemical biosensor for the rapid detection of saxitoxin based on air stable lipid films with incorporated anti-STX using graphene electrodes. Electroanalysis 29:1–9

Campuzano S, Montiel VRV, Torrente-Rodríguez M, Reviejo J, Pingarrón JM (2016) Electrochemical biosensors for food security: allergens and adulterants detection. In: Biosensors for security and bioterrorism applications. Springer, Cham, pp 287–307

Campuzano S, Montiel V-RV, Serafín V, Yáñez-Sedeño P, Pingarrón JM (2020) Cutting-edge advances in electrochemical affinity biosensing at different molecular level of emerging food allergens and adulterants. Biosensors 10(2):10

Chandra P, Koh WCA, Noh HB, Shim YB (2012) In vitro monitoring of i-NOS concentrations with an immunosensor: The inhibitory effect of endocrine disruptors on i-NOS release. Biosens Bioelectron. https://doi.org/10.1016/j.bios.2011.11.027

Chen X, Wu K, Sun Y, Song X (2013) Highly sensitive electrochemical sensor for sunset yellow based on the enhancement effect of alumina microfibers. Sens Actuators B 185:582–586

Choudhary M, Yadav P, Singh A, Kaur S, Ramirez-Vick J, Chandra P, Arora K, Singh SP (2016) CD 59 Targeted ultrasensitive electrochemical immunosensor for fast and noninvasive diagnosis of oral cancer. Electroanalysis 28:2565–2574. https://doi.org/10.1002/elan.201600238

Cui L, Wu J, Ju H (2016) Label-free signal-on aptasensor for sensitive electrochemical detection of arsenite. Biosens Bioelectron 79:861–865

Dai H, Wang N, Wang D, Ma H, Lin M (2016) An electrochemical sensor based on phytic acid functionalized polypyrrole/graphene oxide nanocomposites for simultaneous determination of Cd (II) and Pb (II). Chem Eng J 299:150–155

Daizy M, Tarafder C, Al-Mamun MR, Liu X, Aly Saad Aly M, Khan MZH (2019) Electrochemical detection of melamine by using reduced graphene oxide-copper nanoflowers modified glassy carbon electrode. ACS Omega 4(23):20324–20329

Deka S, Saxena V, Hasan A, Chandra P, Pandey LM (2018) Synthesis, characterization and in vitro analysis of α-Fe_2O_3-$GdFeO_3$ biphasic materials as therapeutic agent for magnetic hyperthermia applications. Mater Sci Eng C 92:932–941. https://doi.org/10.1016/j.msec.2018.07.042

Dou W, Tang W, Zhao G (2013) A disposable electrochemical immunosensor arrays using 4-channel, screen-printed carbon electrode for simultaneous detection of *Escherichia coli O157:H7* and Enterobactersakazakii. Electrochim Acta 97:79–85

Ezhilan M, Gumpu MB, Ramachandra BL, Nesakumar N, Babu KJ, Krishnan UM, Rayappan JBB (2017) Design and development of electrochemical biosensor for the simultaneous detection of melamine and urea in adulterated milk samples. Sens Actuators B 238:1283–1292

Gannavarapu KP, Ganesh V, Thakkar M, Mitra S, Dandamudi RB (2019) Nanostructured diatom-ZrO_2 composite as a selective and highly sensitive enzyme free electrochemical sensor for detection of methyl parathion. Sens Actuators B 288:611–617

Gao X, Gao Y, Bian C, Ma H, Liu H (2019) Electroactivenanoporous gold driven electrochemical sensor for the simultaneous detection of carbendazim and methyl parathion. Electrochim Acta 310:78–85

Guidelines for Drinking-Water Quality, (2011). 4th ed., World Health Organization, Switzerland

Hartati YW, Suryani AA, Agustina M, Gaffar S, Anggraeni A (2019) A gold nanoparticle-DNA bioconjugate–based electrochemical biosensor for detection of Sus scrofamt DNA in raw and processed meat. Food Anal Methods 12(11):2591–2600

Heydari M, Ghoreishi SM, Khoobi A (2019) Chemometrics-assisted determination of Sudan dyes using zinc oxide nanoparticle-based electrochemical sensor. Food Chem 283:68–72

Hoffmann S, Anekwe TD (2013) Making sense of recent cost-of-foodborne-illness estimates, EIB-118. US Department of Agriculture, Economic Research Service, Washington, DC

Hou L, Jiang L, Song Y, Ding Y, Zhang J, Wu X, Tang D (2016a) Amperometric aptasensor for saxitoxin using a gold electrode modified with carbon nanotubes on a self-assembled monolayer, and methylene blue as an electrochemical indicator probe. Microchim Acta 183:1971–1980

Hou L, Ding Y, Zhang L, Guo Y, Li M, Chen Z, Wu X (2016b) An ultrasensitive competitive immunosensor for impedimetric detection of microcystin-LR via antibody-conjugated enzymatic biocatalyticprecipitation. Sens Actuators B 23:363–370

Huang H, Chen T, Liu X, Ma H (2014) Ultrasensitive and simultaneous detection of heavy metal ions based on three-dimensional graphene-carbon nanotubes hybrid electrode materials. Anal Chim Acta 852:45–54

Huang X, Wei S, Yao S, Zhang H, He C, Cao J (2019) Development of molecularly imprinted electrochemical sensor with reduced graphene oxide and titanium dioxide enhanced performance for the detection of toltrazuril in chicken muscle and egg. J Pharm Biomed Anal 164: 607–614

Jing S, Zheng H, Zhao L, Qu L, Yu L (2017) Electrochemical sensor based on poly (sodium 4-styrenesulfonate) functionalized graphene and Co_3O_4 nanoparticle clusters for detection of amaranth in soft drinks. Food Anal Methods 10(9):3149–3157

Jun-shi C (2009) A worldwide food safety concern in 2008- melamine-contaminated infant formula in China caused urinary tract stone in 290,000 children in China. Chin Med J (Engl) 122(3): 243–244

Kadir Abdul MK, Tothill IE (2010) Development of an electrochemical immunosensor for fumonisins detection in foods. Toxins 2:382–398

Karimi A, Husain SW, Hosseini M, Azar PA, Ganjali MR (2018) Rapid and sensitive detection of hydrogen peroxide in milk by enzyme-free electrochemiluminescence sensor based on a polypyrrole-cerium oxide nanocomposite. Sens Actuators B 271:90–96

Khalil I, Yehye WA, Muhd Julkapli N, Ibn Sina AA, Rahmati S, Basiruna WJ, Seyfoddinf A (2020) Dual platform based sandwich assay surfaceenhanced Raman scattering DNA biosensor for the sensitive detection of food adulteration. Analyst 145:1414–1426

Kundu M, Bhardwaj H, Pandey K, Krishnan P, Kotnala R. K, Sumana G (2019) Development of electrochemical biosensor based on CNT–Fe_3O_4 nanocomposite to determine formaldehyde adulteration in orange juice. J Food Sci Technol 56(4): 1829-1840

Li F, Wang X, Sun X, Guo Y (2017) An aptasensor with dsDNA for rapid and highly sensitive detection of kanamycin in milk. RSC Adv 7(62):38981–38988

Li S, Zhang C, Wang S, Liu Q, Feng H, Ma X, Guo J (2018) Electrochemical microfluidics techniques for heavy metal ions detection. Analyst 143:4230–4246

Lian W, Liu S, Yu J, Li J, Cui M, Xu W, Huang J (2013) Electrochemical sensor using neomycin-imprinted film as recognition element based on chitosan-silver nanoparticles/graphene-multiwalled carbon nanotubes composites modified electrode. Biosens Bioelectron 44:70–76

Lim SA, Ahmed MU (2016) A label free electrochemical immunosensor for sensitive detection of porcine serum albumin as a marker for pork adulteration in raw meat. Food Chem 206:197–203

Lin Y, Lin Y, Tang D, Chen G, Tang D (2015) Simple and sensitive detection of aflatoxin B1 within five minute using a non-conventional competitive immunosensing mode. Biosens Bioelectron 74:680–686

Liu B, Tang D, Zhang B, Que X, Yang H, Chen G (2013) Au (III)-promoted magnetic molecularly imprinted polymer nanospheres for electrochemical determination of streptomycin residues in food. Biosens Bioelectron 41:551–556

Liu G, Zhang Y, Guo W (2014) Covalent functionalization of gold nanoparticles as electronic bridges and signal amplifiers towards an electrochemical immunosensor for botulinum neurotoxin typeA. Biosens Bioelectron 61:547–533

Liu J, Moakhar R. S, Perumal A. S, Roman H. N, Mahshid S, Wachsmann-Hogiu S (2020) An AgNP-deposited commercial electrochemistry test strip as a platform for urea detection. Sci Rep 10(1):1-11

Luong JH, Lam E, Male KB (2014) Recent advances in electrochemical detection of arsenic in drinking and groundwaters. Anal Methods 6:6157–6169

Ma H, Sun J, Zhang Y, Bian C, Xia S, Zhen T (2016) Label-free immunosensor based on one-step electrodeposition of chitosan-gold nanoparticles biocompatible film on Au microelectrode for determination of aflatoxinB_1 in maize. Biosens Bioelectron 80:222–229

Mahato K, Kumar S, Srivastava A, Maurya PK, Singh R, Chandra P (2018) Electrochemical immunosensors: fundamentals and applications in clinical diagnostics. In: Handbook of immunoassay technologies. https://doi.org/10.1016/B978-0-12-811762-0.00014-1

Mandli J, Fatimi IE, Seddaoui N, Amine A (2018) Enzyme immunoassay (ELISA/immunosensor) for a sensitive detection of pork adulteration in meat. Food Chem 255:380–389

Mani G, Rajaji U, Wang SF, Chang YJ, Ramalingam RJ, Chan CY (2020) Investigation of sonochemically synthesized sphere-like metal tungstate nanocrystals decorated activated carbon sheets network and its application towards highly sensitive detection of arsenic drug in biological samples. J Taiwan Inst Chem Eng 114:211–219

Mansouri M, Khalilzadeh B, Barzegari A, Shoeibi S, Isildak S, Bargahi N, Omidi Y, Dastmalchi S, Rashidi MR (2020a) Design a highly specific sequence for electrochemical evaluation of meat adulteration in cooked sausages. Biosens Bioelectron 150:111916

Mansouri M, Fathi F, Jalili R, Shoeibie S, Dastmalchie S, Khataee A, Rashidia MR (2020b) SPR enhanced DNA biosensor for sensitive detection of donkey meat adulteration. Food Chem 331: 127163

Marzuki N, Bakar FA, Salleh AB, Heng LY, Yusof NA, Siddiquee S (2012) Electrochemical biosensor immobilization of formaldehyde dehydrogenase with nafion for determination of formaldehyde from Indian mackerel (*Rastrelliger kanagurta*) fish. Curr Anal Chem 8(4): 534–542

Mert S, Bankoğlu B, Özkan A, Atar N, Yola ML (2018) Electrochemical sensing of ractopamine by carbon nitride nanotubes/ionic liquid nanohybrid in presence of other β-agonists. J Mol Liq 254: 8–11

Migliorini FL, Sanfelice RC, Mercante LA, Andre RS, Mattoso LH, Correa DS (2018) Urea impedimetric biosensing using electrospunnanofibers modified with zinc oxide nanoparticles. Appl Surf Sci 443:18–23

Misaghpour F, Shabani-Nooshabadi M (2018) An electrochemical sensor for analysis of food red 17 in the presence of tartrazine in food products amplified with CdO/rGO Nanocomposite and 1,3-dipropylimidazolium bromide. Food Anal Methods 11:646–653

Montiel V, Gutiérrez ML, Torrente-Rodríguez RM, Povedano E, Vargas E, Reviejo AJ, Pingarrón JM (2017) Disposable amperometric polymerase chain reaction-free biosensor for direct detection of adulteration with horsemeat in raw lysates targeting mitochondrial DNA. Anal Chem 89(17):9474–9482

Munir A, Shah A, Piro B (2018) Development of a selective electrochemical sensing platform for the simultaneous detection of Tl^{+}, Cu^{2+}, Hg^{2+}, and Zn^{2+} ions. J Electrochem Soc 165:B399–B406

Naik TSK, Saravanan S, Sri Saravana KN, Utkarsh P, Praveen CR (2020) A non-enzymatic urea sensor based on the nickel sulfide/graphene oxide modified glassy carbon electrode. Mater Chem Phys 245:122798

Narayanan J, Mukesh KS, Ponmariappan S, Sarita MS, Sanjay U (2015) Electrochemical immunosensor for botulinum neurotoxin type-E using covalently ordered graphene nanosheets modified electrodes and gold nanoparticles-enzyme conjugate. Biosens Bioelectron 69:249–256

Nascimento CF, Santos PM, Pereira-Filho ER, Rocha RP (2017) Recent advances on determination of milk adulterants. Food Chem 221:1232–1244

Pal M, Meenu M (2020) Food adulteration: a global public health concern. Food Drink Indust:38–40

Pramanik D, Dey SG (2011) Active site environment of heme-bound amyloid βpeptide associated with Alzheimer's disease. J Am Chem Soc 133(1):81–87

Promphet N, Rattanarat P, Rangkupan R, Chailapakul O, Rodthongkum N (2015) An electro chemical sensor based on graphene/polyaniline/polystyrene nanoporous fibers modified electrode for simultaneous determination of lead and cadmium. Sensors Actuators B: Chem 207: 526–534

Rajaji U, Murugan K, Chen SM, Govindasamy M, Tse-Wei C, Lin PH, Lakshmi Prabha P (2019) Graphene oxide encapsulated 3D porous chalcopyrite ($CuFeS_2$) nanocomposite as an emerging electrocatalyst for agro-hazardous (methyl paraoxon) detection in vegetables. Compos Part B Eng 160:268–276

Ramesh R, Puhazhendi P, Kumar J, Gowthaman MK, D'Souza SF, Kamini NR (2015) Potentiometric biosensor for determination of urea in milk using immobilized Arthrobacter creatinolyticus urease. Mater Sci Eng C 49:786–792

Rattanarat P, Dungchai W, Cate D, Volckens J, Chailapakul O, Henry CS (2014) Multilayer paper-based device for colorimetric and electrochemical quantification of metals. Anal Chem 86: 3555–3562

Ravi AK, Punnakkal N, Vasu SP, Nair BG, Satheesh Babu TG (2020) Manganese dioxide based electrochemical sensor for the detection of nitro-group containing organophosphates in vegetables and drinking water samples. J Electroanal Chem 859:113841

Rebechi SR, Vélez MA, Vaira S, Perotti MC (2016) Adulteration of Argentinean milk fats with animal fats: detection by fatty acids analysis and multivariate regression techniques. Food Chem 192:1025–1032

Regasa M B, Soreta T R, Femi O E, Praveen C. R, Saravanan S (2020) Novel multifunctional molecular recognition elements based on molecularly imprinted poly (aniline-co-itaconic acid) composite thin film for melamine electrochemical detection. Sensing Bio-Sensing Res 27: 100318

Rodrigues SA, Clésia CN (2013) Development of a simple method for the determination of lead in lipstick using alkaline solubilization and graphite furnace atomic absorption spectrometry. Talanta 105:272–277

Rodrigues JA, Rodrigues CM, Almeida PJ, Valente IM, Gonçalves LM, Compton RG, Barros AA (2011) Increased sensitivity of anodic stripping voltammetry at the hanging mercury drop electrode by ultracathodic deposition. Anal Chim Acta 701(2):152–156

Ruiyi L, Qianfang X, Zaijun L, Xiulan S, Junkang L (2013) Electrochemical immunosensor for ultrasensitive detection of microcystin-LR based on graphene–gold nanocomposite/functional conducting polymer/gold nanoparticle/ionic liquid composite film with electrodeposition. Biosens Bioelectron 44(15):235–240

Şenocak A, Köksoy B, Akyüz D, Koca A, Klyamer D, Basova T, Durmuş M (2019) Highly selective and ultra-sensitive electrochemical sensor behavior of 3D SWCNT-BODIPY hybrid material for eserine detection. Biosens Bioelectron 128:144–150

Shah A (2020) A novel electrochemical nanosensor for the simultaneous sensing of two toxic food dyes. ACS Omega 5(11):6187–6193

Shalini Devi KS, Anusha N, Raja S, Senthil Kumar A (2018) A new strategy for direct electrochemical sensing of a organophosphorus pesticide, triazophos, using a coomassie brilliant-blue dye surface-confined carbon-black-nanoparticle-modified electrode. ACS Appl Nano Mater 1(8):4110–4119

Shang L, Zhao F, Zeng B (2014) 3D porous graphene-porous PdCu alloy nanoparticles-molecularly imprinted poly (para-aminobenzoic acid) composite for the electrocatalytic assay of melamine. ACS Appl Mater Interfaces 6(21):18721–18727

Singh AK, Singh M, Verma N (2020) Electrochemical preparation of Fe_3O_4/MWCNT-polyaniline nanocomposite film for development of urea biosensor and its application in milk sample. J Food Meas Charact 14(1):163–175

Syed Imran AM, Konstantinos G, Yiannaka A (2020) Cornhusker economics. Economic Impacts of Food Fraud Agric Econ:29. (https://agecon.unl.edu/cornhusker-economics/2020/economic-impacts-food-fraud.pdf)

Tang D, Zhong Z, Niessner R, Knopp D (2009) Multifunctional magnetic bead-based electrochemical immunoassay for the detection of aflatoxin B1 in food. Analyst 134:1554–1560

Tang D, Tang J, Su B, Chen G (2010) Ultrasensitive electrochemical immunoassay of staphylococcal enterotoxin b in food using enzyme-nanosilica-doped carbon nanotubes for signal amplification. J Agric Food Chem 58:10824–10830

Thangarasu R, Victor VD, Alagumuthu M (2019) MnO_2/PANI/rGO–A modified carbon electrode based electrochemical sensor to detect organophosphate pesticide in real food samples. Anal Bioanal Electrochem 11(4):427–447

Tripathy S, Ghole AR, Deep K, Vanjari S, Singh SG (2017) A comprehensive approach for milk adulteration detection using inherent bio-physical properties as 'Universal Markers': towards a miniaturized adulteration detection platform. Food Chem 217:756–765

Tvorynska S, Josypčuk B, Barek J, Dubenska L (2019) Electrochemical behavior and sensitive methods of the voltammetric determination of food azo dyes amaranth and allura red AC on amalgam electrodes. Food Anal Methods 12(2):409–421

Verma S, Choudhary J, Singh KP, Chandra P, Singh SP (2019) Uricase grafted nanoconducting matrix based electrochemical biosensor for ultrafast uric acid detection in human serum samples. Int J Biol Macromol. https://doi.org/10.1016/j.ijbiomac.2019.02.121

Walsh GK, Salaün P, Van den Berg CM (2010) Arsenic speciation in natural waters by cathodic stripping voltammetry. Anal Chim Acta 662:1–8

Wang L, Chen W, Xu D, Shim BS, Zhu Y, Sun F, Liu L, Peng C, Jin Z, Xu C, Kotov NA (2009) Simple, rapid, sensitive, and versatile swnt-paper sensor for environmental toxin detection competitive with ELISA. Nano Lett 9(12):4147–4152

Wang B, Luo B, Liang M, Wang A, Wang J, Fang Y, Chang Y, Zhi L (2011) Chem Nanoscale 3: 5059–5066

Wang Q, Fang J, Cao D, Li H, Su K, Hu N, Wang P (2013) Impedimetric immunosensor based on gold nanoparticles modified graphene paper for label-free detection of *Escherichia coli* O157: H7. Biosens Bioelectron 49:492–498

Wang D, Hu W, Xiong Y, Xu Y, Li CM (2015a) Multifunctionalized reduced graphene oxide-doped polypyrrole/ pyrrolepropylic acid nanocomposite impedimetric immunosensor to ultra-sensitively detect small molecular aflatoxin B_1. Biosens Bioelectron 63:185–189

Wang Q, Kaiqi S, Liang H, Ling Z, Tianxing W, Liujing Z, Ning H, Ping W (2015b) A novel and functional assay for pharmacological effects of marinetoxins, saxitoxin and tetrodotoxin by cardiomyocyte-based impedance biosensor. Sens Actuators B 209:828–837

Wang Y, Wang L, Huang W, Zhang T, Hu X, Perman JA, Ma S (2017a) A metal-organic framework and conducting polymer based electrochemical sensor for high performance cadmium ion detection. J Mater Chem A 5:8385–8393

Wang J, Wang Y, Cui M, Xu S, Luo X (2017b) Enzymelessvoltammetric hydrogen peroxide sensor based on the use of PEDOT doped with Prussian Blue nanoparticles. Microchim Acta 184(2): 483–489

Wang P, Li H, Hassan MM, Guo Z, Zhang ZZ, Chen Q (2019) Fabricating an Acetylcholinesterase modulated UCNPs-Cu^{2+} fluorescence biosensor for ultrasensitive detection of organophosphorus pesticides-diazinon in Food. J Agric Food Chem 67:4071–4079

Wang C, Song Q, Liu X, Zhu X (2020) Development of electrochemical sensor based on graphene oxide electrode modified by silver-doped ZnO nanorods for detection of carbamate pesticide in food. Int J Electrochem Sci 15:5623–5631

Wei Q, Zhao Y, Du B, Wu D, Cai Y, Mao K, Li H, Xu C (2011) Nanoporous Pt-Ru alloy enhanced nonenzymatic immunosensor for ultrasensitive detection of microcystin-LR. Adv Funct Mater 21:4193–4198

Wei Y, Gao C, Meng FL, Li HH, Wang L, Liu JH, Huang XJ (2012) SnO_2/reduced graphene oxide nanocomposite for the simultaneous electrochemical detection of cadmium (II), lead (II), copper (II), and mercury (II): an interesting favorable mutual interference. J Phys Chem C 116:1034–1041

Wei J, Li SS, Guo Z, Chen X, Liu JH, Huang XJ (2016) Adsorbent assisted *in situ* electrocatalysis: an ultra-sensitive detection of As(III) in water at Fe_3O_4 Nanosphere densely decorated , with Au nanoparticles. AnalChem 88:1154–1161

Xi H, Chen X, Cao Y, Xu J, Ye C, Deng D, Huang G (2020) Electrochemical determination of formaldehyde via reduced AuNPs@ PPy composites modified electrode. Microchem J 156: 104846

Xia S, Tong J, Bian C, Sun J, Li Y (2018) Microsensors and systems for water quality determination. In: Huang QA (ed) Micro electro mechanical systems. Micro/nano technologies. Springer, Singapore

Xiaoyan Q, Yan X, Zhao L, Huang Y, Wang S, Liang X (2019) A facile label-free electrochemical aptasensor constructed with nanotetrahedron and aptamer-triplex for sensitive detection of small molecule: saxitoxin. Electroanal Chem 858:113805

Xiong X, Shi X, Liu Y, Lu L, You J (2018) An aptamer-based electrochemical biosensor for simple and sensitive detection of staphylococcal enterotoxin B in milk. Anal Methods 10(3):365–370

Xu M, Wang R, Li Y (2016) An electrochemical biosensor for rapid detection of *E. coli* O157:H7 with highly efficient bifunctional glucose oxidase-polydopamine nano composites and Prussian blue modified screen-printed interdigitated electrode. Analyst 141(5441)

Xu S, Lin G, Zhao W, Wu Q, Luo J, Wei W, Zhu Y (2018) Necklace-like molecularly imprinted nanohybrids based on polymeric nanoparticles decorated multiwalled carbon nanotubes for highly sensitive and selective melamine detection. ACS Appl Mater Interfaces 10(29): 24850–24859

Xuan X, Hossain MF, Park JY (2018) A fully integrated and miniaturized heavy-metal-detection sensor based on micro-patterned reduced graphene oxide. Sci Rep 6(1):1–8

Yang G, Zhao F (2015) Electrochemical sensor for chloramphenicol based on novel multiwalled carbon nanotubes@ molecularly imprinted polymer. Biosens Bioelectron 64:416–422

Yang X, Xipeng Z, Xian Z, Ying Q, Mei L, Xiao L, Chaorui L, Yingli L, Huiming X, Jingfu Q (2015) a highly sensitive electrochemical immunosensor for fumonisin B1 detection in corn using single-walled carbon nanotubes/chitosan. Electroanalysis 27:2679–2687

Yang M, Jeong SW, Chang SJ, Kim KH, Jang M, Kim CH, Bae NH, Sim GS, Kang T, Lee SJ, Choi BG (2016) Flexible and disposable sensing platforms based on newspaper. ACS Appl Mater Interfaces 8:34978–34984

Yao Y, Liu Y, Yang Z (2016) Highly sensitive electrochemical sensor for the food toxicant Sudan I based on a glassy carbon electrode modified with reduced graphene oxide decorated with Ag-Cu nanoparticles. Microchim Acta 183:3275–3283

Yola ML, Atar N (2017) Electrochemical detection of atrazine by platinum nanoparticles/carbon nitride nanotubes with molecularly imprinted polymer. Ind Eng Chem Res 56(27):7631–7639

Yu Y, Zhao H, Dong G, Yang R, Li L, Liu Y Zhang W (2015) Discrimination of milk adulterated with urea using voltammetric electronic tongue coupled with PCA-LSSVM. Int J Electrochem Sci 10(12):10119-10131

Yukirda J, Kongsittikulb P, Qinc J, Chailapakuld O, Rodthongkum N (2018) ZnO@graphene nanocomposite modified electrode for sensitive and simultaneous detection of Cd (II) and Pb (II). Synth Met 245:251–259

Yun M, Choe JE, You JM, Ahmed MS, Lee K, Üstündağ Z, Jeon S (2015) High catalytic activity of electrochemically reduced graphene composite toward electrochemical sensing of Orange II. Food Chem 169:114–119

Zaijun L, Zhongyun W, Xiulan S, Yinjun F, Peipei C (2010) A sensitive and highly stable electrochemical impedance immunosensor based on the formation of silica gel–ionic liquid biocompatible film on the glassy carbon electrode for the determination of aflatoxin B1 in bee pollen. Talenta 80:1632–1637

Zeng L, Lei P, Dazhi W, Baoguo Y (2018) Electrochemical sensors for food safety. In: Mózsik G, Figler M (eds) Nutrition in health and disease—our challenges now and forthcoming time. IntechOpen. https://doi.org/10.5772/intechopen.82501

Zhang D, Zhang J, Li M, Li W, Aimaiti G, Tuersun G, Chu Q (2011) A novel miniaturised electrophoretic method for determining formaldehyde and acetaldehyde in food using 2-thiobarbituric acid derivatisation. Food Chem 129(1):206–212

Zhang D, Ouyang S, Cai M, Zhang H, Ding S, Liu D, Cai P, Le Y, Hu QN (2020a) FADB-China: a molecular-level food adulteration database in China based on molecular fingerprints and similarity algorithms prediction expansion. Food Chem 327:127

Zhang Y, Wang W, Lin Z, Liu B, Zhou X (2020b) Dual-output toehold-mediated strand displacement amplification for sensitive homogeneous electrochemical detection of specie-specific DNA sequences for species identification. Biosens Bioelectron 161:112256

Zhaoling J, Lifang Z, Chong Z, Xiao Z, Feiqun X (2014) SPE–UPLC–UV method for the determination of Toltrazuril and its two metabolite residues in chicken and porcine tissues. Chromatographia 77(23–24):1705–1712

Zhou W, Li C, Sun C, Yang X (2016) Simultaneously determination of trace Cd^{2+} and Pb^{2+} based on L-cysteine/graphene modified glassy carbon electrode. Food Chem 192:351–357

Potential of Nanotechnology in Food Analysis and Quality Improvement

8

Anuradha Saini, Divyani Panwar, Parmjit S. Panesar, and Pranjal Chandra

Abstract

Nanotechnology has reformed the food sector with producing better-quality food products through its contribution in functional foods development, food nanopackaging, and nanodevices for food analysis. The existing techniques such as culture-based techniques, sensory analysis, and GC techniques for food analysis are time consuming, cumbersome, and labour intensive. To overcome these drawbacks, nanotechnology is nowadays applied to develop techniques that show more accurate and precise results, which is important for maintaining food quality. Nanotechnology in food analysis is used to detect toxins, adulterants, pathogens, sugar, and antioxidants using nanodevices like nanosensors. Furthermore, nanotechnology can also be applied in food packaging and processing domain to sense food spoilage as well as improve food quality. This chapter delivers comprehensive information about the value and potential of nanotechnology for food analysis, packaging, and quality improvement in the food processing domain.

Keywords

Nanotechnology · Food analysis · Food packaging · Nanosensors · Quality

A. Saini · D. Panwar · P. S. Panesar (✉)
Food Biotechnology Research Laboratory, Department of Food Engineering and Technology, Sant Longowal Institute of Engineering and Technology, Longowal, Punjab, India

P. Chandra
Laboratory of Bio-Physio Sensors and Nanobioengineering, School of Biochemical Engineering, Indian Institute of Technology (BHU) Varanasi, Varanasi, Uttar Pradesh, India

P. Chandra, P. S. Panesar (eds.), *Nanosensing and Bioanalytical Technologies in Food Quality Control*, https://doi.org/10.1007/978-981-16-7029-9_8

8.1 Introduction

Nanotechnology is the "science of manipulating or fabricating the materials on molecular or atomic scale". It is also described as "the design, production, utilization of devices, structures, and systems by dealing with the size and shape of material between 1 and 100 nm known as nano-scale" (Neethirajan and Jayas 2011). Based on the dimensional characteristics, nanoengineering materials are categorized into zero (e.g. nanoclusters), one- (e.g. nanotubes, nanorods), two- (e.g. nano-thin films), and three-dimensional (dendrimers, nanocomposites) structures. Richard Feynman first introduced the ideology related to nanotechnology in 1959, whereas Norio Taniguchi created the term nanotechnology in 1974 (Taniguchi 1974).

In nanotechnology, two methodologies are used to synthesize nanomaterials. In general, "top down" methodology refers to the physical processes which include milling, grinding, slicing, and converting substance into nano-sized particles, while the "bottom up" methodology is about assembling of substance from the smallest particle: atom by atom, molecule by molecule (Rai and Ingle 2015) as described in Fig. 8.1.

In food sector, nanotechnology have numerous novel applications, including the development of environment-friendly packaging materials containing nanomaterials, sensing of food spoilage using smart packaging, and improving quality of food products through nutrient-loaded nanomaterials having different functional properties. Furthermore, biological activity of high value compounds has been improved by encapsulating these compounds in nanoscale spheres. Different barrier characteristics can be improved by adding nanomaterials in the packaging films to extend the food products' shelf life (Valdés et al. 2009).

Nanotechnology has a vital role to enhance the food products' shelf life, production of food with improved quality, and prevention of food from contamination. As compared to macroscale systems, nanoscale materials offer unique characteristics such as chemical, physical, and biological characteristics along with higher surface-volume ratio (Kumar et al. 2019). For example, gold, titanium dioxide, and silver nanoparticles were used as potential antimicrobial in food storage containers, air filters, deodorants, etc. (Parisi et al. 2015; El-Temsah and Joner 2012). Apart from this, nanosensors and nanomaterials provide information regarding the condition and nutritional status of food with improved security through improved microbial detection. This chapter provides a summarized role and importance of nanotechnology in food industry such as food analysis to analyse the food (e.g. adulterants, toxins, sugars, antioxidants), food nanopackaging for sensing the food spoilage and improving the food quality, and delivery of bioactive compounds for improving the functional quality of food products. Table 8.1 shows various nanomaterials used in the analysis of different components of food products.

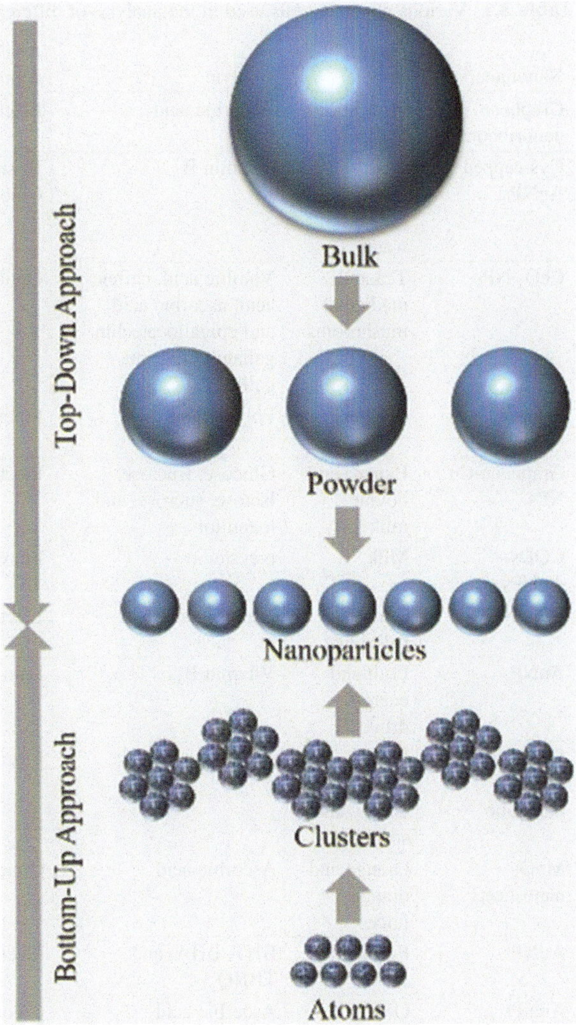

Fig. 8.1 Top-up and bottom-down approach for fabrication of nanoparticles. (Adopted from Sasidharan et al. 2019)

8.2 Potential of Nanotechnology in Food Industry

Nanotechnology is emerging as a promising technology by contributing significantly to revolutionize the food processing industry by various means. For instance, protection of functional ingredients from degradation, bioactive fortification, modulate the food structures and textures, improvement of food quality, and identification and deactivation of microbiological and biochemical modifications through food nanopackaging systems (Tadros et al. 2004; Mason et al. 2006; Steinvil et al. 2016; Sonneville-Aubrun et al. 2004).

Table 8.1 Various nanomaterials used in the analysis of different components of food products

Nanomaterials	Food sample	Analyte	Analysis approach	References
Graphene nanoribbons	Fruit juice	Ascorbic acid	Electrochemical	Yang et al. (2013)
Cys-capped AgNPs	Peas, grapes, tomato, water	Vitamin B_1	Plasmonic Colorimetric	Khalkho et al. (2020)
CeO_2 NPs	Tea and medicinal mushrooms	Vanillic acid, caffeic acid, ascorbic acid, and epigallocatechin gallate, quercetin, gallic acid	Optical	Sharpe et al. (2013)
CDs/Cu^{2+}	Tea	Polyphenols	Fluorescence	Wei et al. (2020a, b)
Graphene-Cu NPs	Banana and bovine milk	Glucose, fructose, lactose, sucrose, and mannitol	Electrochemical	Chen et al. (2012)
CQDs–AuNPs	Milk	L-cysteine	Fluorescence	Chen et al. (2020)
CdO NPs	Canned fruit juice	Ascorbic acid	Electrochemical	Gopalakrishnan et al. (2018)
AuNPs	Fruit and energy drinks	Vitamin B_{12}	Immunodipstick	Selvakumar et al. (2013)
Single-walled carbon nanotube	Honey, fruit juices, energy and soft drinks	D-fructose	Electrochemical	Antiochia et al. (2013)
MnO_2 nanosheets	Orange and orange juice	Ascorbic acid	Optical	He et al. (2018)
AuNPs	Edible oils	BHT, BHA, and TBHQ	Electrochemical	Lin et al. (2013)
AuNPs	Orange juice	Ascorbic acid	Colorimetric	Wei et al. (2020a, b)
Au nanocage	Green tea	Total antioxidants (in terms of GA equivalents)	Colorimetric	Wang et al. (2018a, b)
Cu NPs-based ink	Soft drinks	Glucose, sucrose, and fructose	Electrochemical	Terzi et al. (2017)
Magnetic iron nanoparticle (Fe_3O_4)	Orange juices	*p*-coumaric acid	Electroanalytical	Şenocak (2020)

In food sector, two main applications, which are based on food nanostructured ingredients (nanomaterials), and food nanosensing (nanodevices) are being utilized for enhanced food safety and quality evaluation, rapid sampling of chemical and biological contaminants, nanoencapsulation of functional compounds for improved delivery, controlled release, solubilization and bioavailability in food systems (Ravichandran 2010).

Various functions are performed using nanomaterials to uplift the food industry in terms of taste, colour, and sensory attributes, shelf-life extension, bioavailability, and functionality enhancement of various bioactive compounds. Moreover, the food packaging and food analysis are also an important aspect of food industry that is being improved using nanomaterials (Bajpai et al. 2018). Nanopackaging is the smart technology that maintains the food products' safety and quality all over the food supply chain. Further, these particles are used in the food analysis, food processing, and food packaging to enhance the functionality, short time analysis, and enhance the shelf life of product.

8.2.1 Nanotechnology in Food Analysis

Food analysis is defined as the "discipline dealing with food products in terms of composition, safety, traceability, quality, nutritional, and sensory value". In other words, it is "the study of analytical procedures that can be used to characterize the food properties along with their ingredients" and can also ensure the food quality and support the production of innovative food products/technologies. In recent years, considerable attention is being given on traceability of food products from production to distribution. Therefore, rapid food testing is required, as the conventional microbiological detection method for the identification of foodborne pathogens was no longer effective due to expense of time, complexity, and analysis cost. Hence, rapid quality control methods in a routine laboratory have been adopted to meet the local and global food regulations (Raju and Yoshihisa 2002). For example, biosensors, mainly electrochemical and optical biosensors with nanomaterials have been continuously growing in food sector, which can be considered as a good alternative to traditional analytical methods due to their fast, simple, and cheap multi-detection (Goriushkina et al. 2009; Roy et al. 2019; Purohit et al. 2020). Figure 8.2 shows the schematic diagram of main elements of a general nanobiosensor that has been developed for various applications.

Nanosensors have promising applications in food industry for quality assessment, shelf life monitoring, indicating microbial, residual, and toxin contamination in food, which have been discussed in detail below:

8.2.1.1 Detection of Sugars

Sugars are important structural compounds extensively distributed in both plant and animal kingdoms. Moreover, these have a substantial role in explaining the pathways and processes for synthesis of valuable organic biological compounds. The detection of sugars is commonly being carried out in food industries, biochemical and clinical laboratories. Therefore, it is important to analyse the real-time monitoring of carbohydrates in the fermentative processes to obtain the higher yield of the product. In biosensors, the utilization of AuNPs with pyranose oxidase

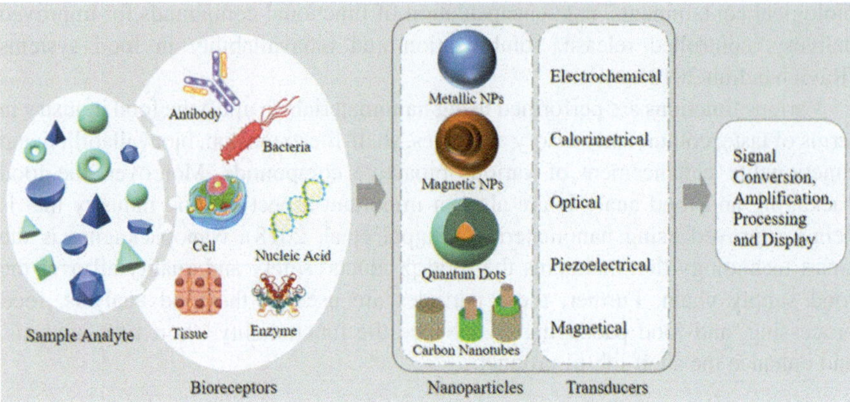

Fig. 8.2 Schematic diagram of main elements of a nanobiosensor. (Adapted from Srivastava et al. 2018)

showed high affinity and good biocompatibility towards the D-glucose that was applied on the fruit juices such as peach, pomegranate, and orange. In addition, it showed stable and fast responses in biosensing of D-glucose due to retention of protein enzyme activity (Ozdemir et al. 2010). Various nanomaterials have been utilized in the formation of nanobiosensors, such as multi-walled carbon nanotubes (MWCNT), gold nanoparticles (AuNPs), single wall carbon nanotubes (SWCNT), Fe_3O_4 magnetic nanoparticles (MNPs), and ZnS-coated CdSe nanoparticle (CdSe@ZnS NPs), which have been utilized to analyse sugars of food products (Antiochia et al. 2004; Sandros et al. 2006; Wei and Wang 2008; Tominaga et al. 2009; Ozdemir et al. 2010).

Nanomaterial-based system was created to detect glucose using the exfoliated graphite nanoparticles with Pd and Pt nanoparticles. These nanoparticles increase the electroactive space of electrode and decrease the overpotential significantly to identify the hydrogen peroxide (Lu et al. 2008). The biosensing system based on unimolecular protein–CdSe@ZnS nanoparticle assembly was created to detect maltose in food (Sandros et al. 2006). The use of carbon nano-tubes (CNTs) as transducer nanomaterials to determine fructose in honey has been reported and results depicted that detection of fructose in real samples showed high current sensitivity. Also, these nanomaterials obtained lower detection limit than conventional solid and paste electrode (Antiochia et al. 2004).

8.2.1.2 Detection of Antioxidants

Antioxidants are the compounds often present in foods to inhibit or delay the oxidative damage to cells of organisms. Several chromatographic and spectroscopic techniques have been created to detect antioxidants in food products. However, focus has currently been shifted to the use of nanosensors for antioxidant analysis as these have been accepted as more sensitive and reliable method as compared to others. The sensors based on Au-gr have shown a lower limit of detection and a higher sensitivity towards the ascorbic acid in fruit juices as compared to the gold nanoparticles-based sensor formulated by Turkevich method (Brainina et al. 2020). Various sensitive luminescent systems were developed to detect the ascorbic acid in a few fruit juices. These systems were based on the utilization of copper oxide nanoparticles (SDS−CuONPs), silver nanoparticles (SDS−AgNPs SDS), and gold nanoparticles (SDS−AuNPs) in the presence of micellar medium sodium dodecyl sulphate that showed a lower detection limit of 5×10^{-8} M, 1.0×10^{-10} M, and 5×10^{-7} M, respectively, which indicated that all suggested approaches can be exploited to detect ascorbic acid in commercial fruit juices (Al-onazi et al. 2020).

8.2.1.3 Determination of Other Analytes

Apart from the above-mentioned compounds, several other analytes like vitamins (e.g. B_{12}, B_1), protein (e.g. L-cysteine) etc. have also been determined by nanosensors. The detection of vitamin B_1 in food (grapes, peas, and tomatoes) and environmental water samples was done through Cys-capped AgNPs (sensor), which was established on the measurement of red shift of localized surface plasmon resonance (LSPR) band of sensor in the region of 200–800 nm. This sensor in colorimetry assay is low cost, simple, and suitable to detect vitamin B_1 in food as well as environmental water samples (Khalkho et al. 2020). The silver nanoparticles (AgNPs) as a nanoprobe have also been used to determine arginine (Arg) and histidine (His) in fruit and vegetable samples. This method was established on the hypochromic shift measurement of LSPR absorption band at 400 nm for His and 395 nm for Arg (Shrivas et al. 2021). The detection of vitamin C was determined using voltammetric method in the presence of vitamin B_6, which was established on the usage of ZrO_2 nanoparticle/ionic liquids carbon paste electrode (ZrO_2/NPs/IL/CPE). The modified electrode determined the vitamin B_6 and vitamin C in food samples efficiently as compared to other methods. The detection limits for vitamin C and vitamin B_6 were 0.009 and 0.1 μM, respectively (Baghizadeh et al. 2015).

8.2.1.4 Detection of Pathogens

Food poisoning is one of the major diseases, which occurs due to foodborne pathogens like viruses, bacteria, viruses, and parasites present in food products. The very familiar pathogens of food origin are bacteria (*Salmonella, Clostridium, Campylobacter, Listeria*), viruses (noroviruses, rotaviruses, hepatitis A viruses, and astroviruses), and parasites (*Cryptosporidium, Cyclospora, Giardia, Toxoplasma*). These can be categorized as per the specific food consumed (FDA 2010). For instance, *Salmonella* spp. (meat, poultry, and eggs), *Campylobacter* spp. (raw or undercooked poultry), *Shigella* spp. (meat and unpasteurized milk), *Clostridium* spp. (home-canned foods), *Yersinia* spp., *Vibrio* spp., *Staphylococcus* spp., *Clostridium* spp., *Listeria* spp., and *Bacillus* spp., are associated with soft cheese, vegetables, uncooked meats, unpasteurized milk, and vegetables that cause different foodborne diseases (Rasooly and Herold 2006). Several viruses such as norovirus damage the human body and parasites like *G. lamblia* and *Cryptosporidium* responsible for 71% waterborne diseases.

According to the reports of WHO's Foodborne Disease Burden Epidemiology Reference Group (FERG), viruses, bacteria, and parasites affected approximately 48 million people yearly (Labib et al. 2016). Most of the outbreaks were due to pathogenic microorganisms like *Salmonella* spp., *Campylobacter* spp. and *Escherichia coli* (WHO 2015). In addition, under consideration of mortality rate and estimated cases of foodborne illness related to bacteria, viruses and parasites, the estimated values for *Salmonella*, *Listeria*, *Campylobacter* and *E. coli* O157:H7 are 31%, 20%, 5%, and 3% respectively, which causes food-related deaths (IFT 2004). Most of the foodborne and waterborne diseases are triggered by bacteria. The conventional culture-based methods need discrete biochemical validation for the pathogenic bacteria identification in isolated colonies although these are complicated, labour-intensive, and time-consuming methods (Ellis and Goodacre 2001). In addition, other methods such as electrochemical assays (Kim et al. 2006; Koubova et al. 2001), immunological based biosensors involving enzyme-linked immunosorbent assay (ELISA), and immunoblotting (Brewster and Mazenko 1998; Valdivieso-Garcia et al. 2001) have overcome these limitations, but these are not always the ideal choices due to the lack of specificity and sensitivity (Thomas et al. 1991). Therefore, it is important to take the advantage of nanoscale approach in food industry, which includes shorter analysis time, multi-analyte analysis, and low functional cost due to less use of reagents, compact, safe, and environmentally friendly devices (Valdés et al. 2009).

A quartz crystal microbalance (QCM) DNA sensor was created to detect *Escherichia coli* O157:H7 that showed high sensitivity and a lower detection limit (267×10^2 cfu/mL) than previous QCM DNA sensors (Mao et al. 2006). An AuNPs Surface Plasmon Resonance (SPR) assay was created to detect *Salmonella* and *Escherichia coli* in raw food samples. The system detected 11.7×10^3 and 7.4×10^3 cfu/mL for *Salmonella* and 17 cfu/mL and 57 cfu/mL for *E. coli* in hamburger and cucumber extracts within 80 min (Vaisocherová-Lísalová et al. 2016). The impedance biosensor was created that is related to an interdigitated array gold microelectrode combined with magnetic nanoparticle-antibody conjugate to detect *E. coli* O157:H7 in real sample (e.g. ground beef). The improved sensitivity was reported owing to the creation of clusters among cells and conjugates (Varshney and Li 2007). Furthermore, a rapid gold nanoparticle (GNP) colorimetric assay for the genome DNA of *Salmonella enterica* and *Listeria monocytogenes* detection based on the polymerase chain reaction (PCR) through thiol-labelled primers was developed and the limit of detection (LOD) was 0.013 ng mL^{-1} and 0.015 ng mL^{-1}, respectively (Fu et al. 2013a, b). A recent study has reported the use of CDS@ZIF-8 (cadmium sulphide QD within zeolitic imidazolate framework-8) as an amplification tag for the electrochemical immune detection of *E. coli* O157:H7 and the detection limit (3 CFU/mL) was increased by 16 times using ZIF-8 than ZIF-8 free sensor (Zhong et al. 2019). Another study reported the detection of *Salmonella enterica* in skimmed milk using the AuNPs amplification along with magneto-immunorecognition at the detection limit of 143 cells/mL indicating that this method was better than conventional methods in terms of higher sensitivity and shorter analysis time (Afonso et al. 2013).

8.2.1.5 Detection of Toxins

Food toxins are natural materials that include a huge range of molecules, produced via fungi, plants, bacteria, or algae metabolism having detrimental effect on humans, even at low level of doses. These are distributed into three major classes: algal toxins, mycotoxins, and plant/bacteria toxins (Dridi et al. 2017). For instance, group of toxic compounds (e.g. aflatoxins) found in food spoiled by *Aspergillus parasiticus and Aspergillus flavus*. Also, organic synthetic compounds like bisphenol A (BPA) can leach into the food through the lining of food and beverages cans.

Aflatoxin M1 detection in milk was conducted using gold nanoparticles-based immunochromatographic strip. The colourless zone was found on the strip when milk sample was contaminated with aflatoxin M1 whereas red colour band appears in the absence of aflatoxin M1 (Wang et al. 2011). A binary colour quantum dots encoded with frit-based immunoassay was created to detect aflatoxin M1 in milk visually and the cut-off value was 0.02 μg/kg (Jiang et al. 2017). In 2004, the detection of sterigmatocystin (secondary metabolite) was done using enzyme electrode along with MWNT. It showed the linear detection ranged from $8.32 \times 10^{(-5)}$ to $66.56 \times 10^{(-5)}$ mg/mL and detection limit was $8.32 \times 10^{(-5)}$ mg/mL with response time of 10 s (Yao et al. 2004).

Mycotoxin is a poisonous fungal metabolite that possesses high toxicity at low level of exposure. Due to the presence of mycotoxin, approximately 25% of the grain is contaminated worldwide and even at low concentration, it can induce significant health issues such as cancer, liver disease, kidney disease, and death (Cheli et al. 2008). Among mycotoxins, ochratoxin A is the most ample food contaminating nephrotoxic toxin generated through *Aspergillus ochraceus* and *Penicillium verrucosum*. In a study, a novel approach of designing optical sensing technique based on aptamer recognition for colorimetric sensing of ochratoxin A was developed which was assembled using CeO_2 particles along with ochratoxin A-specific ssDNA aptamers. It can identify as low as 0.15×10^{-9} M ochratoxin A resulting in higher sensitivity, selectivity, and activity (Bülbül et al. 2016).

The most dominant marine toxin is palytoxin that was detected by the electrochemiluminescence-based nanobiosensor, and it was able to quantify the palytoxin in mussels along with the limit of quantification (LOQ) of 2.2 μg/kg of mussel meat (Zamolo et al. 2012).

8.2.1.6 Detection of Adulterants

Food adulteration is commonly defined as "the adding or subtracting any substance to or from food that affects the natural composition and quality of food substance". It can be categorized into two groups like intentional and incidental adulteration. The intentional adulteration includes cautious addition of substandard constituents into food to gain greater profits whereas the incidental adulteration arises due to the addition of foreign substances as a result of negligence, ignorance, or improper facilities (Jha et al. 2016; ASTA 2004). Examples of food adulterants include sawdust, starches, urea, caustic soda, vegetable oil, etc.

Adulterant's detection in various food samples is the important parameter for food industries and regulatory authorities owing to rise in occurrences of various foodborne diseases. In today's scenario, the processed foods are attaining more consideration because of the changing lifestyle of people that requires fast and simple methods for finding the causative agents. Different techniques involve physicochemical methods (e.g. anthrone test, DNS test, iodine test), chromatographic (e.g. ion-exchange chromatography, liquid chromatography), and immunoassays (e.g. enzyme immunoassay, immunoprecipitation, fluorescence immunoassays), etc. However, owing to the limitations (time consuming, laborious sample preparation) of these techniques, there is a serious requisite for upgraded equipment to use onsite and online detection of adulterants from consuming adulterated foods. The inclusion of nanotechnology-based techniques in food analysis will be a better option as compared to the traditional techniques.

The common adulterant for milk is melamine that is mixed to increase the protein content. Various studies have detected the melamine in milk through nanomaterials as shown in Table 8.2. The LSPR sensor chip was created to detect melamine using AuNPs. The high sensitivity showed by sensors when applied on the commercial milk powder (Oh et al. 2019). Other examples of adulterants are food dyes and Sudan I is the most popular as food colouring agents. The detection of Sudan I using electrochemiluminescent (ECL) immunosensor that was built with nanocrystallines of palladium/aurum and quantum dots (e.g. CdSe@CdS) as signal bioprobe. This sensor has detection limit of 0.3 pg mL^{-1} and exhibits high sensitivity and good stability (Wang et al. 2018a, b).

In a recent study, the detection of sibutramine in food supplements was done using gold nanoparticles (AuNPs) and a smartphone. The results were in alignment with those of UV–Vis spectrophotometry having a relative error ranged from −9.3 to 1.7% (Chaisiwamongkhol et al. 2020). Meat authentication was studied by a colorimetric immunoassay that was constructed using magnetic nanoparticles (solid substrate) and polydopamine (coating polymer). Among different meat species (chicken, turkey, lamb, pork, and beef), the pork showed the highest immunoassay specificity. Furthermore, this method delivers a sensitive in-field test for verifying halal meat authenticity (Seddaoui and Amine 2020).

8.2.2 Nanotechnology in Food Packaging

Food packaging is mainly aimed at storing of food product, providing information of ingredients along with nutritional value and protection of food from external influences (Coles et al. 2003). Food packaging safeguards them from physical (e.g. mechanical damage), biological (e.g. microorganisms, pathogens, insects, rodents), and environmental factors (e.g. light, oxygen, relative humidity) to retain the quality of products. Moreover, the functions such as tamper indication, convenience, and traceability are important to meet the consumer's desires. It can be categorized into primary, secondary, tertiary, and quaternary (Robertson 2006). However, food industries have always been seeking new approaches in technologies

Table 8.2 Various nanomaterials used to detect pathogens, toxins, and contaminants in food products

Nanomaterials	Food sample	Target element	Detection method	References
Pathogens				
ZIF-8 metal organic framework containing Cd QDs	Milk	*E .coli*	Electrochemical	Zhong et al. (2019)
Dendritic superparamagnetic iron oxide nanoparticles	Milk	*Salmonella*	Low field nuclear magnetic resonance	Li et al. (2020)
AuNPs	Artificial contaminated food samples	*Salmonella enterica* and *Listeria monocytogenes*	Colorimetric	Fu et al. (2013a, b)
AuNPs	Hamburger and cucumber extracts	*E. coli* and *Salmonella* sp.	Optical	Vaisocherová-Lísalová et al. (2016)
Multicolour UCNPs–MNPs–aptamers	Food matrixes (e.g. milk and shrimp)	*S. aureus, V. parahaemolyticus, S. typhimurium*	Luminescence bioassay	Wu et al. (2014)
Silica nanoprobes	Milk	*Brucella* spp.	Fluorescence	Vyas et al. (2015)
AuNPs label	Skimmed milk	*Salmonella*	Electrochemical	Afonso et al. (2013)
Cysteine-loaded nanoliposomes and AuNPs	Apple juice, milk, and ground beef	*Salmonella, Listeria, and E. coli* O157	Liposome-amplified plasmonic immunoassay (LAPIA)	Bui et al. (2015)
Au-coated magnetic nanoparticles (AuMNPs)	–	*Staphylococcus aureus*	Surface-enhanced Raman spectroscopy	Wang et al. (2016)
Toxins				
MWCNT	Baby feeding bottles	BPA	Electrochemical	Anirudhan et al. (2018)
AuNPs	Spiked beer	Ochratoxin A	Enhanced SPR	Evtugyn et al. (2013)
Gold nanoprobes	Spiked milk	Aflatoxin M1	Dynamic light scattering	Zhang et al. (2013a, b)
SWCNT and MWCNT	Mussel meat	Palytoxin	Electrochemiluminescence	Zamolo et al. (2012)
CeO_2 NPs	Milk	Mycotoxin (ochratoxin)	Optical	Bülbül et al. (2016)
AuNPs loaded on MWCNT	–	BPA	Electrochemical	Messaoud et al. (2017)
Adulterants				

(continued)

Table 8.2 (continued)

Nanomaterials	Food sample	Target element	Detection method	References
GNRs/GO	Tomato sauce	Sudan I	Electrochemical	Wang et al. (2018a, b)
AuNPs	Food supplement products	Sibutramine	Colorimetric	Chaisiwamongkhol et al. (2020)
AuNPs	Milk powder	Melamine	Optical	Oh et al. (2019)
MWCNT	Hot chili	Sudan I	Electrochemical	Gan et al. (2008)
ParaNitroaniline (p-NA)-modified silver nanoparticles	Milk	Melamine	Optoelectronic	Kalpana et al. (2020)
Pd/Au nanocrystallines (Pd/Au CSNs)	Chili sauce	Sudan I	Electrochemical	Wang et al. (2018a, b)
AgNPs	Milk	Melamine	Interference biosynthesis	Moshahary and Mishra (2021)
Au nanocolloid	Tomato sauce and chili powder	Sudan I	Optical	Wang et al. (2013)
CdSe@CdS QDs	Chili powder	Sudan I	Electrochemical	Wang et al. (2018a, b)
Chemical residues/contaminants				
Au@AgNPs	Milk	Thiram and dicyandiamide	Surface-enhanced Raman spectroscopy	Hussain et al. (2020)
CuO nanostructure	Cabbage and spinach extract	Chlorpyrifos, fenthion, and methyl parathion	Electrochemical	Tunesi et al. (2018)
AuNPs	Fish	Histamine	Colorimetric aptamer assay	Lerga et al. (2020)
Au@Ag NPs	Milk	Kanamycin	Surface-enhanced Raman spectroscopy	Jiang et al. (2019)
Au@Ag-TGANPs	Milk	Thiabendazole and ferbam	Surface-enhanced Raman spectroscopy	Hussain et al. (2021)

for improving the food packaging functions for better food quality. For example, industries are producing nanomaterials to overcome different problems such as microbial contamination and low physical-chemical stability of food during storage, non-biodegradable nature (e.g. plastic polymers), and single function-oriented packaging materials that can be improved using nanotechnology to increase the shelf life of food product.

In industries, various provisions have been used in the packaging sector such as wooden crates, cellophane wrappings, cardboard boxes, and various plastic polymers (polypropylene, LDPE, HDPE, PET, PVC, and PS) although these provisions have their own advantages or disadvantages (Robertson 2006). These materials have various functional properties, which includes antimicrobial activity, oxygen scavenging capacity, biodegradability, and barrier properties that can be enhanced by the nanotechnology.

Furthermore, nanomaterials used to enhance the functional attributes of food packaging that have been introduced the new definitions of the active, improved, and intelligent/smart packaging. Generally, food packaging based on nanotechnology can be classified (Duncan 2011; Silvestre et al. 2011) as follows:

- "Intelligent/smart nano-packaging" is developed for sensing and monitoring various food changes, for instance, sensing particular pathogens or gases that are developed during food spoilage. For example, time-temperature indicators (TTI), quality indicators, oxygen sensors, nanosensors (e.g. microbial growth), electronic tongue and nose, holograms, and light-emitting diodes.
- "Active nano-packaging" uses nanomaterials for interaction with the product directly or the environment to permit better safety of food product. This type of packaging performs various functions such as oxygen or UV scavengers, emitters, microbial blocking, and regulating or buffering. For example, silver nanoparticles, gold nanoparticles, and metal oxides nanoparticles.
- "Improved nano-packaging" includes the nanomaterials, which are mixed with polymer matrix to improve various properties of packaging material, and can be categorized as under:
  (a) Physical improved packaging: It involves the nanomaterials, which improve the properties like gas barrier, moisture stability, durability, temperature, and flexibility. For example, carbon nanotubes, metal oxides nanoparticles, nanoclays.
  (b) Biochemical improved packaging: It includes the bio-nanomaterials that improve the different attributes such as edibility, low-waste, biocompatibility, and eco-friendly nature. For example, bio-nanocomposites.

The application of functional nanomaterials can enhance the various physicochemical properties of food packaging that includes moisture and temperature stability, durability, flexibility, and barrier properties. For example, the incorporation of nanocomposites formed with CMC was strengthened using $TiO_2$and sodium montmorillonite nanomaterials enhanced the density, moisture content, and glass transition temperature slightly; however, decrease in water vapour permeability was observed (Fathi Achachlouei and Zahedi 2018). In active packaging, the functions can be enriched by nanomaterials, e.g. antioxidative, antimicrobial, UV protective. The functional properties like antioxidant and antimicrobial activity have been improved by essential oil-loaded biopolymeric nanocarriers, which showed the suitability of the material used in active food packaging owing to inhibiting properties of microbial development in various food products (Rehman et al. 2020). Furthermore, the smart packaging is also being used to detect small organic molecules, gases, active stage, product quality and authenticity through freshness indicators, integrity indicators, temperature, and time indicators. The indicator (shikonin-based pH-sensitive colour indicator), which was used to observe the pork and fish freshness during storage, showed the interaction between the change in indicator's colour and change in sample's pH thus, it can be used to observe the packaged food quality in real-time (Ezati et al. 2021).

Different nanopackagings like improved, active, and bio-based nanopackaging have been used to maintain the quality of food product and protect from various environmental conditions via nanomaterials. However, smart food packaging has been used to sense and protect from food spoilage. For example, constrained release of preservative, detection of pathogen, and/or incorporation of biosensors/sensors at nanoscale were done to extend the shelf life and monitor the food during the storage and transportation (Sozer and Kokini 2009; Scrinis and Lyons 2007).

Recent trends in food packaging propose the applications of nanotechnology to offer a variety of improved applications through functional nanomaterials. The present objective of food packaging is to build a food packaging system having multiple properties that have no negative interrelation in between the constituents (Han et al. 2018). Many international corporations such as Tetra Laval International S.A., Amcor Ltd., Sonoco Products Company, and Chevron Phillips Chemical Company are carrying out the production of nanotechnology-based packaging materials that can improve food safety along with the shelf-life extension (Primožič et al. 2021).

8.2.2.1 Sensing the Food Spoilage Using Smart Nanopackaging

The most extensively explored application of NPs is sensing that is based on the changes in physicochemical attributes of NPs, determined through target analytes (Carrillo-Carrión et al. 2014; Polo et al. 2013; Xia et al. 2010; Elghanian et al. 1997; Cho et al. 2012).

Nanosensors are generally classified as intravenous sensors that evaluate the state of food within packaging, and extraneous sensors measure the outside condition of food package. The most important nanosensing approach is colour changing of metallic NPs solutions in the presence of specific target. For instance, a colorimetric sensor based on plasmonic NPs. Temperature is a vital extrinsic parameter, which affects the quality of food significantly, and it is important to know the temperature during production, storage, and transportation of food products. The examples of patented time-temperature indicators are Timestrip Complete, Timestrip Plus® (−20 °C to 86 °F), and 3 M MonitorMark (15–31 °C). These indicators were built with a grouping of sensor, response element that can sense the temperature breach, and generated response in the form of different colours (e.g. red, white, and blue).

For advanced quality control, time-temperature indicators are the current requirement of food industry; however, these are inadequate because of the high cost and poor programmability. Through nanotechnology, it has become possible to develop the cost-efficient and kinetically programmable TTI protocol fabricated from plasmonic nanocrystals. It was used to mimic the degradative mechanisms, track perishables, and specify the quality of product using sharp-contrast multicolour changes (Zhang et al. 2013a, b).

Different nanosensors (e.g. freshness sensors, use-by indicators, leak detection systems, time/temperature indicators) are used to quantify the changes regarding the oxygen concentration, carbon dioxide concentration, temperature change, and detection of pathogen microorganisms. For example, a nanosensor known as *OxyDot*® used to check the dissolved oxygen from sealed drink, and packaged food product commercially. A company named TEMPTIMES Corporation has designed a time-temperature indicator known as *Fresh Check,* which is fixed on outside packaging of food products and provides the details related to product quality by monitoring temperature variations throughout the storage period. Another product named Toxin Guard has been manufactured by Toxin Alert, Canada to detect microorganisms that include *Campylobacter* sp., *Salmonella* sp., *Listeria* sp., and *E. coli*. Moreover, a ripeness indicator called RipeSense® labels was constructed on the colour changes from red to orange and lastly into the yellow, which is used to detect the volatile compounds, released by the ripened food (Srivastava et al. 2018).

8.2.2.2 Improving the Food Products Using Active Nanopackaging

Nanomaterial-based active packaging has been used to enhance the food quality that can be helpful in reduction of serious environmental problems, which is caused due to discarded (approx. 1.3 billion tons/year) food. These are formed using bioactive NPs having antimicrobial and/or antioxidant properties. Mostly biopolymers and traditional polymers used for the incorporation of nanomaterials include clays and organically modified clays, bioactive inorganic metallic and metal oxides NPs, etc. The metallic NPs are produced using electrochemical techniques, chemical, biological, and physical methods. These NPs have gained consideration as they utilize natural sources (e.g. plants and plant extracts) that have high concentration of bioactive metabolites. For instance, the formation of gallic acid/AuNPs using SiO_2NPs were produced, which enables the exploitation of beneficial properties of gallic acid (e.g. antioxidant) and antimicrobial activities of AuNPs simultaneously (Rattanata et al. 2016). The improvement in attributes such as tensile strength, glass transition temperature, melting point, and lower water vapour permeability were obtained through bio-based films incorporated with potato starch-TiO_2NPs/MMT nanocomposites (Oleyaei et al. 2016).

8.2.2.3 Enhancing the Food Quality Using Improved Nanopackaging

The incorporation of nanomaterials in improved packaging has been used to enhance the physical, mechanical, biochemical attributes such as gas barrier, moisture stability, temperature, mechanical strength, biocompatibility, and edibility, through metallic nanoparticles, nanoclays, carbon nanotubes, and bio-nanocomposites. Different nanoparticles or nanocomposites were incorporated in polymers for packaging applications in food industry, where clay nanoparticle composites can be used in the proportion of up to 5% (w/w). The main role to incorporate these types of nanomaterials was to improve functional properties such as reduction in carbon dioxide and oxygen permeation up to 80–90% (Brody 2007). The single layer food packaging films were developed using organically modified montmorillonite nanoclays. With the addition of appropriate level of nanoclay (3% w/w), a significant increment was observed in the mechanical, optical, thermal, and barrier attributes of the formed EVOH/clay nanocomposite films as compared to the films without added montmorillonite clay nanoparticles (Kim and Cha 2014). In a recent study, a pure starch packaging film was produced with a temperature-aided electrospinning method that consists of starch nanofibers. Various properties such as thermal stability, ultralow hydrophobicity were improved using stearic acid (STA) coating having superhydrophobic properties (Cai et al. 2021).

8.2.3 Nanotechnology for Delivery of Bioactive Compounds

Nano-based materials have potential applications in the delivery and targeted distribution of bioactive compounds in value-added food products. Various nanocarriers have been used to improve the functional attributes of high value compounds. Basically, these nanocarriers can be divided into five categories based on the application of different ingredients or equipment for their synthesis and are used as improved delivery systems (Jafari 2017a, b). Nanocarriers are lipid-based (nanostructured phospholipid carriers, nanoemulsions, nano-lipid carriers), special equipment-based (nanospray, dryer, nanofluidics, electrospinning), nature-inspired (cyclodextrins, caseins), and biopolymers that involve complex biopolymer nanoparticles, single biopolymer nanoparticles, nano-gels, and nanotubes/nanofibrils (Assadpour and Jafari 2018). The various nanocarriers have been used to encapsulate bioactive compounds are discussed below:

8.2.3.1 Lipid-Based Nanocarriers

Lipid-based nanocarriers are synthesized by oil, solvents, and surfactants. The role of lipids in the production of lipid-based nanocarriers is important to ameliorate the bioavailability of high value compounds through improvement in stability and constrained release of high value compounds in GIT and also assists the absorption of functional compounds in epithelial cells (Esfanjani et al. 2018). These nanocarriers are found in various forms like single (W/O and O/W), double (O/W/O and W/O/W), structural and pickering nanoemulsions (Jafari et al. 2017; Akhavan et al. 2018; Saini et al. 2020). **Nanoemulsions** refer to an isotropically clear dispersion of two non-miscible liquids, such as water and oil at nanoscale. These are stabilized through interfacial film of surfactant molecules that includes various types of surfactants (McClements 2005) such as ionic (e.g. DATEM, SLS,

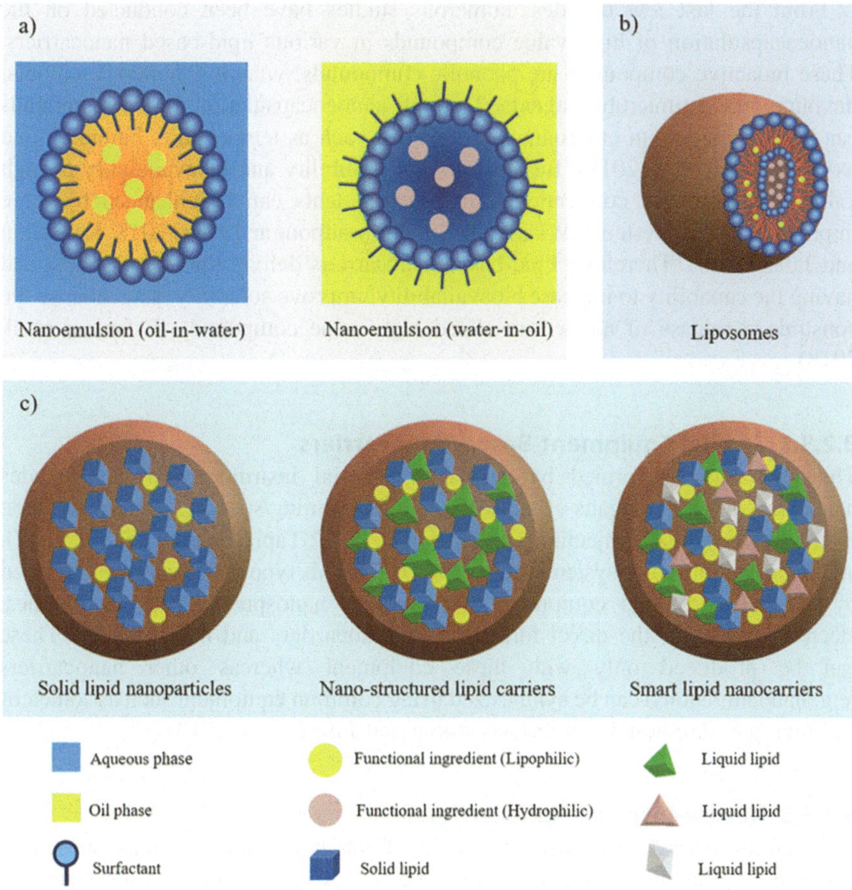

Fig. 8.3 Schematic representation of lipid-based nanocarriers (**a**) Nanoemulsions, (**b**) liposomes (**c**) nano-lipid carriers (Source: Saini et al. 2020)

CITREM), zwitterionic surfactants (e.g. lecithin), and non-ionic (e.g. sucrose monopalmitate, sorbitan monooleate). Nanoemulsions are formulated by three main components that include aqueous phase, oil, and surfactant/cosurfactant (McClements and Rao 2011). Various aspects to consider throughout the synthesis of nanoemulsions are ultralow interfacial tension and must have the flexible interface to support the formation of nanoemulsions (McClements and Rao 2011). **Nanoliposomes** are the lipid-based nanocarriers that are formulated by oil, phospholipids, and surfactants. These nanocarriers have important property to encapsulate the hydrophilic (central cavity) as well as lipophilic (membrane) compounds in it. The last group of this class is known as new generation nanocarriers (nano-lipid carriers) that include nanostructured lipid carriers, solid lipid nanocarriers, and smart lipid nanocarriers (Fig. 8.3).

From the last few decades, numerous studies have been conducted on the nanoencapsulation of high value compounds in various lipid-based nanocarriers. These bioactive compounds are phenolic compounds, vitamins, minerals, colours, flavours, and antimicrobial agents. Through nanoencapsulation, these compounds can be protected from environment conditions such as temperature, humidity, and oxygen. (Saini et al. 2019). Besides, the low solubility and bioavailability of high value compounds in conventional food supplements can be enhanced by these improved delivery systems (Walia et al. 2019; Assadpour and Jafari 2018; Katouzian and Jafari 2016). Therefore, lipid-based nanocarriers deliver high surface area and having the capability to increase bioavailability, improve solubility, and enhance the constrained release of nanoencapsulated high value compounds (Esfanjani et al. 2018).

8.2.3.2 Special Equipment-Based Nanocarriers

The next class is formed by special/professorial instruments which includes nanospray dryer (Arpagaus et al. 2017), electrospinning/spraying (single injection nozzle, coaxial double injection) (Ghorani et al. 2017; Tapia-Hernández et al. 2017), and micro/nanofluidics systems (Ran et al. 2017). This type of nanocarrier is formed by specially designed equipment that involves nanospray dryer, nanofluidics, electrospinning for the direct formation of nanocarriers and nanoparticles. These can be produced only with these equipment whereas other nanocarriers (e.g. nanoemulsions) can be synthesized using common equipment such as sonicator and high pressure homogenizer (Assadpour and Jafari 2018, 2019).

8.2.3.3 Nature-Inspired Nanocarriers

The nature-inspired nanocarriers are one of the key classes of nanocarriers as nanovehicles that already exist in nature and acts as nanoencapsulation systems for bioactive compounds that involve amylose nanostructures, caseins, and cyclodextrins. For instance, casein is present naturally as spherical micelles with nanoparticles having size in the range of 50–200 nm in the cow's milk. Caseins have similarity with copolymers due to its well-balanced hydrophilic and hydrophobic regions that may self-assemble to produce nanosized carriers along with good thermal stability. Therefore, these natural nanoparticles like caseins, starch granules, etc. are used in the nanoencapsulation of functional compounds, which have hydrophobic or sensitive properties (Esfanjani et al. 2018; Assadpour and Jafari 2019; Jafari 2017b; Haratifar and Guri 2017; Gharibzahedi and Jafari 2017).

8.2.3.4 Biopolymer Nanoparticles

This class of nanocarriers can be synthesized using single biopolymer nanoparticles (Sadeghi et al. 2017), complexation of two unlike charge on surface of biopolymers (Ghasemi et al. 2017, 2018Hosseini et al. 2017), nanofibrils or nanotubes of whey proteins (Jafari 2017a), and nano-gels formed with specific biopolymers which include chitosan, whey, soy proteins, etc. (Mokhtari et al. 2017; Abaee et al. 2017).

8.2.3.5 Miscellaneous Nanocarriers

Mostly, this class of nanocarriers is used in the pharmaceutical fields and availability of information data for nanoencapsulation of functional food ingredients is meagre. Miscellaneous nanocarriers are the nanocarriers that are not covered in any above-mentioned classes. These types of nanocarriers include the nanoparticles formulated from nanostructured surfactants such as niosomes, inorganic nanoparticles, chemical polymers, cubosomes, and nanocrystals.

8.2.4 Summary and Future Prospects

The rapid food testing is required for maintaining the food quality in real-world applications. However, conventional microbiological analysis techniques for foodborne pathogens are time consuming and have high analysis cost. Sensors based on nanotechnology have great possibilities to provide the rapid quality control methods in a routine laboratory, which will meet the local and global food regulations. On the other hand, nanotechnology in food packaging and processing can be used to sense the food spoilage, better food quality, and improve the shelf life by incorporation of nanomaterials. Bioactive compounds are sensitive towards the change in light, pH, oxygen, heat, etc. and nano-based delivery systems are used to overcome this problem, which is not possible with conventional delivery system efficiently. Further, attention should be given on the cost-effective rapid and real-time applications for food analysis in food industries at large scale.

References

Abaee A, Mohammadian M, Jafari SM (2017) Whey and soy protein-based hydrogels and nanohydrogels as bioactive delivery systems. Trends Food Sci Technol 70:69–81

Afonso AS, Pérez-López B, Faria RC, Mattoso LH, Hernández-Herrero M, Roig-Sagués AX, Maltez-da Costa M, Merkoçi A (2013) Electrochemical detection of Salmonella using gold nanoparticles. Biosens Bioelectron 40(1):121–126

Akhavan S, Assadpour E, Katouzian I, Jafari SM (2018) Lipid nano scale cargos for the protection and delivery of food bioactive ingredients and nutraceuticals. Trends Food Sci Technol 74:132–146

Al-onazi WA, Alarfaj NA, El-Tohamy MF, Al-Malki NA (2020) Facile dual enhanced modes of nanoparticles/sodium dodecyl sulfate for luminescent detection of vitamin C in commercial fruit juices. J Anal Chem 75(10):1285–1294

Anirudhan TS, Athira VS, Sekhar VC (2018) Electrochemical sensing and nano molar level detection of Bisphenol-A with molecularly imprinted polymer tailored on multiwalled carbon nanotubes. Polymer 146:312–320

Antiochia R, Lavagnini I, Magno F (2004) Amperometric mediated carbon nanotube paste biosensor for fructose determination. Anal Lett 37(8):1657–1669

Antiochia R, Vinci G, Gorton L (2013) Rapid and direct determination of fructose in food: a new osmium-polymer mediated biosensor. Food Chem 140(4):742–747

Arpagaus C, John P, Collenberg A, Rütti D (2017) Nanocapsules formation by nano spray drying. In: Jafari SM (ed) Nanoencapsulation technologies for the food and nutraceutical industries. Academic Press, pp 346–401

Assadpour E, Jafari SM (2018) A systematic review on nanoencapsulation of food bioactive ingredients and nutraceuticals by various nanocarriers. Crit Rev Food Sci Nutr. https://doi.org/10.1080/10408398.2018.1484687

Assadpour E, Jafari SM (2019) Nanoencapsulation. In: Amparo LR, Rovira MJ, Sanz MM, Gomez-Mascaraque LG (eds) . Nanomaterials for food applications, Elsevier, pp 35–61

Baghizadeh A, Karimi-Maleh H, Khoshnama Z, Hassankhani A, Abbasghorbani M (2015) Voltammetric sensor for simultaneous determination of vitamin C and vitamin B_6 in food samples using ZrO_2 nanoparticle/ionic liquids carbon paste electrode. Food Anal Methods 8: 549–557

Bajpai VK, Kamle M, Shukla S, Mahato DK, Chandra P, Hwang SK, Kumar P, Huh YS, Han YK (2018) Prospects of using nanotechnology for food preservation, safety, and security. J Food Drug Anal 26(4):1201–1214

Brainina KZ, Bukharinova MA, Stozhko NY, Sokolkov SV, Tarasov AV, Vidrevich MB (2020, 1800) Electrochemical sensor based on a carbon veil modified by phytosynthesized gold nanoparticles for determination of ascorbic acid. Sensors 20(6)

Brewster JD, Mazenko RS (1998) Filtration capture and immunoelectrochemical detection for rapid assay of Escherichia coli O157: H71. J Immunol Methods 211:1–8

Brody AL (2007) Case studies on nanotechnologies for food packaging. Food Technol 61:102–107

Bui MP, Ahmed S, Abbas A (2015) Single-digit pathogen and attomolar detection with the naked eye using liposome-amplified plasmonic immunoassay. Nano Lett 15(9):6239–6246

Bülbül G, Hayat A, Andreescu S (2016) ssDNA-functionalized nanoceria: a redox-active aptaswitch for biomolecular recognition. Adv Healthc Mater 5(7):822–828

Cai J, Zhang D, Zhou R, Zhu R, Fei P, Zhu ZZ, Cheng SY, Ding WP (2021) Hydrophobic interface starch nanofibrous film for food packaging: From bioinspired design to self-cleaning action. J Agric Food Chem 69(17):5067–5075

Carrillo-Carrión C, Nazarenus M, Paradinas SS, Carregal-Romero S, Almendral MJ, Fuentes M, Pelaz B, del Pino P, Hussain I, Clift MJ, Rothen-Rutishauser B (2014) Metal ions in the context of nanoparticles toward biological applications. CurrOpin Chem Eng 4:88–96

Chaisiwamongkhol K, Labaidae S, Pon-in S, Pinsrithong S, Bunchuay T, Phonchai A (2020) Smartphone-based colorimetric detection using gold nanoparticles of sibutramine in suspected food supplement products. Microchem J 158:105273

Cheli F, Pinoti L, Campagnoli A, Fusi E, Rebuci R, Baldi A (2008) Mycotoxin analysis, mycotoxin producing fungi assays and mycotoxin toxicity bioassays in food mycotoxin monitoring and surveillance. Ital J Food Sci 20(4):447–462

Chen Q, Zhang L, Chen G (2012) Facile preparation of graphene-copper nanoparticle composite by in situ chemical reduction for electrochemical sensing of carbohydrates. Anal Chem 84(1): 171–178

Chen Y, Qin X, Yuan C, Wang Y (2020) Switch on fluorescence mode for determination of L-cysteine with carbon quantum dots and Au nanoparticles as a probe. RSC Adv 10(4): 1989–1994

Cho ES, Kim J, Tejerina B, Hermans TM, Jiang H, Nakanishi H, Yu M, Patashinski AZ, Glotzer SC, Stellacci F, Grzybowski BA (2012) Ultrasensitive detection of toxic cations through changes in the tunnelling current across films of striped nanoparticles. Nat Mater 11(11): 978–985

Coles R, McDowell D, Kirwan MJ (eds) (2003) Food packaging technology. CRC Press

Dridi F, Marrakchi M, Gargouri M, Saulnier J, Jaffrezic-Renault N, Lagarde F (2017) Nanomaterial-based electrochemical biosensors for food safety and quality assessment. In: Grumezescu AM (ed) Nanobiosensors. Academic Press, pp 167–204

Duncan TV (2011) Applications of nanotechnology in food packaging and food safety: barrier materials, antimicrobials and sensors. J Colloid Interface Sci 363(1):1–24

Elghanian R, Storhoff JJ, Mucic RC, Letsinger RL, Mirkin CA (1997) Selective colorimetric detection of polynucleotides based on the distance-dependent optical properties of gold nanoparticles. Science 277(5329):1078–1081

Ellis D, Goodacre R (2001) Rapid and quantitative detection ofthe microbial spoilage of muscle foods: current status and futuretrends. Trends Food Sci Technol 12:414–424

El-Temsah YS, Joner EJ (2012) Impact of Fe and Ag nanoparticles on seed germination and differences in bioavailability during exposure in aqueous suspension and soil. Environ Toxicol 27(1):42–49

Esfanjani AF, Assadpour E, Jafari SM (2018) Improving the bioavailability of phenolic compounds by loading them within lipid-based nanocarriers. Trends Food Sci Technol 76:56–66

Evtugyn G, Porfireva A, Stepanova V, Kutyreva M, Gataulina A, Ulakhovich N, Evtugyn V, Hianik T (2013) Impedimetric aptasensor for ochratoxin A determination based on Au nanoparticles stabilized with hyper-branched polymer. Sensors 13(12):16129–16145

Ezati P, Bang YJ, Rhim JW (2021) Preparation of a shikonin-based pH-sensitive color indicator for monitoring the freshness of fish and pork. Food Chem 337:127995

Fathi Achachlouei B, Zahedi Y (2018) Fabrication and characterization of CMC-based nanocomposites reinforced with sodium montmorillonite and TiO_2 nanomaterials. Carbohydr Polym 199:415–425

Food and Drug Administration (2010). Available: http://www.fda.gov/Food/default.htm

Fu Z, Zhou X, Xing D (2013a) Rapid colorimetric gene-sensing of food pathogenic bacteria using biomodification-free gold nanoparticle. Sens Actuators B Chem 182:633–641

Fu Z, Zhou X, Xing D (2013b) Sensitive colorimetric detection of Listeria monocytogenes based on isothermal gene amplification and unmodified gold nanoparticles. Methods 64(3):260–266

Gan T, Li K, Wu K (2008) Multi-wall carbon nanotube-based electrochemical sensor for sensitive determination of Sudan I. Sens Actuators B Chem 132(1):134–139

Gharibzahedi SMT, Jafari SM (2017) Nanocapsule formation by cyclodextrins. In: Jafari SM (ed) Nanoencapsulation technologies for the food and nutraceutical industries. Academic Press, pp 187–261

Ghasemi S, Jafari SM, Assadpour E, Khomeiri M (2017) Production of pectin-whey protein nanocomplexes as carriers of orange peel oil. Carbohydr Polym 177:369–377

Ghasemi S, Jafari SM, Assadpour E, Khomeiri M (2018) Nanoencapsulation of d-limonene within nanocarriers produced by pectin-whey protein complexes. Food Hydrocoll 77:152–162

Ghorani B, Alehosseini A, Tucker N (2017) Nanocapsule formation by electrospinning. In: Jafari SM (ed) Nanoencapsulation technologies for the food and nutraceutical industries. Academic Press, pp 264–319

Gopalakrishnan A, Sha R, Vishnu N, Kumar R, Badhulika S (2018) Disposable, efficient and highly selective electrochemical sensor based on Cadmium oxide nanoparticles decorated screen-printed carbon electrode for ascorbic acid determination in fruit juices. Nano-Struct Nano-Objects 16:96–103

Goriushkina TB, Soldatkin AP, Dzyadevych SV (2009) Application of amperometric biosensors for analysis of ethanol, glucose, and lactate in wine. J Agric Food Chem 57(15):6528–6535

Han JW, Ruiz-Garcia L, Qian JP, Yang XT (2018) Food packaging: a comprehensive review and future trends. Compr Rev Food Sci Food Saf 17:860–877

Haratifar S, Guri A (2017) Nanocapsule formation by caseins. In: Jafari SM (ed) Nanoencapsulation technologies for the food and nutraceutical industries. Academic Press, pp 140–164

He L, Wang F, Chen Y, Liu Y (2018) Rapid and sensitive colorimetric detection of ascorbic acid in food based on the intrinsic oxidase-like activity of MnO_2 nanosheets. Luminescence 33(1): 145–152

Hosseini SM, Ghiasi F, Jahromi M (2017) Nanocapsule formation by complexation of biopolymers. In: Jafari SM (ed) Nanoencapsulation technologies for the food and nutraceutical industries. Academic Press, pp 447–492

Hussain A, Sun DW, Pu H (2020) Bimetallic core shelled nanoparticles (Au@ AgNPs) for rapid detection of thiram and dicyandiamide contaminants in liquid milk using SERS. Food Chem 317:126429

Hussain A, Pu H, Hu B, Sun DW (2021) Au@ Ag-TGANPs based SERS for facile screening of thiabendazole and ferbam in liquid milk. Spectrochim Acta A Mol BiomolSpectrosc 245: 118908
IFT (2004) Bacteria associated with foodborne diseases. A Scientific Status Summary of the Institute of Food Technologists, Chicago
Jafari SM (2017a) An overview of nanoencapsulation techniques and their classification. In: Jafari SM (ed) Nanoencapsulation technologies for the food and nutraceutical industries. Academic Press, pp 1–34
Jafari SM (2017b) An Introduction to nanoencapsulation techniques for the food bioactive ingredients. In: Jafari SM (ed) Nanoencapsulation of food bioactive ingredients. Academic Press, pp 1–62
Jafari SM, Paximada P, Mandala I, Assadpour E, Mehrnia MA (2017) Encapsulation by nanoemulsions. In: Jafari SM (ed) Nanoencapsulation technologies for the food and nutraceutical industries. Academic Press, pp 36–73
Jha SN, Jaiswal P, Grewal MK, Gupta M, Bhardwaj R (2016) Detection of adulterants and contaminants in liquid foods—a review. Crit Rev Food Sci Nutr 56(10):1662–1684
Jiang W, Beloglazova NV, Luo P, Guo P, Lin G, Wang X (2017) A dual-color quantum dots encoded frit-based immunoassay for visual detection of aflatoxin M1 and pirlimycin residues in milk. J Agric Food Chem 65(8):1822–1828
Jiang Y, Sun DW, Pu H, Wei Q (2019) Ultrasensitive analysis of kanamycin residue in milk by SERS-based aptasensor. Talanta 197:151–158
Kalpana R, Devasena T, Sudha S (2020) Optoelectronic method of test for melamine adulteration in milk using paranitroaniline modified silver nanoparticles. In: 2020 IEEE 20th International Conference on Nanotechnology (IEEE-NANO), 29 July 2020
Katouzian I, Jafari SM (2016) Nano-encapsulation as a promising approach for targeted delivery and controlled release of vitamins. Trends Food Sci Technol 53:34–48
Khalkho BR, Kurrey R, Deb MK, Shrivas K, Thakur SS, Pervez S, Jain VK (2020) L-cysteine modified silver nanoparticles for selective and sensitive colorimetric detection of vitamin B_1 in food and water samples. Heliyon 6(2):e03423
Kim SW, Cha SH (2014) Thermal, mechanical, and gas barrier properties of ethylene–vinyl alcohol copolymer-based nanocomposites for food packaging films: Effects of nanoclay loading. J Appl Polym Sci 131(11):40289
Kim HJ, Bennetto HP, Halablab MA, Choi C, Yoon S (2006) Performance of an electrochemical sensor with different types of liposomal mediators for the detection of hemolytic bacteria. Sens Actuators B Chem 119:143–149
Koubova V, Brynda E, Karasova L, Škvor J, Homola DJ, Tobiška P, Rošický J (2001) Detection of foodborne pathogens using surface plasmon resonance biosensors. Sens Actuators B Chem 74: 100–105
Kumar A, Purohit B, Maurya PK, Pandey LM, Chandra P (2019) Engineered nanomaterial assisted signal-amplification strategies for enhancing analytical performance of electrochemical biosensors. Electroanalysis 31(9):1615–1629
Labib M, Sargent EH, Kelley SO (2016) Electrochemical methods for the analysis of clinically relevant biomolecules. Chem Rev 116:9001–9090
Lerga TM, Skouridou V, Bermudo MC, Bashammakh AS, El-Shahawi MS, Alyoubi AO, O'Sullivan CK (2020) Gold nanoparticle aptamer assay for the determination of histamine in foodstuffs. Microchim Acta 187(8):1–9
Li T, Jin L, Feng K, Yang T, Yue X, Wu B, Ding S, Liang X, Huang G, Zhang J (2020) A novel low-field NMR biosensor based on dendritic superparamagnetic iron oxide nanoparticles for the rapid detection of *Salmonella* in milk. LWT 133:110149
Lin X, Ni Y, Kokot S (2013) Glassy carbon electrodes modified with gold nanoparticles for the simultaneous determination of three food antioxidants. Anal Chim Acta 765:54–62

Lu J, Do I, Drzal LT, Worden RM, Lee I (2008) Nanometal-decorated exfoliated graphite nanoplatelet based glucose biosensors with high sensitivity and fast response. ACS Nano 2(9):1825–1832

Mao X, Yang L, Su X, Li Y (2006) A nanoparticle amplificationbased quartz crystal microbalance DNA sensor for detection ofEscherichia coliO157:H7. BiosensBioelectron 21:1178–1185

Mason TG, Wilking JN, Meleson K, Chang CB, Graves SM (2006) Nanoemulsions: formation, structure, and physical properties. J Phys Condens Matter 18(41):R635–R666

McClements D (2005) Food Emulsions: Principles, Practices, and Techniques. CRC Press, Boca Raton, FL, USA

McClements DJ, Rao J (2011) Food-Grade Nanoemulsions: Formulation, fabrication, properties, performance, biological fate, and potential toxicity. Crit Rev Food Sci Nutr 51:285–330

Messaoud NB, Ghica ME, Dridi C, Ali MB, Brett CM (2017) Electrochemical sensor based on multiwalled carbon nanotube and gold nanoparticle modified electrode for the sensitive detection of bisphenol A. Sens Actuators B Chem 253:513–522

Mokhtari S, Jafari SM, Assadpour E (2017) Development of a nutraceutical nano-delivery system through emulsification/internal gelation of alginate. Food Chem 229:286–295

Moshahary S, Mishra P (2021) Synthesis of silver nanoparticles (AgNPs) using culinary banana peel extract for the detection of melamine in milk. J Food Sci Tech 58(2):797–804

Neethirajan S, Jayas DS (2011) Nanotechnology for the food and bioprocessing industries. Food Bioprocess Tech 4:39–47

Oh SY, Lee MJ, Heo NS, Kim S, Oh JS, Lee Y, Jeon EJ, Moon H, Kim HS, Park TJ, Moon G (2019) Cuvette-type LSPR sensor for highly sensitive detection of melamine in infant formulas. Sensors 19(18):3839

Oleyaei SA, Almasi H, Ghanbarzadeh B, Moayedi AA (2016) Synergistic reinforcing effect of TiO_2 and montmorillonite on potato starch nanocomposite films: thermal, mechanical and barrier properties. CarbohydrPolym 152:253–262

Ozdemir C, Yeni F, Odaci D, Timur S (2010) Electrochemical glucose biosensing by pyranose oxidase immobilized in gold nanoparticle-polyaniline/AgCl/gelatin nanocomposite matrix. Food Chem 119(1):380–385

Parisi C, Vigani M, Rodríguez-Cerezo E (2015) Agricultural nanotechnologies: what are the current possibilities? Nano Today 10(2):124–127

Polo E, del Pino P, Pelaz B, Grazu V, Jesus M (2013) Plasmonic-driven thermal sensing: ultralow detection of cancer markers. Chem Commun 49(35):3676–3678

Primožič M, Knez Ž, Leitgeb M (2021) (Bio) Nanotechnology in food science—Food packaging. Nano 11(2):292

Purohit B, Vernekar PR, Shetti NP, Chandra P (2020) Biosensor nanoengineering: Design, operation, and implementation for biomolecular analysis. Sensors Int 1:100040

Rai M, Ingle A (2015) Role of nanotechnology in agriculture with special reference to management of insect pests. Applied MicrobiolBiot 94(2):287–293

Raju, K. V. R., Yoshihisa, O (2002) Report of the APO. Seminar on Quality Control for Processed Food held in the Republic of China, 2002; 02-AG-GE-SEM-02

Ran R, Sun Q, Baby T, Wibowo D, Middelberg APJ, Zhao CX (2017) Multiphase microfluidic synthesis of micro- and nanostructures for pharmaceutical applications. Chem Eng Sci 169:78–96

Rasooly A, Herold KE (2006) Biosensors for the analysis of food- and waterborne pathogens and their toxins. J AOAC Int 89:873–883

Rattanata N, Klaynongsruang S, Leelayuwat C, Limpaiboon T, Lulitanond A, Boonsiri P, Chio-Srichan S, Soontaranon S, Rugmai S, Daduang J (2016) Gallic acid conjugated with gold nanoparticles: antibacterial activity and mechanism of action on foodborne pathogens. Int J Nanomedicine 11:3347–3356

Ravichandran R (2010) Nanotechnology applications in food and food processing: innovative green approaches, opportunities and uncertainties for global market. Int J Green Nanotechnol: Phys Chem 1(2):P72–P96

Rehman A, Jafari SM, Aadil RM, Assadpour E, Randhawa MA, Mahmood S (2020) Development of active food packaging via incorporation of biopolymeric nanocarriers containing essential oils. Trends Food Sci Technol 101:106–121

Robertson GL (2006) Food packaging: principles and practice. CRC Press, Boca Raton

Roy S, Malode SJ, Shetti NP, Chandra P (2019) Modernization of biosensing strategies for the development of lab-on-chip integrated systems. In: Krishnaraj RN, Sani RK (eds) Bioelectrochemical interface engineering. John Wiley & Sons, pp 325–342

Sadeghi R, Mehryar L, Karimi M, Kokini J (2017) Nanocapsule formation by individual biopolymer nanoparticles. In: Jafari SM (ed) Nanoencapsulation technologies for the food and nutraceutical industries. Academic Press, pp 404–446

Saini A, Panesar PS, Bera MB (2019) Valorization of fruits and vegetables waste through green extraction of bioactive compounds and their nanoemulsions-based delivery system. Bioresour Bioprocess 6(1):1–2

Saini A, Panwar D, Panesar PS, Bera MB (2020) Encapsulation of functional ingredients in lipidic nanocarriers and antimicrobial applications: a review. Environ Chem Lett:1–28

Sandros MG, Shete V, Benson DE (2006) Selective, reversible, reagentless maltose biosensing with core–shell semiconducting nanoparticles. Analyst 131(2):229–235

Sasidharan S, Raj S, Sonawane S, Sonawane S, Pinjari D, Pandit AB, Saudagar P (2019) Nanomaterial synthesis: chemical and biological route and applications. In: Pottathara YB, Thomas S, Kalarikkal N, Grohens Y, Kokol V (eds) Nanomaterials synthesis: design, fabrication and applications. Elsevier, pp 27–51

Scrinis G, Lyons K (2007) The emerging nano-corporate paradigm: nanotechnology and the transformation of nature, food and agri-food systems. IJSAF 15(2):22–44

Seddaoui N, Amine A (2020) A sensitive colorimetric immunoassay based on poly (dopamine) modified magnetic nanoparticles for meat authentication. LWT 122:109045

Selvakumar LS, Ragavan KV, Abhijith KS, Thakur MS (2013) Immunodipstick based gold nanosensor for vitamin B_{12} in fruit and energy drinks. Anal Methods 5(7):1806–1810

Şenocak A (2020) Fast, simple and sensitive determination of coumaric acid in fruit juice samples by magnetite nanoparticles-zeolitic imidazolate framework material. Electroanalysis 32(10): 2330–2339

Sharpe E, Frasco T, Andreescu D, Andreescu S (2013) Portable ceria nanoparticle-based assay for rapid detection of food antioxidants (NanoCerac). Analyst 138(1):249–262

Shrivas K, Naik W, Kumar D, Singh D, Dewangan K, Kant T, Yadav S, Jaiswal N (2021) Experimental and theoretical investigations for selective colorimetric recognition and determination of arginine and histidine in vegetable and fruit samples using bare-AgNPs. Microchem J 160:105597

Silvestre C, Duraccio D, Cimmino S (2011) Food packaging based on polymer nanomaterials. Prog Polym Sci 36(12):1766–1782

Sonneville-Aubrun O, Simonnet JT, L'alloret F (2004) Nanoemulsions: a new vehicle for skincare products. Adv Colloid Interf Sci 108:145–149

Sozer N, Kokini JL (2009) Nanotechnology and its applications in the food sector. Trends Biotechnol 27(2):82–89

Srivastava AK, Dev A, Karmakar S (2018) Nanosensors and nanobiosensors in food and agriculture. Environ Chem Lett 16(1):161–182

Steinvil A, Zhang YJ, Lee SY, Pang S, Waksman R, Chen SL, Garcia-Garcia HM (2016) Intravascular ultrasound-guided drug-eluting stent implantation: an updated meta-analysis of randomized control trials and observational studies. Int J Cardiol 216:133–139

Tadros T, Izquierdo P, Esquena J, Solans C (2004) Formation and stability of nano-emulsions. Adv Colloid Interf Sci 108:303–318

Taniguchi N (1974) On the basic concept of nanotechnology. In: Proceeding of the International Conference on Production Engineering, Tokyo, pp 18–23

Tapia-Hernández JA, Rodríguez-Félix F, Katouzian I (2017) Nanocapsule formation by electrospraying. In: Jafari SM (ed) . Nanoencapsulation technologies for the food and nutraceutical industries, Academic Press, pp 320–345

Terzi F, Zanfrognini B, Ruggeri S, Dossi N, Casagrande GM, Piccin E (2017) Amperometric paper sensor based on Cu nanoparticles for the determination of carbohydrates. Sens Actuators B Chem 245:352–358

Thomas EJ, King RK, Burchak J, Gannon VP (1991) Sensitive and specific detection of *Listeria monocytogenes* in milk and ground beef with the polymerase chain reaction. Appl Environ Microbiol 57:2576

Tominaga M, Nomura S, Taniguchi I (2009) D-Fructose detection based on the direct heterogeneous electron transfer reaction of fructose dehydrogenase adsorbed onto multi-walled carbon nanotubes synthesized on platinum electrode. Biosens Bioelectron 24(5):1184–1188

Tunesi MM, Kalwar N, Abbas MW, Karakus S, Soomro RA, Kilislioglu A, Abro MI, Hallam KR (2018) Functionalised CuO nanostructures for the detection of organophosphorus pesticides: a non-enzymatic inhibition approach coupled with nano-scale electrode engineering to improve electrode sensitivity. Sens Actuators B Chem 260:480–489

Vaisocherová-Lísalová H, Víšová I, Ermini ML et al (2016) Low-fouling surface plasmon resonance biosensor for multi-step detection of foodborne bacterial pathogens in complex food samples. Biosens Bioelectron 80:84–90

Valdés MG, González AC, Calzón JA, Díaz-García ME (2009) Analytical nanotechnology for food analysis. Microchim Acta 166(1–2):1–9

Valdivieso-Garcia A, Riche E, Abubakar O, Waddell TE, Brooks BW (2001) A double antibody sandwich enzyme-linked immunosorbent assay for the detection of *Salmonella* using biotinylated monoclonal antibodies. J Food Protect 64:1166–1171

Varshney M, Li Y (2007) Interdigitated array microelectrodebased impedance biosensor coupled with magnetic nanoparticle–antibody conjugates for detection of *Escherichia coli* O157:H7 in food samples. Biosens Bioelectron 22:2408–2414

Vyas SS, Jadhav SV, Majee SB, Shastri JS, Patravale VB (2015) Development of immunochromatographic strip test using fluorescent, micellar silica nanosensors for rapid detection of B. abortus antibodies in milk samples. Biosens Bioelectron 70:254–260

Walia N, Dasgupta N, Ranjan S, Ramalingam C, Gandhi M (2019) Methods for nanoemulsion and nanoencapsulation of food bioactives. Environ Chem Lett. https://doi.org/10.1007/s10311-019-00886-w

Wang JJ, Liu BH, Hsu YT, Yu FY (2011) Sensitive competitive direct enzyme-linked immunosorbent assay and gold nanoparticle immunochromatographic strip for detecting aflatoxin M1 in milk. Food Control 22:964–969

Wang J, Wang Z, Liu J, Li H, Li QX, Li J, Xu T (2013) Nanocolloidal gold-based immuno-dip strip assay for rapid detection of Sudan red I in food samples. Food Chem 136(3-4):1478–1483

Wang J, Wu X, Wang C, Rong Z, Ding H, Li H, Li S, Shao N, Dong P, Xiao R, Wang S (2016) Facile synthesis of Au-coated magnetic nanoparticles and their application in bacteria detection via a SERS method. ACS Appl Mater Interfaces 8(31):19958–19967

Wang C, Hu L, Zhao K, Deng A, Li J (2018a) Multiple signal amplification electrochemiluminescent immunoassay for Sudan I using gold nanorods functionalized graphene oxide and palladium/aurum core-shell nanocrystallines as labels. Electrochim Acta 278:352–362

Wang Y, Zhang P, Fu W, Zhao Y (2018b) Morphological control of nanoprobe for colorimetric antioxidant detection. Biosens Bioelectron 122:183–188

Wei H, Wang E (2008) Fe_3O_4 magnetic nanoparticles as peroxidase mimetics and their applications in H_2O_2 and glucose detection. Anal Chem 80(6):2250–2254

Wei Q, Liu T, Pu H, Sun DW (2020a) Development of a fluorescent microwave-assisted synthesized carbon dots/Cu^{2+} probe for rapid detection of tea polyphenols. J Food Process Eng 43(7):e13419

Wei S, Wang X, Pang B, Li H, Shi X, Zhao C, Li J, Wang J (2020b) Analyte-triggered autoacceleration of 4-mercaptophenylboronic acid-mediated aggregation of silver nanoparticles for facile and one-step ratiometric colorimetric method for detection of ascorbic acid. Microchem J 158:105122

WHO Global burden of foodborne diseases (2015)

Wu S, Duan N, Shi Z, Fang C, Wang Z (2014) Simultaneous aptasensor for multiplex pathogenic bacteria detection based on multicolorupconversion nanoparticles labels. Anal Chem 86(6): 3100–3107

Xia F, Zuo X, Yang R, Xiao Y, Kang D, Vallée-Bélisle A, Gong X, Yuen JD, Hsu BB, Heeger AJ, Plaxco KW (2010) Colorimetric detection of DNA, small molecules, proteins, and ions using unmodified gold nanoparticles and conjugated polyelectrolytes. Proc Natl Acad Sci 107(24): 10837–10841

Yang Y, Zhou J, Zhang H, Gai P, Zhang X, Chen J (2013) Electrochemical evaluation of total antioxidant capacities in fruit juice based on the guanine/graphene nanoribbon/glassy carbon electrode. Talanta 106:206–211

Yao DS, Wen SM, Liu DL, Xie CF, Bai Y, Ran YH (2004) The primary study on the detection of sterigmatocystin by biologic enzyme electrode modified with the multiwall carbon nanotubes. Sheng Wu Gong Cheng Xue Bao 20(4):601–606

Zamolo VA, Valenti G, Venturelli E, Chaloin O, Marcaccio M, Boscolo S, Castagnola V, Sosa S, Berti F, Fontanive G, Poli M (2012) Highly sensitive electrochemiluminescentnanobiosensor for the detection of palytoxin. ACS Nano 6(9):7989–7997

Zhang C, Yin AX, Jiang R, Rong J, Dong L, Zhao T, Sun LD, Wang J, Chen X, Yan CH (2013a) Time–Temperature indicator for perishable products based on kinetically programmable Ag overgrowth on Au nanorods. ACS Nano 7(5):4561–4568

Zhang Z, Lin M, Zhang S, Vardhanabhuti B (2013b) Detection of aflatoxin M1 in milk by dynamic light scattering coupled with superparamagnetic beads and gold nanoprobes. J Agric Food Chem 61(19):4520–4525

Zhong M, Yang L, Yang H, Cheng C, Deng W, Tan Y, Xie Q, Yao S (2019) An electrochemical immunobiosensor for ultrasensitive detection of *Escherichia coli* O157: H7 using CdS quantum dots-encapsulated metal-organic frameworks as signal-amplifying tags. Biosens Bioelectron 126:493–500

Advancements in Molecular Techniques for the Detection of Foodborne Pathogens

9

Reena Kumari, Nitish Sharma, Sanjukta Samurailatpam, Amit Kumar Rai, and Sudhir P. Singh

Abstract

Advancements in the survival mechanisms of foodborne pathogens have confronted several challenges for the storage and preservation of food products, which, when consumed, results in unwellness or sickness. One of the significant challenges in the field of food science and technology is the accuracy of the detection of foodborne pathogens. The detection of foodborne pathogens involves different processes, some of which are based on the detection by using the microorganism's genetic material (DNA/RNA) or by using the specific enzyme or peptide sequence. Modern microbiology and biotechnology offering FBP detection techniques involve nucleotide-based methods (e.g., PCR, AFLP, RFLP, and FISH), signal-based methods (e.g., biosensors), and immunological methods (ELISA and MALDI-TOF/MS). The application of nanotechnology and bioinformatics is gaining popularity in the recent past to improve the accuracy in identification of bacterial toxins and viruses. Different molecular identification methods applied to detect different food pathogens, challenges, and prospects have been discussed in this chapter.

Keywords

Pathogens · Molecular techniques · PCR · Signal-based methods

R. Kumari · S. Samurailatpam · A. K. Rai
Institute of Bioresources and Sustainable Development, Sikkim Centre, Tadong, Gangtok, Sikkim, India

N. Sharma · S. P. Singh (✉)
Center of Innovative and Applied Bioprocessing, SAS Nagar, Mohali, Punjab, India
e-mail: sudhirsingh@ciab.res.in

P. Chandra, P. S. Panesar (eds.), *Nanosensing and Bioanalytical Technologies in Food Quality Control*, https://doi.org/10.1007/978-981-16-7029-9_9

9.1 Introduction

The interdependence of microorganisms for their growth and survival is often linked to different food products, as described earlier by Hippocrates in 460 BC (Bintsis 2017). Some food products are the obvious environment for microbial growth, like the interrelation of fermentation associated generally regarded as safe (GRAS) organisms, which benefit human health (Rai et al. 2017; Ayyash et al. 2020). Nonetheless, innumerable microorganisms contribute to food spoilage as well and detriment human health. Microorganisms involved in food contamination and illness are recognized as foodborne pathogens (FBP) (Bintsis 2017). These FBPs are accountable for millions of foodborne illnesses, diseases, and even deaths either directly *via* infectious biological agents or indirectly *via* the production of bacterial/viral toxins, mycotoxins, or toxic metabolites in the system (Sekse et al. 2017). As per reports of Centers for Disease Control and Prevention (CDC) in the United States (Zeng et al. 2016) and European Union in 2015 (Bintsis 2017), Australia, Netherlands, France, New Zealand, and India, FBPs have affected health and economy at a global level. FBPs like *Staphylococcus aureus*, *Listeria monocytogenes*, *Campylobacter* species, Hepatitis E virus, norovirus, *Salmonella* species (Foddai and Grant 2020), *Clostridium botulinum*, *Escherichia coli*, *Exophiala dermatitidis*, *Bacillus cereus*, *Candida* species, *Toxoplasma gondii*, *Trichosporon asahii*, *Yersinia enterocolitica*, Hepatitis A virus, etc., are reported to be associated with contamination of a wide range of food items, exerting adverse effect on human health (Rai et al. 2016; Zeng et al. 2016; Bintsis 2017; Zhang et al. 2019; Kumar et al. 2019).

This global morbidity and mortality rate caused by FBPs has persistently influenced socio-economic growth and the immune health of the forthcoming generations. In a recent study, Bhist et al. (2021) have reported the effect of different seasons on the growth of specific FBPs. It was concluded that bacterial growth (of *Staphylococcus aureus*, *Salmonella* spp., *Bacillus* species, *Vibrio* sp., *E. coli*, etc.) was higher in summers, whereas winter favored the growth of viruses like rotavirus and norovirus. Therefore, it becomes an essential service to identify the FBPs. FBPs are detected by microbiological, biochemical, and molecular techniques. Molecular biology techniques offer a better accuracy to identify pathogens because of the specific and speedy detection of the microorganisms. The conventional methods of FBP detection are often arduous and questionable with precision, especially the detection of viral pathogens. In order to combat the challenges confronted in food quality, safety, and hygiene assessment, it is important to select an appropriate technique, keeping in mind the selectivity and specificity of the FBP detection process. The emergence of various FBP strains has stimulated researchers worldwide to develop advanced techniques for their identification. The rapid detection molecular techniques for FBPs include polymerase chain reaction (PCR), upgraded techniques for enhanced specificity of pathogens, development of bio- and nano-sensors, immunological and sequencing techniques. Table 9.1 comprehensively summarizes the applied molecular techniques for detection of FBPs. A detail on such techniques is discussed in the following sections.

Table 9.1 Applied molecular techniques for detection of FBPs

Molecular technique	Food pathogen detected	Reference
Polymerase Chain Reaction (PCR)	*Salmonella* spp.	Wang et al. (2018)
	Listeria monocytogenes	Liu et al. (2019)
	Escherichia coli	Naravaneni and Jamil (2005)
	Vibrio cholera	Shangkuan et al. (1995)
	Vibrio vulnificus	Brauns et al. (1991)
	Vibrio parahaemolyticus	Liu et al. (2019)
	Staphylococcus aureus	Liu et al. (2019)
	Yersinia spp.	Liu et al. (2019)
	Proteus mirabilis	Zhang et al. (2020)
	Enterococcus faecalis, *Vibrio fluvialis*, *E. coli*, *Shigella* spp., *Campylobacter jejuni*, *Proteus mirabilis*, *Salmonella enterica*, β-*Streptococcus hemolyticus*	Liu et al. (2019)
	Trypanosoma cruzi	de Oliveira et al. (2019)
	Norovirus, Hepatitis A virus	Li et al. (2018a, b)
Loop mediated isothermal amplification	*Salmonella* spp.	Sayad et al. (2016), Mori and Notomi (2009)
	Bacillus stearothermophilus	Umesha and Manukumar (2016)
	E. coli	Oh et al. (2016)
	Yersinia pseudotuberculosis	Horisaka et al. (2004)
	Verotoxin producing *Enterococci*, *Listeria* spp.	Mori and Notomi (2009)
	Campylobacter jejuni, *Vibrio parahaemolyticus*, *Staphylococcus aureus*, *Aeromonas caviae*	Shi et al. (2010)
	Campylobacter spp.	Shi et al. (2010)
Single-stranded conformation polymorphism PCR	*Rastolnia solanacearum*	Umesha et al. (2012)
	Bacillus cereus	Kim et al. (2016)
	E. coli	Kim et al. (2016)
	Salmonella spp.,	Kim et al. (2010, 2016)
	Clostridium perfringens	Oh et al. (2012), Kim et al. (2016)
	Campylobacter jejuni	Oh et al. (2008), Oh et al. (2012)
	Campylobacter coli	Kim et al. (2016)
	Listeria monocytogenes	Oh et al. (2012), Kim et al. (2016)
	Vibrio parahaemolyticus	Oh et al. (2012), Kim et al. (2016)
	Yersinia enterocolitica	Kim et al. (2016)

(continued)

Table 9.1 (continued)

Molecular technique	Food pathogen detected	Reference
	Enterobacter sakazaki, *Staphylococcus aureus*	Oh et al. (2012)
	Cronobacter sakazaki, *Enterococcus* spp., *Shigella* spp., *Staphylococcus aureus*, *Vibrio vulnificus*	Kim et al. (2010, 2016)
Ligase chain reaction PCR (LCR-PCR)	*Mycobacterium tuberculosis*, *Chlamydia trachomatis*, *Neisseria gonorrhoneae*	Fakruddin et al. (2013)
	Different species of *Staphylococcus*, *Campylobacter* spp., *Mycoplasma* spp. *Samlonella* spp.	Cremonesi et al. (2009)
	Vibrio spp.	Cariani et al. (2012)
Restriction fragment length polymorphism	*Staphylococcus aureus*	Paillard et al. (2003)
	Listeria spp.	Paillard et al. (2003)
	E. coli	Fields et al. (1997)
	Campylobacter jejuni	Arguello et al. (2018)
Amplified fragment length polymorphism	*Fuarium mangiferae*, *Fuarium proliferatum*, *Fuarium sacchari*, *Fuarium sterilihyphosum*, *Fuarium subglutinans*	Parisi et al. (2010), Umesha and Manukumar (2016)
	Campylobacter spp.	Schouls et al. (2003), Johnsen et al. (2006)
	Salmonella spp.	Torpdahl et al. (2005), Lindstedt et al. (2000)
	Bacillus cereus	Ripabelli et al. (2000)
Random amplified polymorphic DNA	*Photobacterium damsilae*	Magarino et al. (2000)
	Listeria spp.	Aguado et al. (2001)
	Salmonella spp.	Shangkuan and Lin (1998)
	E. coli	Darkazanli et al. (2018)
	Enterococcus lactis	Braiek et al. (2018)
Nucleic acid sequence based amplification	*Salmonella* spp.	Zhai et al. (2019), Bodulev and Sakharov (2020)
	Listeria monocytogenes, *Campylobacter jejuni*, Rotavirus, *Bacillus* spp., Hepatitis A virus	Aslan et al. (2020)
Rolling circle amplification	*E. coli*	Zhong and Zhao (2018), Li et al. (2021)
	Listeria monocytogenes	Zhan et al. (2020)
	Salmonella typhimurium	Zhong and Zhao (2018)
	Vibrio parahaemolyticus	Song et al. (2018, 2019)
	Cronobacter spp.	Liu et al. (2020)
	C. perfringens	Milton et al. (2021)
	Staphylococcus aureus	Yang et al. (2019)
	Shigella spp.	Wang et al. (2017)

(continued)

Table 9.1 (continued)

Molecular technique	Food pathogen detected	Reference
Strand displacement amplification	*E. coli*	Zhong and Zhao (2018)
	Salmonella spp.	Wang et al. (2020a, b, c), Zhang et al. (2021)
	Staphylococcus aureus	Cai et al. (2019)
Fluorescent in-situ hybridization	*Salmonella* spp.	Aslan et al. (2020), Salimi et al. (2020)
	Listeria spp.	Rocha et al. (2019), Aslan et al. (2020)
	Yersinia spp., *Pseudomonas* spp., *Enterobacteriaceae*, *Helicobacter pylori*, *E. coli*, *B. cereus*, *Mycobacterium avium*, *Campylobacter* spp.,	Rohde et al. (2017), Dias and Rathyanaka (2018)
Peptide nucleic acid FISH	*Listeria monocytogenes*, *Vibrio* spp.	Shan et al. (2018)
	Salmonella spp.	Adebowale et al. (2020)
DNA microarray	*Salmonella* spp., *E. coli* *Listeria monocytogenes*	Li et al. (2017)
	Norovirus	Hu et al. (2015)
	E. coli, *Brucella* spp., *Legionella pneumophila*	Ranjbar et al. (2017)
	E. coli, *Campylobacter jejuni*,	Liu et al. (2017)
	E. coli, *Bacillus* spp.,	Aslan et al. (2020)
	Shigella spp.	Ranjbar et al. (2017), Aslan et al. (2020)
	Salmonella spp.	Aslan et al. (2020)
	Vibrio cholera	Ranjbar et al. (2017)
	Enterobactericeae	Zhu et al. (2007)
Multiplex PCR	*Yersinia enterocolitica*	Tao et al. (2020)
	Staphylococcus aureus	Chen et al. (2012), Wei et al. (2018)
	Listeria monocytogenes	Du et al. (2020), Tao et al. (2020)
	Vibrio cholera, *Vibrio parahaemolyticus*, *Clostridium botulinum*, *Bacillus cereus*	Tao et al. (2020)
	E. coli	Du et al. (2020), Tao et al. (2020)
	Salmonella spp.	Du et al. (2020), Tao et al. (2020)
Signal based techniques	*Listeria monocytogenes*, *E. coli*, *Campylobacter jejuni*, *Salmonella* spp.	Saravanan et al. (2020), Zhang et al. (2020)
Immunological techniques	*Campylobacter* spp.	Khan et al. (2018), Aslan et al. (2020), Saravanan et al. (2020)

(continued)

Table 9.1 (continued)

Molecular technique	Food pathogen detected	Reference
	Salmonella spp.	Jenikova et al. (2000)
	Staphylococcus spp., *Listeria monocytogenes*	Umesha and Manukumar (2016)
Pulse field gel electrophoresis	*E. coli*, *Listeria* spp.	Umesha and Manukumar (2016)
	Salmonella spp.	Ha et al. (2020); Taheri et al. (2018)
	E. coli	Akindolire and Ateba (2018)
Ribotyping	*Listeria* spp.	Rao and Arora (2020)
Repetitive extragenic palindromic PCR	*E. coli*	Deng et al. (2018)
	Salmonella spp.	Deng et al. (2018)
	Campylobacter jejuni	Yadav et al. (2017)
	Vibrio parahaemolyticus	Sadeghi et al. (2019)
Multilocus sequencing typing	*Streptococcus pneumonia*	Umesha and Manukumar (2016)
	Campylobacter jejuni	Hsu et al. (2020)
	E. coli	Ramdan et al. (2020)
	Vibrio parahaemolyticus	Escalona et al. (2017)
	Staphylococcus aureus	Yang et al. (2018)
	Bacillus cereus	Yang et al. (2017)
DNase treated DNA PCR	*E. coli*, *Salmonella enterica*, *Listeria monocytogenes*, *Vibrio parahaemolyticus*	Umesha and Manukumar (2016)

9.2 Molecular Methods for Detection of FBPs

9.2.1 Basic Methods

The molecular techniques applied in FBP detection include use of genetic material for sequencing and identification. These techniques are not only time-saving but also show specificity to the identification of FBPs.

9.2.1.1 Cyclic Amplification Techniques

Polymerase Chain Reaction (PCR)

The PCR technique is frequently being used for FBP detection for the last two to three decades. This is a gold standard technique offering promising results and so is followed as a primary laboratory technique. In this technique, amplification of the specific nucleic acid sequence of the targeted FBP is done. This technique is preferred due to lesser error, higher accuracy, rapidity, and capacity to identify FBP even in minimal quantity in the food sample. In PCR process, the DNA polymerase is used to amplify a specific nucleic acid sequence. *Taq* DNA polymerase and *Pfu* DNA polymerase isolated from *Thermus aquaticus* and *Pyrococcus furiosus*, respectively, are frequently used enzymes in PCR (Valasek and Repa 2005).

This technique has been applied in the detection of FBPs from different food and water sources. *Salmonella* spp. has been periodically reported to be identified by PCR method (Rahn et al. 1992; Eichelberg et al. 1994). The detection of *Salmonella* spp. is reported to be more sensitive by PCR method as compared to ELISA (Enzyme-Linked Immunosorbent Assay) technique (Murphy et al. 2007). Similarly, PCR-based detection technique is considered to be a benchmark for detecting *Shigella* spp. (Hartman et al. 1990; Frankel et al. 1990), *Listeria monocytogenes* (Furrer et al. 1991), *E. coli* (Tsai et al. 1993; Naravaneni and Jamil 2005), *Vibrio cholera*, *V. cholerae* biotype O1 (Shangkuan et al. 1995), *Vibrio vulnificus* (Brauns et al. 1991), *Vibrio parahaemolyticus* (Tada et al. 1992), *S. aureus* (Wilson et al. 1991), and *Yersinia* spp. (Ibrahim et al. 1992). In the twenty-first century, this technique was used for the rapid detection of FBPs, *viz.* in the detection of enteric pathogens, foodborne bacteria (*Enterobacter sakazakii*, *S. aureus*, *Proteus mirabilis*, *E. coli*, and *Shigella* spp.) in milk samples (Park et al. 2018; Zhu et al. 2018; Zhang et al. 2020).

Rapid quantification of nucleic acid copies is also possible by quantitative PCR (qPCR) or real-time PCR. The identification of common FBPs including *Enterococcus faecalis*, *Vibrio fluvialis*, *Y. enterocolitica*, *L. monocytogenes*, *E. coli*, *Shigella* spp. *S. aureus*, *Campylobacter jejuni*, *P. mirabilis*, *Salmonella enterica*, *Streptococcus hemolyticus*, and *Vibrio parahaemolyticus*, has been reported by real-time PCR (Liu et al. 2019). Real-time PCR has been considered as a sensitive and rapid process in FBP detection, such as emetic and diarrheal *B. cereus* in food samples (Fricker et al. 2007). The detection and quantification of transcripts of FBPs by qPCR are one of the best techniques with respect to sensitivity (Umesha and Manukumar 2016). qPCR has been applied to identify and quantify *Salmonella typhimurium* nucleic acid sequence (Wang et al. 2018), hemolysin from *Vibrio parahaemolyticus* (Ward and Bej 2006) and *Trypanosoma cruzi* (de Oliveira et al. 2019) (Fig. 9.1).

Loop-Mediated Isothermal Amplification

Loop-mediated isothermal amplification (LAMP) offers the benefits of detection with additional efficiency, rapid detection, ease of sample preparation, uncomplicated visualization, specificity, and less sensitivity to contaminants or inhibitory substances (Notomi et al. 2000). LAMP requires a set of four (and/or six) primers to recognize six specific DNA sites and only one type of DNA polymerase enzyme, preferably Bst DNA polymerase larger fragment (Mori and Notomi 2009; Fakruddin et al. 2013). These primer sets are designed in such a way that both the ends of inner primers lack AT-rich regions, temperature range (Tm) for each domain is restricted to 55–65 °C, length of the amplified region is limited to 200 base pairs (bp), and the distance between two regions should be limited to 40–60 bp. The target gene gets amplified within 1 h at 65 °C, and further extension of the amplified gene is processed to the continuing cycles (Nagamine et al. 2002; Ohtsuka et al. 2005; Oh

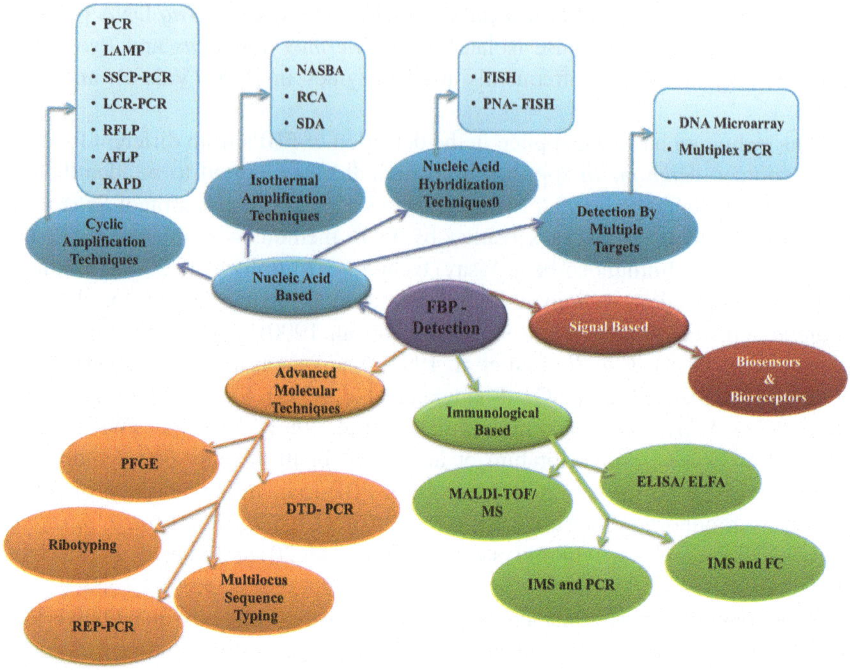

Fig. 9.1 Methods applied in foodborne pathogen detection. (*PCR* Polymerase chain reaction, *LAMP* loop-mediated isothermal amplification, *SSCP-PCR* single-stranded conformation polymorphism PCR, *LCR-PCR* ligase chain reaction PCR, *RFLP* restriction fragment length polymorphism, *AFLP* amplified fragment length polymorphism, *RAPD* random-amplified polymorphic DNA, *NASBA* nucleic acid sequence-based amplification, *RCA* rolling circle amplification, *SDA* strand displacement amplification, *FISH* fluorescent in situ hybridization, *PNA-FISH* peptide nucleic acid–FISH, *ELISA* enzyme-linked immunosorbent assay/ ELFA- enzyme-linked fluorescent assay, *MALDI-TOF/MS* matrix-assisted laser desorption ionization—time of flight/mass spectrometry, *IMS* immunomagnetic separation, *FC* flow cytometry, *PFGE* pulse field gel electrophoresis, *REP-PCR* repetitive extragenic palindromic-PCR, *DTD- PCR* Dnase-treated DNA (DTD) PCR

et al. 2016; Umesha and Manukumar 2016). Accuracy in the LAMP detection process is increased when used along with reverse transcription (RT-LAMP), even in a small sample quantity. It can also detect viruses and amplify DNA from RNA bases (Mori and Notomi 2009). Different software developed for LAMP primer designing are:

- LAMP designer (Premier Biosoft International, Palo Alto, Calif, USA). http://www.premierbiosoft.com/tech_notes/Loop-Mediated-Isothermal-Amplification.html
- Eiken Chemical Co. Ltd. Tokyo, Japan. Primer Explorer V4 Software. http://primerexplorer.jp/e/v4_manual/index.html

FBPs detected by LAMP method are *Salmonella* spp. (Wang et al. 2008; Sayad et al. 2016), *Salmonella* subsp. *enterica*, *Salmonella enteritidis*, and *Salmonella enterica* subsp. *arizonae* (Kudo et al. 2005), *S. enterica* (Ohtsuka et al. 2005), *Bacillus stearothermophilus* (Umesha and Manukumar 2016), *E. coli*, *Salmonell typhimurium*, *Vibrio parahaemolyticus* (Oh et al. 2016), and *Yersinia pseudotuberculosis* (Horisaka et al. 2004). LAMP kits have also been developed to detect *Enterococci* (which produces verotoxin), *Salmonella*, *Legionella*, *Campylobacter*, and *Listeria* (Mori and Notomi 2009). The modified LAMP methods, *viz.* multiplex LAMP, RT-LAMP, real-time reverse transcription LAMP, in situ LAMP have also been applied to identify and detect *Campylobacter jejuni*, *Vibrio parahaemolyticus*, *Staphylococcus aureus*, *E. coli*, *Campylobacter coli*, and *Aeromonas caviae* (Shi et al. 2010).

Single-Stranded Conformation Polymorphism PCR
Single-stranded conformation polymorphism (SSCP) is an efficient molecular technique in FBP detection (Orita et al. 1989). It is a sensitive and precise assay with fewer efforts and high accuracy of microbial species identification. This technique was initially designed to find out the minute differences between species and subspecies by genotyping approach (Oh et al. 2008). In this assay, amplification, denaturation, and electrophoresis are three major basic steps, where a unique conformation is formed from ssDNA (single-stranded DNA) fragments. The primary sequence is so unique that change in even a single base can alter the conformation, which SSCP can observe. The susceptibility of analysis by SSCP relies on the size of the DNA fragment, its GC content, type of mutation, DNA concentration, gel (bead) size, the content of gel matrix composition, buffer composition, run time and temperature, pH and ionic strength of the buffer during electrophoresis (Umesha et al. 2012). *Ralstonia solanacearum* from tomato was reported to be identified by SSCP assay (Umesha et al. 2012). The detection of *E. coli*, *B. cereus*, *Clostridium perfringens*, *Salmonella enterica*, and *C. jejuni* has been reported by SSCP PCR method (Oh et al. 2008). SSCP followed by capillary electrophoresis method has been applied to detect common food pathogens such as *L. monocytogenes*, *B. cereus*, *E. coli*, *E. enterocolitica*, *V. parahaemolyticus*, *Enterobacter sakazakii*, *S. aureus*, *Clostridium perfringens*, and *Salmonella enterica* (Oh et al. 2012). The usage of SSCP–PCR technique has been established for the identification of *Campylobacter coli*, *B. cereus*, *Yersinia enterocolitica*, *Enterococcus* spp., *Salmonella* spp., *Cronobacter sakazakii*, *Shigella* spp., *S. aureus*, *Vibrio vulnificus*, *L. monocytogenes*, *V. parahaemolyticus*, *E. coli*, and *C. perfringens* (Kim et al. 2010, 2016). SSCP in combination with multiplex ligation-dependent probe amplification has been reported to detect *C. jejuni*, *E. coli*, *L. monocytogenes*, and *C. perfringens*. The *gyr*A gene mutation in *C. jejuni* was also detected by SSCP (Hakanen et al. 2002).

Ligase Chain Reaction PCR
Ligase chain reaction (LCR), also recognized as ligase amplification reaction (LAR), is a cyclic DNA amplification method that requires DNA as a template (Shi et al.

2010). It somewhat resembles PCR in the amplification process and requirement of a thermal cycler; however, it has more specificity and amplifies only the probe molecule instead the amplicon. The enzymes required to carry LCR assay are—a thermostable DNA polymerase and ligase. In this assay, four oligonucleotides are used as primers, where two probes with closer proximity and complementary sequence are ligated together. The resulting nucleotide sequences are ligated to form one nucleotide using DNA ligase by a subsequent exponential amplification (Umesha and Manukumar 2016; Gibriel and Ola Adel 2017). The amplification of the target sequence (up to 10^6 folds) gets completed in approximately 20–30 cycles. The amplified product can easily be detected by enzyme-linked immunosorbent assay (ELISA) or gel electrophoresis techniques. This technique is less applied in FBP identification because of its disadvantage in detecting dead microorganisms. However, clinical samples are often used to spot *Mycobacterium tuberculosis*, *Neisseria gonorrhoeae*, and *Chlamydia trachomatis* (Fakruddin et al. 2013; Umesha and Manukumar 2016).

There are several modified versions of LCR technique used for FBP identification. Ligase detection reaction (LDR) PCR is a redesigned form of LCR technique that amplifies the DNA by linear amplification process and has only two oligonucleotides instead of one. LDR has higher specificity due to the low dissociation temperature (T_d) of the used oligonucleotides. It offers better sensitivity to the identification of a microorganism, as the probes are very specific to the target (Umesha and Manukumar 2016; Gibriel and Ola Adel 2017). The detection of pathogens like *S. aureus*, *Streptococcus agalactiae*, *Streptococcus dysgalactiae*, *Streptococcus parauberis*, *Streptococcus pyogenes*, *Streptococcus equi*, *Streptococcus bovis*, *Streptococcus canis*, *Nonaureus staphylococci*, *Campylobacter* spp. *Mycoplasma* spp., and *Salmonella* spp. from milk has been reported to be done by LDR assay (Cremonesi et al. 2009). The sensing of *Vibrio* spp. from shellfish has also been reported using LDR assay (Cariani et al. 2012).

Restriction Fragment Length Polymorphism

Restriction fragment length polymorphism (RFLP) is a simple and rapid approach to detecting FBP where the genomic DNA is disentangled by the restriction endonuclease followed by gel electrophoresis helps in visualizing and tracing the size or mass change in the DNA fragment. Small DNA fragments are formed after digestion by the restriction enzyme at fixed sites. The size and mass of these fragments are compared in RFLP (Umesha and Manukumar 2016). RPLF assay was found helpful in detecting *Listeria* spp. and *S. aureus* (Paillard et al. 2003), *E. coli* (Fields et al. 1997), *C. jejuni* from poultry (Sierra-Arguello et al. 2018). It could also identify the point mutations critical for macrolide resistance in *Campylobacter* spp. (Vacher et al. 2003). The genes (*ltr*A, *ltr*B, and *ltr*C) for Listeriosis causative *L. monocytogenes*, responsible for growth at a lower temperature (4 °C), have been identified using RFLP method (Zheng and Kathariou 1995). The analysis of *inl*A polymorphism by PCR-RFLP was found a promising method to screen *Listeria monocytogenes* strains that are unable to invade Caco-2 cells. This gene expresses the truncated internalin, which helps the intracellular pathogen to get attached to the epithelial cells, followed by invasion (Rousseaux et al. 2004).

Amplified Fragment Length Polymorphism

Similar to RFLP, amplified fragment length polymorphism (AFLP) is also a genotyping amplification process where DNA fragments are digested by specific restriction endonuclease and further amplified. The DNA is completely unbound, broken down into fragments and then ligated by a double-stranded oligonucleotide adapter, which is complementary to the restriction site sequence. These adapters restrict the restoration of the original restriction site because of their specific designing. The specific amplified fragment is visualized after electrophoresis. AFLP aids in the identification of contamination sources *viz.* live stocks. When used in combination with an automated laser fluorescence analyzer, it identifies the organisms even at the strain level. The detection of disease-causing pathogens in mango, e.g., *Fusarium mangiferae, Fusarium proliferatum, Fusarium sacchari, Fusarium sterilihyphosum,* and *Fusarium subglutinans,* has been reported to be detected by AFLP technique (Umesha and Manukumar 2016). The high-resolution genotyping of *L. monocytogenes* from food has been made by AFLP in combination with multilocus sequence typing method (Parisi et al. 2010). The identification of *C. jejuni* has been established by using AFLP (Siemer et al. 2005). The transmission routes and genetic diversity of *Campylobacter* spp. from broiler farms have been studied using AFLP (Johnsen et al. 2006). The population structure (subgroups and strain detection) study by AFLP has been done to observe the minor differences between the pathogenic strains of *Salmonella* spp. (Torpdahl and Ahrens 2004; Lindstedt et al. 2000; Torpdahl et al. 2005). The differences in the subspecies by epidemiological typing of *B. cereus* (isolated from cooked chickens and cooked rice causing food poisoning) has been studied by AFLP (Ripabelli et al. 2000). AFLP in association with multilocus sequence typing has been used for genotyping using a strain of *C. jejuni* (isolated from poultry sources). The specific gene sequences have been successfully visualized to see the closest relativity between the species (Schouls et al. 2003).

Random Amplified Polymorphic DNA Technique

As the name suggests, the random amplified polymorphic DNA (RAPD) technique, is useful in identifying the organism by PCR analysis of random and primers under low stringency. This technique is rapid and easy to perform. Short sequence primers from DNA of the known sources are used, and the amplified product is observed by gel electrophoresis. There is a high possibility of having a complementary sequence of the primers to the DNA used, resulting in the production of variable fragments that might be specific to certain fungal pathogenic strains (Adzitey et al. 2013b; Umesha and Manukumar 2016). The identification of *Photobacterium damselae* subsp. *piscicida* (previously *Pasteurella piscicida*) strains from different geographical regions has been done by RAPD assay (Magarino et al. 2000). The cross-contamination in ready-to-eat processed food products (cooked meat and smoked salmon) by *L. monocytogenes* has been done by RAPD, in combination with serotyping (Aguado et al. 2001). The RAPD application has also been made to differentiate *Listeria innocua* from *L. monocytogenes* (Czajka et al. 1993) and *Salmonella typhi* from other *Salmonella* spp. (Shangkuan and Lin 1998). *E. coli* from green leafy vegetables (lettuce, and spinach) growing near the river beds have been detected with its genetic diversity using RAPD (Darkazanli et al. 2018). RAPD was used to study the intraspecific genetic diversity of *Enterococcus lactis* 4CP3, an enterotoxin producer isolated from raw beef meat (Braiek et al. 2018).

9.2.1.2 Isothermal amplification techniques

Nucleic Acid Sequence-Based Amplification

Nucleic acid sequence-based amplification (NASBA) is commonly called transcription-mediated replication and/or 3SR because of its ability to perform self-sustained sequence replication (3SR). This assay is rapid (1–2 h of completion), cost-effective, and has fewer contamination chances in detecting viable microbes. NASBA is an isothermal replication process where a constant temperature of 41 °C is uniformly maintained throughout the assay, unlike PCR. This technique is used to amplify nucleic acids (specifically RNA) without using a thermal cycler. The assay is achieved by utilization of a club of three enzymes, namely RNase H (to digest the DNA-RNA hybrid form and produce ssDNA), avian myeloblastosis virus (AMV) reverse transcriptase, and T7 DNA-dependent RNA polymerase. In NASBA amplification process, manifold ssRNA copies from DNA are produced in approximately 20 cycles, unlike PCR where binary amplification per cycle is done. The temperature at which this assay is performed keeps the DNA in dsDNA form; hence, the amplification substrate cannot be replaced. The amplified product is visualized by electrophoresis (Fakruddin et al. 2013; Umesha and Manukumar 2016). Several rapid detection kits have been developed following this technique, *viz. Salmonella* spp. detection and differentiation (Zhai et al. 2019; Bodulev and Sakharov 2020). This technique has been successfully used to spot *L. monocytogenes*, *C. jejuni*, rotavirus, *Bacillus* spp., and hepatitis A virus (Aslan et al. 2020).

Rolling Circle Amplification

Rolling circle amplification (RCA) carries out the amplification of a circular template into multi-fold copies (10^9) with the help of Ø29 DNA polymerase. This assay produces ssDNA from DNA or RNA and can be performed both in the solid and solution phase at a constant temperature. When the amplified product is linear, it is called linear RCA or single primer RCA. However, the sensitivity of the test is refined by using a reverse primer. DNA polymerase extends the final product of linear RCA (ssDNA) after a reverse primer (complementary to the strand polymerized by a single oligonucleotide primer) is added to the reaction mixture. The final DNA template (dissociated ssDNA) is hybridized with a reverse primer, and the cycle is extended to produce branched or ramified DNA. The reaction is stopped when all ssDNA get converted to dsDNA. This type of RCA amplification process is called ramification amplification (RAM) or hyper-branched RCA or cascade RCA (Fakruddin et al. 2013; Umesha and Manukumar 2016; Zhong and

Zhao 2018). This assay has been applied to detect *E. coli* (Jiang et al. 2017; Zhong and Zhao 2018; Li et al. 2021), *L. monocytogenes* (Zhong and Zhao 2018; Zhan et al. 2020), *S. typhimurium* (Zhong and Zhao 2018), *V. parahaemolyticus* (Teng et al. 2017; Song et al. 2018, 2019), *Cronobacter* spp. from powdered infant formula food samples (Liu et al. 2020) and from milk samples (Liu et al. 2020), *C. perfringens* from meat (Milton et al. 2021), *S. aureus* from meat (Hao et al. 2017) and from milk (Yang et al. 2019), and *Shigella* spp. from vegetable salad (Wang et al. 2017).

Strand Displacement Amplification

Strand displacement amplification (SDA) is a highly efficient, rapid, and manageable isothermal amplification process (Walker et al. 1992; Zhong and Zhao 2018; Aslan et al. 2020). This assay can be performed at a higher temperature (37–70 °C). The essential components required for SDA are four oligonucleotide primers (bumper primers B1 and B2 with adjacent primers S1 and S2) having Hinc II exonuclease restriction site (5′-GTTGAC-3′), exo-Klenow (*E. coli* DNA polymerase with an exonuclease-deficient restriction sites), and a DNA template. The assay occurs in two phases: production of DNA template with restriction site and target DNA amplification (Fakruddin et al. 2013; Umesha and Manukumar 2016; Zhong and Zhao 2018). SDA has been applied to detect *Salmonella enteritidis* and *E. coli* (Zhong and Zhao 2018), specificity and sensitivity of *Salmonella* spp. (Zhang et al. 2021), *S. aureus* (Cai et al. 2019), and *S. typhimurium* (Zhang et al. 2016; Li et al. 2019; Wang et al. 2020a, b, c).

9.2.1.3 Nucleic Acid Hybridization Techniques

The upgradation of techniques in the FBP detection process also employs methods where PCR amplification can be skipped, and time can be saved. The techniques can rapidly identify the closely related DNA sequences in two different populations and do not require enrichment of the medium. Instead, a labeled DNA probe (labeled with either a radioactive or non-radioactive marker), complementary to the target sequence (DNA/RNA), is enough to carry out the assay. The probe is usually a dsDNA oligonucleotide. The non-amplification methods that are easy to perform in the laboratories include fluorescent in situ hybridization (FISH), peptide nucleic acid–FISH, line-probe assay (LPA), and hybridization protection assay (HPA). The specific targets can be detected by fluorescent, calorimetric, and chemiluminescence signals (Umesha and Manukumar 2016; Aslan et al. 2020).

Fluorescent In Situ Hybridization

Fluorescent in situ hybridization (FISH) is chiefly a microscopic technique used for FBP detection. In FISH assay, the target is marked with a fluorescent probe at the 5′ end, targeting the rRNA that can be visualized in a fluorescent microscope. The labeling of the probe (15–25 oligonucleotide sequence) is done either by direct labeling (i.e., by using a fluorescent nucleotide) or indirect labeling (i.e., by combining the probe with a reporter molecule that is defined by a fluorescent antibody or any other affinity molecule). FISH has four major steps, including fixation and permeabilization of specimens, probe and target nucleotide hybridization, unbounding the probe by washing and visualization, and interpretation of the result. A fluorescent labeled 16 s rRNA probe does the detection. The target stained cells can be detected by epifluorescence microscopy after the hybridization (Umesha and Manukumar 2016; Dias and Rathyanaka 2018; Aslan et al. 2020). Cross-contamination by *Salmonella* spp. in chickens has been studied and reported by Alzaabi and Khan et al. (2017). FISH, in association with flow cytometry, has been applied for rapid detection of *Salmonella* spp. (Umesha and Manukumar 2016). The detection of *S. enterica* and *Listeria* spp. in food samples have been reported (Aslan et al. 2020). FISH has been used for the identification of FBPs like spiked *Salmonella* spp. (from vegetables, tomato, pork sausage, barley plant, and sweet corn roots), *Yersinia* spp. (from minced pork meat), *Pseudomonas* spp. (from milk), *E. coli* (from ground beef milk), *Enterobacteriaceae* (from milk), *Helicobacter pylori* (from bovine milk), *L. monocytogenes* (from mozzarella cheese, smoked salmon and Julienne cabbage), *B. cereus* (from milk and barley plant), *Mycobacterium avium* (from portable water), *Campylobacter* spp. (from drinking water and chicken products), *E. coli* (from Japanese seafood called *ikura* and minced chicken meat), using rRNA as target probe (Dias and Rathyanaka 2018). Other studies for detection of FBP from food samples include identification of pathogenic strains of *L. monocytogenes* (Rocha et al. 2019), *Salmonella* spp. (Salimi et al. 2020), and *Yersinia* spp. (Rohde et al. 2017).

Peptide Nucleic Acid–FISH

Peptide nucleic acid–FISH (or PNA-FISH) was originally described by Nielsen and his colleagues in 1991 at the University of Copenhagen, Denmark. PNA is a DNA/RNA analog (a pseudopeptide DNA mimic), where N-(2-aminoethyl) glycine-repeating units replace the sugar-phosphate backbone and polyamide chain is covalently linked to nucleobases by carboxymethyl spacer. PNA is an achiral and a non-ionic molecule, resistant to degradation by enzymes (Nielson et al. 1991; Saadati et al. 2019). It is preferred over DNA probes due to higher stability of PNA-RNA or PNA-DNA hybrid (because of the absence of electrostatic repulsions and no charge on PNA as PNA backbone does not have charged phosphate group) than DNA-DNA hybrid, resistance to degradable enzymes, resistance to nucleases or proteases, high Tm, and more specificity and the easy entrance of PNA in the bacterial cells (because of its relative hydrophobicity). This assay can be accessed at a low salt concentration that promotes destabilization of rRNA–rRNA interaction, improving the accessibility to the target site out of reach with the standard RNA or DNA probe (Umesha and Manukumar 2016; Rocha, 2018; Zhao and Wu 2020; Adebowale et al. 2020). This technique has been used for the detection of *L. monocytogenes*, *Vibrio* spp. (Shan et al. 2018), and *Salmonella* spp. (Adebowale et al. 2020).

9.2.1.4 Detection by Multiple Targets

The advancements in molecular methods of FBP detection brought the easier process and cost-effectiveness. Techniques like broad-range PCR, DNA microarray, multiplex PCR can perform detection of more than one microorganism.

DNA Microarray

DNA microarray is one of the most advanced, flexible, and specific detection techniques for FBP detection (Ranjbar et al. 2017). The word "microarray" was first introduced by Schena in (1995). The technique of DNA microarray allows the study of the expression of multiple genes at one time (Campos et al. 2012). The term "microarray" refers to a two-dimensional high-density matrix of DNA fragments, printed or synthesized on glass or silicon slide (chip) in a fixed order (Umesha and Manukumar 2016). Microarray is a specific ssDNA pattern (sequence) that is immobilized on a slide or chip. The basic principle for DNA microarray is complementary base pairing hybridization of ssDNA and its detection. The array can be picked up by binding the identification targets like proteins, nucleotides, cDNA, antibodies, peptides, carbohydrates, aptamers to distinct regions called "microarray" (Aslan et al. 2020). This chip diagnostic technique has basically two divisions: dry lab (in silico/bioinformatics part), where probe designing is done, and wet lab (in vitro) that includes nucleic acid extraction and hybridization.

In the dry lab, designing of the specific probe is done, and in wet lab work, probe spotting, probe labeling, hybridization, and labeling are done. The International Nucleotide Sequence Database Collaboration (INSDSC): http://www.insdc.org/ is a multifunctional database. However, although limited, but easily accessible sequence databases are also available for achieving *in silico* work, including European Nucleotide Archive (ENA): www.ebi.ac.uk/ena, National Center for Biotechnology Information (NCBI): http://www.ncbi.nlm.nih.gov/, and DNA Data Bank of Japan (DDBJ): http://www.ddbj.nic.ac.jp/. For prokaryotes, including archaea and bacteria, the available database is Rapid Annotations using Subsystems Technology (RAST): http://rast.nmpdr.org/, SEED: http://pubseed.theseed.org/, PathoSystems Resource Integration Center (PATRIC): http://www.patricbrc.org/ (for bacteria), Virus Pathogen Database and Analysis Resource (ViPR): http://ViPRbrc.org, Influenza Research Database (IRD): www.fludb.org, and Atlas of Biological Databases and Tools (ABDT): http://bis.zju.edu.cn/DaTo/ . The most effective tool for proper designing of microarray to date has been reported to be the BLAST (Basic Local Alignment Search Tool): https://blast.ncbi.nlm.nih.gov/Blast.cgi, PanSeq: http://lfz.corefacility.ca/panseq/, and GView: https://server.gview.ca/. Further, the analysis of the designed probe can be done on OligoAnalyzer: http://edu.idtdna.com/calc/analyzer .

In the wet lab work, probe spotting can be achieved with the help of robotic spotters/printer. There are three major groups of array platforms, namely glass, microwell, and micropillar. Glass is used most frequently. In hybridization, the target DNA or protein molecule is analyzed by a labeled and hybridized identification probe on the array, where the probe's position on the array is called spot or future. The advancements in microarray technology include mutation analysis, comparative genomic analysis, gene expression array, and multiple species component array. Microarray, when applied in combination with pulsed-filed gel electrophoresis (PFGE) and multilocus sequence typing (MLST) gives high specificity and discrimination. The first application of microarray in FBP detection was for enteric pathogens (Campos et al. 2012; Umesha and Manukumar 2016; Ranjbar et al. 2017;

Li et al. 2017; Liu et al. 2017; Behzadi and Ranjbar, 2019; Aslan et al. 2020). DNA microarray has been applied in the detection of *Salmonella* spp., *E. coli*, *L. monocytogenes* (Li et al. 2017), norovirus (Hu et al. 2015), *E. coli*, *Shigella boydii*, *S. typhi*, *S. typhimurium*, *Shigella flexneri*, *Shigella sonnei*, *Brucella* spp. *Legionella pneumophila*, *V. cholera*, *Shigella dysenteriae* (Ranjbar et al. 2017), *C. jejuni*, and *S. enterica* (Liu et al. 2017). FBPs belonging to *Shigella* spp., *Salmonella* spp., *E. coli*, and *Bacillus* spp. Strains have been reported to be identified using microarray technology (Aslan et al. 2020). Multiplex PCR (mPCR)-based microarray technique has also been applied in the detection of point mutation of drug-resistant genes (extended-spectrum β-lactamases ESBL and plasmid-mediated AmpC β-lactamases) in *Enterobacteriaceae* (Zhu et al. 2007).

Multiplex PCR

The multiplex PCR allows the concomitant identification and amplification of numerous gene sequences from FBPs. Its cost-effectiveness, rapidity, specificity, less sample preparation time, detection of viable pathogens, and requirement of fewer enzymes and reagents prioritize it from other techniques (Lantz et al. 1998; Umesha and Manukumar 2016; Tao et al. 2020). This technique was reported for the first time in 1988 and applied in studying mutation for tracking the gene deletion of Duchenne muscular dystrophy (DMD) locus. The assay has also been applied in detection of virus molecules and antimicrobial gene determinants. The assay requires a template DNA, Taq polymerase, PCR buffer mixture, and dNTPs. The overall conditions of the mPCR resemble PCR (temperature and electrophoresis, etc.) to get the final product (Chamberlain et al. 1998; Zhu et al. 2007). In mPCR, two or more primer sets are used in the same PCR reaction for amplification of different targets. The mPCR requires a higher concentration of $MgCl_2$ as compared to conventional PCR (Umesha and Manukumar 2016). The application of mPCR has been reported for the identification of *Y. enterocolitica* from pork (Lantz et al. 1998), *Salmonella* spp., *S. aureus*, *E. coli* O157:H7 and *L. monocytogenes* in milk (Wei et al. 2018; Sheng et al. 2018). An mPCR method was developed that can detect common foodborne pathogens including *S.* enterica, *V. parahaemolyticus*, *S. aureus* *L. monocytogenes*, *E. coli* O157:H7, *S. flexneri*, *V. cholerae*, *C. botulinum* type A, *B. cereus*, *Y. enterocolitica*, and *C. perfringens* Alpha toxin (Tao et al. 2020). Further, a multiplexed real-time PCR assay method was developed to detect *V. parahaemolyticus*, which is associated with outbreaks due to consumption of different seafoods (Ward and Bej, 2006). Several reports are available on detection of *L. monocytogenes*, one of the common foodborne pathogens in food products, by this approach (Zhang et al. 2020; Feng et al. 2020; Du et al. 2020; Chen et al. 2012).

9.2.2 Advanced Molecular Techniques

9.2.2.1 Pulse Field Gel Electrophoresis

Pulse field gel electrophoresis (PFGE) technique is useful in the separation of larger DNA molecules. The assay accesses the molecular grouping, genotypic characterization, and helps to identify the clonal strains of the same organism (de Melo et al. 2021). This process was first reported by Schwartz and Cantorto, where they progressed with size-dependent electrophoretic separation without using molecular sieves. The technique utilizes electrical pulse field to relax and separate the DNA in the gel. In order to have new orientations and obtain larger DNA fragments, the direction of the electric field is altered intermittently. The time required for DNA orientation is directly proportional to the size of the DNA and the electric field applied, which results in the separation of DNA with maximum electrophoretic mobility. Therefore, the orientation of molecules allows them to raptate through the gel matrix without any size dependency (Umesha and Manukumar 2016). Briefly, PFGE offers a method to produce DNA fingerprints from larger DNA fragments and helps in detailed and pinpoint distinction of closely related species based on their genetic relatedness. The assay is very sensitive and has been applied in the detection and differentiation of 55 *Salmonella* spp. strains from mechanically deboned chickens (Ha et al. 2020).

Because of its ability to separate larger molecules, PFGE is applied in the molecular analysis of genes and genomes of mammalian cells and microbes. This technique has been widely used at Nebraska Public Health Laboratory (NPHL) for studying the molecular epidemiology of FBPs (Rao and Arora 2020). Due to frequent FBP outbreaks, a similar database system (that collects molecular subtyping information of FBPs) has been developed in China to track the foodborne diseases and associated pathogens, named as TraNet. TraNet was established in 2013, and its role is quite similar to PulseNet in the United States (Li et al. 2018a, b). PFGE has been used for detecting *Salmonella* spp., *E. coli*, and *Listeria* spp. (Umesha and Manukumar 2016), *Salmonella* spp. from poultry (Ashrafudoulla et al. 2021), *Y. enterocolitica* bio-serotype 4/O:3 from pork (Martins et al. 2020), thermotolerant *C. jejuni* from poultry (Zbrun et al. 2020), *S. enterica* serovar Minnesota from poultry (de Melo et al. 2021), and *Salmonella* Serovar Infantis from poultry flocks (Taheri et al. 2018). This technique was helpful in detecting the virulent *E. coli* O157:H7 strain from the fecal samples of cattle farm beef (Akindolire and Ateba, 2018).

9.2.2.2 Ribotyping

In the process of FBP detection, ribotyping plays a vital role in identifying as well as classifying the microbes based on structural differences of gene sequences in rRNA. Each ribosomal operon in ribotyping, the polycistronic operon, is consisted of three major consensus genes that encode structural rRNA molecules (5S, 16S, and 23S). The 16S rRNA is considered ideal for bacterial ribotyping because of its conserved region. This method comprises the digestion of the genomic DNA into smaller fragments with the help of restriction enzymes (endonucleases) and the electrophoretic separation of these smaller fragments (1–30 kb), followed by Southern blotting for hybridization with a radiolabeled ribosomal operon probe for decoding the rRNA gene sequences. Further processing involves autoradiography that allows the visualization of the bands with the ribosomal operon consensus sequence (Umesha and

Manukumar 2016; Rao and Arora 2020). The multiplicity of rRNA operon in the microbe can be inferred by the number of fragments produced in ribotyping.

The different ribotyping techniques modified to date are named according to the technology introduced, *viz.* PCR ribotyping, automated ribotyping, PCR ribotyping followed by restriction endonuclease subtyping, long PCR ribotyping, and amplified rRNA gene restriction analysis (ARDRA). A species of *Listeria* has been differentiated by automated ribotyping (Rao and Arora 2020). Ribotyping has also been successfully applied to distinct the lactic acid bacteria (LAB) from rainbow trout (spoiled vacuum packaged "gravad" by Lyhs et al. (2002).

9.2.2.3 Repetitive Extragenic Palindromic PCR

Repetitive extragenic palindromic PCR (REP-PCR) is another advanced molecular technique. With this method, FBP detection is supposed to be more accurate, and it has a higher differentiation ability between species. In this technique, a regulatory sequence called REP sequence, also called palindromic unit located on the untranslated operon region, is focused as a target. These types of sequences are found in *E. coli*, *Salmonella*, and *Pseudomonas*. The REP sequences play a major role in transcription termination, stabilizing RNA viability, and chromosomal maintenance. These sequences also assist in encoding the gene sequences for respiration, degradation, biosynthesis, and regulation. This technique has been successfully applied for the detection of FBPs like *E. coli*, *Salmonella* spp. (Deng et al. 2018), (Adzitey et al. 2013a, b), *C. jejuni* (Yadav et al. 2017), and *V. parahaemolyticus* (Sadeghi et al. 2019).

9.2.2.4 Multilocus Sequencing Typing

Multilocus sequence typing (or MST) is one of the prime molecular techniques in FBP detection process. This is a nucleotide-based typing method that uses housekeeping genes data to provide the sequence type for finding the intraspecific genetic relationship between bacterial and fungal species. In this method, rapid sequencing is done to check the allelic variants in conserved genes for characterization, subtyping, and classification of the species. Initially, PCR is done to amplify the housekeeping genes for sequence analysis, followed by the comparison of the individual genes. MST starts with the production of nucleotide sequences from the target DNA region. Further, in order to achieve the data typing, the significant steps included are data summary, allele assignment, lineage assignment, sequence typing assignment, and estimation of recombinants. Approximately 450–500 bp of internal fragments are essential to carry out the MST process. The different allele numbers are assigned to the different sequences of the bacterial species. Likewise, for each isolate, the alleles at each locus define the upcoming sequence type. The generated nucleotide sequences are compared on programs like sequence typing analysis and retrieval system (STARS), DiscoverIR (Licor, UK), Staden Package (a complete set for assembly of the DNA sequence, editing, and analysis), and NCBI BLAST. After allele assigning, the data is entered in the MST website (http://www.mlst.net) to get the sequence type. This is followed by grouping the sequence types into clonal complexes by their similarity to central allelic profile (i.e., the genotype), further identified by computer-based statistical methods (Umesha and Manukumar 2016; Rao and Arora 2020). This advancement has been successfully applied in detecting pathogens like *Streptococcus pneumonia* (Umesha and Manukumar 2016), *L. monocytogenes* IVb serogroup from meat (Lachtara et al. 2021), *C. jejuni* (Hsu et al. 2020), *E. coli* (Yu et al. 2018; Ramdan et al. 2020), *V. parahaemolyticus* (Escalona et al. 2017), *S. aureus* (Yang et al. 2018), and *B. cereus* (Yang et al. 2017).

9.2.2.5 DNase-Treated DNA (DTD) PCR

In this molecular technique, FBPs can be detected by PCR-based techniques in which PCR is followed by the treatment with DNaseI enzyme. This technique has been demonstrated to be successful for the detection by amplifying the DNA from dead cells also, without causing any harm to the DNA of live cells. FBPS like *E. coli*, *L. monocytogenes*, *S. enterica*, and *V. parahaemolyticus* have been successfully detected by this technique (Umesha and Manukumar 2016).

9.2.3 Signal-Based Techniques

The biosensors are the analytical devices that aid the visual detection of FBP in signals, where electrical signals are converted from biological signals. Basically, biosensors have three major parts: a biological molecule, a device for converting them into signals, and a detection system. Hence, biosensors are a combined form of biological elements and transducers. In this process, the transducers and bioreceptors recognize biological elements like cells, tissues, organelles, microbe, enzyme, antibody, nucleic acid, aptamers, bacteriophage, or any biomimic and convert the recognition to a measurable electric signal in the form of electrochemical: impedance, amperometric and potentiometric, optical: light scattering, surface plasmon resonance (SPR) and fiber optics, surface acoustic and piezoelectric, thermometric, electrical, magnetic, or their combinations. Because of the strong bonding with the antigens and high specificity, antibodies are frequently used as biorecognition elements (Chandra et al., 2012; Choudhary et al., 2016; Deka et al., 2018; Mahato et al., 2018; Verma et al., 2019).

Depending upon the use of a type of measurable signals, the biosensors are named piezoelectric, bioluminescence, electrochemical, voltammetric, colorometric, potentiometric, impedimetric, fluorescent, label-based or label-free, and optical biosensors. The first report of FBP detection by biosensoring was the detection of tobacco mosaic virus (TMV) and cowpea mosaic virus (CMV). The detection of pathogenic *L. monocytogenes*, *C. jejuni*, and *E. coli* has also been done by using immunoassay by applying highly dispersed carbon particles. Umesha et al. (2016) have reported an artificial cell-based biosensor by applying a liposome-doped silica nanocomposite, which imitates a whole-cell method to detect Listeriolysin O produced by *L. monocytogenes*. The biosensor detects the Listeriolysin O and confirms the presence of toxin by *L. monocytogenes*. The lectin-based array biosensor has been used to detect and differentiate FBPs, including *S. aureus*, *B. cereus*, *Proteus vulgaris*, *E. coli*, and *Enterobacter aerogenes*. More specificity to the FBP detection has been achieved by including the use of spike proteins. *S. aureus, E. coli,* and *B. anthracis* have also been detected by using a phage as biorecognition elements.

In the detection process of FBPs by biosensors, the transducers play a significant role. The transducers use optical oriented detection methods, *viz.* dispersion, reflection, refraction, fluorescence, chemiluminescence, Raman (surface-enhanced Raman scattering or SERS), infrared, or phosphorescence. SPR and fluorescence spectrometry have been considered highly sensitive transducing methods. SPR has been applied in the detection of *L. monocytogenes* and *Salmonella* spp. Similarly, an amperometric detection method has been applied for detecting *Salmonella, E. coli, C. jejuni*, and *L. monocytogenes*, where an ample amount of current is produced at a constant potential between the working and reference electrodes. However, conventional amperometric biosensors contain three electrode cells (working, reference, and auxiliary electrodes), a voltage source, and a device for measuring voltage and current (Umesha and Manukumar 2016; Arora et al. 2018; Saravanan et al. 2020; Zhang et al. 2020).

9.2.4 Immunological Techniques

The detection of FBPs using immunological techniques has also been achieved. The successful detection of FBPs by enzyme-linked immunosorbent assay (ELISA) and MALDI-TOF/MS (matrix-assisted laser desorption ionization-time of flight/mass spectrometry) operation systems are still being reported periodically. The principle behind the immunological techniques is the recognition of the binding of an antigen (Ag) to the specific target antibody (Ab) (Ag-Ab complex) that causes change in color (Umesha and Manukumar 2016). The specific Ag binds to the specific Ab at a specific site called epitope. Another more sensitive immunological technique than ELISA is the ELFA (enzyme-linked fluorescent assay). Some of the modern-day advanced immune techniques applied for the detection of FBPs are immunofluorescence, immune-electrophoresis, immune-diffusion, and radioimmunoassay. For the detection of *Campylobacter* spp., the immunoassays have played a crucial role. The ELISA technique is acknowledged as the most reliable technique for mycotoxins (Alfatoxin M1 from milk, T2 in wheat flour, fumonisin B1, AFB1, and DON in maize) diagnosis. ELISA, in combination with immunomagnetic beads, has been utilized for the visual detection of bacterial pathogens (Khan et al. 2018; Aslan et al. 2020; Saravanan et al. 2020). Immunomagnetic separation method in synchronization with PCR assay has also been employed for detecting *Salmonella* spp. (Jenikova et al. 2000). MALDI-TOF/MS has been reported to find out enterotoxin B from *Staphylococcus* spp. and *L. monocytogenes*, using immunomagnetic separation and flow cytometry (Umesha and Manukumar 2016).

9.2.5 Smart Devices for FBP Detection

The modern-day biosensors also apply smartphones as tools for detection and 3D printing of pathogens and biochemicals (Ding et al. 2018; Mahato et al. 2018; Purohit et al. 2020a, b). FBPs like *E. coli*, *S. aureus*, and *S. enteritidis* have been reported to be detected by smartphone-based optical detection techniques (Ding et al. 2018). The principle behind detection is the endpoint lateral flow immune-assay technique, where a single strip is divided into four parts—sample loading, conjugate formation, a wicking membrane (for conjugate-analyte complex formation), and a waste collector (Ding et al. 2018). The smartphone-based imaging devices for increasing equitable food quality and longevity have also been one of the emerging technologies by which FBPs (specifically *E. coli* in milk) and food allergens are targeted (Banik et al. 2021). The application of biosensors is time-saving and cost-effective, which can be applied to portable devices like smartphones. This minimizes the labor and, therefore, has potential application in food technologies.

9.3 Conclusions and Perspectives

PCR-based techniques are one of the most accepted and widely applied detection techniques among different molecular methods. However, some of the techniques have limitations, complicated to handle, some methods need large number of samples, and some are limited only to detecting either viral or bacterial pathogens. Techniques like LCR-PCR have contamination risks and error chances due to variable copy numbers in plasmid having LCR target. Multiplex PCR is limited to the visualization of amplicons of similar length in low quantity in agarose gel. Among the DNA-sequencing techniques, whole-exome sequencing (WES) is expansive. Immuno-based methods are also restricted to the requirement of the specific antibody. Some PCR-based techniques require very specifically designed probes. The use of biosensors need highly purified samples, which again are expansive and time taking.

In recent advancements, biosensors and nanoparticles are emerging technological approaches for FBP detection. Nanoparticles contribute in signal amplification in the detection process even from a single cell. Similarly, magnetic-based techniques in association with microfluidics can be helpful for rapid, error-free, and portable detection methods. One of the recent advances in FBP detection is the use of CRISPER Cas9 coupled with isothermal amplification (Sun et al. 2020). The limit of sensitivity for this method has been overcome by coupling it with other gene amplification methods like DETECTR (or DNA endonuclease-targeted CRISPER Trans reporter), which binds RPA with Cas12a or CrRNA-based ssDNA-FQ reported cleavage. The use of DNA apta sensors is also one of the newly introduced molecular techniques that can be further improvized for the specificity of FBP detection.

It is noteworthy that techniques like microarray, next-generation PCR, metagenomic sequencing, whole-genome sequencing, and nanotechnology contribute a promising role in the FBP detection process and bring it from "farm to fork." Hence, the application of modern-day biotechnological methods is tremendous and vital to be carried forward to handle the health-related issues associated with FBPs.

References

Adebowale OO, Goh S, Good L (2020) The development of species-specific antisense peptide nucleic acid method for the treatment and detection of viable *Salmonella*. Heliyon 6(6):e04110

Adzitey F, Huda N, Ali GRR (2013a) Molecular techniques for detecting and typing of bacteria, advantages and application to foodborne pathogens isolated from ducks. 3 Biotech 3(2):97–107

Adzitey F, Ali GRR, Huda N, Ahmed R (2013b) Genotyping of *Salmonella* strains isolated from ducks and their environments in Penang, Malaysia using Repetitive Extragenic Palindromic (REP). J Microbiol Biotechnol Food Sci 3(1):87–93

Aguado V, Vitas AI, Gracia I (2001) Random amplified polymorphic DNA typing applied to the study of cross-contamination by *Listeria monocytogenes* in processed food products. J Food Prot 64(5):716–720

Akindolire MA, Ateba CN (2018) Use of pulsed field gel electrophoresis genetic typing for tracing contamination with virulent *Escherichia coli* O157:H7 in beefcattle producing farms. Gene Rep 13:59–65

Alzaabi SE, Khan MA (2017) A study on foodborne bacterial cross-contamination during fresh chicken preparation. Arab J Nutr Exercise 2(1):128–138

Arora S, Ahmed N, Khubber S, Siddiqui S (2018) Detecting food borne pathogens using electrochemical biosensors: an overview. Micromachines 10(4):222

Ashrafudoulla M, Na KW, Byun KH, Kim DH, Yoon JW, Mizan MFR, Kang I, Ha S (2021) Isolation and characterization of *Salmonella* spp. from food and food contact surfaces in a chicken processing factory. Poult Sci 101234

Aslan H, Ekinci A, Aslan I (2020) Nucleic acid-based methods in the detection of foodborne pathogens. In: Natural remedies for pest, disease and weed control. Chapter 13, pp 143–161

Ayyash M, Olaimat A, Al-Nabulsi A, Liu SQ (2020) Bioactive properties of Novel probiotic *Lactococcus lactis* fermented camel sausages: cytotoxicity, angiotensin converting enzyme inhibition, antioxidant capacity, and antidiabetic activity. Food Sci Anim Resources 40(2): 155–171

Banik S, Melanthota SK, Arbaaz VJM, Kadambalithaya VM, Iftak H, Dutta S, Mazumdar N (2021) Recent trends in smartphone-based detection for biomedical applications: a review. Anal Bioanal Chem 413:2389–2406

Behzadi P, Ranjbar R (2019) DNA microarray technology and bioinformatic web services. Acta Microbiol Immunol Hung 66(1):19–30

Bintis T (2017) Food pathogens. AIMS Microbiol 3(3):529–563

Bisht A, Kamble MP, Choudhary P, Chaturvedi K, Kohli G, Juneja VK, Sehgal S, Taneja NK (2021) A surveillance of food borne disease outbreaks in India: 2009–2018. Food Control 121: 107630

Bodulev IL, Sakharov Y (2020) Isothermal nucleic acid amplification techniques and their use in bioanalysis. Biochem Mosc 85(2):147–166

Braiek OB, Smaoui S, Ennouri K, Hani K, Ghrairi T (2018) Genetic analysis with random amplified polymorphic DNA of the multiple Enterocin-producing *Enterococcus lactis* 4CP3 strain and its efficient role in the growth of *Listeria monocytogenes* in raw beef meat. Lett Appl Microbiol 47(3):153–157

Brauns LA, Hudson MC, Oliver JD (1991) Use of the polymerase chain reaction in detection of culturable and nonculturable *Vibrio vulnificus* cells. Appl Environ Microbiol 57(9):2651–2655

Cai R, Yin F, Zhang Z, Tian Y, Zhou N (2019) Functional chimera aptamer and molecular beacon based fluorescent detection of *Staphylococcus aureus* with strand displacement-target recycling amplification. Anal Chim Acta 1075:128–136

Campos GL, Suarez JVM, Urda MA, Alonso VL (2012) Microarray detection and characterization of bacterial foodborne pathogens. SpringerBriefs in Food, Health and Nutrition Series Springer

Cariani A, Piano A, Consolandi C, Severgnini M, Castiglioni B, Caredda G, Candela M, Serratore P, De Bellis G, Tinti F (2012) Detection and characterization of pathogenic vibrios in shellfish by a ligation detection reaction-universal array approach. Int J Food Microbiol 153: 474–482

Chamberlain JS, Gibbs RA, Ranier JE, Nguyen PN, Caskey CT (1998) Deletion screening of the Duchenne muscular dystrophy locus via multiplex DNA amplification. Nucleic Acids Res 16(23):11141–11156

Chandra P, Koh WCA, Noh HB, Shim YB (2012) In vitro monitoring of i-NOS concentrations with an immunosensor: the inhibitory effect of endocrine disruptors on i-NOS release. Biosens Bioelectron. https://doi.org/10.1016/j.bios.2011.11.027

Chen J, Tang J, Liu J, Cai Z, Bai X (2012) Development and evaluation of a multiplex PCR for simultaneous detection of five foodborne pathogens. J Appl Microbiol 112(4):823–830

Choudhary M, Yadav P, Singh A, Kaur S, Ramirez-Vick J, Chandra P, Arora K, Singh SP (2016) CD 59 Targeted ultrasensitive electrochemical immunosensor for fast and noninvasive diagnosis of oral cancer. Electroanalysis 28:2565–2574. https://doi.org/10.1002/elan.201600238

Cremonesi P, Pisoni G, Severgnini M, Consolandi C, Moroni P, Raschetti M, Castiglioni B (2009) Pathogen detection in milk samples by ligation detection reaction-mediated universal array method. J Dairy Sci 92(7):3027–3039

Czajka J, Bsat N, Piani M, Russ W, Sultana K, Weidmann M, Whitaker R, Bati CA (1993) Differentiation of *Listeria monocytogenes* and *Listeria innocua* by 16S rRNA genes and intraspecies discrimination of *Listeria monocytogenes* strains by random amplified polymorphic DNA polymorphisms. Appl Environ Microbiol 59(1):304–308

Darkazanli M, Kiseleva I, Darkazanli K (2018) Genetic diversity of *E. coli* O157:H7 isolated from some leafy greens, irrigated by Aleppo River, using random amplified polymorphic DNA (RAPD) marker. Russ Agric Sci 44(2):146–152

de Melo RT, Cardoso TDR, Peres PABM, Braz RF, Monteiro GP, Rossi DA (2021) *Salmonella enterica* Serovar minnesota biofilms, susceptibility to biocides, and molecular characterization. Pathogens 10(5):581

de Oliveira AC, Soccol VT, Rogez H (2019) Prevention methods of foodborne Chagas disease: disinfection, heat treatment and quality control by RT-PCR. Int J Food Microbiol 301:34–40

Deka S, Saxena V, Hasan A, Chandra P, Pandey LM (2018) Synthesis, characterization and in vitro analysis of α-Fe_2O_3-$GdFeO_3$ biphasic materials as therapeutic agent for magnetic hyperthermia applications. Mater Sci Eng C 92:932–941. https://doi.org/10.1016/j.msec.2018.07.042

Deng C, Lv X, Li J, Liu Y, Du G, Amaro RL, Liu L (2018) Synthetic Repetitive Extragenic Palindromic (REP) sequence as an efficient mRNA stabilizer for protein production and metabolic engineering in prokaryotic cells. Biotechnol Bioeng 116(1):5–18

Dias PGI, Rathyanaka RMUSK (2018) Fluorescence in situ hybridization (FISH) in food pathogen detection. Int J Mol Biol 3(3):143–149

Ding X, Mauk MG, Yin K, Kadimisetty K, Liu C (2018) Interfacing pathogen detection with smartphones for point-of-care applications. Anal Chem 91(1):655–672

Du J, Wu S, Niu L, Li J, Zhao D, Bai Y (2020) A gold nanoparticles-assisted multiplex PCR assay for simultaneous detection of *Salmonella typhimurium*, *Listeria monocytogenes* and *Escherichia coli* O157:H7. Anal Methods 12:212–217

Eichelberg K, Ginocchio CC, Galan JE (1994) Molecular and functional characterization of the *Salmonella typhimurium* invasion Genes *inv*B and *inv*C: homology of InvC to the FoF1 ATPase family of proteins. J Bacteriol 176(15):4501–4510

Escalona NG, Jolley KA, Reed E, Urtaza JM (2017) Defining a core genome multilocus sequence typing scheme for the global epidemiology of *Vibrio parahaemolyticus*. J Clin Microbiol 55(6): 1682–1697

Fakruddin M, Mannan KSB, Chowdhary A, Mazumdar RM, Hossain MN, Islam S, Chowdhary MA (2013) Nucleic acid amplification: Alternative methods of polymerase chain reaction. J Pharm Bioallied Sci 5(4):245–252

Feng Y, Yao H, Chen S, Sun X, Yin Y, Jiao X (2020) Rapid detection of hypervirulent serovar 4h *Listeria monocytogenes* by multiplex PCR. Front Microbiol 11:1309

Fields PI, Blom K, Hughes HJ, Helsel LO, Feng P, Swaminathan B (1997) Molecular characterization of the gene encoding H antigen in *Escherichia coli* and development of a PCR-restriction fragment length polymorphism test for identification of *E. coli* O157:H7 and O157:NM. J Clin Microbiol 35(5):1066–1070

Foddai ACG, Grant IR (2020) Methods for detection of viable foodborne pathogens: current state-of-art and future prospects. Appl Microbiol Biotechnol 104:4281–4288

Frankel G, Riley L, Giron JA, Valmassoi J, Friedmann A, Strockbine N, Falkow S, Schoolnik GK (1990) Detection of *Shigella* in feces using DNA amplification. J Infect Dis 161:1252–1256

Fricker M, Messelhauber U, Busch U, Scherer S, Ehling-Schulz M (2007) Diagnostic real-time PCR assays for the detection of emetic *Bacillus cereus* strains in foods and recent food-borne outbreaks. Appl Environ Microbiol 73(6):1892–1898

Furrer B, Candrian U, Hoefelein C, Luethy J (1991) Detection and identification of *Listeria monocytogenes* in cooked sausage products and in milk by in vitro amplification of haemolysin gene fragments. J Appl Bacteriol 70:372–379

Gibriel AA, Ola Adel O (2017) Advances in ligase chain reaction and ligation based amplifications for genotyping assays; detection and applications. Mutat Res/Rev Mutat Res 773:66–90

Ha AJW, Perez LGS, Kim TJ, Mizan MFR, Nahar S, Park SH, Chun HS, Ha SD (2020) Identification and characterization of *Salmonella* spp. in mechanically deboned chickens using pulsed-field gel electrophoresis. Poult Sci 100(3):100961

Hakanen A, Jalava J, Kotilainen P, Jousimies-Somer H, Siitonen A, Huovinen P (2002) *gyr*A Polymorphism in *Campylobacter jejuni*: detection of *gyr*A mutations in 162 *C. jejuni* isolates by single-strand conformation polymorphism and DNA sequencing. Antimicrob Agents Chemother 46(8):2644–2647

Hao L, Gu H, Duan N, Wu S, Ma X, Xia Y, Tao Z, Wang Z (2017) An enhanced chemiluminescence resonance energy transfer aptasensor based on rolling circle amplification and WS2 nanosheet for *Staphylococcus aureus* detection. Anal Chim Acta 959:83–90

Hartman AB, Venkatesan M, Oaks EV, Buysse JM (1990) Sequence and molecular characterization of a multicopy invasion plasmid antigen gene, *ipa*H, of *Shigella flexneri*. J Bacterial 172(4): 1905–1915

Horisaka T, Fujita K, Iwata T, Nakadai A, Okatani AT, Horikita T, Taniguchi T, Honda E, Yokomizo Y, Hayashidani H (2004) Sensitive and specific detection of *Yersinia pseudotuberculosis* by loop-mediated isothermal amplification. J Clin Microbiol 42(11):5349–5352

Hsu CH, Harrison L, Mukharjee S, Strain E, McDermott P, Zhang Q, Zhao S (2020) Core genome multilocus sequence typing for food animal source attribution of human *Campylobacter jejuni* infections. Pathogens 9(7):532

Hu Y, Yan H, Mammel M, Chen H (2015) Sequence-independent amplification coupled with DNA microarray analysis for detection and genotyping of noroviruses. AMB Express 5(1):69

Ibrahim A, Liesack W, Stackebrandt E (1992) Polymerase chain reaction-gene probe detection system specific for pathogenic strains of *Yersinia enterocolitica*. J Clin Microbiol 30(8): 1942–1947

Jenikova G, Pazlarova J, Demnerova K (2000) Detection of *Salmonella* in food samples by the combination of immunomagnetic separation and PCR assay. Int Microbiol 3:225–229

Jiang Y, Zou S, Cao X (2017) A simple dendrimer-aptamer based microfluidic platform for *E. coli* O157:H7 detection and signal intensification by rolling circle amplification. Sensors Actuators B Chem 251:976–984

Johnsen G, Kruse H, Hofshagen M (2006) Genetic diversity and description of transmission routes for *Campylobacter* on broiler farms by amplified-fragment length polymorphism. J Appl Microbiol 101(5):1130–1139. https://doi.org/10.1111/j.1365-2672.2006.02995.x

Khan JA, Rathore RS, Abulreesh HH, Qais FA, Ahmad I (2018) Cultural and immunological methods for the detection of *Campylobacter jejuni*: a review. Indian J Biotechnol Pharm Res 6(3):4–10

Kim AR, Oh MH, Seol KH, Shin GW, Jung GY, Oh S (2010) Parallel analysis of 7 food-borne pathogens using capillary electrophoresis-based single-strand conformation polymorphism. Food Sci Biotechnol 19(6):1441–1447

Kim SY, Chung B, Chang JH, Jung GY, Kim HW, Park BY, Oh SS, Oh MH (2016) Simultaneous identification of 13 foodborne pathogens by using capillary electrophoresis–single strand conformation polymorphism coupled with multiplex ligation-dependent probe amplification and its application in foods. Foodborne Pathog Dis, 13(10):566-574

Kudo YH, Yoshino M, Kojima T, Ikedo M (2005) Loop-mediated isothermal amplification for the rapid detection of *Salmonella*. FEMS Microbiol Lett 253(1):155–161

Kumar J, Sharma N, Kaushal G, Samurailatpam S, Sahoo D, Rai AK, Singh SP (2019) Metagenomic insights into the taxonomic and functional features of *Kinema*, a traditional fermented Soybean product of Sikkim Himalaya. Front Microbiol 10:1744

Lachtara B, Osek J, Wieczorek K (2021) Molecular typing of *Listeria monocytogenes* IVb serogroup isolated from food and food production environments in Poland. Pathogens 10(4):482

Lantz PG, Knutsson R, Blixt Y, Al- Soud WA, Borch E, Radstrom P (1998) Detection of pathogenic *Yersinia enterocolitica* in enrichment media and pork by a multiplex PCR: a study of sample preparation and PCR-inhibitory components. Int J Food Microbiol 45(2):93–105

Li B, Patel IR, Tall BD, Elkins CA (2017) Advancements in microarray utility for detection and tracking of foodborne microbes in the genomic era. Adv Tech Biol Med 5(3):1000239

Li D, Butot S, Zuber S, Uyttendaele M (2018a) Monitoring of foodborne viruses in berries and considerations on the use of RTPCR methods in surveillance. Food Control 89:235–240

Li W, Wu S, Fu P, Liu J, Han H, Bai L, Pei X, Li N, Liu X, Gou Y (2018b) National molecular tracing network for foodborne disease surveillance in China. Food Control 88:28–32

Li S, Liu S, Xu Y, Zhang R, Zhao Y, Qu X, Wang Y, Huang J, Yu J (2019) Robust and highly specific fluorescence sensing of *Salmonella typhimurium* based on dual-functional phi29 DNA polymerase-mediated isothermal circular strand displacement polymerization. Analyst 144: 4795–4802

Li S, Jiang Y, Yang X, Lin M, Dan H, Zou S, Cao X (2021) In situ rolling circle amplification surface modifications to improve *E. coli* O157:H7 capturing performances for rapid and sensitive microfluidic detection applications. Anal Chim Acta 1150:338229

Lindstedt BA, Heir E, Vardund T, Kapperud G (2000) Fluorescent amplified-fragment length polymorphism genotyping of *Salmonella enterica* subsp. *enterica* serovars and comparison with pulsed-field gel electrophoresis typing. J Clin Microbiol 38(4):1623–1627

Liu H, Dong H, Chen Z, Lin L, Chen H, Li S, Deng Y (2017) Magnetic nanoparticles enhanced microarray detection of multiple foodborne pathogens. J Biomed Nanotechnol 13(10): 1333–1343

Liu Y, Cao Y, Wang T, Dong Q, Li J, Niu C (2019) Detection of 12 common food-borne bacterial pathogens by TaqMan real-time PCR using a single set of reaction conditions. Front Microbiol 10:00222

Liu J, Zhan Z, Liang T, Xie G, Aguilar ZP, Xu H (2020) Dual-signal amplification strategy: universal asymmetric tailing-PCR triggered rolling circle amplification assay for fluorescent detection of *Cronobacter* spp. in milk. J Dairy Sci 103(4):3055–3065

Lyhs U, Korkeala H, Bjorkroth J (2002) Identification of lactic acid bacteria from spoiled, vacuum-packaged 'gravad' rainbow trout using ribotyping. Int J Food Microbiol 72(1-2):147–153

Magarino B, Toranzo AE, Barja JL, Romalde JL (2000) Existence of two geographically-linked clonal lineages in the bacterial fish pathogen *Photobacterium damselae* subsp. *piscicida* evidenced by random amplified polymorphic DNA analysis. Epidemiol Infect 125:213–219

Mahato K, Kumar S, Srivastava A, Maurya PK, Singh R, Chandra P (2018) Electrochemical immunosensors: fundamentals and applications in clinical diagnostics. In: Handbook of immunoassay technologies. https://doi.org/10.1016/B978-0-12-811762-0.00014-1

Martins BTF, de Azevedo EC, Yamatogi RS, Call DR, Nero AL (2020) Persistence of Yersinia enterocolitica bio-serotype 4/O:3 in a pork production chain in Minas Gerais. Brazil Food Microbiol 94:103660

Milton AAP, Momin KM, Ghatak S, Priya GB, Angappan M, Das S (2021) Development of a novel polymerase spiral reaction (PSR) assay for rapid and visual detection of *Clostridium perfringens* in meat. Heliyon 7(1):e05941

Mori Y, Notomi T (2009) Loop-mediated isothermal amplifi cation (LAMP): a rapid, accurate, and cost-effective diagnostic method for infectious diseases. J Infect Chemother 15(2):62–69

Murphy NM, McLauchlin J, Ohai C, Grant KA (2007) Construction and evaluation of a microbiological positive process internal control for PCR-based examination of food samples for *Listeria monocytogenes* and *Salmonella enterica*. Int J Food Microbiol 120:110–119

Nagamine K, Hase T, Notomi T (2002) Accelerated reaction by loop-mediated isothermal amplification using loop primers. Mol Cell Probes 16:223–229

Naravaneni R, Jamil K (2005) Rapid detection of food-borne pathogens by using molecular techniques. J Med Microbiol 54(1):51–54

Nielsen PE, Egholm M, Berg RH, Buchardt O (1991) Sequence-selective recognition of DNA by strand displacement with a thymine-substituted polyamide. Science 254(5037):1497–1500. https://doi.org/10.1126/science.1962210

Notomi T, Okayama H, Masubuchi H, Yonekawa T, Watanabe K, Amino N, Hase T (2000) Loop—mediated isothermal amplification of DNA. Nucleic Acid Res 28(12):E63

Oh MH, Park YS, Paek SH, Kim HY, Jung GY, Oh S (2008) A rapid and sensitive method for detecting foodborne pathogens by capillary electrophoresis-based single-strand conformation polymorphism. Food Control 19(11):1100–1104

Oh MH, Hwang HS, Chung B, Paik HD, Han S, Kang SM, Ham JS, Kim HW, Seol KH, Jang A, Jung GY (2012) Simultaneous detection of 10 foodborne pathogens using capillary electrophoresis-based single strand conformation polymorphism. Korean J Food Sci Anim Resour 32(2):241–246

Oh SJ, Park BH, Jung JH, Choi G, Lee DC, Kim DH, Seo TS (2016) Centrifugal loop-mediated isothermal amplification microdevice for rapid, multiplex and colorimetric foodborne pathogen detection. Biosens Bioelectron 75:293–300. https://doi.org/10.1016/j.bios.2015.08.052

Ohtsuka K, Yanagawa K, Takatori K, Hara-Kudo Y (2005) Detection of *Salmonella enterica* in naturally contaminated liquid eggs by loop-mediated isothermal amplification, and characterization of *Salmonella* isolates. Appl Environ Microbiol 71(11):6730–6735

Orita M, Iwahana H, Kanazawa H, Hayashi K, Sekiya T (1989) Detection of polymorphisms of human DNA by gel electrophoresis as single-strand conformation polymorphisms. Proc Natl Acad Sci U S A 86(8):2766–2770

Paillard D, Dubois V, Duran R, Nathier F, Guittet C, Caumette P, Quentin C (2003) Rapid identification of *Listeria* species by using restriction fragment length polymorphism of PCR-amplified 23S rRNA gene fragments. Appl Environ Microbiol:6386–6392

Parisi A, Latorre L, Normanno G, Miccolupo A, Fraccalvieri R, Lorussuo V, Santagada G (2010) Amplified fragment length polymorphism and multi-locus sequence typing for high-resolution genotyping of *Listeria monocytogenes* from foods and the environment. Food Microbiol 27: 101–108

Park YM, Lim SY, Shin SJ, Kim CH, Jeong SW, Shin SY, Bae NH, Lee SJ, Na J, Jung GY, Lee TJ (2018) A film-based integrated chip for gene amplification and electrochemical detection of pathogens causing foodborne illnesses. Sensors Actuators B Chem 329:129130

Purohit B, Mahato K, Kumar A, Chandra P (2020a) Smartphone assisted personalized diagnostics devices and wearable sensors. Curr Opin Biomed Eng 13:42–50

Purohit B, Vernekar PR, Shetti NP, Chandra P (2020b) Biosensor nanoengineering: Design, operation, and implementation for biomolecular analysis. Sensors Int 1:100040

Rahn K, De Grandis SA, Clarke RC, McEwen SA, Galan JE, Ginocchio C, Curtis R III, Gyles CL (1992) Amplification of an *inv*A gene sequence of *Salmonella typhimurium* by polymerase chain reaction as a specific method of detection of *Salmonella*. Mol Cell Probes 6(4):271–279

Rai AK, Kumari R, Sanjukta S, Sahoo D (2016) Production of bioactive protein hydrolysate using the yeasts isolated from soft *chhurpi*. Bioresour Technol 219:239–245

Rai AK, Sanjukta S, Chourasia R, Bhat I, Bhardwaj PK, Sahoo D (2017) Production of bioactive hydrolysate using protease, β-glucosidase and α-amylase of *Bacillus* spp. isolated from *kinema*. Bioresour Technol 235:358–365

Ramdan H, Jackson CR, Frye JG, Hiott LM, Samir M, Awad A, Woodley TA (2020) Antimicrobial resistance, genetic diversity and multilocus sequence typing of *Escherichia coli* from humans, retail chicken and ground beef in Egypt. Pathogens 9(5):357

Ranjbar R, Behzadi P, Najafi A, Roudi R (2017) DNA microarray for rapid detection and identification of food and water borne bacteria: from dry to wet lab. Open Microbiol J 11:330–338

Rao S, Arora K (2020) Recent trends in molecular techniques for food pathogen detection. In: Chemical analysis of food. Academic Press, pp 177–285

Ripabelli G, McLauchlin J, Mithani V, Threlfall EJ (2000) Epidemiological typing of *Bacillus cereus* by amplified fragment length polymorphism. Lett Appl Microbiol 30(5):358–363

Rocha RJA (2018) Optimization of Peptide Nucleic Acid Fluorescence in situ Hybridization (PNA-FISH) for the identification of microorganisms in food matrices

Rocha R, Sousa JM, Cerqueira L, Vieira MJ, Almeida C, Azevedo NF (2019) Development and application of peptide nucleic acid fluorescence *in situ* Hybridization for the specific detection of *Listeria monocytogenes*. Food Microbiol 80:1–8

Rohde A, Hammerl JA, Appel B, Dieckmann R, Al Dahouk S (2017) Differential detection of pathogenic *Yersinia* spp. by fluorescence in situ hybridization. Food Microbiol 62:39–45

Rousseaux S, Olier M, Lemaitre JP, Piveteau P, Guzzo J (2004) Use of PCR-restriction fragment length polymorphism of *inl*A for rapid screening of *Listeria monocytogenes* strains deficient in the ability to Invade Caco-2 cells. Appl Environ Microbiol 70(4):2180–2185

Saadati A, Hassanpour S, de la Guardia M, Mosafer J, Hashemzaei M, Mokhtarzadeh A, Baradaran B (2019) Recent advances on application of peptide nucleic acids as a bioreceptor in biosensors development. TrAC Trends Anal Chem 114:56–68

Sadeghi S, Thong KL, Chai LC (2019) Pre-enrichment step, incubation temperature and type of selective media affect the pathogenic *Vibrio parahaemolyticus* detection efficiency in frozen prawns. J Consumer Protect Food Safety 14(4):355–364

Salimi G, Mousavi ZE, Kiani H (2020) Efficiency of fluorescence in situ hybridization (FISH) method for the rapid detection of *Salmonella* in minced lamb meat: Method analysis and optimization. J Microbiol Methods 175:105989

Saravanan A, Kumar PS, Hemavathy RV, Jeevanantham S, Kamalesh R, Sneha S, Yaashikaa PR (2020) Methods of detection of food-borne pathogens: a review. Environ Chem Lett 19:189–207

Sayad AA, Ibrahim F, Uddin SM, Pei KX, Mohktar MS, Madou M, Thong KL (2016) A microfluidic lab-on-a-disc integrated loop mediated isothermal amplification for foodborne pathogen detection. Sensors Actuators B 227:600–609

Schena M, Shalon D, Davis RW, Brown PO (1995) Quantitative monitoring of gene expression patterns with a complementary DNA microarray. Science 220(5235):467–470

Schouls LM, Reulen S, Duim B, Wagenaar JA, Willems RJL, Dingle KE, Colles FM, Embbden JDAV (2003) Comparative genotyping of *Campylobacter jejuni* by amplified fragment length polymorphism, multilocus sequence typing, and short repeat sequencing: strain diversity, host range, and recombination. J Clin Microbiol 41(1):15–26

Sekse C, Holst-Jensen A, Dobrindt U, Johannessen GS, Li W, Spilsberg B, Shi J (2017) High throughput sequencing for detection of foodborne pathogens. Front Microbiol 8:02029

Shan W, Ke L, Yun F, Huizhen Y, Mingzhe Z, Jiangbing S, Xiaofeng Z (2018) Simultaneous detection of *Listeria monocytogenes* and pathogenic *Vibrio* by dual peptide nucleic acid fluorescence *in situ* hybridization. J Zhejiang Univ 44(6):659–666

Shangkuan YH, Lin HC (1998) Application of random amplified polymorphic DNA analysis to differentiate strains of *Salmonella typhi* and other *Salmonella* species. J Appl Microbiol 85:693–702

Shangkuan YH, Show YS, Wang TM (1995) Multiplex polymerase chain reaction to detect toxigenic *Vibrio cholerae* and to biotype *Vibrio cholerae* O1. J Appl Bacteriol 79(3):264–273

Sheng J, Tao T, Zhu X, Bie X, Lv F, Zhai H, Lu Z (2018) A multiplex PCR detection method for milk based on Novel primers specific for *Listeria monocytogenes* 1/2a serotype. Food Control 86(1):183–190

Shi XM, Long F, Sou B (2010) Molecular methods for the detection and characterization of foodborne pathogens. Pure Appl Chem 82(1):69–79

Siemer BL, Nielsen EM, On SLW (2005) Identification and molecular epidemiology of *Campylobacter coli* isolates from human gastroenteritis, food, and animal sources by amplified fragment length polymorphism analysis and penner serotyping. Appl Environ Microbiol 71(4): 1953–1958

Sierra-Arguello YM, Furian TQ, Perdoncini G, Moraes HLS, Salle CTP, Rodrigues LB, dos Santos LR, Gomes MJP, do Nascimento VP (2018) Fluoroquinolone resistance in *Campylobacter jejuni* and *Campylobacter coli* from poultry and human samples assessed by PCR-restriction fragment length polymorphism assay. PLoS One 13(7):e0199974

Song S, Wang X, Xu K, Li Q, Ning L, Yang X (2018) Selection of highly specific Aptamers to *Vibrio Parahaemolyticus* using cell-SELEX powered by functionalized graphene oxide and rolling circle amplification. Anal Chim Acta 1052:153–162

Song S, Wang X, Xu K, Xia G, Yang X (2019) Visualized detection of *Vibrio parahaemolyticus* in food samples using dual functional aptamers and cut-assisted rolling circle amplification. J Agric Food Chem 67(4):1244–1253

Sun X, Wang Y, Zhang L, Liu S, Zhang M, Wang J, Ning B, Peng Y, He J, Hu Y, Gao Z (2020) A CRISPR–Cas9 triggered two–step isothermal amplification method for *E. coli* O157:H7 detection based on metal–organic framework platform. Anal Chem 92(4):3032–3041

Tada J, Ohashi T, Nishimura N, Shirasaki Y, Ozaki H, Fukushima S, Takano J, Nishibuchi M, Takeda Y (1992) Detection of the thermostable direct hemolysin gene (*tdh*) and the thermostable direct hemolysin-related hemolysin gene (*trh*) of *Vibrio parahaemolyticus* by polymerase chain reaction. Mol Cell Probes 6(6):477–487

Taheri H, Peighambari SM, Shahcheraghi F, Solgi H (2018) Pulse-Field Gel Electrophoresis (PFGE) of *Salmonella* Serovar infantis isolates from poultry. Iran J Vet Med 12(3):187–197

Tao J, Liu W, Ding W, Han R, Shen Q, Xia Y, Zhang Y, Sun W (2020) A multiplex PCR assay with a common primer for the detection of eleven foodborne pathogens. J Food Sci 85(3):744–754

Teng J, Ye Y, Yao L, Yan C, Cheng K, Xue F, Pan D, Li B, Chen W (2017) Rolling circle amplification based amperometric aptamer/immuno hybrid biosensor for ultrasensitive detection of *Vibrio parahaemolyticus*. Microchim Acta 184:3477–3485

Torpdahl M, Ahrens P (2004) Population structure of *Salmonella* investigated by amplified fragment length polymorphism. J Appl Microbiol 97(3):566–573

Torpdahl M, Skov MN, Sandvang D, Baggesen DL (2005) Genotypic characterization of *Salmonella* by multilocus sequence typing, pulsed-field gel electrophoresis and amplified fragment length polymorphism. J Microbiol Methods 63:173–184

Tsai YL, Palmer CJ, Sangermano LR (1993) Detection of *Escherichia coli* in sewage and sludge by polymerase chain reaction. Appl Environ Microbiol 59(2):353–357

Umesha S, Manukumar HM (2016) Advanced molecular diagnostic techniques for detection of food-borne pathogens; current applications and future challenges. Crit Rev Food Sci Nutr 58(1): 84–104

Umesha S, Chandan S, Swamy LN (2012) Colony PCR–single strand confirmation polymorphism for the detection of *Ralstonia solanacearum* in tomato. Int J Integr Biol 13(1):45–51

Vacher S, Menard A, Bernard E, Megraud F (2003) PCR-Restriction fragment length polymorphism analysis for detection of point mutations associated with macrolide resistance in *Campylobacter* spp. Antimicrob Agents Chemother 47(3):1125–1128

Valasek MA, Repa JJ (2005) The power of real-time PCR. Adv Physiol Educ 29:151–159

Verma S, Choudhary J, Singh KP, Chandra P, Singh SP (2019) Uricase grafted nanoconducting matrix based electrochemical biosensor for ultrafast uric acid detection in human serum samples. Int J Biol Macromol. https://doi.org/10.1016/j.ijbiomac.2019.02.121

Walker GT, Fraiser MS, Schram JL, Little MC, Nadeau JG, Malinowski DP (1992) Strand displacement amplification—an isothermal, in vitro DNA amplification technique. Nucleic Acid Res 20(7):1691–1696. https://doi.org/10.1093/nar/20.7.1691

Wang L, Shi L, Alam MJ, Geng Y, Li L (2008) Specific and rapid detection of foodborne *Salmonella* by loop-mediated isothermal amplification method. Food Res Int 41:69–74

Wang Z, Yang Q, Zhang Y, Meng Z, Ma X, Zhang W (2017) Saltatory Rolling Circle Amplification (SRCA): a Novel nucleic acid isothermal amplification technique applied for rapid detection of *Shigella* spp. in vegetable salad. Food Anal Methods 11(2):1–10

Wang M, Yang J, Gai Z, Huo S, Zhu J, Li J, Wang R, Xing S, Shi G, Shi F, Zhang L (2018) Comparison between digital PCR and real-time PCR in detection of *Salmonella typhimurium* in milk. Int J Food Microbiol 266:251–256

Wang X, Luo Z, Xie Q, Huang Z, Wu M, Duan Y (2020a) Toehold-mediated strand displacement reaction formation of three—way junction DNA structure combined with nicking enzyme signal amplification for highly sensitive colorimetric detection of *Salmonella Typhimurium*. Anal Chim Acta 1139:138–145

Wang Y, Ke Y, Liu W, Sun Y, Ding X (2020b) A one-pot toolbox based on Cas12a/crRNA enables rapid foodborne pathogen detection at attomolar level. ACS Sensors 5(5):1427–1435

Wang Y, Zhao T, He X, Ke Y, Liu W, Zou D (2020c) Multiplex real-time SYBR Green I PCR assays for simultaneous detection of 15 common enteric pathogens in stool samples. Mol Cell Probes 53:101619

Ward LN, Bej AK (2006) Detection of *Vibrio parahaemolyticus* in shellfish by use of multiplexed real-time PCR with TaqMan fluorescent probes. Appl Environ Microbiol 72(3):2031–2042

Wei C, Zhong J, Hu T, Zhao X (2018) Simultaneous detection of *Escherichia coli* O157:H7, *Staphylococcus aureus* and *Salmonella* by multiplex PCR in milk. 3 Biotech 8(1):76

Wilson IG, Cooper JE, Gilmour A (1991) Detection of enterotoxigenic *Staphylococcus aureus* in dried skimmed milk: use of the polymerase chain reaction for amplification and detection of staphylococcal enterotoxin genes *ent*B and *ent*C1 and the thermonuclease gene *nuc*. Appl Environ Microbiol 57(6):1793–1798

Yadav R, Yadav J, Gahlot K, Purva M, Deora A, Kumar P, Nathawat P, Rathore NS, Maherchandani S, Kashyap SK (2017) Repetitive extragenic palindromic-PCR (REP-PCR) typing of *Campylobacter jejuni* isolated from poultry. Vet Pract 18(2):160–162

Yang Y, Yu X, Zhan L, Chen J, Zhang Y, Zhang J, Chen H, Zhang Z, Zhang Y, Lu Y, Mei L (2017) Multilocus sequence type profiles of *Bacillus cereus* isolates from infant formula in China. Food Microbiol 62:46–50. https://doi.org/10.1016/j.fm.2016.09.007

Yang X, Yu S, Wu Q, Zhang J, Wu S, Rong D (2018) Multilocus sequence typing and virulence-associated gene profile analysis of *Staphylococcus aureus* isolates from retail ready-to-eat food in China. Front Microbiol 9:197

Yang Q, Zhang Y, Li S, Lu X, Yuan Y, Zhang W (2019) Saltatory rolling circle amplification for sensitive visual detection of *Staphylococcus aureus* in milk. J Dairy Sci 102(11):9702–9710

Yu F, Chen X, Zheng S, Han D, Wang Y, Wang R, Wang B, Chen Y (2018) Prevalence and genetic diversity of human diarrheagenic *Escherichia coli* isolates by multilocus sequence typing. Int J Infect Dis 67:7–13

Zbrun MV, Rossler E, Sotto LP, Rosmini MR, Sequeira GJ, Frizzo LS, SignoriniML (2020) Molecular epidemiology of *Campylobacter jejunii* solates from the broiler production chain: first report of MLST profiles in Argentina. Rev Argent Microbiol, 53(1):59-63

Zeng D, Chen Z, Jiang Y, Xue F, Li B (2016) Advances and challenges in viability detection of foodborne pathogens. Front Microbiol 7:01833

Zhai L, Liu H, Chen Q, Lu Z, Zhang C, Lv F, Bie X (2019) Development of a real-time nucleic acid sequence–based amplification assay for the rapid detection of *Salmonella* spp. from food. Braz J Microbiol 50(1):255–261. https://doi.org/10.1007/s42770-018-0002-9

Zhan Z, Li H, Liu J, Xie G, Xiao F, Wu X, Aguilar Z, Xu H (2020) A competitive enzyme linked aptasensor with rolling circle amplification (ELARCA) assay for colorimetric detection of *Listeria monocytogenes*. Food Control 107:106806

Zhang P, Liu H, Li X, Ma S, Men S, Wei H, Cui J, Wang H (2016) Label-free fluorescent direct detection of live *Salmonella typhimurium* using cascade triple trigger sequences-regenerated strand displacement amplification and hairpin template-generated -scaffolded silver nanoclusters. Biosens Bioelectron 87:1044–1049

Zhang Z, Zhou J, Du X (2019) Electrochemical biosensors for detection of foodborne pathogens. Micromachines 10:222

Zhang R, Belwal T, Li L, Lin X, Xu Y, Luo Z (2020) Nanomaterial-based biosensors for sensing key foodborne pathogens: advances from recent decades. Compr Rev Food Sci Food Saf 19(4): 1465–1487

Zhang Y, Li ST, Tian JJ, Li K, Du Z, Xu WT (2021) Universal linker polymerase chain reaction-triggered strand displacement amplification visual biosensor for ultra-sensitive detection of *Salmonella*. Talanta 222:121575

Zhao X, Wu C (2020) Recent advances in peptide nucleic acids for rapid detection of foodborne pathogens. Food Anal Methods 13(10):1956–1972

Zheng W, Kathariou S (1995) Differentiation of epidemic-associated strains of *Listeria monocytogenes* by restriction fragment length polymorphism in a gene region essential for growth at low temperatures (48 °C). Appl Environ Microbiol 61(12):4310–4314

Zhong J, Zhao X (2018) Isothermal amplification technologies for the detection of foodborne pathogens. Food Anal Methods 11(22):1543–1560

Zhu LX, Zhang ZW, Liang JD, Wang C, Du N, Zhang Q, Mitchelson K, Cheng J (2007) Multiplex asymmetric PCR-based oligonucleotide microarray for detection of drug resistance genes containing single mutations in *Enterobacteriaceae*. Antimicrob Agents Chemother 51(10): 3707–3371

Zhu L, Zhang Y, He P, Zhang Y, Wang Q (2018) Simultaneous detection of foodborne bacteria in milk by microchip electrophoresis combined with multiplex PCR amplification. J Chromatogr B 1093–1094:141–146

Aptamer-Based Technologies in Foodborne Pathogen Detection

10

Li Yao, Jun Teng, and Wei Chen

Abstract

Food safety is a severe problem facing the global public health system. Foodborne diseases caused by foodborne pathogens or their toxins are one of the main burdens of public health, which seriously hinder the global social and economic development. Therefore, the establishment of highly sensitive detection method is the primary task of prevention and control of foodborne pathogenic bacteria pollution. Traditional detection methods of pathogenic bacteria mainly rely on precision instruments, and these methods have high sensitivity and excellent accuracy, but time-consuming and tedious operation steps limit its application in on-site detection. Immunoassay and polymerase chain reaction (PCR) can solve the above problems to a certain extent, but the cost of obtaining antibodies is high, and PCR needs complex DNA extraction process. The emergence of aptamers has greatly overturned this detection limit. Aptamers are DNA or RNA sequences with a length of about 25–80 bases that bind to the targets with high affinity and specificity as antibodies, and it was obtained by the method called systematic evolution of ligands by exponential enrichment (SELEX). Aptamers can specifically bind to their target, in addition, they are more stable and easier to be prepared than antibodies, which make them widely used in the field of detection. So far, aptamer has been applied in various pathogen detection technologies, such as ELISA, fluorescence, electrochemical, surface-enhanced Raman scattering (SERS), which greatly promotes the development of rapid detection of pathogenic bacteria.

L. Yao · W. Chen (✉)
School of Food and Biological Engineering, Hefei University of Technology, Hefei, China

J. Teng
Guangzhou College of Technology and Business, Guangzhou, China

P. Chandra, P. S. Panesar (eds.), *Nanosensing and Bioanalytical Technologies in Food Quality Control*, https://doi.org/10.1007/978-981-16-7029-9_10

Keywords

Food safety · Foodborne pathogens · Detection · Aptamer · SELEX

10.1 Introduction

10.1.1 Pathogenic Bacteria in Food

Pathogenic bacteria refer to the microorganisms that can cause diseases, also known as pathogenic microorganisms. Generally speaking, pathogenic bacteria refer to bacteria in pathogenic microorganisms. People infected with foodborne pathogens mainly have nausea, vomiting, abdominal pain, abdominal spasm, diarrhea and other gastrointestinal symptoms. The common pathogenic bacteria in food are Salmonella, *Staphylococcus aureus*, *Vibrio parahaemolyticus*, *Listeria monocytogenes*, *Escherichia coli* O157: H7, and so on. Although different foodborne pathogens have different infection sources, most of them are meat products, aquatic products, egg products, dairy products, fruits, and vegetables and other fresh foods.

10.1.1.1 Salmonella

Salmonella is a large group of Gram-negative bacteria that parasitize in human and animal intestines. Its biochemical reaction and antigen structure are similar. It is a large genus of Enterobacteriaceae. It was named after Daniel Elmer salmon (D.V.M., 1850–1914), the first veterinary doctor in the United States, who first isolated Salmonella cholerae from diseased pigs in 1884. Salmonella has no spore and capsule. Most of the bacteria have flagella and pili all over the body. They have motility and can grow on simple medium. The optimum pH and temperature for Salmonella growth were 6.5–7.5 and 35–37 °C. Enterobacteriaceae selective identification medium is often used for isolation and culture of Salmonella. Bile salt and brilliant green in the medium have less inhibitory effect on Salmonella than other Enterobacteriaceae, so it can be used to prepare Enterobacteriaceae selective medium. Most Salmonella strains form colorless colonies because they do not ferment lactose.

Salmonella is widely distributed in animal foods such as poultry, eggs, milk, beef, and pork, as well as fresh fruits, vegetables (Mahon et al. 1997), water (Mahon et al. 1997), reptiles (Friedman et al. 1998), and even human skin (Stone et al. 1993). Some *Salmonella* serotypes have a certain host specificity, such as human is the main host of *Salmonella typhi*, cattle are the main host of *Salmonella Dublin*, poultry is the main host of *Salmonella pullorum*, pigs are the main host of Salmonella cholerae, but *Salmonella typhimurium* has no host specificity.

According to the British Public Health Laboratory Service report, a batch of eggs containing Salmonella were imported from Spain to the UK, causing the most serious disease infection in the history of the UK. Only in September 2002, at least 250 British people were infected with Salmonella enteritidis because of eating the eggs, resulting in two deaths and 10 hospitalizations. Once the human body ingests food containing a large amount of *Salmonella*, it will cause bacterial infection, and then food poisoning will occur under the action of toxins.

10.1.1.2 *Listeria monocytogenes*

Listeria monocytogenes is an important foodborne pathogen, which is widely distributed in nature (Nightingale et al. 2005). Murray et al. first isolated the pathogen from rabbits and guinea pigs with sepsis in 1924 (Murray et al. 1926), and finally named it *L. monocytogenes* by Pirie in 1940.

Listeria monocytogenes is a type of Gram-positive facultative anaerobic bacterium with no spores and flagella. This kind of bacteria is psychrophilic and suitable for growth in the pH range of 4–9. It can grow and reproduce at low temperature (such as 4 °C storage temperature) (Farber and Peterkin 1991). And, there are seven species of Listeria: *L. monocytogenes*, *L. seeligeri*, *L. grayi*, *L. ivanvii*, *L. innocua*, *L. welshimeri* and *L. murrayi*. *Listeria monocytogenes* is widely distributed in nature. Food poisoning is mainly caused by oral infection of food (mainly animal products) contaminated by animal and human feces. *Listeria monocytogenes* outbreaks occur frequently all over the world, which has gradually attracted wide attention. *Listeriosis* caused by *monocytogenes* has a high mortality rate for human beings, especially in people with low immunity, the mortality rate is as high as 30%.

10.1.1.3 *Vibrio parahaemolyticus*

Vibrio parahaemolyticus was first isolated from a food poisoning patient in Japan by Fujino et al. In 1953, it was named *Pasteurella haemolyticus*. In 1958, because of its salt tolerance, it was identified as halophilic bacteria by Takikawa et al. (Kourany 1983). Until 1963, sakazaki et al. named it *Vibrio parahaemolyticus*, and it has been used up to now.

Vibrio parahaemolyticus is a Gram-negative bacterium, it belongs to the genus Vibrio of Vibrio family, and it is a halophilic polymorphous bacillus. They were divided into 13 serogroups according to the O antigen of *Vibrio parahaemolyticus*. The shape of *Vibrio parahaemolyticus* is arc, rod, or filamentous, no spore and capsule exist around the bacterial cell. The arrangement of bacteria is irregular under the microscope, occasionally in pairs. Most of the bacteria have single flagella in liquid medium, which is helpful to the movement. *Vibrio parahaemolyticus* can grow periflagella in solid medium, it is a kind of aerobic or microaerophilic bacteria. This kind of bacteria has low nutritional requirements and can grow in common nutrient agar or peptone solution. The suitable growth pH condition is 5–10, and the optimum pH is from 7.2 to 8.2, while the optimum temperature range is 5–10 °C, but some experiments showed that it could still survive for 12 days at −34 °C when in the fish. It is not heat resistant and will be killed at 90 °C for 1 min, besides, it is not acid resistant and will die in 1% acetic acid or 50% vinegar for 1 min. In addition, it can't grow without salt, 3.5% of NaCl in the medium is the most suitable concentration, but it can't grow when the concentration is higher than 8%.

Vibrio parahaemolyticus is one of the main pathogens causing food poisoning in coastal areas. The food containing *Vibrio parahaemolyticus* are mainly aquatic products and salted products. It is easy to be detected in oysters, crabs, squid, jellyfish, fish, scallops, and other aquatic products. Octopus and squid are the most vulnerable foods to be contaminated by *Vibrio parahaemolyticus*, and the probability of carrying the bacteria can be as high as 100% In addition, *Vibrio parahaemolyticus* is often detected in eggs, meat, and vegetables, and the probability of *Vibrio parahaemolyticus* in flies near seaports and fish shops is also high. Pathogenic *Vibrio parahaemolyticus* has strong toxicity and pathogenicity to human and animals. Its pathogenic factors mainly include direct thermolysin (TDH) and direct thermolysin-related hemolysin (TRH). These toxic factors have hemolytic activity, enterotoxin and lethal effect, and can cause many adverse symptoms such as food poisoning, reactive arthritis, and heart disease (Nishibuchi and Kaper 1995). The main symptoms of clinical patients infected with *Vibrio parahaemolyticus* include acute diarrhea, abdominal pain, vomiting and watery stool, who with low immunity may have symptoms of spasticity, dehydration, and acidosis. It can be seen that *Vibrio parahaemolyticus* is harmful to human beings. Therefore, it is particularly important to detect it as soon as possible.

10.1.1.4 *Staphylococcus aureus*

Staphylococcus aureus belongs to the genus Staphylococcus. The typical *Staphylococcus aureus* is spherical and arranged in clusters under the microscope. It has no spores, flagella, and most of them have no capsule. *Staphylococcus aureus* is widely distributed on the surface of the environment, animals, and human body, and it is difficult to be removed. *Staphylococcus aureus* has high salt tolerance and can grow in broth containing 10–15% of NaCl.

The colonies of *Staphylococcus aureus* on the plate were thick, glossy, round and convex in shape, with a diameter of about 1–2 mm. A transparent hemolytic ring was formed around the colony of blood plate. *Staphylococcus aureus* has high salt tolerance and can grow in broth containing 10–15% of NaCl. It can decompose glucose, maltose, lactose, and sucrose, producing acid but not gas. Methyl red reaction of *Staphylococcus aureus* is positive while its V-P reaction is weakly positive. Many strains of the *Staphylococcus aureus* have the ability to decompose arginine, hydrolyze urea, reduce nitrate, and liquefy gelatin. *Staphylococcus aureus* has strong resistance, it is low sensitive to sulfonamides, but highly sensitive to penicillin and erythromycin. It has high sensitivity to basic dyes, and its growth process can be inhibited immediately by 1/100,000 of gentian violet solution.

Staphylococcus aureus is widely distributed in nature. It can grow and reproduce in foods such as dairy products, meat products, fish, and canned foods. It can produce many enterotoxins (SETs) related to virulence and pathogenicity. These toxins can cause serious diseases such as human food poisoning, toxic shock syndrome, osteomyelitis, necrotizing pneumonia, and endocarditis, and cause infection and death in humans and many animals (Cowie et al. 2005). *Staphylococcus aureus* enterotoxin is a worldwide health problem. Food poisoning caused by *Staphylococcus aureus* enterotoxin accounts for 33% of the total bacterial food poisoning in the United States, and the problem is more serious in Canada, accounting for 45%. The contamination rate of *Staphylococcus aureus* to raw meat is 11%, while reaches 16% to unprocessed milk, and about 3% to processed food.

10.1.1.5 *Escherichia coli*

Escherichia coli (*E. coli*) is named after a German bacteriologist because it was isolated by him in 2005. It is classified in Enterobacteriaceae and belongs to the genus Escherichia. *Escherichia coli* is small, simple in shape, and has a cell wall. It is a single-celled microorganism that reproduces based on dichotomy. The bacteria have no spores, and most of the strains have motility. Its structure is basically divided into part of the cell membrane in the cell wall, cytoplasm, some sugar coats outside the cell wall of the nucleoplasty, flagella, fimbriae and sex hairs, and the cell wall, and it belongs to Gram-negative bacilli. On the tryptone yeast extract medium, the appearance of *E. coli* colonies is smooth, round, and colorless.

Infants and newborn animals, often a few hours after birth, there are *E. coli* from the mouth into their digestive tract and in the rear end of the digestive tract, after a large number of reproductions, these bacteria will exist for all over the host's life, constitute the main part of the intestinal flora, and has an important physiological role. On the other hand, when the body is poor of resistance or other tissues and organs invaded by *E. coli*, it will form conditional pathogenic bacteria and cause extraintestinal infection subsequently. Some serotypes of *E. coli* can produce specific fimbriae antigen, toxin, or have specific invasiveness, causing gastrointestinal tract infection, urinary tract infection, meningitis, etc. of people or animals, these *E. colis* are pathogenic *E. coli*.

Escherichia coli is usually excreted from human and animal bodies with feces and is widely spread in nature. Therefore, once *E. coli* is detected in food, it means that there is direct or indirect fecal contamination. So, in terms of hygiene, *E. coli* is used as a microbiological indicator of fecal pollution in drinking water, milk, or food. Moreover, because the survival time of *E. coli* in the outside world is similar to that of some major intestinal pathogens, its appearance may also indicate the existence of some intestinal pathogens, such as Salmonella and Shigella. Therefore, *E. coli* is internationally recognized as an indicator of health monitoring. At present, the traditional detection methods of *Escherichia coli* include plate dilution method, multi-tube fermentation method, filter membrane method, and so on. Although these methods have high accuracy, some disadvantages of long culture time, poor specificity, and complex operation are still non-negligible. Therefore, it is of great significance to develop sensitive and efficient detection methods for *Escherichia coli*.

10.1.2 Detection of Pathogenic Bacteria

Infectious diseases caused by foodborne pathogenic bacteria has always been a main threat to public health. The Centers for Disease Control and Prevention (CDC) estimated that nearly 9.4 million case of foodborne illness occurred every year in the United States. Just in 2013, the frequent outbreak of foodborne diseases up to a total of 19,056 infections, 4200 hospitalizations, and 80 deaths in America. Furthermore, due to the poor medical conditions, the occurrence of foodborne diseases is even highly frequent in many developing countries. Preliminary stage, the detection methods for foodborne pathogens mainly include traditional microbial detection technology, analysis based on instrument, immunological detection methodm and molecular biology technology.

Traditional microbial detection technology is mainly based on the distinction of physiological and biochemical characteristics of microorganisms. Generally, it requires a series of complex steps including enrichment culture, isolation and culture, biochemical test, serological test, etc. This technology is recognized as the authoritative method for the detection of foodborne pathogens in food because of its admirable stability. Although the method does not require experimental equipment, the experimental operation required is complex and time-consuming, it takes about 3–5 days from sampling to identification, besides, the sensitivity and specificity are limited, so it does not meet the requirements for rapid on-site testing.

For the instrumental analysis of pathogenic bacteria, there are mainly two ways: gas chromatography (GC) and high performance liquid chromatography (HPLC). The detection principle is mainly based on the difference of the chemical components of pathogens themselves and the distinction of the metabolites they produce. We can determine the specific chemical marker components of pathogenic microorganisms through chromatographic analysis of bacterial metabolites in body fluids, which can be applied in assisting in the diagnosis and detection of pathogens. And gas chromatography is more widely used among instrument detection of pathogenic bacteria. Instrumental analysis is easy to operate as well as has outstanding sensitivity, besides, the detection results are reliable and accurate. However, the equipment applied are relatively costly, which lead to the unsuitable on-site detection.

The development of nanomaterials (Chandra and Prakash 2020) and biorecognition elements (Purohit et al. 2020) application, greatly promote the application of biosensors in biomolecule detection, and the biosensors contains immunobiosensors, nucleic acid biosensors, optical biosensors, and electrochemical biosensors (Chandra et al. 2012; Choudhary et al. 2016; Deka et al. 2018; Mahato et al. 2018; Verma et al. 2019). Immunoassay for foodborne pathogens detection is mainly based on specific binding between antibodies and corresponding antigens like proteins, polysaccharides, and other molecules on cell surface. The commonly applied methods include enzyme-linked immunosorbent assay (ELISA), enzyme-linked fluorescence assay (ELFA), time-resolved fluorescence immunoassay (TrFIA) (Watanabe et al. 2002), polymerase immunoassay (Stamm et al. 1981), chemiluminescence immunoassay (CIA) (Karsunke et al. 2009), etc. Brigmon group constructed an enzyme-linked immunosorbent assay using monoclonal antibody (ASCII) to detect the *Salmonella enteritidis* in environmental samples (Brigmon et al. 1992). The minimum detection limit was 10^5 cfu/mL, and there was no cross-reaction with other 31 *Salmonella* strains, which showed excellent selectivity. Hochel et al. established an indirect competitive ELISA for the detection of *Campylobacter jejuni* in food with a detection limit of 50 cfu/μL (Hochel et al. 2004). A multi-channel sandwich chemiluminescence immunoassay for simultaneous detection of *Escherichia coli*, *Yersinia coli*, *Salmonella typhoid*, and *Listeria monocytogenes* were constructed by Magliulo (2007), the detection limits were

10^4–10^5 cfu/mL, and the result can compete with the traditional slab method. Besides, electrochemical enzyme immunoassay was also a method to improve sensitivity effectively (Jenkins et al. 1988). Based on the significant sensitivity, this kind of sensors is the most recommended approach for point-of-care applications (Suman and Chandra 2021). Immunoassay has the advantages of high specificity, high sensitivity, and easy observation, but it takes a long time to prepare antibody, which is also a costly process.

Polymerase chain reaction (PCR), including conventional PCR, multiplex PCR (m-PCR), real-time PCR, and reverse transcription PCR (RT-PCR), is the most widely used molecular biological technology to detect bacteria. Usually, conventional PCR is applied to detect single pathogenic bacteria, while multiplex PCR often applied in detecting three or more kinds of foodborne pathogenic bacteria simultaneously. Xu Yiping et al. designed three pairs of specific primers for multiplex PCR detection method according to the sequence of *Salmonella invA* gene, *Escherichia coli phoA* gene, and *Staphylococcus aureus nuc* gene, and the detection limits were 10.2 pg, 10.2 pg, and 102.0 pg, respectively. Jongsoo et al. used multiplex PCR to analyze the contamination of *E. coli*, *Listeria monocytogenes*, and *Salmonella typhimurium* in wheat, and obtain detection limits of 56, 1800, and 54 cfu/mL, respectively (Kim et al. 2006). Real-time fluorescent quantitative PCR is to monitor the amplification process of genes in real time, as fluorescent groups were added to the PCR reaction system, and the fluorescence signal will increase with the amplification process, so we can monitor the amplification information instantly through the accumulation of fluorescence intensity. Finally, quantitatively analyze the unknown template through the standard curve, without the need for agar after the conventional PCR amplification reaction glycogel electrophoresis experiment. Based on the specific *O* gene of *Listeria monocytogenes*, Liu Zhongmin and others designed a pair of specific primers, using SYBR Green as the fluorophore, established a real-time fluorescent quantitative PCR method for the detection of *Listeria monocytogenes*, with a detection limit of 8 cfu/mL. Similarly, Jiang Luyan designed two pairs of specific primers and established a method based on the Taq Man probe to detect these two pathogenic bacteria simultaneously based on the *gyrB* gene of *Vibrio parahaemolyticus* and the *coa* gene of *Staphylococcus aureus*. Reverse transcription polymerase chain reaction (RT-PCR) is a project to amplify messenger RNA (mRNA). As mRNA is unstable and easy to be degraded after the cell death, so it can be detected only in living bacterial cells. As a result, his method can be applied in distinguishing dead bacteria from living bacteria. In general, the method based on molecular biology technology to detect foodborne pathogens has great advantages in shortening the detection time and simplifying the detection procedure. At the same time, it also has certain defects, for instance, it needs to extract total bacterial DNA or RNA in early stage, besides, the false-positive rate is relatively high.

In addition, the common situation in life is, multiple bacterial pathogens may coexist in the same food sample but typically occur at different low concentration levels. Therefore, the challenge now is to develop rapid, sensitive, and specific methods that have the ability of simultaneously detecting multiple pathogens. Traditional culture-based methods are preceded by an enrichment process to raise the number of bacteria to meet the detection level. For the simultaneous detection of different kinds of pathogen, differences in growth requirements and growth rates have to be taken into accounts. As an alternative, some biological amplification methods such as multiplex polymerase chain reaction (PCR) or real-time PCR detection can also be considered for the simultaneous detection of five, six, or even more pathogens (Kramer et al. 2009). However, the traditional bacteria detection methods require PCR amplification and/or cell culturing, which are slow, time-consuming, and laborious have limited suitability on-site analyses. In order to meet this demand, several other detection methods have been developed. For instance, a paper-based radial flow chromatographic immunoassay (RFCI) employing gold nanoparticles (AuNPs) as chromatic agents was developed for the detection of *Escherichia coli* O157:H7 in whole milk (Luo et al. 2019), and the detection limit for target pathogenic bacteria in whole milk is as low as 10^3 cfu/mL. In addition, paper-based colorimetric sensors, such as lateral flow assay (LFA), have great potential as a point-of-care diagnostics platform that can instantly identify the presence of pathogenic microorganisms in food samples (Kong et al. 2017). Furthermore, a new cotton swab-based detection system that involved integrating bacterial collection, preconcentration, and detection on Q-tips was developed (Alamer et al. 2018). The platform is based on a sandwich assay that can detect different pathogens visually by color changes. However, the antibodies which play key roles in the above methods have defects which cannot be ignored, such as high production cost and short shelf half-life because of low stability at high temperatures and pH changes.

From what has been discussed above, there are more or less some drawbacks for the above-mentioned methods to detect pathogenic bacteria, these limitations call for the emergence of a new molecular recognition probes, which can not only maintain the affinity and specificity of antibody, but also overcome many defects in the preparation and use of antibody. Once this expectation can be realized, it will be of great significance to improve the detection technology of foodborne pathogens. Based on this demand, oligonucleotide aptamers, this new type of recognition element came into being, and gradually became a research hotspot. The physical and chemical properties of the aptamer make it more suitable for on-site testing. Aptamers can maintain their stability under different temperatures and pH ranges; therefore, they have an extensive range of assay advantage.

10.1.3 SELEX Methods for Aptamers

Aptamers are DNA or RNA sequences with a length of about 25–80 bases that bind to the targets with high affinity and specificity as antibodies, and they provide a variety of advantages compared with antibodies. Aptamers can target small metal ion, amino acids, organic molecules, proteins, viruses, bacteria, whole cells, and animals, etc. Since the oligonucleotide itself can be folded into diverse three-dimensional structure (Fig. 10.1), they can bind to the target with high specificity and affinity, and its action pattern is similar to that of antibody, so it is also regarded

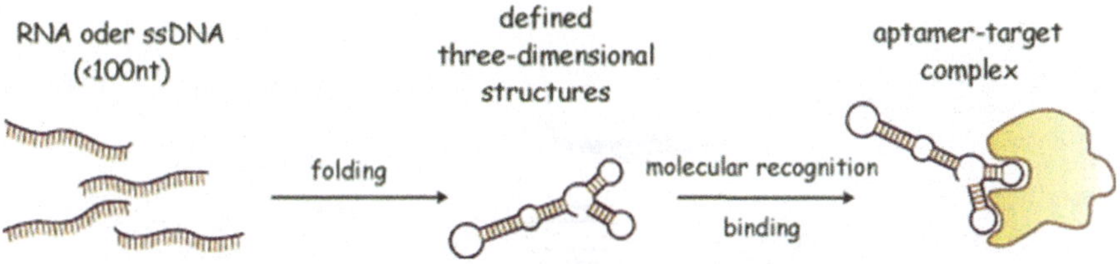

Fig. 10.1 Schematic representation of the functionality of aptamers (Wang et al. 2019)

as a nucleic acid or chemical similar to an antibody. But aptamers are not equivalent to antibody absolutely. Aptamers have many other unique advantages, such as excellent thermal stability, less batch variation, low immunogenicity, and low cost. Demonstrates considerable advantages in purification, cell tracing, biomarker discovery, biosensing, clinical diagnosis, drug delivery, etc. Further, the most important property of aptamers rests with the feasibility by which these oligonucleotide sequences can be easily modified and engineered into aptamer–drug conjugates, which facilitates their clinical applications.

Aptamers are screened from synthetic short-strand nucleic acid libraries by a method called systematic evolution of ligands by exponential enrichment (SELEX) in vitro, which was established by Tuerk and Gold, and Ellington and Szostak in 1990. The basic principle of this technology lies in the correlation between single-stranded oligonucleotide bases. Interactions often form many spatial conformations, such as hairpins, pseudoknots, pockets, or G-quadruplexes, through spatial conformation matching, enrichment of bases in the sequence, hydrogen bonding between charged groups or electrostatic interactions, etc., high-affinity and high-specific binding with the target molecule achieved. Based on the random DNA library, which has a huge capacity and a rich variety of nucleic acid spatial conformations, it is possible to screen and obtain high-affinity and high-specificity aptamers from any target. Combined with PCR in vitro amplification technology, the oligonucleotide sequences that specifically bind to the target are enriched exponentially. At the same time, through strict control of the screening conditions, with several or dozens of rounds of screening and enrichment, the final obtained one or a set of oligonucleotide aptamers that bind to the target substance with high affinity and specificity.

10.1.3.1 Conventional SELEX Methods

The conventional technical route of SELEX technology is as follows (Fig. 10.2). First step, construct and artificially synthesize random oligonucleotide library and primers. The sequence length of the random region is about 30–60 nt, and the two ends are fixed sequences, which are convenient for primer annealing and PCR amplification; the random library can be an RNA library or a single-stranded DNA (ss DNA) library. There is sufficient evidence to show whether to use an RNA library or a single-stranded DNA library to screen a target molecule. Experiments have used both RNA library and single-stranded DNA library to screen the same target molecule, and the results show that the affinity and specificity of the selected

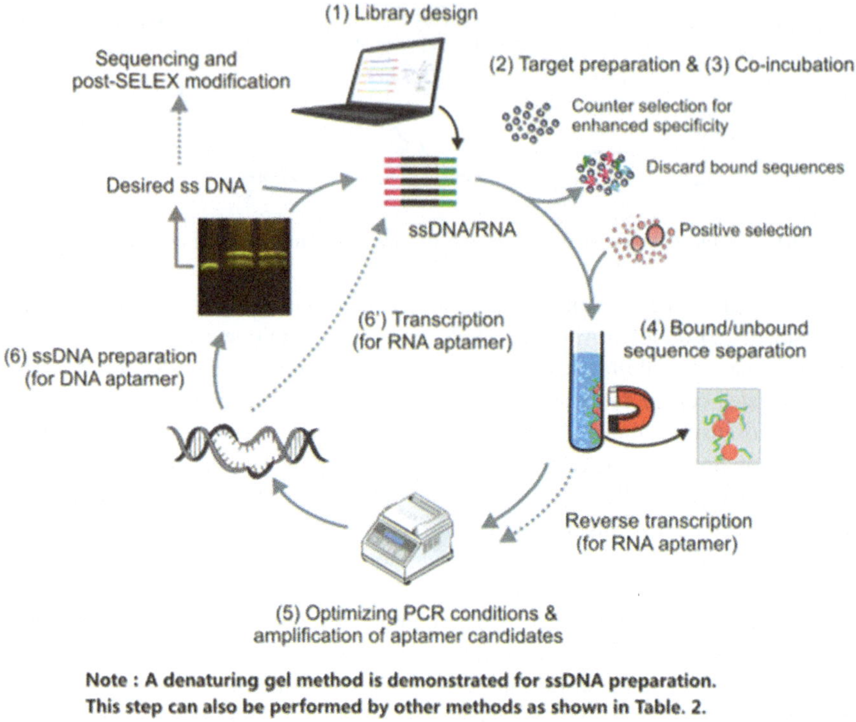

Fig. 10.2 Illustration of the key steps of a typical SELEX protocol. Reproduced with permission from (Stoltenburg et al. 2007)

oligonucleotide aptamers are similar. From the point of view of the molecular structure of oligonucleotides, in addition to G-C and A-U pairing in single-stranded RNA molecules, there are also G-U variant base pairs, so it is easier to form diverse spatial structures and may be more conducive to binding with target molecules. This may be the reason why most SELEX literature uses RNA libraries. However, compared with RNA molecules, DNA is more stable, and the production cost is low, so it has an advantage for in vitro diagnosis and treatment. The second step, the target is incubated with the single-stranded oligonucleotide library for a certain period of time under certain buffer conditions; The third step, the separation and purification of the specific oligonucleotide molecule bound to the target molecule. Use nitrocellulose filter membrane method, microplate or affinity chromatography to separate and obtain the oligonucleotides that bind to the target molecule, and elute and purify them; the fourth step, PCR amplification and enrichment, and separation. The oligonucleotides that specifically bind to the target molecule are amplified by PCR, and the resulting double-stranded DNA (dsDNA) is used as the library template for the next round of screening; if an RNA random library is used, reverse transcription is required firstly and DNA is then amplified into dsDNA, which subjected to in vitro transcription and the next round of screening. The fifth step, the amplified dsDNA product is prepared by heat denaturation or alkali denaturation; The sixth step, as the library for the second round of screening, repeat the above screening and enrichment process, repeating 8–15 rounds of screening; the seventh step, clone and sequence the products of the last round of enrichment; the eighth step, the sequencing results. The sequences are analyzed for homology and secondary structure, and they are grouped according to the analysis results; the ninth step, select representative sequences from each group, examine their affinity and specificity with the target, and fit the dissociation constant curve. As a result, the sequence which binds to the target with high affinity and specificity will be selected as the aptamer of the target.

10.1.3.2 Other Types of SELEXS

In recent years, in addition to conventional aptamer screening methods, other pathogenic bacteria aptamer screening methods have also been established, and these methods mainly include the following.

Ultrafiltration SELEX

Ultrafiltration SELEX mainly refers to nitrocellulose membrane ultrafiltration SELEX, this method is mainly applied to the screening of protein target molecular aptamers. The target protein is first adsorbed on the nitrocellulose membrane, then the random library is added to a certain binding buffer and incubated for a certain period of time, and then the membrane is washed with a washing buffer. The random single strands bound to the protein are trapped on the membrane, and the random single strands that are not bound to the target protein pass through the filter membrane, and finally the random single strands bound to the target protein on the membrane are eluted, so as to achieve the purpose of separation (Gopinath 2007). Tuerk, the founder of SELEX screening aptamers, was the first to use nitrocellulose membrane filtration to screen aptamers for phage T4 DNA polymerase, making this method a classic method for aptamer screening, especially for protein target molecules. The application is particularly extensive.

Centrifugal Precipitation SELEX

Centrifugal precipitation SELEX is a method mainly based on incubating the target and the random library in binding buffer for a certain period of time. The random single strands that can bind to the target would be precipitated with the target by centrifugation, at the same time, the random single strands that cannot bind to the target will be removed as they are freed in the supernatant, in this way, the purpose of separation achieved. This method is easy to operate and is a time-saving process, and it is widely used due to the high enrichment efficiency. Cao et al. applied the centrifugal precipitation SELEX method to gain a set of aptamers that specifically bind to the whole bacteria of *Staphylococcus aureus* (Cao et al. 2009). This set of aptamers contains five aptamers that all specifically bind to *Staphylococcus aureus*, and the combination of these aptamers greatly improves the affinity and specificity of recognition.

Nitrocellulose Membrane Filtration SELEX

This method is mainly applied to the screening of aptamers for protein target molecules. Firstly, the target protein is adsorbed on the nitrocellulose membrane, and then the random library is added to incubate in a certain binding buffer for a period of time. Then, the membrane is washed with the flushing buffer. It is important to emphasize that a random single chain binding to the target protein was designed firstly for the purpose of separation, and the chain bound with the protein is intercepted on the membrane, while the random single chain not bound with the target protein passes through the filter membrane, and finally elutes on the membrane. Tuerk, who developed SELEX-based aptamer screening method, firstly used nitrocellulose membrane filtration method to screen aptamers for phage T4 DNA polymerase, making this method a classic method for aptamer screening, especially in the screening of protein target molecules. Hijiri Hasegawa et al. used nitrocellulose membrane to screen human vascular endothelial growth factor (VEGF165) aptamer, which provided a favorable means for cancer diagnosis (Hasegawa et al. 2008).

Affinity Chromatography SELEX

Affinity chromatography SELEX is a method mainly fixes the target on the chromatographic column. The main principle is: firstly, the target is fixed on the chromatographic column, and the gene library is incubated long enough with the target in the column. Due to the interaction difference between the target and oligonucleotides, the nucleic acids with strong binding ability are captured on the chromatographic column, and the other nucleic acids with weak binding ability are lost. In this way, the nucleic acids with strong binding ability with the target molecule are separated Oligonucleotides were screened out. This method has wide applicability that not only for large molecules but also for small molecules. Ellington et al. established affinity columns by fixing dyes on agarose beads and obtained RNA aptamers that can specifically recognize seven dyes. Abbas Ali Imani Fooladi utilizes the affinity chromatography system to select novel ssDNA aptamers for the detection of staphylococcal enterotoxin B (SEB) (Hedayati et al. 2016). To use this method, a dsDNA library-based standard SEB protein as the target need to build firstly, and affinity chromatography matrix in microfuge tubes are also included. Then, the specific ssDNA aptamers were isolated by SELEX program and purified by ethanol precipitation. For the purified aptamer, ELISA was used for affinity-binding test and specificity detection. The results showed that three of the 12 readable sequences were selected as the most suitable aptamers because of their good affinity and specificity for SEB. Through 12 rounds of SELEX screening, a set of ssDNA aptamers with good selectivity for SEB were obtained (Fig. 10.3). However, this method also has some shortcomings (1) The target molecule needs to be immobilized on the chromatographic column, and the immobilization of the target molecule will cover some of the sites that can interact with oligonucleotides; (2) A large-volume chromatographic column must be used in the SELEX process, as a result, it is a process that takes a lot of samples because each round of screening

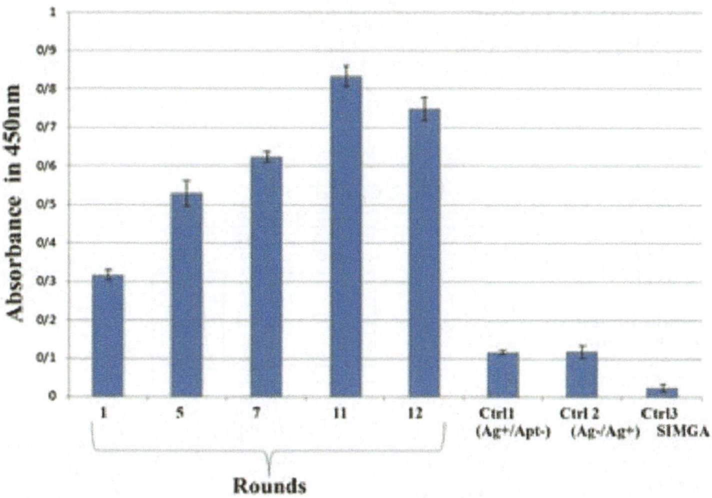

Fig. 10.3 Binding affinity test for different rounds of SELEX (Hedayati et al. 2016)

requires a few milliliters of the target and oligonucleotides. (3) It takes 2–4 weeks to complete the screening for wasting samples, so it is a time-consuming approach.

Capillary Electrophoresis SELEX

Capillary electrophoresis SELEX (CE-SELEX) is a SELEX method that efficiently separates target-bound oligonucleotides and untargeted oligonucleotides by electrophoresis in free solution. CE-SELEX overcomes limitation of classic SELEX technology that cannot effectively separate the bound and free nucleic acid. In CE-SELEX, the random nucleic acid library was incubated with the target in free solution for a period of time. After that, the incubation mixture injected into the capillary is separated by high pressure. Because the migration ability of the nucleic acid binding to the target is different from that of the non-binding sequence, different components can be collected according to this principle. Moreover, in free solution, the natural conformation of the target and nucleic acid can be maintained by the binding of the target and library. Making the aptamers obtained more suitable for application in real samples. CE-SELEX also greatly shortened the screening process, which could be completed by 2–4 rounds of screening. The Kd value of the obtained aptamers could be as low as 180 pM. So far, a series of aptamers have been screened by CE-SELEX method; Michael T. Bowser has obtained a high affinity ssDNA aptamer for HIVRT using CE-SELEX (Fig. 10.4). At present, the number of AIDS cases in the world is still growing at an alarming rate. ssDNA aptamers with picomolar affinity for HIVRT will become a potential choice.

Microfluidic SELEX

Microfluidic SELEX (M-SELEX) is a new SELEX method that separates the oligonucleotides that have been bound to the target through a driven microfluid in

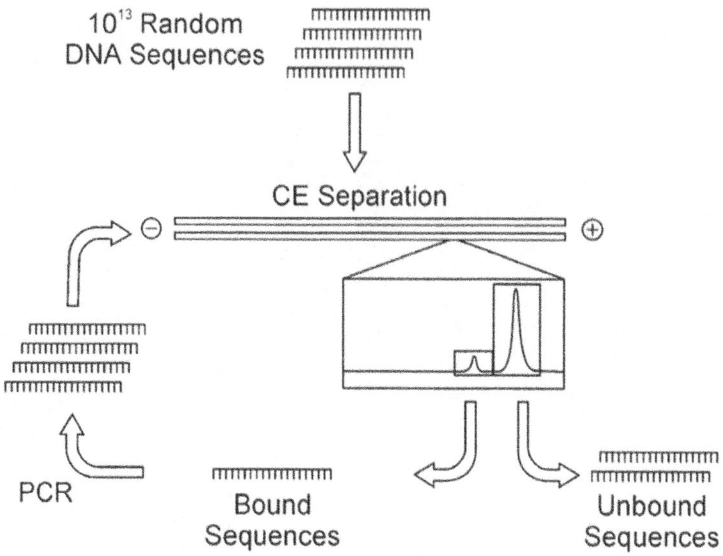

Fig. 10.4 Schematic of CE-SELEX (Mosing et al. 2005)

the microfluidic channel of the chip. Multiple functional modules are integrated on a microfluidic chip, which can realize the rapid automatic SELEX and improve the screening efficiency. First, the chip is small in size, which increases the rigor of screening when processing small amount of reagent. And can reduce the amount of reagents and samples, thereby reducing screening costs. Secondly, in the chip fluid channel, the fluid flow mode can be controlled, which can better remove weakly bound or unbound nonspecific nucleic acids, greatly reducing the capacity of the library. Then, the microfluidic chip can increase the surface area to volume ratio, further increase the surface tension, significantly improve the separation efficiency, effectively reduce the screening rounds; in this way, the entire screening time is saved. Furthermore, the chip is more integrated and automated, which is another big advantage compared to traditional SELEX. Since Hyberger et al. proposed "a microfluidic SELEX prototype" in 2005 (Hybarger et al. 2006), combining microfluidic chip with SELEX for the first time, microfluidic SELEX has developed rapidly, and a variety of SELEX methods based on microfluidic have been established subsequently, such as sol-gel microfluidic SELEX, beads-based microfluidic SELEX. In particular, magnetic beads-based microfluidic SELEX has showed excellent separation ability, the magnetic bead-based M-SELEX method is mainly to fix the target on the surface of the magnetic microspheres in covalent or non-covalent manner, and then pass it into the microfluidic chip to form a region of separation and enrichment. The specific nucleic acid will bind to the target when fluid flow, nonspecific nucleic acids and weakly bound nucleic acids are removed. Finally, the nucleic acid molecules specifically bound to the target on the surface of the magnetic microspheres were eluted and separated for PCR expansion. Soh et al.

developed a microfluidic SELEX based on magnetic beads in 2009 (Fig. 10.5). In the micro scale, superior screening efficiency was achieved due to the unique physical phenomenon. The specific aptamers for *Botulinum neurotoxin* type A (BoNT/a-rlc) can be screened by only one round of screening. The Kd range is 34–86 nM. In the same year, Soh et al. proposed a new method to replace the above separation method. They used micromagnetic separation (MMS) chip to remove unbound ssDNA, only three rounds were needed to screen specific aptamers with high affinity for streptavidin. The magnetic bead recovery was as high as 99.5%, and the dissociation constant ranged from 25 to 65 nmol / L.

Whole-Cell SELEX

Cell SELEX (cell SELEX) directly interacts with a random library in the SELEX process of intact cells and screens out oligonucleotides bound to target molecules on the cell surface. A variety of marker protein molecules are distributed on the cell surface. Using the difference in protein expression between target cells and non-target cells (such as normal cells), cell SELEX screens out the aptamers of target cell marker protein molecules to obtain a group of aptamers that can specifically recognize this target cell. Since cell-SELEX uses living cells as the target, the aptamers screened out can interact with the natural conformation of the target protein. They can be used for positioning and tracing of living cells, targeted drug delivery, and targeted treatment of diseases. In addition, under the condition that the type and structure of the target protein on the cell surface are not yet clear, specific aptamers can also be screened through cell SELEX. Using this aptamer as a bait, the target protein can be "capture" from the cell membrane for identification, thereby discovering new biomarkers.

Some studies have shown that the use of whole-cell targets in SELEX process is actually faster, easier, and more repeatable than the use of purified target molecules, and it is more successful in generating functional aptamers with binding affinity to living cells. In view of this reason, Lee-Ann Jaykus selected and characterized biotinylated DNA aptamers with binding affinity to *Salmonella typhimurium* using a whole-cell SELEX approach (Dwivedi et al. 2013). As a proof-of-concept, these aptamers were then used to capture and concentrate *Salmonella typhimurium* cells for direct detection using qPCR.

Using an improved and rapid whole-cell SELEX method, Shylaja Ramlal's team developed a live whole-cell aptamer for enteric Salmonella serotype *Salmonella typhimurium* using four other Salmonella serotypes and other bacterial cells as negative selection targets (Fig. 10.6). The aptamer obtained by this method has high specificity and meets the requirements of practical detection.

Genomic SELEX

Genome SELEX is a method that uses the entire genome of a specific organism as a SELEX library, the genomic DNA library used in genomic SELEX is different from the traditional SELEX using chemically synthesized library. This is a method to screen out the natural recognition sequence of biological active molecules (Teng et al. 2016). Genome SELEX is particularly suitable for exploring the relationship

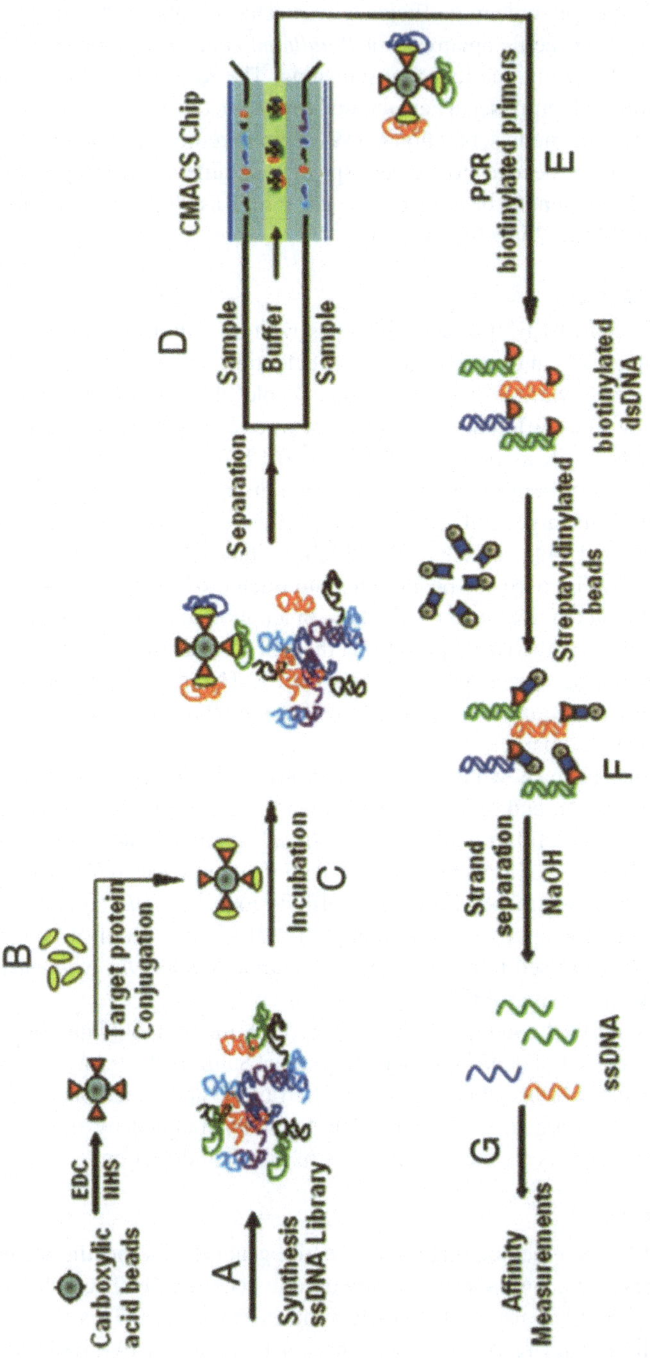

Fig. 10.5 Overview of the M-SELEX process (Lou et al. 2009)

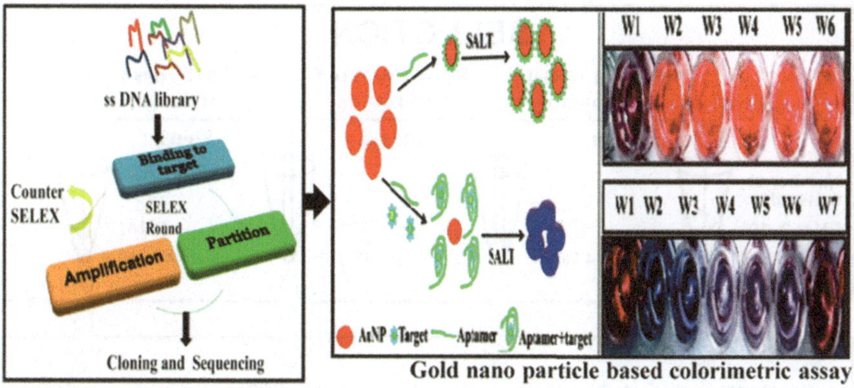

Fig. 10.6 Selection and Characterization of Aptamers Using a Modified Whole Cell Bacterium SELEX for the Detection of Salmonella enterica Serovar Typhimurium (Lavu et al. 2016)

between biologically active molecules and nucleic acids, or looking for RNAs that are rarely expressed, such as transcripts of silent regions of genomic heterochromatin, and RNA expressed at a specific stage of the cell cycle. As a new method, genomic SELEX has shown great advantages in the analysis of intracellular gene regulation and metabolic regulation. For example, Shimada used genomic SELEX technology to study the regulation of *E. coli* peptidoglycan (PG) degradation and found that the transcription factor YcJZ is the expression inhibitor of the initial enzyme in the PG degradation pathway.

However, although numerous studies demonstrated that the SELEX process generates better aptamers to bacterial cells, the repetitive enrichment steps also elevate the cost and time associated to this process and require extremely large amounts of the target cells for aptamer isolation. To overcome the drawbacks of SELEX, new aptamer selection processes have been introduced to reduce some steps of SELEX, it was non-SELEX. The traditional SELEX technology has to apply PCR and other technologies to enrich the nucleic acid sequences that can bind to the target after elution and separation, and the amplified products will be applied to the next round of screening process, so as to improve the proportion of aptamer sequences in the library, then finally select the aptamer sequences with low dissociation constant. Unlike SELEX, non-SELEX does not require PCR and other amplification steps. After 2–3 times of isolation, analysis, and screening steps, the aptamer sequence directly obtained.

Berezovski et al. firstly proposed the concept of non-SELEX technology and obtained h-Ras protein aptamers by this technology (Fig. 10.7). They applied the non-equilibrium capillary electrophoresis of equilibrium mixtures (NECCM) in the screening of aptamers, using equilibrium mixture of h-Ras protein and aptamer library as samples, after three rounds of separation and analysis, the nucleic acid sequence that can specifically bind to h-Ras protein was obtained.

Subsequently, non-SELEX technology has been developed and applied in aptamer screening. Ashley et al. used non-equilibrium capillary electrophoresis to

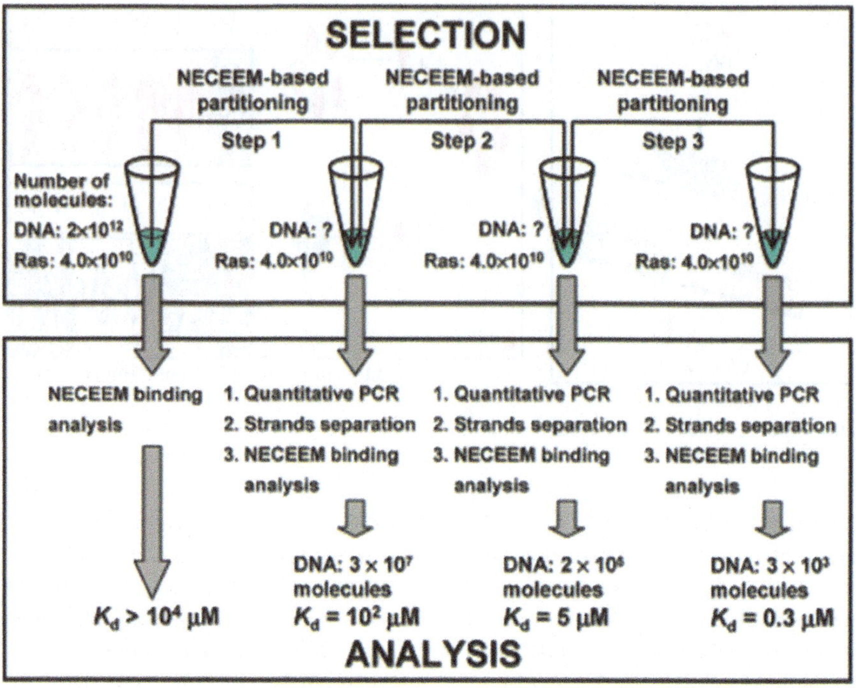

Fig. 10.7 Non-SELEX selection of aptamers with three steps of NECEEM-based partitioning (Berezovski et al. 2006)

perform negative screening with lysozyme, trypsinogen, chymotrypsinogen A, and myoglobin as targets. After three rounds of non-SELEX process, an aptamer that can specifically bind to bovine catalase was found. Yu et al. used a mathematical reasonable model to analyze the effect of protein concentration and separation efficiency, on the non-SELEX screening process, and provided a reference for subsequent researchers' experimental design.

Then, Byoung Chan Kim et al. propose a rapid method to isolate bacterial cell-specific DNA aptamers that does not require repeated rounds of elution and amplification of bound probes (Fig. 10.8). They used *E. coli* as a model to implement the process, through repetitive centrifugation-based partitioning, the DNA bound to and unbound bacterial cells are separated, Finally, in this method, the DNA pools bound to cells (*E. coli*) are amplified only once prior to cloning, while SELEX needs repetitive rounds of binding, elution, and amplification.

Compared with the SELEX screening technology, which usually has a screening cycle of 1–3 months, the non-SELEX screening technology can complete the entire aptamer screening process within a few days or even hours, greatly shortening the screening cycle. However, since the latter needs to be screened with the help of capillary electrophoresis, there should be significant changes in electrophoretic mobility before and after screening the target substance bound to the nucleic acid

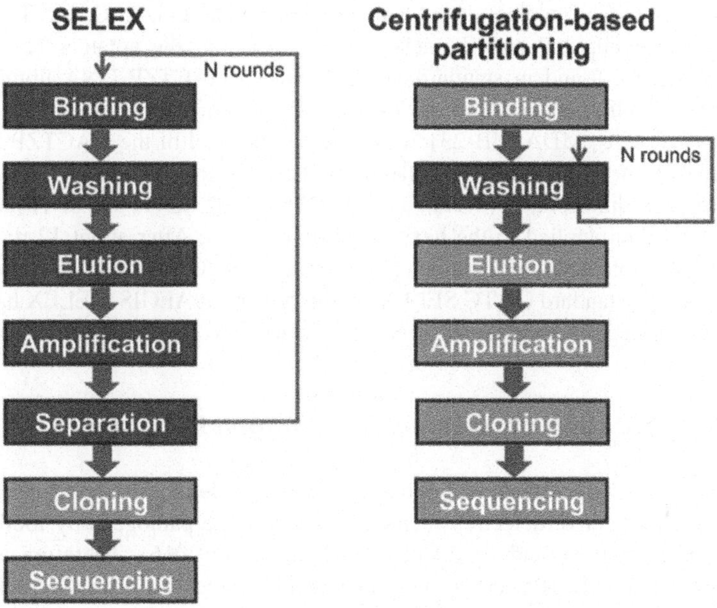

Fig. 10.8 Schematic representation of the centrifugation-based partitioning process in comparison to Cell-SELEX of ssDNA aptamers (Kim et al. 2020)

sequence, so it has certain limitations in application, as a result, non-SELEX technology is mostly used for the aptamer of macromolecular screening.

10.1.3.3 Advanced SELEX methods

Aptamer Selection Express

In addition to these common SELEX methods, there are some special aptamer methods, aptamer selection express (ASExp) is one of them (Fan et al. 2008). Johnathan L. Kiel group mainly used a double-stranded (ds)-DNA library to interact with target, after one of the single-stranded (ss)-DNA bound to the target, the random ss-DNA will bind with magnetic beads for fast separation. They have applied ASExp to select aptamers against different types of targets successfully, and the targets contain *B. anthracis* spores, *B. thuringiensis* spores, *MS-2 bacteriophage*, and ovalbumin. Although the ASExp has cross-reaction problems, but it provides a much faster and easier SELEX approach because it requires a small amount of target, so, compared with SELEX, it costs less. As a result, ASExp is a promising rapid selection method for aptamers.

Artificially Expanded Genetic Information Systems-SELEX

Another special advanced aptamer SELEX method called artificially expanded genetic information systems-SELEX (AEGIS–SELEX). AEGIS are unnatural forms of DNA because in this system (Sefah et al. 2014), aptamers are built from six different kinds of nucleotides. Apart from the standard G, A, C, and T, AEGIS involves nonstandard P and Z nucleotides, as a result, the aptamer has a nitro functionality not found in standard DNA. Firstly, a GACTZP DNA library with two primer binding sequences flanking a 20-nt random region. Then, the line of breast cancer cells, MDA-MB-231 was added in the pool of the GACTZP library, and the DNA survivors would be collected and amplified by PCR with a mixture of six nucleotide triphosphates (dGTP, dATP, dCTP, dTTP, dZTP, and dPTP), and the product was then applied to the next round of selection. After about 12 rounds of selection, the aptamer can be obtained. Generally, 15–20 rounds of selection were required for the standard GATC SELEX, so this advanced AEGIS–SELEX has great potential to speed up the aptamer screening process.

10.2 Application of Aptamer in Pathogen Detection

In recent years, the outbreak of foodborne diseases has led to an increasing demand for rapid and sensitive detection methods for foodborne pathogens by food safety inspection institutions and food industry (Jiang et al. 2016). Therefore, a large number of studies have been carried out to find a rapid and sensitive method to detect foodborne pathogens (lazcka et al. 2007). Despite these efforts, traditional pathogen detection methods and culture-based methods are still widely used to identify microbial pathogens in different environmental media. However, problems still exist, such as cumbersome procedure, long time required and the skill to detect viable but not culturable (VBNC) state of bacteria.

Over the years, many aptamer-based detection methods have been developed, but their applications in public health and food safety are limited. The main reason may be the complexity of the methods because these methods involve various techniques in the process of sample preparation and detection, such as sample extraction, purification, enrichment, and separation (Pitcher and Fry 2000; Stevens and Jaykus 2004).

Aptamers, single-stranded DNA or RNA oligonucleotides selected in vitro with an aptamer, a single-stranded DNA or RNA oligonucleotide (Tuerk and Gold 1990), obtained in vitro by exponential ligand enriched phylogenetic selection (SELEX), can bind to the target analyte in a specific three-dimensional conformation (Ellington and Szostak 1992). It is considered as "chemical antibody" because of its sequence specificity and target-binding function, which provides a high affinity for them to form higher order structures (Ren et al. 2017). Compared with antibodies, aptamers provide a variety of unique advantages. For example, aptamers are characterized by simple synthesis, low immunogenicity, stability, high sensitivity, low price, flexible chemical modification and low immunogenicity (Breaker 1997). In addition, aptamers can be modified with a variety of signal tags such as tumor cells, bacteria, virus, proteins, some small molecules (ATP), and even metal ions (Kim et al. 2012; Medley et al. 2011; Rhinehardt et al. 2015). These advantages make aptamer a promising candidate for aptamer-based sensors (Drolet et al. 1996; Mayer 2009). Drolet in 1996 firstly used aptamer to detect human vascular endothelial growth factor (Ferreira et al. 2008). Since then, multifarious aptamer-based biosensors have been developed, including ELISA (Mayer 2009), fluorescence (Shen et al. 2016), surface plasma resonance (SPR) (Bai et al. 2013), electrochemical (Hai et al. 2014), colorimetry (Wang et al. 2014), surface enhanced Raman scattering (SERS) (Yao et al. 2017) and flow cytometry sensors for a wide range of targets (Wu et al. 2014).

10.2.1 ELISA Detection

Enzyme-linked immunosorbent assay (ELISA) refers to the combination of soluble antigen or antibody with polystyrene and other solid-phase carrier, and the qualitative and quantitative detection of immune response by specific combination of antigen and antibody. It is a technology that combines enzyme-linked antibody with corresponding substrate catalytic reaction. ELISA for quantitative determination of IgG was published by Engvall and Perlmann in 1971 (Engvall and Perlmann 1971), which led to the development of enzyme-labeled antibody technology for antigen localization in 1966 into a method for the determination of trace substances in liquid specimens. Since then, ELISA develops rapidly and is used broadly. In particular, all kinds of ELISA kits are made for the detection of different targets due to a variety of advantages including easy operation, rapidly, untrained, and sensitivity. The basic principle of this method is: ① the antigen or antibody is bound to the surface of a solid phase carrier and maintain its immune activity, ② the antigen or antibody is linked to an enzyme to form an enzyme-labeled antigen or antibody, which retains both its immune activity and enzyme activity. At the time of measurement, after adding the substrate for the enzyme reaction, the substrate is catalyzed by the enzyme to turn into colored products. The amount of the product is directly related to the amount of the tested substance in the sample, so it can be analyzed qualitatively or quantitatively according to the depth of the color reaction. Chen et al. did the related work that contains aptamer selection for *Campylobacter jejuni* (*C. jejuni*) and the use of selected aptamer for *C. jejuni* detection by ELISA (Chen et al. 2020). Herein, authors name this ELISA method hetero-sandwich assay where aptamer and antibody are used to recognize target bacterium simultaneously (Fig. 10.9). And the assay had a great sensitive detection limit and good specificity.

10.2.2 Fluorescence Detection

Fluorescence detection is based on the fluorescent materials that were broadly used (Vigneshvar et al. 2016), including quantum dots (QDs), fluorescent microsphere, and so on. If the molecule is excited to the excited state and then returns to the ground state, the material will emit fluorescence. Fluorescent materials are commonly used as indicators in biological analysis. With the development of nanotechnology in recent years, fluorescent nanomaterials have been paid more and more attention and play an increasingly important role in biological analysis (Yao et al. 2014). Among the fluorescent nanomaterials, QDs offer a tremendous advantage

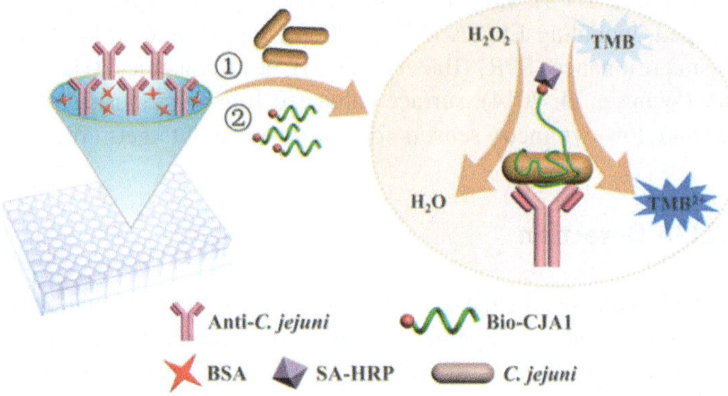

Fig. 10.9 The application for hetero-sandwich-based detection of *C. jejuni* (Chen et al. 2020)

over conventional fluorophores in terms of quantum efficiency and strong fluorescence intensities, as well as single excitation, with multicolored emissions based on different QDs size (Riegler et al. 2008).

Fluorescence resonance energy transfer (FRET) is a typical and widely used method for fluorescence detection. However, the conventional fluorescence analysis method is seriously affected by its autofluorescence interference, which limits the practical application of the technology. Wang et al. (2017) describes a time-resolved fluorescence resonance energy transfer (TR-FRET)-based adaptive sensor for the identification of *Salmonella typhimurium*. The authors used nanoparticles as energy donor and carboxyl fluorescein (FAM)-labeled complementary oligonucleotide (cDNA) as receptor. The detection scheme is based on the hybridization between aptamer and cDNA. When there are no *S. typhimurium* in the solution, photonic energy is transferred from NPs to FAM, and then it would get a 520 nm emission. On the contrary, when there are *S. typhimurium* existing in the solution, the FRET would decrease due to the aptamer interaction with bacteria. Based on the specific recognition ability of aptamers and the strong fluorescence property of nanoparticles, this method has high sensitivity and selectivity for the detection of *Salmonella typhimurium*. In addition to quantum dots and fluorescent dyes, fluorescent nanoparticles are also commonly used in the detection of pathogenic bacteria (Chung et al. 2015). In this study, the aptamer-conjugated fluorescent nanoparticles are combined with the photofluidic detection platform, as a result, the real-time and rapid detection of *Escherichia coli* was realized.

In another report, Kurt et al. developed a dual-excitation sensing method using aptamer-modified quantum dots and upconverting nanoparticles (Kurt et al. 2016). Figure 10.10 shows the multiple detection using the double excitation luminescence method. Here, we introduce the coupling of magnetic beads with aptamer-functionalized luminescent nanoparticles. The magnetic beads were modified with short DNA sequences that can be partially complementary to the aptamer sequences and were used to separate the analyte-free conjugates for fluorescent measurement.

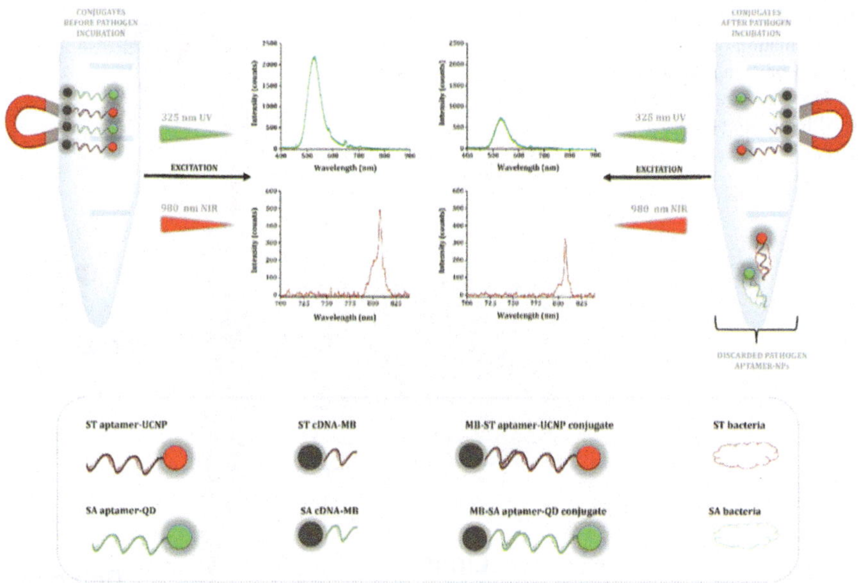

Fig. 10.10 Illustration of the dual-excitation luminescence strategy for multiplexed detection (Kurt et al. 2016)

The quantum dots were excited by 325 nm UV light, and the upconversion nanoparticles were excited by 980 nm near-infrared light, and the multiple detection of food model pathogens, *Salmonella typhi* and *Staphylococcus aureus* were achieved, with detection limits of 16 cfu/mL and 28 cfu/mL, respectively.

10.2.3 Surface Plasmon Resonance Detections

Surface plasmon resonance (SPR) is a new technology developed in the 1990s. SPR sensors are widely used in various fields because of its appealing and promising, which detect the interaction between ligand and analyte on biosensor chip. Among all the biosensor platforms, surface plasmon resonance (SPR) is a label-free optical detection technique. Besides, SPR-based sensors display versatile advantages, such as specificity, sensitivity, simplicity, rapid analysis, and providing the kinetic reaction process in real time.

Recently, plasmonic nanoparticle-based localized surface plasmon resonance (LSPR) biosensors have attracted the attention of many researchers. Even though LSPR biosensors have many advantageous traits, most LSPR biosensors have low reproducibility, and it is difficult to immobilize large areas of plasmonic-active nanoparticles while reducing the detection limit. Therefore, Seo Yeong Oh et al. (2017) introduced a gold nanoparticle-aptamer-based LSPR analytical method for the rapid detection of *Salmonella typhimurium*. As shown in Fig. 10.11, it is the schematic representation of the detection of bacteria using the LSPR sensing chip.

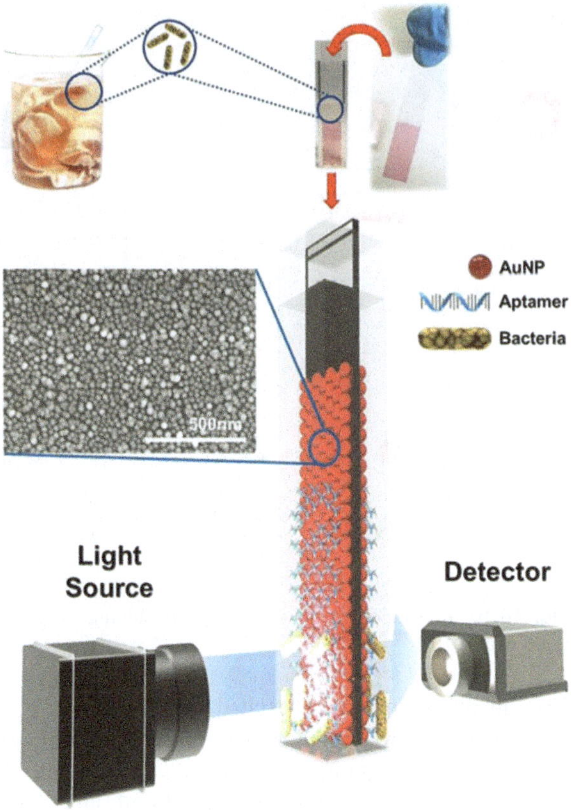

Fig. 10.11 Schematic representation of the detection of bacteria using the localized surface plasmon resonance (LSPR) sensing chip (Oh et al. 2017)

The aptamer-based LSPR chip was fabricated by the self-assembly of AuNPs attached to a glass substrate slide. Next, bacterial aptamer was modified on the surface of chip. Finally, sample was analyzed for the sensing signals by using a spectrophotometer.

Based on LSPR, Seung Min Yoo et al. (2015) developed an easily constructed, aptamer-based LSPR sensor for simple, rapid, sensitive, specific, and multiple detection of bacterial species (Fig. 10.12). In this study, they designed a multi-point gold-capped nanoparticle array (MG-NPA) chip, which was successfully used to detect and identify three different bacteria simultaneously. The detection limit of LSPR sensor based on aptamer is 30 cfu/mL. The combination of LSPR sensor and aptamer has unique advantages. One is that the MG-NPA structure results in sensitivity and reproducibility. Another is that the specific aptamer functionalized on the MG-NPA structure surface allows the simple and rapid detection of bacterial species. The other is that a single chip with the multiple spots can allow us to simultaneously identify different bacterial species in a single assay.

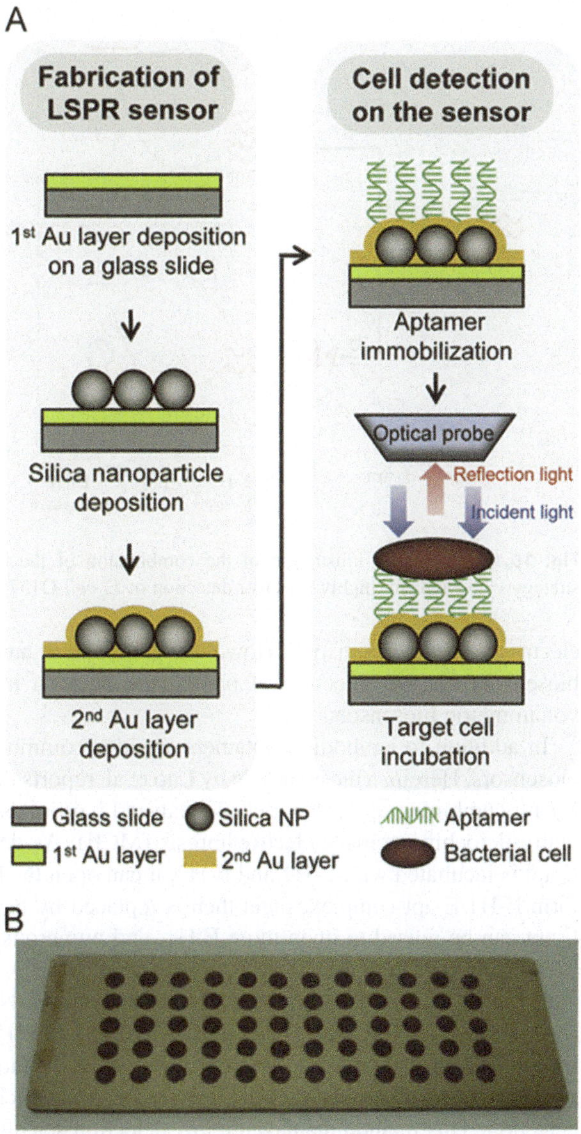

Fig. 10.12 Aptamer-immobilized LSPR sensor system (Yoo et al. 2015)

10.2.4 Electrochemical Biosensors

Because of their several advantages such as high sensitivity, selectivity, miniaturized, fast response time, simple operation and low-cost, electrochemical sensors have received considerable interest. So, the number of reports published on electrochemical biosensors is increasing. Electrochemical biosensors have different types according to different sensing principles, including electrochemical impedance spectroscopy (EIS), amperometric detection, voltammetry, and other

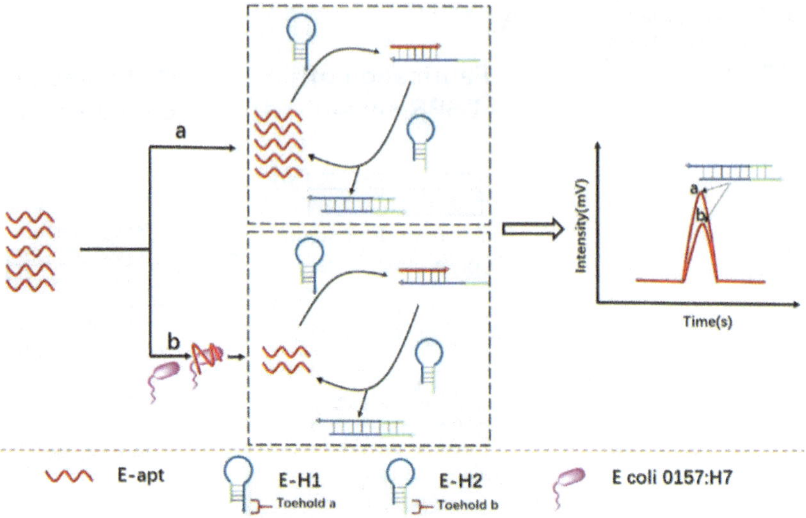

Fig. 10.13 Schematic illustration of the combination of the CHA-based signal amplification strategy with MCE for highly sensitive detection of *E. coli* O157:H7 (Xu et al. 2017)

electrochemical biosensors. However, most used and reported electrochemical biosensors for the detection of pathogenic bacteria in food are impedimetric or voltammetric biosensors.

In addition to antibodies, aptamers are also commonly used in electrochemical biosensors. Herein, a recent article by Luo et al. reports the detection of *E. coli* O157:H7 by combining aptamer-induced catalyzed hairpin assembly circle amplification with microchip capillary electrophoresis (MCE). As depicted in Fig. 10.13, when E-apt is incubated with E-H1 and E-H2, it can open the hairpin structure of E-H1 to form E-H1/E-apt complexes and then is replaced by hairpin E-H2. As a result, the E-apt can be reused to open more E-H1, and numerous E-H1/E-H2 complexes are formed through the DNA circuit (Scheme 1a). Finally, the E-H1/E-H2 complexes with longer sequences can be separated with other DNA strands by MCE, which show strong signal in MCE detection. While *E. coli* O157:H7 is added to the E-apt solution, a portion of aptamer would bind with it and lead to a decrease in the quantity of E-apt as well as E-H1/E-H2 complexes, which show weak signal in MCE detection. This method displayed a low detection sensitivity with 75 cfu/mL. Then, another research (Xu et al. 2017) by Xu et al. that reports the determination of *E. coli* in licorice extract using an aptamer-based biosensor over a gold electrode. Besides, an aptamer-conjugated silver nanoparticles electrochemical sensor with dual-aptamer based was also developed (Abbaspour et al. 2015), with the assembling of the AgNPs, the detection sensitivity for *Staphyococcus aureus* was improved effectively.

10.2.5 Lateral Chromatography Test Strips

Lateral flow analysis (LFA) is a fast, on-site detection method based on chromatographic simplified sensor format. In recent years, the lateral chromatographic strip biosensor has attracted extensive attention because of its portability, fast detection time, cost-effectiveness, easy to use, and other advantages. Disposable biosensor detection platform is a kind of user-friendly equipment, with portable size, provides the possibility of on-site visual analysis, and can be easily operated without special training. At present, strip biosensors are often used as point-of-care tools in food, medical, diagnostic, and environmental fields.

Aptamers and antibodies are the common recognition elements of band biosensors. Generally, the strip detection is completed on the prefabricated nitrocellulose strip. Firstly, the prefabricated reagent is sprayed on the nitrocellulose. After drying, the liquid sample can activate the signal mechanism. This kind of sensor is usually colorimetric or fluorescent method. Colloidal Gold (CG), Blue Latex Particle, and Red Quantum dot (QD) were used as the signal reporter in the detection, respectively.

Although there are many advantages of lateral chromatography test strips, they are often developed to detect a single target per assay (Berlina et al. 2013). Thus, in order to detect multiple analytes on a single strip, several lateral flow immunoassays (LFIAs) multiple detection techniques have been developed, which further provides an opportunity to improve the speed and reduce the cost by analyzing multiple targets simultaneously. However, these LFIAs could have nonspecific binding and crossover reactions, which can lead to false-positive results. These may be caused by the use of antibodies. Compared to antibody, the aptamer is easier to modify and tag, which is employed in lateral flow aptamer assays (LFAAs). Based on this, Jin et al. (2018) developed a smart phone portable multiple detection device based on lateral flow aptamer analysis (Fig. 10.14). This device mainly uses aptamer-functionalized polychromatic upconversion nanoparticles as probes, which can accurately detect mercury ion, ochratoxin A, and Salmonella on the test strip with high sensitivity. In detection zone, this strip biosenser achieved the test based on the principle of nucleic acid hybridization. And finally, a smartphone was used to read the fluorescent signals.

10.2.6 Surface-Enhanced Raman Scattering Detection

In 1928, Raman scattering was discovered by C. V. However, weak effect of the Raman scattering greatly limits the application and development of Raman spectroscopy. Whereafter, Fleischmann and Jeanmaire et al. have systematically studied how to improve the intensity of Raman scattering signal by changing the surface roughness of the substrate through systematic experimental study, which is called the surface-enhanced Raman scattering (SERS). Subsequently, SERS technology developed rapidly to be a very powerful analysis tool and has been applied in various fields. Of course, the use of SERS to detect pathogen has attracted great attention.

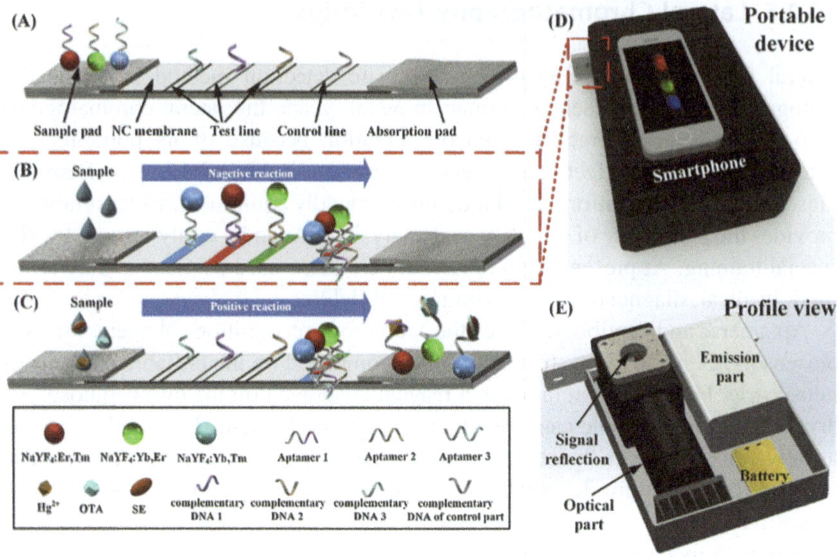

Fig. 10.14 Schematic illustration of LFAA for simultaneous multiple targets detection (Jin et al. 2018)

Then, a magnetically assisted surface-enhanced Raman scattering (SERS) biosensor based on aptamer recognition was established (Wang et al. 2015), with the application of good dispersity superparamagnetic magnetic NPs and a novel SERS tag, the sensor obtained a satisfactory detection limit for *Staphylococcus aureus*. Another aptamer-based SERS detection method is shown in Fig. 10.15 (Pang et al. 2019). In the schematic illustration, Fig. 10.1a shows the synthetic process of vancomycin-SERS tags (Van-SERS tag) that is on the surface of AuNPs modifying Van. The synthetic process of aptamer-Fe_3O_4@Au MNPs is illustrated in Fig. 10.8b. Briefly, Fe_3O_4@Au MNPs were synthesized firstly. Then, the Fe_3O_4@Au MNPs get surface carboxylation by sonicating in a 11-mercapto-undecanoic acid (MUA) ethanol solution for 2 h. After that, COOH-Fe_3O_4@Au MNPs were dispersed in MES buffer with EDC to activate the carboxyl. Finally, mixing the NH_2-aptamer and Fe_3O_4@Au, aptamer would be fixed on the surface of Fe_3O_4@Au MNPs through of amino-carboxyl condensation. Figure 10.8c shows the operating procedure of this strategy for *S. aureus* detection that is based on aptamer recognition.

10.2.7 Flow Cytometry Detection

Flow cytometry (FCM) is a high-throughput analysis technology, which is mainly used to quantitatively analyze single cell or other biological particles by monoclonal antibodies at the cellular molecular level. Light source, fluid flow pathway, signal detection and transmission, and data analysis system are the main components of

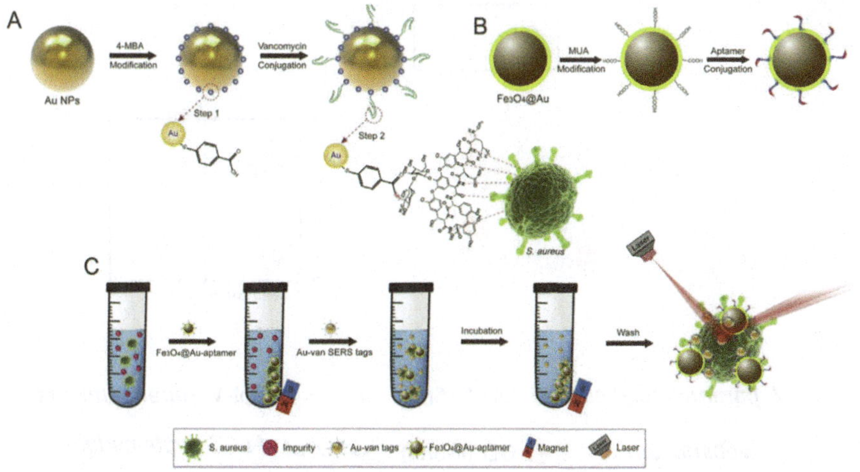

Fig. 10.15 Schematic illustration of (**a**) the synthesis of Au-Van SERS tags, (**b**) the synthesis of aptamer-modified Fe_3O_4@Au MNPs, and (**c**) the operating procedure for *S. aureus* detection via the dual-recognition SERS biosensor (Pang et al. 2019)

FCM. It is widely used in oncology, immunology, toxicology and other medical fields, and also used in the quantitative analysis of bacterial cell subsets. Moreover, it can also be used to analyze heterogeneous cell mixtures and characterize cells. And we can use the proper fluorescent markers or probes that are antibodies conjugated to dyes or fluorescent phycobiliproteins and small molecules that bind to nucleic acids to characterize the same type of cells that possess different biochemical or biophysical properties.

As we all know, fluorescent materials (e.g., organic fluorophores, fluorescent proteins, or QDs) would be employed in the flow cytometry detection. Among the fluorescent materials, semiconductor QDs provide more advantages to be employed in biological labeling. The broad excitation spectra and narrow emission spectra of QDs allow for simultaneous excitation and detection. This will make it possible to detect different targets using multicolor QDs in a single multiplex analysis. It is difficult to modify antibodies with a variety of fluorescent dyes or other tags, and the activity of antibodies may be affected. So, we can use aptamers selected by SELEX instead of antibodies as biosensing elements because of their high affinity and specificity towards targets. A paper published by Duan et al. describes a dual functional platform for the simultaneous detection of two pathogenic bacteria using quantum dots as fluorescence markers coupled with aptamers as the molecular recognition element by flow cytometry (Duan et al. 2013). In Fig. 10.16, we can see that QD 535 nm and QD 585 nm were used to detect *Vibrio parahaemolyticus* and *Salmonella typhimurium*.

Therefore, the application of suitable aptamers is promising for the sensitive detection of pathogenic bacteria in food matrix. The combination of aptamers and QDs used as biomolecule probes also provides a potential platform for detection of pathogens in flow cytometry.

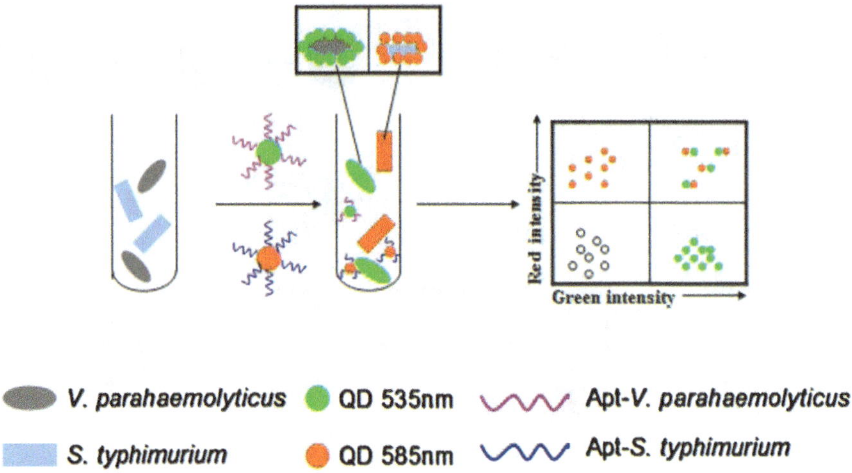

Fig. 10.16 Conceptual scheme of the flow cytometric assay for *Vibrio parahaemolyticus* and *S. typhimurium* detection (Duan et al. 2013)

References

Abbaspour A, Norouz-Sarvestani F, Noori A, Soltani N (2015) Aptamer-conjugated silver nanoparticles for electrochemical dual-aptamer-based sandwich detection of *Staphylococcus aureus*. Biosens Bioelectron 68:149–155

Alamer S, Eissa S, Chinnappan R, Herron P, Zourob M (2018) Rapid colorimetric lactoferrin-based sandwich immunoassay on cotton swabs for the detection of foodborne pathogenic bacteria. Talanta 185:275–280

Bai Y, Feng F, Zhao L, Wang C, Wang H, Tian M, Qin J, Duan Y, He X (2013) Aptamer/thrombin/aptamer-AuNPs sandwich enhanced surface plasmon resonance sensor for the detection of subnanomolar thrombin. Biosens Bioelectron 47:265–270

Berezovski M, Musheev M, Drabovich A, Krylov SN (2006) Non-SELEX selection of aptamers. J Am Chem Soc 128(5):1410–1411

Berlina AN, Taranova NA, Zherdev AV, Vengerov YY, Dzantiev BB (2013) Quantum dot-based lateral flow immunoassay for detection of chloramphenicol in milk. Anal Bioanal Chem 405(14):4997–5000

Breaker RR (1997) DNA aptamers and DNA enzymes. Curr Opin Chem Biol 1(1):26–31

Brigmon RL, Zam SG, Bitton G, Farrah SR (1992) Detection of Salmonella enteritidis in environmental samples by monoclonal antibody-based ELISA. J Immunol Methods 152(1):135–142

Cao X, Li S, Chen L, Ding H, Xu H, Huang Y, Li J, Liu N, Cao W, Zhu Y, Shen B, Shao N (2009) Combining use of a panel of ssDNA aptamers in the detection of *Staphylococcus aureus*. Nucleic Acids Res 37(14):4621–4628

Chandra P, Prakash R (2020) Nanobiomaterial engineering concepts and their applications in biomedicine and diagnostics: concepts and their applications in biomedicine and diagnostics. Springer

Chandra P, Koh WCA, Noh HB, Shim YB (2012) In vitro monitoring of i-NOS concentrations with an immunosensor: the inhibitory effect of endocrine disruptors on i-NOS release. Biosens Bioelectron. https://doi.org/10.1016/j.bios.2011.11.027

Chen W, Teng J, Yao L, Xu J, Liu G (2020) Selection of specific DNA Aptamers for hetero-sandwich-based colorimetric determination of *Campylobacter jejuni* in food. J Agric Food Chem 68(31):8455–8461

Choudhary M, Yadav P, Singh A, Kaur S, Ramirez-Vick J, Chandra P, Arora K, Singh SP (2016) CD 59 Targeted ultrasensitive electrochemical immunosensor for fast and noninvasive diagnosis of oral cancer. Electroanalysis 28:2565–2574. https://doi.org/10.1002/elan.201600238

Chung J, Kang JS, Jurng JS, Jung JH, Kim BC (2015) Fast and continuous microorganism detection using aptamer-conjugated fluorescent nanoparticles on an optofluidic platform. Biosens Bioelectron 67:303–308

Cowie SE, Ma I, Lee SK, Smith RM, Hsiang YN (2005) Nosocomial MRSA infection in vascular surgery patients: impact on patient outcome. Vasc Endovasc Surg 39(4):327–334

Deka S, Saxena V, Hasan A, Chandra P, Pandey LM (2018) Synthesis, characterization and in vitro analysis of α-Fe_2O_3-$GdFeO_3$ biphasic materials as therapeutic agent for magnetic hyperthermia applications. Mater Sci Eng C 92:932–941

Drolet DW, Moon-McDermott L, Romig TS (1996) An enzyme-linked oligonucleotide assay. Nat Biotechnol 14(8):1021–1025

Duan N, Wu S, Yu Y, Ma X, Xia Y, Chen X, Huang Y, Wang Z (2013) A dual-color flow cytometry protocol for the simultaneous detection of *Vibrio parahaemolyticus* and *Salmonella typhimurium* using aptamer conjugated quantum dots as labels. Anal Chim Acta 804:151–158

Dwivedi HP, Smiley RD, Jaykus L-A (2013) Selection of DNA aptamers for capture and detection of *Salmonella typhimurium* using a whole-cell SELEX approach in conjunction with cell sorting. Appl Microbiol Biotechnol 97(8):3677–3686

Ellington AD, Szostak JW (1992) Selection in vitro of single-stranded DNA molecules that fold into specific ligand-binding structures. Nature 355(6363):850–852

Engvall E, Perlmann P (1971) Enzyme-linked immunosorbent assay (ELISA) quantitative assay of immunoglobulin G. Immunochemistry 8(9):871–874

Fan M, McBurnett SR, Andrews CJ, Allman AM, Bruno JG, Kiel JL (2008) Aptamer selection express: a novel method for rapid single-step selection and sensing of aptamers. J Biomol Tech 19(5):311–319

Farber JM, Peterkin PI (1991) *Listeria monocytogenes*, a food-borne pathogen. Microbiol Rev 55(3):476–511

Ferreira CSM, Papamichael K, Guilbault G, Schwarzacher T, Gariepy J, Missailidis S (2008) DNA aptamers against the MUC1 tumour marker: design of aptamer–antibody sandwich ELISA for the early diagnosis of epithelial tumours. Anal Bioanal Chem 390(4):1039–1050

Friedman CR, Torigian C, Shillam PJ, Hoffman RE, Heltze D, Beebe JL, Malcolm G, DeWitt WE, Hutwagner L, Griffin PM (1998) An outbreak of salmonellosis among children attending a reptile exhibit at a zoo. J Pediatr 132(5):802–807

Gopinath SCB (2007) Methods developed for SELEX. Anal Bioanal Chem 387(1):171–182

Hai H, Yang F, Li J (2014) Highly sensitive electrochemiluminescence "turn-on" aptamer sensor for lead(II) ion based on the formation of a G-quadruplex on a graphene and gold nanoparticles modified electrode. Microchim Acta 181(9):893–901

Hasegawa H, Sode K, Ikebukuro K (2008) Selection of DNA aptamers against VEGF165 using a protein competitor and the aptamer blotting method. Biotechnol Lett 30(5):829–834

Hedayati CM, Amani J, Sedighian H, Amin M, Salimian J, Halabian R, Imani Fooladi AA (2016) Isolation of a new ssDNA aptamer against staphylococcal enterotoxin B based on CNBr-activated sepharose-4B affinity chromatography. J Mol Recogn 29(9):436–445

Hochel I, Viochna D, Škvor J, Musil M (2004) Development of an indirect competitive ELISA for detection of *Campylobacter jejuni* subsp. jejuni O:23 in foods. Folia Microbiol 49(5):579–586

Hybarger G, Bynum J, Williams RF, Valdes JJ, Chambers JP (2006) A microfluidic SELEX prototype. Anal Bioanal Chem 384(1):191–198

Jenkins SH, Heineman WR, Halsall HB (1988) Extending the detection limit of solid-phase electrochemical enzyme immunoassay to the attomole level. Anal Biochem 168(2):292–299

Jiang Y, Zou S, Cao X (2016) Rapid and ultra-sensitive detection of foodborne pathogens by using miniaturized microfluidic devices: a review. Anal Methods 8(37):6668–6681

Jin B, Yang Y, He R, Park YI, Lee A, Bai D, Li F, Lu TJ, Xu F, Lin M (2018) Lateral flow aptamer assay integrated smartphone-based portable device for simultaneous detection of multiple targets using upconversion nanoparticles. Sensors Actuators B Chem 276:48–56

Karsunke XYZ, Niessner R, Seidel M (2009) Development of a multichannel flow-through chemiluminescence microarray chip for parallel calibration and detection of pathogenic bacteria. Anal Bioanal Chem 395(6):1623

Kim J, Demeke T, Clear RM, Patrick SK (2006) Simultaneous detection by PCR of *Escherichia coli, Listeria monocytogenes* and *Salmonella typhimurium* in artificially inoculated wheat grain. Int J Food Microbiol 111(1):21–25

Kim B, Jung IH, Kang M, Shim HK, Woo HY (2012) Cationic conjugated polyelectrolytes-triggered conformational change of molecular beacon aptamer for highly sensitive and selective potassium ion detection. J Am Chem Soc 134(6):3133–3138

Kim HR, Song MY, Chan Kim B (2020) Rapid isolation of bacteria-specific aptamers with a non-SELEX-based method. Anal Biochem 591:113542

Kong M, Shin JH, Heu S, Park J-K, Ryu S (2017) Lateral flow assay-based bacterial detection using engineered cell wall binding domains of a phage endolysin. Biosens Bioelectron 96:173–177

Kourany M (1983) Medium for isolation and differentiation of *Vibrio parahaemolyticus* and *Vibrio alginolyticus*. Appl Environ Microbiol 45(1):310

Kramer M, Obermajer N, Bogovič Matijašić B, Rogelj I, Kmetec V (2009) Quantification of live and dead probiotic bacteria in lyophilised product by real-time PCR and by flow cytometry. Appl Microbiol Biotechnol 84(6):1137–1147

Kurt H, Yüce M, Hussain B, Budak H (2016) Dual-excitation upconverting nanoparticle and quantum dot aptasensor for multiplexed food pathogen detection. Biosens Bioelectron 81: 280–286

Lavu PSR, Mondal B, Ramlal S, Murali HS, Batra HV (2016) Selection and characterization of Aptamers using a modified whole cell bacterium SELEX for the detection of Salmonella enterica serovar typhimurium. ACS Comb Sci 18(6):292–301

Lazcka O, Campo FJD, Muñoz FX (2007) Pathogen detection: a perspective of traditional methods and biosensors. Biosens Bioelectron 22(7):1205–1217

Lou X, Qian J, Xiao Y, Viel L, Gerdon AE, Lagally ET, Atzberger P, Tarasow TM, Heeger AJ, Soh HT (2009) Micromagnetic selection of aptamers in microfluidic channels. Proc Natl Acad Sci 106(9):2989

Luo K, Ryu J, Seol I-H, Jeong K-B, You S-M, Kim Y-R (2019) Paper-based radial chromatographic immunoassay for the detection of pathogenic bacteria in milk. ACS Appl Mater Interfaces 11(50):46472–46478

Magliulo M, Simoni P, Guardigli M, Michelini E, Luciani M, Lelli R, Roda A (2007) A rapid multiplexed chemiluminescent immunoassay for the detection of *Escherichia coli* O157:H7, Yersinia enterocolitica, *Salmonella typhimurium*, and *Listeria monocytogenes* pathogen bacteria. J Agric Food Chem 55(13):4933–4939

Mahato K, Kumar S, Srivastava A, Maurya PK, Singh R, Chandra P (2018) Electrochemical immunosensors: fundamentals and applications in clinical diagnostics. In: Handbook of immunoassay technologies

Mahon BE, Pönkä A, Hall WN, Komatsu K, Dietrich SE, Siitonen A, Cage G, Hayes PS, Lambert-Fair MA, Bean NH, Griffin PM, Slutsker L (1997) An international outbreak of salmonella infections caused by Alfalfa sprouts grown from contaminated seeds. J Infect Dis 175(4): 876–882

Mayer G (2009) The chemical biology of Aptamers. Angew Chem Int Ed 48(15):2672–2689

Medley CD, Bamrungsap S, Tan W, Smith JE (2011) Aptamer-conjugated nanoparticles for cancer cell detection. Anal Chem 83(3):727–734

Mosing RK, Mendonsa SD, Bowser MT (2005) Capillary electrophoresis-SELEX selection of Aptamers with affinity for HIV-1 reverse transcriptase. Anal Chem 77(19):6107–6112

Murray EGD, Webb RA, Swann MBR (1926) A disease of rabbits characterised by a large mononuclear leucocytosis, caused by a hitherto undescribed bacillus Bacterium monocytogenes (nsp). J Pathol Bacteriol 29(4):407–439

Nightingale KK, Windham K, Wiedmann M (2005) Evolution and molecular phylogeny of *Listeria monocytogenes*; isolated from human and animal listeriosis cases and foods. J Bacteriol 187(16):5537

Nishibuchi M, Kaper JB (1995) Thermostable direct hemolysin gene of *Vibrio parahaemolyticus*: a virulence gene acquired by a marine bacterium. Infect Immun 63(6):2093–2099

Oh SY, Heo NS, Shukla S, Cho HJ, Vilian ATE, Kim J, Lee SY, Han YK, Yoo SM, Huh YS (2017) Development of gold nanoparticle-aptamer-based LSPR sensing chips for the rapid detection of *Salmonella typhimurium* in pork meat. Sci Rep 7(1):10130

Pang Y, Wan N, Shi L, Wang C, Sun Z, Xiao R, Wang S (2019) Dual-recognition surface-enhanced Raman scattering(SERS)biosensor for pathogenic bacteria detection by using vancomycin-SERS tags and aptamer-Fe_3O_4@Au. Anal Chim Acta 1077:288–296

Pitcher DG, Fry NK (2000) Molecular techniques for the detection and identification of new bacterial pathogens. J Infect 40(2):116–120

Purohit B, Vernekar PR, Shetti NP, Chandra P (2020) Biosensor nanoengineering: design, operation, and implementation for biomolecular analysis. Sensors Int 1:100040

Ren X, Gelinas AD, von Carlowitz I, Janjic N, Pyle AM (2017) Structural basis for IL-1α recognition by a modified DNA aptamer that specifically inhibits IL-1α signaling. Nat Commun 8(1):810

Rhinehardt KL, Srinivas G, Mohan RV (2015) Molecular dynamics simulation analysis of anti-MUC1 Aptamer and Mucin 1 peptide binding. J Phys Chem B 119(22):6571–6583

Riegler J, Ditengou F, Palme K, Nann T (2008) Blue shift of CdSe/ZnS nanocrystal-labels upon DNA-hybridization. J Nanobiotechnol 6(1):7

Sefah K, Yang Z, Bradley KM, Hoshika S, Jiménez E, Zhang L, Zhu G, Shanker S, Yu F, Turek D, Tan W, Benner SA (2014) In vitro selection with artificial expanded genetic information systems. Proc Natl Acad Sci 111(4):1449

Shen H, Wang J, Liu H, Li Z, Jiang F, Wang F-B, Yuan Q (2016) Rapid and selective detection of pathogenic bacteria in bloodstream infections with Aptamer-based recognition. ACS Appl Mater Interfaces 8(30):19371–19378

Stamm WE, Cutter BE, Grootes-Reuvecamp GA (1981) Enzyme immunoassay for detection of antibody-coated bacteria. J Clin Microbiol 13(1):42–45

Stevens KA, Jaykus L-A (2004) Bacterial separation and concentration from complex sample matrices: a review. Crit Rev Microbiol 30(1):7–24

Stoltenburg R, Reinemann C, Strehlitz B (2007) SELEX—a (r)evolutionary method to generate high-affinity nucleic acid ligands. Biomol Eng 24(4):381–403

Stone A, Shaffer M, Sautter RL (1993) Salmonella poona infection and surveillance in a neonatal nursery. Am J Infect Control 21(5):270–273

Suman P, Chandra P (2021). Immunodiagnostic technologies from laboratory to point-of-care testing

Teng J, Yuan F, Ye Y, Zheng L, Yao L, Xue F, Chen W, Li B (2016) Aptamer-based technologies in foodborne pathogen detection. Front Microbiol 7:1426–1426

Tuerk C, Gold L (1990) Systematic evolution of ligands by exponential enrichment: RNA ligands to bacteriophage T4 DNA polymerase. Science 249(4968):505

Verma S, Choudhary J, Singh KP, Chandra P, Singh SP (2019) Uricase grafted nanoconducting matrix based electrochemical biosensor for ultrafast uric acid detection in human serum samples. Int J Biol Macromol

Vigneshvar S, Sudhakumari CC, Senthilkumaran B, Prakash H (2016) Recent advances in biosensor technology for potential applications—an overview. Front Bioeng Biotechnol 4:11

Wang J-C, Wang Y-S, Rang W-Q, Xue J-H, Zhou B, Liu L, Qian Q-M, Wang Y-S, Yin J-C (2014) Colorimetric determination of 8-hydroxy–2′-deoxyguanosine using label-free aptamer and unmodified gold nanoparticles. Microchim Acta 181(9):903–910

Wang J, Wu X, Wang C, Shao N, Dong P, Xiao R, Wang S (2015) Magnetically assisted surface-enhanced raman spectroscopy for the detection of *Staphylococcus aureus* based on Aptamer recognition. ACS Appl Mater Interfaces 7(37):20919–20929

Wang X, Niazi S, Yukun H, Sun W, Wu S, Duan N, Hun X, Wang Z (2017) Homogeneous time-resolved FRET assay for the detection of *Salmonella typhimurium* using aptamer-modified NaYF4:Ce/Tb nanoparticles and a fluorescent DNA label. Microchim Acta 184(10):4021–4027

Wang T, Chen C, Larcher LM, Barrero RA, Veedu RN (2019) Three decades of nucleic acid aptamer technologies: lessons learned, progress and opportunities on aptamer development. Biotechnol Adv 37(1):28–50

Watanabe K, Arakawa H, Maeda M (2002) Simultaneous detection of two verotoxin genes using dual-label time-resolved fluorescence immunoassay with duplex PCR. Luminescence: J Biol Chem Luminescence 17(2):123–129

Wu J, Zhu Y, Xue F, Mei Z, Yao L, Wang X, Zheng L, Liu J, Liu G, Peng C, Chen W (2014) Recent trends in SELEX technique and its application to food safety monitoring. Microchim Acta 181(5):479–491

Xu M, Wang R, Li Y (2017) Electrochemical biosensors for rapid detection of *Escherichia coli* O157: H7. Talanta 162:511–522

Yao J, Yang M, Duan Y (2014) Chemistry, biology, and medicine of fluorescent nanomaterials and related systems: new insights into biosensing, bioimaging, genomics, diagnostics, and therapy. Chem Rev 114(12):6130–6178

Yao L, Ye Y, Teng J, Xue F, Pan D, Li B, Chen W (2017) In vitro isothermal nucleic acid amplification assisted surface-enhanced Raman spectroscopic for ultrasensitive detection of *Vibrio parahaemolyticus*. Anal Chem 89(18):9775–9780

Yoo SM, Kim D-K, Lee SY (2015) Aptamer-functionalized localized surface plasmon resonance sensor for the multiplexed detection of different bacterial species. Talanta 132:112–117

Nanomaterials-Based Immunosensors in Food Analysis

11

Nikita Sarawagi, Kalyan Vaid, Jasmeen Dhiman, Treesa Johns, and Vanish Kumar

Abstract

Immunosensors are the type of affinity biosensors that work on the basis of antigen–antibody interactions. These devices can be highly advantageous to develop a portable and point-of-care system for living and non-living food contaminants. The utilization of nanomaterials in the fabrication of immunosensors is found to improve several sensing parameters of an immunosensors. In general, the nanomaterials are used as transducer materials for the fabrication of immunosensors. Till now, nanomaterials are found to show significant improvements in the sensing capabilities of electrochemical, optical, and piezoelectric immunosensors. The use of nanomaterials is found to enhance sensitivity, specificity, rapidness, and portability of the immunosensors. Herein, we have covered the key achievements of nanomaterials for the development of immunosensors. This book chapter covers the important aspects of immunosensors types, nanomaterials-based immunosensors fabrication, and accomplishments of nanomaterials-based electrochemical, optical, and piezoelectric immunosensors for the analysis of food contaminants, e.g., pesticides, pathogens, toxins, pharmaceuticals, allergens, and adulterants.

Keywords

Nanomaterials · Food analysis · Sensors · Bioanalysis · Miniaturization

N. Sarawagi · J. Dhiman · T. Johns · V. Kumar (✉)
National Agri-Food Biotechnology Institute (NABI), S.A.S. Nagar, Punjab, India
e-mail: vanish@nabi.res.in

K. Vaid
National Agri-Food Biotechnology Institute (NABI), S.A.S. Nagar, Punjab, India

Centre for Nanoscience and Nanotechnology, Panjab University, Chandigarh, India

P. Chandra, P. S. Panesar (eds.), *Nanosensing and Bioanalytical Technologies in Food Quality Control*, https://doi.org/10.1007/978-981-16-7029-9_11

11.1 Introduction

These days, contamination in food is very common due to inappropriate agricultural, animal husbandry, food processing, and food storage practices (Njobeh et al. 2009; Nerín et al. 2016; Kumar et al. 2020a, b). For example, the extensive use of pesticides/herbicides in improving the agricultural yield can cause contamination in food crops (Kumar et al. 2020a, b). Likewise, chemicals are used to enhance unnatural growth of animals, food preservation, and disease control can also be part of food in the form of contaminants (Kolok et al. 2018; Falleh et al. 2020; Kumar et al. 2020a, b). In addition to that, the improper storage conditions of the food may lead to the growth of pathogenic microorganisms as biological food contaminants (García-Díaz et al. 2019; Choi et al. 2020). The contaminated food can pose acute to severe threat to human beings or other animals. Specifically, in humans these contaminants can cause irritation, food poisoning, headaches, nausea, diabetes, neurological disorders, asthma, cancer, and even death (Ahmed Adam et al. 2017; Hassan et al. 2017; Vikrant et al. 2018; Agache et al. 2019; Kumar et al. 2020a, b). So, the food samples should be analyzed before consumption for the presence of contaminant to avoid any acute/chronic health issues.

In past, several analytical tools, such as chromatographic techniques, nuclear magnetic resonance, capillary electrophoresis, flow injection calorimetry, and mass spectroscopy have been explored successfully to determine the presence of food contaminants (Patra et al. 2017; Malvano et al. 2020). Interestingly, these techniques are adequately precise and sensitive. However, a good number of limitations are associated with these techniques, e.g., time-consuming methodologies, non-portability, expensive methodologies/instrumentation/ procedures, and requirement of expert manpower, and involvement of multiple pre-treatment steps. All these limitations restricted the current contaminant sensing techniques to develop a point-of-care device in fulfilling the food security demand for common people (Suri et al. 2009; Kumar et al. 2020a, b; Malvano et al. 2020).

In this context, easy, faster, sensitive, cost-effective, and specific sensing is in high demands. The development of immunoassays is one of the most viable ways to achieve the aforementioned sensing goals. In case of an immunosensor, the sensitive and specific interactions between antibody and antigen have been utilized to detect the levels of antigen in the food samples (Kumar et al. 2020a, b). The generated signals are captured and transferred by the transducer to the detection unit. In general, the specificity and sensitivity of an immunosensor is defined by the antigen–antibody (Ag–Ab) interaction and characteristics of the transducer. Since, the inception of radioimmunoassay technique, researchers are working hard to substitute conventional methods of analysis by replacing radioactive material with more promising options (Tian et al. 2012; Mistry et al. 2014). To achieve this, a number of chemicals/molecules (e.g., enzyme, fluorophore, and chromophore) have been explored as a tag for antibodies to sense the analyte molecule, i.e., antigen (Ricci et al. 2007). For example, immunoassay-based pregnancy examination can be achieved by sensing human chorionic gonadotropin via enzyme-coupled antibodies (Makaraviciute and Ramanaviciene 2013). Further advancements in the immunosensors have been achieved by the incorporation of nanomaterials (Mahato et al. 2018; Pottathara et al. 2019; Chandra and Prakash 2020; Kumar et al. 2020a, b; Purohit et al. 2020a, b). In general, nanomaterials have been explored as transducer material to amplify the sensitivity of an immunosensor. In this regard, diverse nanomaterials, such as carbon nanomaterials (e.g., carbon nanotubes; CNTs, graphene-materials, carbon nanorods, and carbon-based nanoparticles), metal-based nanomaterials (e.g., metal nanoparticles, metal oxide nanostructures, and metal sulfide nanomaterials), semiconductor nanostructure (e.g., quantum dots; QDs), and hybrid materials (e.g., metal organic frameworks; MOF) have been tested to enhance the sensing capabilities of immunosensors (Zhang et al. 2016a, b; Patra et al. 2017; Xu et al. 2017; He et al. 2018a, b; Sun et al. 2019; Fan et al. 2020). The use of nanomaterials provided multiple benefits in the development of immunosensors. For instance, nanomaterials were found to enhance sensing parameters (e.g., sensitivity, rapidness, stability, and portability) of the immunosensor (Kumar et al. 2020a, b). Moreover, it was also observed that the use of nanomaterials can enhance antibody loading capacities on the working electrode. On the basis of nanomaterials characteristics, these structures have been employed successfully for the development of diverse immunosensors including optical, electrochemical, and piezoelectric immunosensors (Goud et al. 2018; Bansal et al. 2020; Bhardwaj et al. 2021; Kukkar et al. 2021; Kumar et al. 2021).

In this book chapter, we have provided the in-depth knowledge of all the aspects of nanomaterials-based immunosensors for food contamination analysis. To understand the importance of nanomaterials in the advancements of immunosensors, we have discussed the immunosensing of diverse food contaminants (e.g., biological and non-biological) in diverse food samples, e.g., meat, food grains, dairy products, and poultry products. A discussion on the types of immunosensors with a special focus on the biorecognition molecules, immunosensing methods, and role of nanomaterials is provided in the next section. In Sect. 11.3, the key strategies (e.g., bioconjugation of immunogen and sensing strategies) involved for the fabrication of a nanomaterials-based immunosensors is described. The development, functioning, mechanism, and performances of nanomaterials-based electrochemical, optical, and piezoelectric sensors are discussed in Sect. 11.4. In the last section, conclusion of this chapter is provided with special focus on the research gaps remained and their possible solutions.

11.2 Immunosensors

Immunosensors are one of the most explored affinity biosensors (Mollarasouli et al. 2019). The sensing outcome of immunosensors is specifically based on Ag–Ab interactions. In general, an immunosensor is composed of an immunoreceptor and signal transducer to generate sensing signals for analyte (e.g., food contaminants, clinical samples, and environment pollutants) (Duffy and Moore 2017; Mollarasouli et al. 2019).

On the basis of target analyte, the receptor (or capture probe) can be made up of an antigen or antibody (Banica 2012). Specifically, the immunoreceptor components recognize target analyte by forming an immunochemical complex with analyte molecule (Lara and Perez-Potti 2018). The interactivity in immunocomplex are majorly controlled by non-covalent interactions such as electrostatic, hydrogen bonding, hydrophobic, and van der Waals interaction (Ahmed et al. 2019). Upon the formation of immunocomplex, the signals generated due to affinity immunochemical interactions are monitored with diverse transducers (Malvano et al. 2020). For example, the signals generated in the form of mass variation, upon formation of immunocomplex on the surface of quartz microbalance, can be monitored to know the analyte level in piezoelectric immunosensors. Similarly, in electrochemical sensors, the alterations in the electric charges on the surface of sensing materials, due to formation of immunocomplex, can be used to determine antigen concentrations (Malvano et al. 2020).

A variety of immunosensors have been tested and explored for the determination of diverse analytes. Fascinatingly, immunoassay kits are also commercially available for the detection of various key analytes (e.g., enzyme-linked immunosorbent assay (ELISA)-based kits for the detection of aflatoxin, food allergen, and histamine in fish are commercially available (https://www.biofronttech.com/products/food-safety-prognosis-biotech-histamine-kits/1/24/). With time, several developments witnessed the ascent of immunosensors. After all such development, the modern immunoassays are highly specific, simple, non-destructive, and devoid of complicated sample preparation. Excitingly, the introduction of nanotechnology with immunochemistry not only enhanced the sensing parameters to a great extent, but also helpful in enhancing the stability, simplicity, and portability of the immunosensors (Krishna et al. 2018). A number of nanomaterials found excellent for the immunosensing applications in food sample analysis. On the basis of the type of sensing signals generated, the nanomaterials-based immunosensors can be classified as (1) electrochemical, (2) spectroscopic, and (3) colorimetric immunosensor.

11.2.1 Bioreceptors in Immunosensor

Antibodies are one of the key components in any immunoassay. These are proteinaceous molecules generated in defense of foreign antigen by the immune system of higher organisms (Lara and Perez-Potti 2018). Antibodies are highly specific due to variable domain at terminus of its fragment antigen-binding (Fab) region (Lara and Perez-Potti 2018). This biospecificity of antibodies for its target made them relevant for specific analytic applications. An antibody can be categorized in monoclonal (binds to single epitope of antigen) or a polyclonal (binds to multiple epitopes) (Felix and Angnes 2018). On the basis of extent of sensing application, either of the aforementioned antibody type can be used. In particular, monoclonal antibodies (mAbs) have negligible probability to involve in cross reactivity with nonspecific antigen and are thus better in terms of specificity for immunoassays (Felix and Angnes 2018). In certain situations (e.g., for the determination of pollutant classes), polyclonal antibody are also suitable (Felix and Angnes 2018). In general, the target for antibodies-based bioreceptor is antigens. Till now, a number of immunoassays have been employed for the determination of several antigens (e.g., food contaminants, pollutant, drugs, and allergens) (Banica 2012). At several instances, the antigen were also used as bioreceptor in immunosensors for the detection of antibodies (e.g., detection of specific disease by diagnosing antibodies linked with the disease) (Mollarasouli et al. 2019).

11.2.2 Methods in Immunoassays

The Ag–Ab interaction was explored by Yallow and Berson in year 1959 for analysis purpose in the form of radioimmunoassay (Turner 1997). This technique involves the usage of competitive binding of sample antigen to its specific antibody displacing the previously bound known radiolabeled antigen (Cristea et al. 2015). Immunosensors can be classified as indirect and direct immunosensors, where it allows measurement of immunochemical reaction with and without employing label, respectively (Pei et al. 2013; Cristea et al. 2015). In general, label-free immunosensors are less laborious and avoid interference of the labels in analysis process (Saito et al. 2008; Wang and Park 2020). The non-labeled immunosensors could detect physical changes (e.g., mass in case of piezoelectric sensor) induced by immune-specific interactions (Cristea et al. 2015; Wang et al. 2016; Filik and Avan 2019).

In case of indirect immunosensors, the labeling of antibodies helped in enhancing sensitivity of the sensor by supporting signal generation (Aranda et al. 2018). The common examples of labels used during immunosensing are enzyme, radioactive isotopes, and fluorophores (Wang et al. 2016; Niu et al. 2019). Depending upon the type of labeling, the immunoassays can be categorized as (1) radio immunoassay, (2) chemiluminescent immunoassay, (3) enzyme immunoassay, and (4) fluoroimmunoassay techniques (Cristea et al. 2015). In case of radio immunoassay, the labeling of molecules with radioisotopes allow highly sensitive detection of analyte (Felix and Angnes 2018; Wang and Park 2020). Unfortunately, the hazards associated with the exposure of radioactive materials restricted the use of radio immunoassay.

On the other hand, the enzymes are most common label for immunosensing due to its exceptional signal amplification and specificity (Cristea et al. 2015). ELISA is most widely utilized for enzyme-based immunoassay and is well-known for its speed, specificity, and precision. In brief, the enzyme-tagged antibody is allowed to bind Ag–Ab complex and the unbounded antibodies are washed away. Only in case of attachment between antigen and antibody, a color is observed on subsequent addition of substrate. This happens due to conversion of substrate by enzyme tagged to the antibody (Cristea et al. 2015). The resulted color intensity is used to quantify the analyte levels. The most common enzyme labels used to develop ELISA are horseradish peroxidase (HRP), alkaline phosphatase (ALP), luciferase, catalase

(CAT), and glucose oxidase (Ricci et al. 2007; Cristea et al. 2015). Furthermore, ELISA is equally effective for the determination of antigen or antibody.

11.2.3 Nanomaterials in Immunosensor

The use of nanomaterials is found excellent for the fabrication of label-free immunosensors. In case of nanomaterials-based immunosensors, the molecular tags are usually replaced by nanomaterials (e.g., metal nanoparticles, carbon materials, and semiconductor nanostructures). In general, nanomaterials possess extraordinary properties, e.g., high surface area to volume ratio, high biocompatibility, and elevated redox activity (Lara and Perez-Potti 2018). Additionally, these materials also exhibit exceptional optical, electrical, and electronic properties suitable to replace molecular labels (Wang et al. 2016; Lara and Perez-Potti 2018). In particular, the CNTs, carbon nanorods, graphene-materials, metal-based nanomaterials, and semiconductor QDs have been employed to enhance the sensing capabilities of immunosensors (Zhang et al. 2016a, b; Patra et al. 2017; Xu et al. 2017; He et al. 2018a, b; Sun et al. 2019; Fan et al. 2020). In general, nanomaterials help to generate sensitive sensing signals with respect to the analyte levels. Moreover, they also displayed improved loading/immobilization capacity for recognition molecule by offering large surface and more number of active sites.

On the basis of unique properties of nanomaterials, a good number of nanomaterials have been employed to generate diverse immunosensors. For example, the gold nanoparticles (AuNPs) with tunable surface plasmonic properties are excellent for colorimetric immunosensing (Daliri et al. 2019). Similarly, fluorescent QDs were found suitable for the fabrication of fluorescent immunosensor (Daliri et al. 2019). Note that QDs exhibited high photostability, quantum yield (QE), and narrow/symmetric emission and broad absorption spectrum (Krishna et al. 2018; Pu et al. 2018). Likewise, the metal nanoparticles, e.g., gold, palladium, platinum, and silver nanoparticles (AgNPs) exhibit appropriate redox properties to generate electrochemical signals (Iglesias-Mayor et al. 2019). Interestingly, many nanostructures (e.g., graphene oxide; GO, magnetic nanoparticles, and Pd-Pt nanoparticles) imitate the activity of natural enzymes, e.g., CAT, peroxidase, oxidase, and superoxide dismutase enzymes (Iglesias-Mayor et al. 2019; Niu et al. 2019). Consequently, these nanozymes have the potential to replace short lived and costly natural enzyme in immunosensors (Chang et al. 2017; Zhou et al., 2018; Yao et al. 2020). It should be noted that the nanozymes displayed better stability, sensitivity, cost-effectiveness, and production easiness, when compared to the natural enzymes. Overall, a number of ways are possible to explore nanomaterials in the development of immunosensors. Out of these nanomaterials-based immunosensors, several sensors have been employed in food analysis to determine the contaminant levels. The specific discussion on the performance of nanomaterials-based immunosensors for food analysis is discussed in Sect. 11.4.

11.3 Fabrication of Immunosensor in Food Assay

A biosensor mainly consists of three major units: (1) receptor, (2) transducer, and (3) detector. In general, the interaction between receptor and analyte is converted into measurable signals by the transducer. A receptor molecule can be conjugated to transducer through various immobilization strategies. The type of immobilization strategies and transducer is one of the most crucial components for the fabrication of a biosensor.

11.3.1 Bioconjugation of Immunogens on Nanomaterials

Generally, immobilization method can be selected on the basis of transducer type, biorecognition element, and operational conditions (e.g., temperature, pH, and ionic strength) (Ahmed et al. 2019; Tsekenis et al. 2019). Depending on the type of sensors, immunogen, and sensing material, a number of bioconjugation strategies can be used to conjugate biomolecule to develop immunosensor for food contaminants. These conjugation methods can be majorly classified into three: (1) adsorption, (2) covalent, and (3) affinity interactions (Fig. 11.1) (Welch et al. 2017). The performance of an immunosensor is greatly influenced by (1) accessibility of binding site on the capture probe, (2) specificity of capture probe, (3) sensitivity of the designed probe, and (4) density of antibody immobilized on the sensor surface (Kausaite-Minkstimiene et al. 2010; Tsekenis et al. 2019). Usually, a capture probe in immunosensor can be immobilized in random or specific orientation (Makaraviciute and Ramanaviciene 2013). Physical adsorption is the most common and simplest approach to immobilize antibodies in which antibodies are irregularly attached to the surface of nanomaterials (or material of choice) (Makaraviciute and Ramanaviciene 2013). Despite being simple and easy, this method has some major disadvantages, e.g., no protection in harsh conditions and easy detachability from the

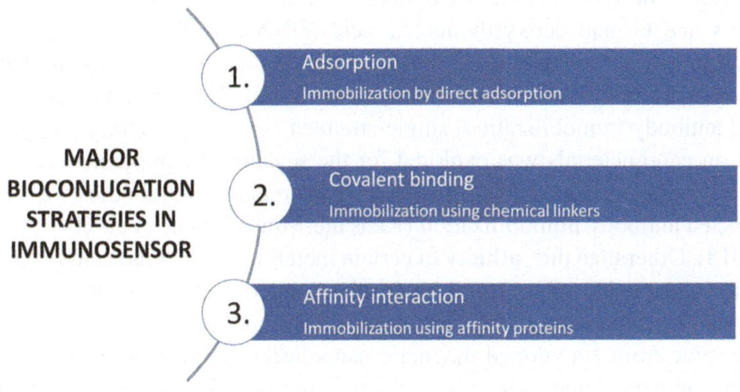

Fig. 11.1 Major bioconjugation strategies for the immobilization of biomolecules on the surface of immunosensors

surface. Consequently, such immunosensor are poorly reproducible in most situations (Makaraviciute and Ramanaviciene 2013; Tsekenis et al. 2019).

The covalent attachment of antibodies is another most explored immobilization approach with high success rate (Jung et al. 2008). In covalent immobilization, a covalent bond is formed between antibody and nanomaterial with the help of a linker molecule (e.g., ethyl(dimethylaminopropyl) carbodiimide/ N-hydroxysuccinimide (EDC/NHS), glutaraldehyde, glutathione, epoxy-lysine, and epoxy-thiol (Baniukevic et al. 2013; Kumar et al. 2020a, b). The EDC/NHS is one of the most efficient covalent immobilization approaches in which the carboxyl or amine group of antibodies are attached with nanomaterials via carbodiimide bond (Welch et al. 2017). To perform EDC/NHS coupling in nanomaterials, their surfaces can be modified by self-assembled monolayer of alkane having different functional groups (Welch et al. 2017; Haddada et al. 2018). These modifications facilitate formation of covalent bond between the surface and antibody (Makaraviciute and Ramanaviciene 2013). Generally, amine and carboxylic group of lysine and aspartic acid/glutamic acid of antibodies are involved in carbodiimide bond formation with nanomaterials surface functionalities, respectively (Welch et al. 2017).

Likewise, the disulfide bond present between thiol group of two cysteine molecule and carbohydrate present on Fc region of antibody can be reduced and oxidized, respectively, for covalent immobilization of antibodies with nanomaterials (Jung et al. 2008; Welch et al. 2017). It has been postulated that the covalent conjugation of antibodies on nanomaterials are highly stable and facilitate denser antibody immobilization on the surface of sensor for the sensitive determination of analytes. Nonetheless, the random distribution of amine and carboxylic groups on antibody can cause their disoriented attachment on sensor surface. A number of studies have witnessed the improved sensing capabilities due to directed antibody immobilization on nanomaterials (Jung et al. 2008; Kausaite-Minkstimiene et al. 2010; Baniukevic et al. 2013).

In a few studies, the use of affinity tag for conjugation of antibodies on nanomaterials is found excellent in site-oriented antibody immobilization (Welch et al. 2017). The most common examples of affinity tags are biotin-streptavidin, protein A and G, and deoxyribonucleic acid (DNA). In biotin-streptavidin-based conjugation of antibodies, the biotin-tagged antibody was immobilized on streptavidin-modified surface (Abdallah et al. 2019). Likewise, in case of DNA directed antibody immobilization, single-stranded (ssDNA) antibody complex conjugated on nanomaterials was explored for the sensing of complementary ssDNA. The affinity of bacterial proteins A and G for Fc region of antibodies was used for site-directed antibody immobilization (Kausaite-Minkstimiene et al. 2010; Haddada et al. 2018). Other than this, affinity of certain metals toward amino acids of antibody (e.g., histidine and cysteine) and molecular printed technology have also been employed for the oriented immobilization of antibody (Welch et al. 2017). For instance, zinc from zinc-doped magnetic nanoclusters form strong bond with thiol group of fragmented antibody, which can be employed for pathogen detection (Pal et al. 2017).

11.3.2 Sensing Strategies

A wide range of sensing strategies has been explored for the quality and safety assessment of food. On the basis of transduction mechanism, immunosensors can be classified into (1) electrochemical, (2) optical, and (3) piezoelectric immunosensors (Mollarasouli et al. 2019).

In general, electrochemical immunosensors (ECM-IS) are simple and effective tools to sense target contaminant, especially if the target analyte is present in trace levels (Aydin et al. 2019). It is a combination of electrochemical analysis and immunological techniques, which led to the generation of highly sensitive and specific sensing signals (Wang et al. 2013). An ECM-IS is composed of biorecognition element (e.g., antibody), transducer (e.g., electrode), and detector. The electrochemical transducers usually convert the changes associated with bio-recognition event (e.g., interaction of antibodies with antigens) into an electrical signal (e.g., current-, potential-, resistivity-, and impedance-change) (Thakur and Ragavan 2013; Kumar et al. 2015; Kumar et al. 2017a, b; Goud et al. 2018; Kumar et al. 2020a, b). Accordingly, the detector can be selected to determine the intensity of the generated signals (Aydin et al. 2019). Till now, various electrochemical immunosensing techniques have been used to analyze the food samples for the presence of contaminants. The most common of them are voltammetry (e.g., cyclic voltammetry; CV) and amperometry (Capoferri et al. 2018). Fascinatingly, the sensing capabilities of the electrochemical electrodes can be improved by several fold via employing the nanomaterials, e.g., CNTs, graphene materials, nanowires, and metallic nanoparticles (Cho et al. 2018).

The optical immunosensor detect the analyte levels through measuring the variations in the optical signals (e.g., fluorescence, absorbance, and color) of transducer (Daliri et al. 2019). The optical sensing can be performed by measuring fluorescence, absorbance, surface plasmon resonance (SPR), and colorimetric signals (Capoferri et al. 2018). In particular, the SPR mainly depend upon changes in refractive index of the transducer originated due to Ag–Ab interaction. The obtained signals can be quantify to detect different levels of analyte antigen (Ricci et al. 2007). The fluorescence immunosensors generally work on the measurement of decrease/increase in the fluorescence of fluorescent material (Capoferri et al. 2018). The common signals monitored in the fluorescence immunosensor are fluorescent intensity, fluorescence lifetime, and fluorescence resonance energy transfer (FRET) (Bhatnagar et al. 2016; Kempahanumakkagari et al. 2018; Kukkar et al. 2018; Berhanu et al. 2019; Kabir et al. 2019; Kumar et al. 2019). Likewise, chromogenic reactions have been employed for colorimetric immunosensing. Importantly, the colorimetric immuno-detection can also be performed visually or by digital image colorimetry (Suman and Chandra 2020). The nanoparticles-based engineering was found crucial for the development of such types of immunosensor (Mohamad et al. 2019). For instance, the AuNPs-based colorimetric immunosensor used to detect human immunoglobin through aggregation of functionalized AuNPs (Iarossi et al. 2018). Note that the change in color of AuNPs from red to purple was observed due to AuNPs aggregation in the presence of target analyte (Iarossi et al. 2018; Vaid et al.

2020a, b). Especially, AuNPs are very beneficial for colorimetric detection owing to their high extinction coefficient (especially in the visible region) (Zhang et al. 2019). Further, ultraviolet-visible (UV-vis) absorption spectroscopy is also a detection method used in optical immunosensors where absorbance is measured parameter to know the analyte levels (Kłos-Witkowska 2016).

The immune reactions specificity and high sensitivity of quartz crystal are utilized in piezoelectric immunosensing of food samples (Chen et al. 2011). In a piezoelectric immunosensor, antibody or antigen is immobilized on a quartz crystal and upon immunointeraction, a shift in the oscillation frequency was monitored corresponding to the change in mass on quartz crystal or analyte levels (Pohanka 2018). A variety of nanomaterials (such as metal nanoparticles, CNTs, graphene, and semiconductor nanomaterials) have been explored to enhance the performances of abovementioned sensing devices. All the nanomaterials-based achievements are discussed in subsequent sections.

11.4 Food Analysis Using Nanomaterials-Based Immunosensors

Food adulterant and contaminants detection is an international concern to achieve healthy human lifestyle. Purposely, immunosensors have been successfully coupled with different transducers assisted with nanomaterials for better performance. Several food contaminants such as biological toxins, pathogen, food allergen, veterinary drug residues, pesticides, and chemical additives are major concerns in food analysis have been analyzed through immunosensors. As discussed above, on the basis of transducer material, a number of sensing methods are employed to detect signals generated due to immune interactions. Out of all the explored detection techniques, electrochemical, optical, and quartz crystal microbalance (QCM)-based sensing approaches are employed most for nanomaterials-based immunosensors.

11.4.1 Electrochemical Immunosensors (ECM-IS)

In electrochemical sensors, the bioreceptors conjugated on modified/bare-electrode surface are used to examine the analyte. In general, a number of electrodes can function as working electrode for electrochemical immunoassays, e.g., screen-printed electrode (SPE), gold electrode (AuE), glassy carbon electrode (GCE), and screen-printed graphene electrode (SPGE) (Duffy and Moore 2017). The surface architecture and transducer materials (e.g., nanomaterials) is crucial in defining the performance of an ECM-IS (Ahmed et al. 2019). To enhance the performance of an ECM-IS, single as well as composite form of nanostructures can be used to modify/develop working electrode or used as label for ECM-ISs (Campuzano et al. 2020). Usually, nanomaterials were found to improve sensing signals of an ECM-ISs via several ways, such as by enhancing the conductivity of the electrodes, improving the immobilization efficiency, enhancing the charge transfer efficiency, increasing the sensitivity and specificity, and biocompatibility of electrodes (Campuzano et al.

2020). Till now, a good number of nanomaterials have been tested to fabricate sensitive ECM-IS (Pan et al. 2017). Broadly, these nanomaterials can be classified into (1) metal/metal oxide/metal sulfide nanostructures (includes nanostructures made of Au, Ag, Pt, ZnO, Fe_3O_4, MoS_2, and etc.), (2) carbon-based nanomaterials (includes CNT, graphene, and carbon-consisting QDs), and (3) composites of nanomaterials mentioned in the above two categories among themselves or with other potent materials.

11.4.1.1 Metal Nanostructures-Based ECM-IS

AuNPs have been used explicitly as an preferable nanomaterial to modify working electrode due to its extraordinary electrical/optical properties, availability, simplicity (in terms of synthesis), biocompatibility, stability, and structural uniformity (Pan et al. 2017). In a report on AuNPs-based ECM-IS, Alves et al. (2015) prepared an AuNPs coated SPCE for the detection of 1–100 ng/mL peanut allergen (Ara h 6) with limit of detection (LOD) 0.27 ng/mL (Alves et al. 2015). (Note that LOD of a sensor is the lowest level of analyte that can be reliably sensed by it.) The presence of AuNPs on the surface of screen printed carbon electrode (SPCE) facilitated stable immobilization of anti-Ara h 6 through strong affinity interaction of thiol group of antibody with Au. Upon addition of the target analyte (e.g., Ara h 6), analyte molecule was captured between immobilized antibody, i.e., anti-Ara h 6 and biotinylated anti-Ara h 6 (Alves et al. 2015). In subsequent steps, streptavidin-ALP was added. The specific conjugation of biotin and streptavidin led to the attachment of streptavidin-ALP on to biotinylated anti-Ara h 6. The ALP enzyme catalyzed 3-indoxyl phosphate by precipitating Ag^+ into Ag (Fig. 11.2).

The presence Ara h 6 was monitored by the electrochemical signals generated due to catalytic precipitation of silver ions (Alves et al. 2015). In contrast, biotinylated

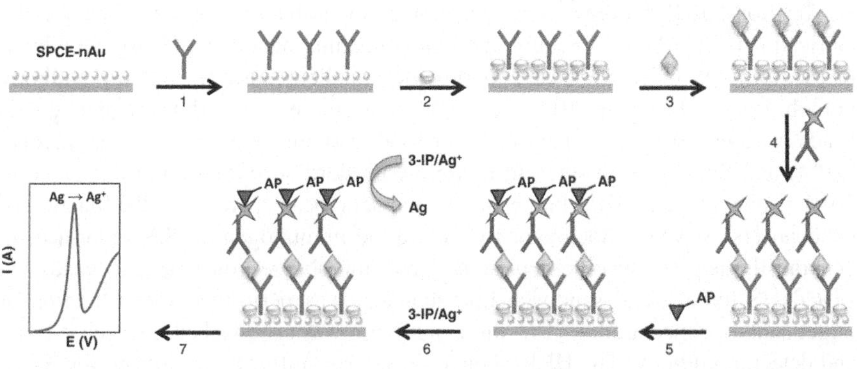

Fig. 11.2 Development and working of AuNPs-based immunosensor for Ara h 6. The SPCE was first modified with AuNPs to immobilize capture antibody (1), followed by the addition of blocking agent (casein) to block the remaining active sites (2), addition of Ara h 6 (3), addition and incubation of biotinylated-antibody (detection antibody probe) (4), Addition and interaction of streptavidin-ALP with detection antibody (5), addition of substrate (e.g., 3-indoxyl phosphate; 3-IP) for ALP and reduction of Ag^+ ions into Ag (6), and voltametric determination by measuring the deposited silver (7). Adapted with permission from Alves et al. (2015)

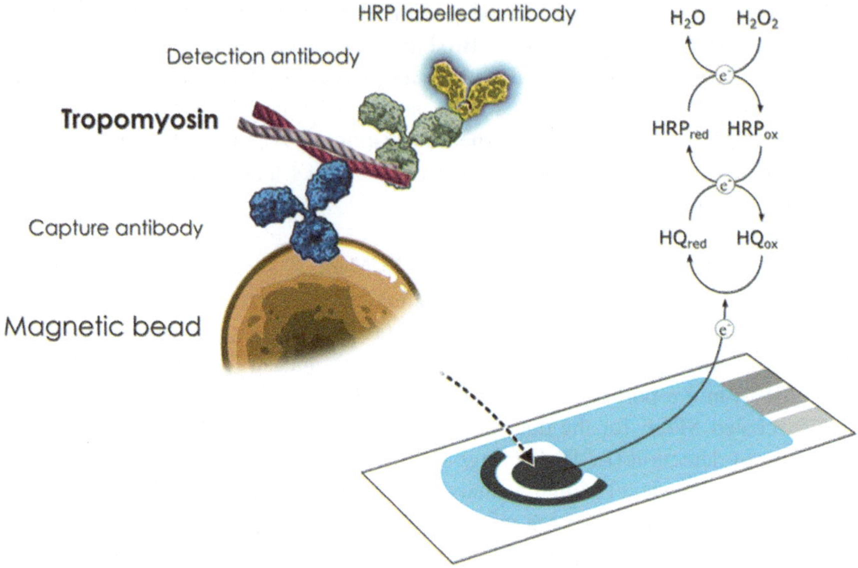

Fig. 11.3 The scheme showing tropomyosin capturing on the working electrode for amperometry detection. Adapted with permission from Angulo-Ibanez et al. (2019)

anti-Ara h 6 was not bound on the electrode surface, when Ara h 6 was absent and therefore, no change in the electrochemical signal was observed. Likewise, gold nanorods (AuNRs) were employed to analyze milk samples for the presence of *S. aureus*. The AuNR-based sensor could detect 1.8×10^2–1.8×10^7 CFU/mL of *S. aureus* with an LOD of 2.4×10^2 CFU/mL (Chen et al. 2016; Han et al. 2020).

The iron-based nanomaterials are another metal-based structure used for the fabrication of ECM-IS. Generally, the iron-containing nanomaterials were explored in the form of magnetic beads to provide excellent surface area for antibody immobilization (Pan et al. 2017). In two different studies, the carboxylated magnetic beads were tested for immunosensing of food allergens, e.g., shrimp tropomyosin (TPM) and Sin a protein (present in yellow mustard seeds) (Angulo-Ibanez et al. 2019; Gamella et al. 2020). Interestingly, magnetic beads performed the function of nanocarriers as well as facilitator to capture the immunogen on SPCE surface. In general, the analyte specific antibodies were immobilized on magnetic beads via EDC/NHS linkage and sandwich immunoassay was performed for evaluation of target analyte concentration. The allergen protein was sandwiched between capture and detector antibody. The HRP labeled secondary antibody was utilized for H_2O_2/hydroquinone (HQ) system based ECM signal for TPM of Shrimp (Angulo-Ibanez et al. 2019; Gamella et al. 2020) (Fig. 11.3).

The above-mentioned magnetic beads immunosensors were found excellent for efficient detection of food allergens. The magnetic beads-based immunosensor displayed a LOD of 0.047 and 0.82 ng/mL for TPM and Sin a protein, respectively (Angulo-Ibanez et al. 2019; Gamella et al. 2020).

11.4.1.2 Carbon Nanomaterials-Based ECM-IS

The carbon nanomaterials are well known for their unique electrochemical properties, high surface area, abundance binding sites for antibody immobilizations, and stability (Zhu 2017; Kour et al. 2020). Furthermore, due to their biocompatibility and catalytic efficiency they are preferential choice to develop ECM-IS (Pan et al. 2017). The most common carbon nanostructure employed for the fabrication of ECM-IS are CNTs and graphene materials (Pan et al. 2017). For instance, single-walled CNTs (SWCNTs) were used for the determination of AFB1 (Zhang et al. 2016a, b), *Staphylococcus aureus* (Bhardwaj et al. 2017), and antibiotic cephalexin (Yu et al. 2020). In another carbon-based immunosensor, multi-walled CNTs (MWCNT) and poly(3,4-ethylenedioxythiophene)-coated SPCE was used for electrochemical sensing of 0–250 ng/mL clenbuterol (Talib et al. 2018a, b). Note that clenbuterol is an illegal drug used in poultry to produce lean meat which can cause serious health issues if consumed. The MWCNT-based clenbuterol immunosensor was developed by immobilizing anti-clenbuterol antibody on the MWCNT-modified SPCE through EDC/NHS linker. During EDC/NHS linkage, the carboxyl groups of MWCNT were activated to interact with the amine group on antibody for the formation of carbamide bond. In the detection step, the HRP-labeled clenbuterol compete with free clenbuterol to bind immobilized anti-clenbuterol antibody. ECM study was performed on the basis of redox activity of HRP (HRP conjugated with antigen). After the addition of substrate (e.g., tetramethylbenzidine), the substrate was reduced due to catalytic activity of conjugated HRP. The exchange of electrons during the catalytic reduction of substrate led to generation of electrochemical signals. Thus, the amount of current produced was proportional to the clenbuterol-HRP bounded to MWCNT-antibody conjugate. Furthermore, the use of nanomaterial is speculated to amplify the sensing signals by improving electron transport to the working electrode. The MWCNT-based immunosensor demonstrated an LOD value of 4.66 ng/mL for clenbuterol (Talib et al. 2018a, b). Working on the similar sensing strategy, GO-based direct competitive assay was used to determine the amount of clenbuterol in beef samples with LOD value of 0.196 ng/mL (Talib et al. 2018a, b). The efficient immobilization of anti-clenbuterol (due to functionality of GO surface) and high conductivity were postulated for the generation of excellent sensing signals in GO-based immunosensor (Talib et al. 2018a, b). It was also observed that the controlled structural defects in reduced GO (rGO) exerted positive impact on its electrochemical sensing properties (Pan et al. 2017). The rGO with controlled defect levels can be beneficial in measuring the sensing signals for 1.2–34 ng/mL gliadin and exhibited an LOD value of 1.2 ng/mL (Chekin et al. 2016). In another report, graphene nanosheets were used for the construction of miniatured electrode to detect saxitoxin lower down to 0.299×10^3 ng/mL (Bratakou et al. 2017).

Apart from 2D morphology of graphene-materials, 0D structures, e.g., graphene QDs (GQD) have also been explored to develop immunosensors for food contaminants (Bhardwaj et al. 2018; Savas and Altintas 2019). The anti-*Y. enterocolitica* immobilized GQD (via EDC/NHS) was also used to modify the working electrode for electrochemical sensing of 1–61.23 $\times$ 10^8 CFU/mL of

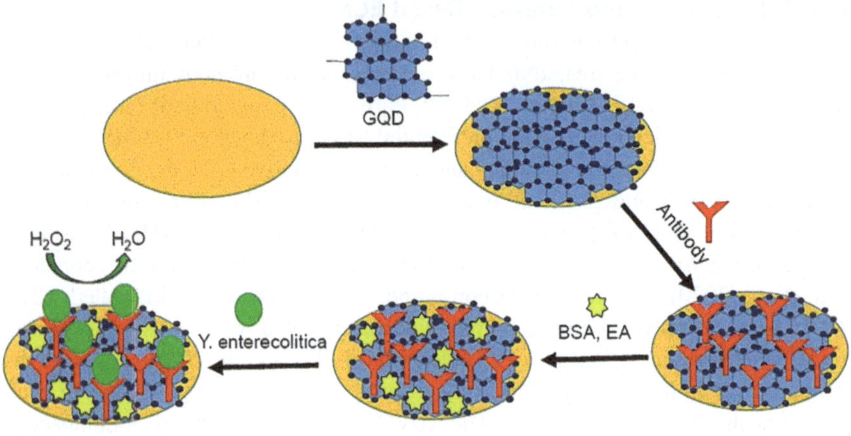

Fig. 11.4 Steps involved in the fabrication of GQDs-based immunosensor for *Y. enterocolitica* (*EA* Ethanolamine, *GQD* Graphene quantum dots). Adapted from Savas and Altintas (2019)

Y. enterocolitica (Savas and Altintas 2019) (Fig. 11.4). Upon addition of test sample, the anti-*Y. enterocolitica* (immobilized on GQDs) captured the *Y. enterocolitica*, which led to the blockade of charge transfer and thus decrease in electrical signal. This sensor can monitor the levels of *Y. enterocolitica* bacteria in spiked milk samples with a good LOD of 5 CFU/mL. The superior quantum confinement and peroxidase like activity of the GQDs were postulated for enhanced performance of GQDs-based sensor (Savas and Altintas 2019).

11.4.1.3 Composite Nanomaterials-Based ECM-IS

The composite form of nanomaterials also displayed excellent characteristics, which can be employed for the development of efficient ECM-IS (Kumar et al. 2017a, b; Kumar et al. 2020a, b; Mahato et al. 2020; Vaid et al. 2020a, b). Recently, conjugation of carbon and metal nanostructure has been tested to sense contaminations levels in diverse food samples (Shukla et al. 2018; Wang et al. 2018a, b; Sun et al. 2019; An et al. 2020; Hong et al. 2020). In particular, mycotoxin T-2 (0.01–100 ng/mL) was successfully detected electrochemically in feed and swine meat with SWCNT/AuNP-based indirect competitive assay (Wang et al. 2018a, b). This sensor was fabricated by immobilizing T-2 ovalbumin antigen analog on SWCNT/AuNP-modified GCE. The testing of antigen was performed by the addition of anti-T-2 antibody, followed by introduction of ALP labeled anti-antibody (secondary antibody). The ALP enzyme attached on the secondary antibodies hydrolyzed the substrate (e.g., α-naphthyl phosphate) to produce electrochemical signals. In the presence of T-2 toxin (in the test sample), T-2 complete with the immobilized T-2 ovalbumin toxin to form T-2: anti-T-2 immunocomplex. (Note that upon washing, the antibodies formed immunocomplex with the T-2 present in test solution was washed away). The formation of immunocomplex with the analyte present in test solution led to the decrease in the number of antibodies attached with the T-2

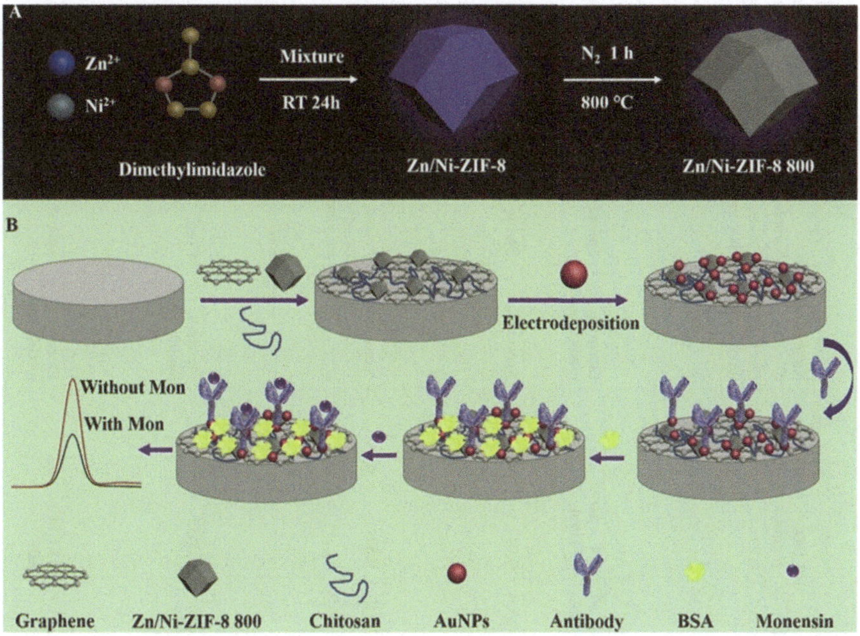

Fig. 11.5 Fabrication of graphene, MOF, and AuNPs-based immunosensor for monensin analysis. (**a**) Synthesis of MOF (Zn/Ni-ZIF-8-800), (**b**) sequential steps of electrode modification and signal generation for monensin. Adapted with permission from Hu et al. (2019)

immobilized on SWCNT/AuNP. Correspondingly, on the basis of interaction between ALP and α-naphthyl phosphate, dropped peak current was observed with respect to the T-2 levels. This SWCNT/AuNP exhibited 0.13 ng/mL LOD for T-2 (Wang et al. 2018a, b).

In addition to the metal nanostructures and carbon nanomaterial composites, composites of metal-organic frameworks with metal and carbon nanomaterials have also been explored for ECM-IS of food contaminants (Hu et al. 2019). In a study on MOF-based composites for ECM-IS, synergic effect of MOF (Zn/Ni-ZIF-8-800), graphene, and AuNPs has been exploited for the diagnosis of 0.25–100 ng/mL monensin in milk samples (Hu et al. 2019).

MOF was used for its selectivity and porosity whereas graphene and AuNPs enhanced the conductivity of electrode. The resultant modified electrode possessing high surface area, high conductivity, and biocompatibility. Using EDC/NHS, anti-monensin antibody was immobilized on hybrid matrix and the ECM signals were generated due to Ag–Ab interaction derived decreased in the current. The decrease in current is attributed to the insulating effect of antigen (Fig. 11.5) (Hu et al. 2019). A few key nanomaterials ECM-based immunosensors developed for food analysis are summarized in Table 11.1.

Table 11.1 Applications of different nanomaterials in ECM-IS for food analysis

No.	Antibodies used	Nanomaterial used	Analyte	Sensor type	Range	Range (ng/mL)	LOD	LOD (ng/mL)	Response time (min)	Nanomaterial role	Sample	Reference
1	Anti-Ara h 6 IgG (capture antibody) and biotinylated Anti-Ara h 6 IgG (detector antibody) (mouse)	AuNPs	Ara h 6 (peanut allergen)	ECM-LSV-SPCE	1–100 ng/mL	1–100	0.27 ng/mL	0.27		Electrode modification	Spiked cookies and chocolate samples	Alves et al. (2015)
2	Anti-*S. aureus*	AuNRs	*S. aureus*	ECM-impedance-GCE	1.8×10–1.8×10^7 CFU/mL	1.8×10^3–1.8×10^7 CFU/mL	2.4×10^2 CFU/mL	2.4×10^2 CFU/mL		Electrode modification	Spiked milk	Han et al. (2020)
3	MNB/Rabbit anti-TPM (capture antibody), HRP/anti-Rabbit Immunoglobulin G (Goat) (detector antibody)	MNB	Shrimp TPM (allergen)	ECM-Amperometric-SPCE			47 pg/mL	0.047	180	Nanocarrier	Shrimp	Angulo-Ibanez et al. (2019)
4	Mouse anti-Sin a (capture antibody), Rabbit anti-Sin a (primary detector antibody), HRP/goat anti-rabbit IgG (secondary detector antibody)	MNB	Sin a (allergen)	ECM-Amperometric-SPCE			0.82 ng/mL	0.82	40–90	Nanocarrier	Peanut, rapeseed, cashew, pine nut, mustard seed extract	Gamella et al. (2020)
5	Anti-AFB1 (primary antibody), anti-mouse immunoglobulin G-AP (secondary antibody)	SWCNT	Aflatoxin B1 (mycotoxin)	ECM-DPV-GCE	0.01–100 ng/mL	0.01–100	3.5 pg/mL	0.0035		Electrode modification	Corn	Zhang et al. (2016a, b)
6	SWCNT/Anti-*S. aureus*	SWCNT	*S. aureus*	ECM-DPV	0.1–2 ng/mL	0.1–2	13 CFU/mL	13 CFU/mL	30	Electrode modification	Spiked milk	Bhardwaj et al. (2017)

7	Rabbit anti-CEX antibody (primary antibody), HRP/goat anti-rabbit-antibody (secondary antibody)	SWCNT	Cephalexin (antibiotic)	ECM-DPV-GCE	1–100 ng/mL	1–100	45.7 ng/mL	45.7		Electrode modification	Milk (yoghurt, milk, milk powder) derived and meat (fish, prawn, beef) products	Yu et al. (2020)
8	Rabbit anti-clenbuterol	MWCNT	Clenbuterol	ECM-SPCE	0–250 ng/mL	0–250	4.66 ng/mL	4.66		Electrode modification	Spiked meat sample	Talib et al. (2018a)
9	Rabbit anti-STX	Graphene	Saxitoxin (Shell fish toxin)	ECM-potentiometric-graphene electrode	1 nM–1 μM	(0.299–299) × 10^3	1 nM	0.299 × 10^3	5–20	Electrode modification	Spiked lake Water and shellfish samples	Bratakou et al. (2017)
10	Rabbit anti-gliadin	rGO	Gliadin (allergen)	ECM-DPV-GCE	1.2–34 ng/mL	1.2–34	1.2 ng/mL	1.2		Electrode modification	Spiked wheat	Chekin et al. (2016)
11	Rabbit anti-clenbuterol	GO	Clenbuterol	ECM-SPCE	5–150 ng/mL	5–150	0.196 ng/mL	0.196		Electrode modification	Spiked milk sample	Talib et al. (2018b)
12	Mouse anti-AFB1	GQD	Aflatoxin B1	ECM-impedance-ITO-coated glass substrate	0.1–1.75 ng/mL (test sample) 0.1–2.0 ng/mL (standard sample)	0.1–1.75 ng/mL (test sample) 0.1–2.0 ng/mL (standard sample)	0.05 ng/gm (test sample) 0.03 ng/mL (standard samples)	0.05 ng/gm (test sample) 0.03 ng/mL (standard samples)		Electrode modification	Maize	Bhardwaj et al. (2018)
13	Anti-*Y. enterocolitica*	GQD	*Y. Enterocolitica*	ECM-Chronoamperometry-AuE	1–61.23 × 10^8 CFU/mL	1–61.23 × 10^8 CFU/mL	5 CFU/mL	5 CFU/mL		Electrode modification and nanozyme	Spiked milk samples	Savas and Altintas (2019)
14	Anti-kidney bean lectin (rabbit)	AuNPs/MWCNT	Kidney bean lectin (allergen)	ECM-DPV-GCE	0.05–100 μg/mL	(0.05–100) × 10^3	0.023 μg/mL	23		Electrode modification	Kidney bean	Sun et al. (2019)
15	Anti-T-2, anti-mouse /ALP (secondary antibody)	AuNPs/CNT	T-2 Toxin (mycotoxin)	ECM-DPV-GCE	0.01–100 μg/L	0.01–100	0.13 μg/L	0.13		Electrode modification	Feed and Swine Meat	Wang et al. (2018a, b)
16	Rabbit anti-BoNT/E (Capture Antibody), Mouse anti-	Graphene and AuNPs	Botulinum neurotoxin-E	ECM-LSV-GCE	10 pg/ml-10 ng/ml	0.01–10	5.0 pg/mL	0.005	65	Electrode modification and signal Amplification	Spiked orange juice and milk	Narayanan et al. (2015)

(continued)

Table 11.1 (continued)

No.	Antibodies used	Nanomaterial used	Analyte	Sensor type	Range	Range (ng/mL)	LOD	LOD (ng/mL)	Response time (min)	Nanomaterial role	Sample	Reference
	BoNT/E (primary detector Antibody), AuNP/ rabbit anti-mouse IgG/ALP (Secondary detector Ab)											
17	Rabbit anti-C. *sakazakii* (rabbit)	GO/AuNP	*C. Sakazakii*	ECM-DPV-GCE	2.0×10^2–2.0×10^7 CFU/mL	2.0×10^2–2.0×10^7 CFU/mL	20 CFU/mL	20 CFU/mL	15	Electrode modification	Spiked infant milk powder	Shukla et al. (2018)
18	Anti β – lactoglobulin	rGO/AuNC	BLG (allergen)	ECM-DPV-SPE	0.01–100 ng/mL	0.01–100	0.08 ng/mL	0.08		Electrode modification	Milk	Hong et al. (2020)
19	Anti-AFM1	GO-/CeO_2	Aflatoxin M1	ECM-DPV-SPE	1.8×10^2–1.8×10^7 CFU/mL	1.8×10^2–1.8×10^7 CFU/mL	0.009 μg/L	0.009		Electrode modification	Milk	An et al. (2020)
20	Mouse anti-monensin	AuNPs/Zn/Ni-ZIF-8-800/graphene	Monensin	ECM-DPV-GCE	0.25–100 ng/mL	0.25–100	0.11 ng/mL	0.11		Electrode modification	Spiked milk	Hu et al. (2019)

Abbreviations: *AuNPs* Gold nanoparticles, *AuNR* Gold nanorods, *AuE* Gold electrode, *MNB* Magnetic nanobeads, *S. aureus Staphylococcus aureus, TPM* Tropomyosin, *HRP* Horseradish peroxidase, *IgG* Immunoglobulin G, *MWCNT* Multi-walled carbon nanotubes, *SWCNT* Single-walled carbon nanotubes, *GQD* Graphene quantum dots, *ALP* Alkaline phosphatase, *AuNC* Gold nanocluster, *GO* Graphene oxide, *rGO* Reduced graphene oxide, *C. sakazakii Cronobacter sakazakii, SPE* Screen-printed electrode, *ITO* Indium tin oxide, *BoNT/E* Botulinum neurotoxin-E, *AFM1* Aflatoxin M1, *ZIF-8* Zeolitic imidazolate framework-8

11.4.2 Optical and Spectroscopy-Based Immunosensors

A large number of nanomaterials exhibited superior optical properties. These materials mainly include metal/metal-oxides nanomaterials, carbon nanomaterials, and composite nanomaterials (Ding et al. 2020). These optical nanomaterials can be employed for the fabrication of optical immunosensors in food analysis. The optical properties of these materials can be assessed by SPR, absorbance, fluorescence, and Raman spectroscopies to determine the level of analyte contaminants (Zhang et al. 2020). This section will cover the role of different nanomaterials used for monitoring food contaminants with discussion on selected recently published reports.

11.4.2.1 Colorimetric and Absorbance-Based Immunosensors

Colorimetric is one of the simplest approaches for on-site analysis of various analyte (Tsagkaris et al. 2021). Analysis through colorimetric immunosensors can be performed in several forms such as microfluidic chip devices, lab-on-a-chip devices, lateral flow assay, and paper-based assays (Tsagkaris et al. 2021). Furthermore, the results can be visualized with naked eyes (Guo et al. 2018) or smartphone-coupled devices/programs (Zheng et al. 2019). In general, noble metal-based nanostructures have been employed most in colorimetric immunosensors for food analysis (Zhang et al. 2020). In particular, metal nanoparticles have significant place in colorimetric immunosensors due to their special ability to display change in color with change in their dispersion state (Zhang et al. 2020). Among all the metal nanostructures, gold nanostructures are most commonly used to develop colorimetric immunosensors. This wonder structures exhibit high extinction coefficient in visible light and functionality-dependent sensitive/selective response to analyte, which give rise the possibility of analyte detection in very low concentration via spectroscopy as well as naked eye (Zhang et al. 2019). The antibody conjugated AuNPs can be utilized for the colorimetric immunosensing of antibiotic residues (Guo et al. 2018). An immunochromatographic strip (ICTS) was fabricated by employing AuNPs for cyproheptadine antibiotic residues examination in pig urine. (Note that cyproheptadine is a prohibited drug usually used for weight gain in animals and its detection in urine can be correlated with its level in pig's body). This ICTS chip was good enough to show 5 ng/mL LOD (Guo et al. 2018). Interestingly, different forms of gold nanostructures can be used to determine four different types of mycotoxins. For example, four mycotoxins (fumonisin B1; FB1, ochratoxin A; OTA, zearalenone; ZEN, and anti-AFB1) were detected simultaneously by different color gold nanostructure, e.g., with nanospheres (red color), nanoflowers (blue color), nanocacti (purple color), and hyperbranched blackbodies morphologies (black color), respectively, for the fabrication of ICTS (Wu et al. 2020). The respective antibodies for the abovementioned gold nanostructure were conjugated through physical absorption. Moreover, the antigen-bovine serum albumin (BSA) was conjugated as a test line on ICTS. The ICTS-based sensing was performed by exposure of the test sample to the strip followed by addition of gold nanostructure-antibody probe. (Note that the ICTS consisted of test (T) and control (C) line and T line was developed by immobilizing antigen on its surface). In the presence of toxins in the

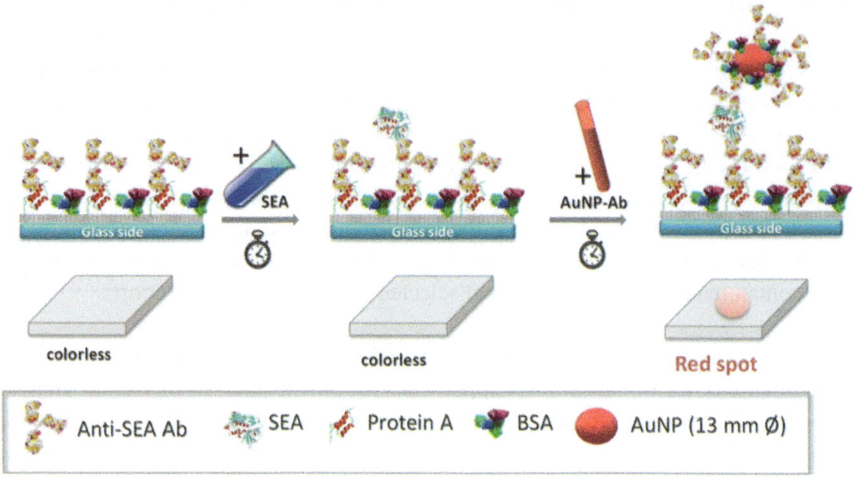

Fig. 11.6 Schematic of AuNPs-based colorimetric immunoassay for SEA. The glass slide was first modified with the anti-SEA followed by the addition of SEA and anti-SEA immobilized on AuNPs. The color as well as absorbance intensity was increased in the presence of SEA is the test sample. Adapted with permission from Zhang et al. (2019)

test samples, the toxin antigens compete with the antigen conjugated on the strip surface to make immunocomplex with gold nanostructure-immobilized antibodies. A decrease in the color intensity of the test line was observed with respect to the increase in the toxin levels in test sample. This ICTS could monitor FB1, OTA, ZEN, and AFB1 with LOD of 3.27, 0.10, 0.70, and 0.06 ng/mL, respectively (Wu et al. 2020). An metal nanoparticles-based ICTS was also developed for the colorimetric detection of 3-[(4-carboxyphenyl) monomethyl] amino-2-oxazolidinone with LOD 0.044 ng/mL (Yu et al. 2018).

In case of absorbance-based immunosensing, UV-vis spectroscopy is majorly employed for quantification of signals. For example, the change in absorbance of AuNPs (ratio of absorbance intensity at 520 and 630 nm) was monitored to detect the OTA levels (Liang et al. 2018; Zhang et al. 2019). Specifically, a sandwich type immunoassay was constructed for the determination of staphylococcal enterotoxin (SEA) by immobilizing anti-SEA on glass slide through affinity protein A (Fig. 11.6) (Zhang et al. 2019). Successively, SEA and AuNP/anti-SEA antibody was added to monitor the levels of SEA. The interaction of immobilized anti-SEA with SEA followed by attachment of AuNP/anti-SEA detection probe induces naked eye visible color in the testing strip, which was further analyzed by UV-vis spectroscopy (at 530 nm). Thus, the color intensity and absorbance of AuNPs increased with an increase in the amount of SEA in test sample. This AuNPs-based sensor exhibited LOD of 1 ng/mL for SEA (Zhang et al. 2019).

In another study, AuNP's tendency of aggregation in cysteine was utilized for amantadine (AMD) drug residues determination in poultry (Yu et al. 2018). The technique involves combination of conventional ELISA with Fenton's reaction to

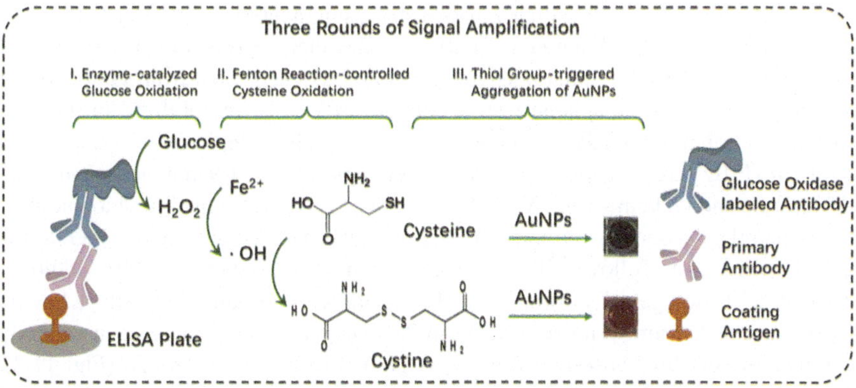

Fig. 11.7 Scheme elucidating process after immunocomplex formation in AuNPs-based immunoassay. (i) glucose oxidation by immobilized glucose oxidase enzyme, (ii) the Fenton reaction triggered cysteine oxidation, and (iii) images of AuNP in aggregated and re-dispersed state. Adapted with permission from Yu et al. (2018)

measure the sensing signals for AMD. Firstly, the microtiter plate was modified by AMD and incubated with mAb 3F2 (antibody against AMD) which was followed by addition of goat anti-mouse IgG labeled with glucose oxidase (gtAm-GOx). AMD in the test sample compete with the attached AMD and displaced surface bound mAb 3F2 antibody, which led to the less binding of secondary antibody gtAm-GOx. The enzyme glucose oxidase labeled on antibody produces H_2O_2 by glucose metabolism and H_2O_2 further oxidizes cysteine. This originally slow process (cysteine oxidation by H_2O_2) was triggered by hydroxyl radicals released though Fenton's reaction where Fe^{2+} catalyze H_2O_2 to form hydroxyl radicals. This leads to re-dispersion of cysteine aggregated AuNPs and cause blue shift in absorbance under visible light. The results were further quantified by calculating ratio of absorbance between 640 nm and 520 nm (A_{640nm}/A_{520nm}) (Fig. 11.7) (Yu et al. 2018).

In another immunosensing studies, Au nanostructures facilitate colorimetric analysis of *Salmonella typhimurium* (LOD: 35 CFU/mL) and alternariol monomethyl ether (mycotoxin) (LOD: 0.16 ng/mL) was performed with magnetic nanoparticles-based immunocomplex separation (Man et al. 2018; Guo et al. 2020a, b).

It was also observed that nanomaterials (such as MOF, platinum nanoparticles, GO, and AuNPs) were very beneficial in improving the sensitivity of ELISA-based immunoassays (He et al. 2018a, b; Xu et al. 2021). These nanomaterials can be employed as antibodies carrier (He et al. 2018a, b) or due to enzyme like activity (Wang et al. 2020a, b; Xu et al. 2021). (Note that a few nanomaterials exhibit enzyme like activity and are much more stable in comparison to enzymes). SiO_2 nanoparticles have good biocompatibility and high surface area. A competitive type immunoassay was tested to examine 2,4-dichlorophenoxyacetic acid pesticide with LOD 0.079 ng/mL. Here, SiO_2-conjugated antigen competes with antigen in the sample to attach with HRP-anti-2,4-D antibody (Wang et al. 2017a, b). In another study, parvalbumin (PV) allergen was detected using nanomaterials assisted ELISA in spiked dish sample (Wang et al. 2020a, b). The allergen protein was captured on magnetic beads through glutaraldehyde linker and was separated from sample using magnetic property of magnetic beads. Anti-frog PV was immobilized on GO via EDC/NHS linker and HRP/anti-mouse IgG2a heavy chain antibody was conjugated with AuNP by direct adsorption. For the detection of PV, PV/magnetic nanobead (PV/MNB) and GO/anti-frog PV antibody were mixed to form the immunocomplex. Subsequently, secondary antibody, AuNP, and HRP labeled anti-mouse IgG2a was added, which was followed by 3,3′',5,5'-Tetramethylbenzidine (TMP) addition. TMP is a chromogenic substrate for HRP catalyst to produce a product having absorbance at 450 nm. With an increase in PV concentration, an increase in antibody conjugation and thus absorbance was observed (LOD = 4.29 ng/mL) (Fig. 11.8). The obtained signals were five times better in comparison to the conventional ELISA. The use of GO and AuNPs was mainly suspected for the enhancement in the sensing signals whereas, MNB are responsible for reducing operation time (Wang et al. 2020a, b). The nanomaterials-based colorimetric sensors are listed in Table 11.2.

11.4.2.2 Surface-Enhanced Raman Spectroscopy (SERS)-Based Immunosensors

The Raman spectroscopy is another efficient tool, which can be utilized to monitor signals generated by the nanomaterials in response to immunointeractions. SERS (an improved version of Raman spectroscopy) is a highly sensitive, portable, non-destructive, and cost-effective method for detection of food contaminants (Guo et al. 2020a, b). Till now, SERS has been successfully tested for the detection of diverse food contaminants, e.g., pesticides (Xu et al. 2020), mycotoxins (Zhang et al. 2020), drug residues (Li et al. 2019; Fan et al. 2020), and pathogens (Chattopadhyay et al. 2019) (Table 11.2). It provides "fingerprint" information of the test sample through its molecular vibration even if present in very low concentration (Guo et al. 2020a, b). SERS is based on the Raman scattering where light is scattered due to inelastic collision between sample and incident light (optical sensor in food analysis) with enhanced signals (due to the presence of metal nano-rough surfaces) (Guo et al., 2020). The physical (e.g., electromagnetic field due to localized SPR; LSPR) and chemical characteristics (e.g., change in polarizability due to charge transfer) of incorporated metal structure (or nanostructure) are usually responsible for signal enhancement in SERB (Neng et al. 2020). Metal nanomaterials (especially gold and silver-based nanostructures) are most explored for the fabrication of SERS-based immunosensor due to their plasmonic resonance properties (Guo et al. 2020a, b). In SERS-based immunosensing, nanomaterials or nanomaterial and Raman reporter-labeled antibodies usually served as Raman nanoprobe to detect analyte. In a report on SERS-based competitive immunoassay, three mycotoxins aflatoxins, e.g., B1 (LOD = 0.061–0.066 μg/kg), ZEN (LOD = 0.53–0.57 μg/kg), and OTA (LOD = 0.26–0.29 μg/kg) were successfully detected in food samples (Li et al. 2018). In this sensor, Raman probe was developed by the attachment of AuNPs with anti-mycotoxin and Raman reporter, e.g.,

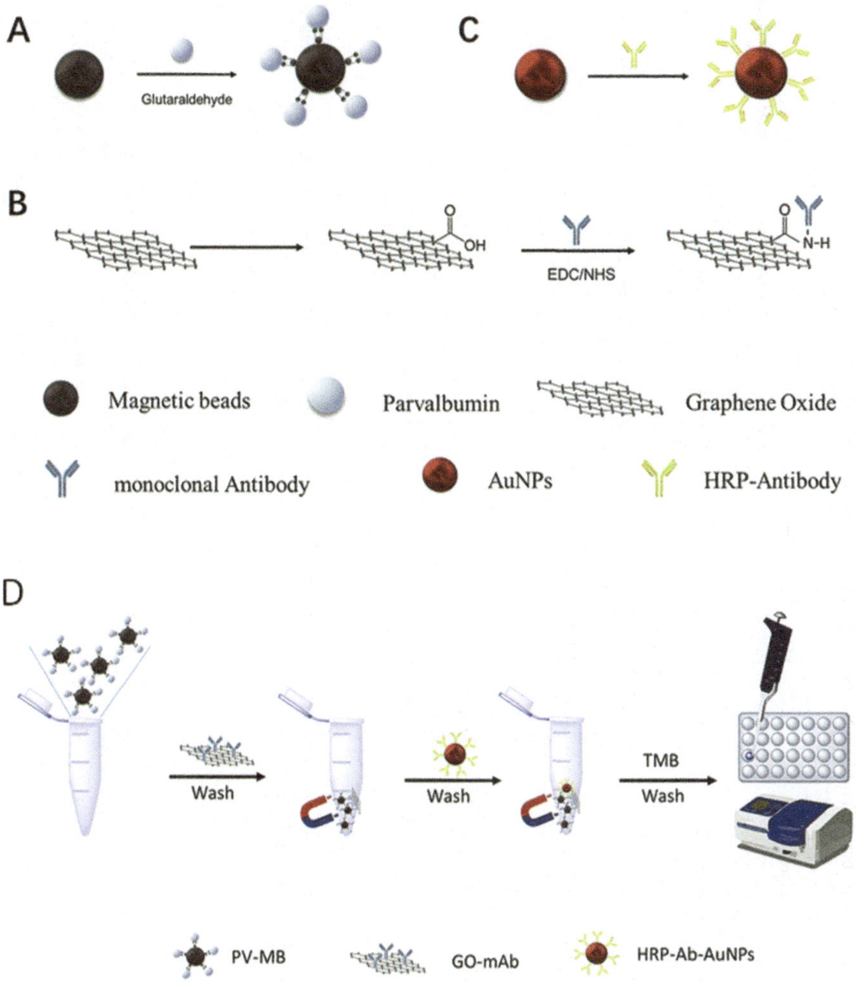

Fig. 11.8 Development and working of graphene and AuNP-based immunosensor. (A) Modification of magnetic beads with parvalbumin (PV-MB), (B) Modification of GO with mAb (anti-frog PV) (GO-mAb), (C) Modification of AuNP with HRP-antibody (HRP-Ab-AuNPs), and (D) schematic detection process. Note: here HRP represent anti-mouse IgG2a heavy chain antibody. Adapted with permission from Wang et al. (2020a, b)

5,5-dithiobis(succinimidyl-2-nitrobenzoate) (DSNB). The sample containing target mycotoxin competed with the capture substrate (immobilized on plate well) for this Raman nanoprobe (Li et al. 2018). Consequently, the corresponding Raman signal variations were analyzed with respect to the mycotoxin levels. Likewise, lateral flow immunochromatographic assay (LFIA) was developed for the SERS-based food analysis (Li et al. 2019). For example, an AuNPs SERS LFIA was developed for 0.12–10 ng/mL ng/mL of colistrin (veterinary antibiotic) (Li et al. 2019). This LFIA

Table 11.2 Applications of different nanomaterials in colorimetric, absorbance, and SERS-based immunosensors for food samples

No.	Antibodies used	Nanomaterial use	Analyte	Linear range	Linear range (ng/mL)	LOD	LOD (ng/mL)	Response time (min)	Sensor type	Nanomaterial role	Sample	Reference
A. Colorimetric and UV-vis-based immunosensors												
1	AuNP/Mice anti-CYP-OVA, goat anti-mouse IgG	AuNPs	Cyproheptadine (antibiotics)	0.16–0.85 ng/mL	0.16–0.85	5 ng/mL	5		Colorimetric-ICTS	Signal amplification	Pig urine	Guo et al. (2018)
2	Mouse anti-FB1, Anti-zearalenone, anti-OTA, Anti-AFB1	AuNS, AuNC, AuNFs, HAuPB	FB1	4–80 ng/mL	4–80	3.27 ng/mL	3.27		Colorimetric-ICTS	Signal generation	Corn	Wu et al. (2020)
			Zearalenone	0.8–40 ng/mL	0.8–40	0.70 ng/mL	0.70					
			OTA	0.2–2 ng/mL	0.2–2	0.10 ng/mL	0.10					
			Aflatoxin B1	0.1–1.25 ng/mL,	0.1–1.25	0.06 ng/mL	0.06					
3	MNP/Murine snti-CPAOZ, MNP/goat anti-mouse antibody	MNP	CPAOZ (antibiotic)			0.044 ng/mL	0.044		Colorimetric-ICTS	Signal generation and amplification	Spiked milk	(Yan et al. 2018)
4	MNP/anti-*E. coli* O157:H7 (capture antibodies) and *E. coli* O157:H7 (detection antibodies)	AuNP MNP	*Escherichia coli* O157:H7	5.0×10^1–5.0×10^4 CFU/mL	5.0×10^1–5.0×10^4 CFU/mL	50 CFU/mL	50 CFU/mL		Colorimetric-microfluidics	Signal generation and immunomagnetic target separation	Spiked Chicken samples	Zheng et al. (2019)
5	Mouse anti-Ara h 2 (capture antibody), HRP/	AuNPs	Ara h 2 (allergen)			1 ng/mL and 0.02 ng/mL	1 and 0.02		Colorimetric-ICTS and UV-vis-ELISA	Signal generation	Spiked sample of nestle fibre diet of corn	Peng et al. (2015)

	goat anti-mouse immunoglobulin G (indirect ELISA) and mouse anti-Ara h 2 (capture antibody), HRP/anti-Ara h 2 (detection antibody) (sandwich ELISA) and mouse anti-Ara h 2 (capture antibody), AuNP/anti-Ara h 2 (detection antibody)										milk for breakfast	
6	Anti-SEA (capture antibody) (rabbit) and AuNP/anti-SEA (detection antibody) (rabbit) HRP/anti-SEA antibody	AuNPs	SEA	10–500 ng/mL	10–500	1 ng/mL	1	90	Colorimetric and UV-vis spectroscopy	Signal generation	Spiked milk	Zhang et al. (2019)

(continued)

Table 11.2 (continued)

No.	Antibodies used	Nanomaterial use	Analyte	Linear range	Linear range (ng/mL)	LOD	LOD (ng/mL)	Response time (min)	Sensor type	Nanomaterial role	Sample	Reference
	(detection antibody)											
7	GOx/anti-mouse IgG (capture antibody) (goat) and Anti-AMD (detection antibody (mouse)	AuNPs	AMD	0–70 μM	0–10.587	0.51 nM	0.07714		Colorimetric-UV-vis spectroscopy	Signal amplification	Chicken sample	(Yu et al. 2018)
8	Anti-OTA	AuNPs	OTA	12.5–150 pg/mL	0.0125–0.150	150 pg/mL	0.15	65	Colorimetric-Plasmonic ELISA-UV	Signal amplification	Spiked corn sample	Liang et al. (2018)
9	Anti-AME (mice)	MNP, AuNP	Alternariol monomethyl ether mycotoxin)	0.08–0.48 ng/mL	0.08–0.48	0.16 ng/mL	0.16		Colorimetric-UV-vis spectroscopy	Immunomagnetic separation Signal amplification	Spiked cherry and orange sample	Man et al. (2018)
10	MNP/anti-*S. typhimurium* (capture antibody), Catalase/anti-*S. typhimurium* (detection antibody)	MNP, AuNRs	*Salmonella typhimurium*	10^1–10^5 CFU/mL	10^1–10^5 CFU/mL	35 CFU/mL	35 CFU/mL	180	Colorimetric-UV-vis spectroscopy	Immunomagnetic separation Signal amplification	Spiked chicken samples	Guo et al. (2020a, b)
11	MP/anti-*Salmonella* Typhimurium (capture	Pt@ZIF-8	*Salmonella typhimurium*	1–50 ng/mL	1–50 ng/mL	11 CFU/mL	11 CFU/mL	150	Colorimetric-UV-vis spectroscopy	Nanocatalysis	Spiked chicken	Wang et al. (2020)

	antibody) And Pt@ZIF-8/anti-*Salmonella* Typhimurium (detection antibody)											
12	Anti-AFB1 (capture antibody) and MOF/anti-AFB1 (detection antibody)	MOF (MIL-88)	Aflatoxin B1	0.01–20 ng/mL	0.01–20	0.009 ng/mL	0.009	60	UV-vis spectroscopy	Nanozyme	Spiked peanut milk and soy milk	Xu et al. (2021)
13	Anti-BLG (capture antibody) and PtNPs/biotinylated anti-BLG (detection antibody)	PtNPs	BLG Allergen	0.49–1.6×10^4 ng/mL	0.49–1.6×10^4	0.12 ng/mL	0.12		UV-vis spectroscopy	Nanocarrier	Milk	He et al. (2018a, b)
14	HRP-Labeled Anti-2,4-D	Nano-SiO_2	2,4-D (herbicide)	1–350 ng/mL	1–350	0.079 ng/mL	0.079		UV-vis spectroscopy	Nanocarrier	Mung bean sprouts and soybean sprout samples.	Wang et al. (2017a, b)
15	GO/Mouse anti-frog PV (primary antibody) AuNPs/Goat anti-mouse IgG2a heavy	GO, AuNPs, MNB	PV	1–50 ng/mL	1–50	4.29 ng/mL	4.29		UV-vis spectroscopy	Nanocarrier, signal amplification, immunomagnetic separation	Spiked fish	Wang et al. (2020a, b)

(continued)

Table 11.2 (continued)

No.	Antibodies used	Nanomaterial use	Analyte	Linear range	Linear range (ng/mL)	LOD	LOD (ng/mL)	Response time (min)	Sensor type	Nanomaterial role	Sample	Reference
	chain antibody (secondary antibody)											
B. SERS immunosensors												
1	AuNPs/DSNB/ anti-mycotoxin (anti-AFB1, anti-ZEA, or anti-OTA antibodies) (Raman probe) Goat anti-mouse IgG (capture antibody)	AuNPs	Aflatoxin B1	1–1000 pg/mL	0.001–1	0.061–0.066 μg/kg	0.061–0.066 μg/kg	60	Microarray-SERS	Signal enhancement	Corn, rice and wheat	Li et al. (2018)
			Zearalenone	10–10,000 pg/mL	0.01–10	0.53–0.57 μg/kg	0.53–0.57 μg/kg					
			OTA (mycotoxins)	5–5000 pg/mL	0.005–5	0.26–0.29 μg/kg	0.26–0.29 μg/kg					
2	Goat anti-mouse antibody (capture antibody) and AuNP/ DTNB/anti-colistin antibody (Raman probe)	AuNPs	Colistin (Antibiotic)	0.12–10 ng/mL	0.12–10	0.10 ng/mL	0.10		SERS-Lateral flow analysis	Signal enhancement	Milk	Li et al. (2019)
3	Goat anti-mouse antibody (capture antibody) and Au@AgNP/ MBA/anti-X	Au@Ag NP-	Aflatoxin B1	0.0014–0.33 ng/mL	0.0014–0.33	0.96 pg/mL	0.96×10^{-3}	20	SERS-Lateral flow analysis	Signal enhancement	Maize	Zhang et al. (2020)
			FB1	0.41–100 ng/mL	0.41–100	0.26 ng/mL	0.26					
			Zearalenone	0.015–3.7 ng/mL	0.015–3.7	6.2 ng/mL	6.2					
			Deoxynivalenol		0.14–33.3		0.11					

	antibody (Raman probe) X = aflatoxin B1, FB1, zearalenone, deoxynivalenol, OTA, T-2 toxin			0.14–33.3 ng/mL		0.11 ng/mL					
			OTA	0.027–6.7 ng/mL	0.027–6.7	15.7 pg/mL	0.0157				
			T-2 toxin (Mycotoxins)	0.014–3.3 ng/mL	0.014–3.3	8.6 pg/mL	0.0086				
4	Goat anti-mouse antibody (capture antibody), Au@Ag NP/DTNB/ anti-tetracycline antibody (Raman probe), Au@Ag NP/MBA/anti-penicillin antibody (Raman probe)	Au@AgNPs	Tetracycline	0.02–11 ng/mL	0.02–11	0.015 ng/mL	0.015	20	SERS-Lateral flow analysis	Signal enhancement	Milk	Fan et al. (2020)
			Penicillin (antibiotics)	0.02–11 ng/mL	0.02–11	0.010 ng/mL	0.010					
5	AuNF@Ag–MBA/anti-BB antibody (Raman probe) and Goat anti-rabbit IgG (detector antibody)	AuNF@Ag	Brombuterol			0.5 pg/mL	0.5×10^{-3}	15	SERS-Lateral flow analysis	Signal enhancement	Swine meat and urine samples.	Fu et al. (2017)
6	MNB/anti-OTA antibody	MNB, AuNPs	OTA	1 pg/mL to1000 pg/mL	0.001–1	0.61 pg/mL	0.61×10^{-3}		SERS	Surface support and signal enhancement	Wine	Ding et al. (2020)
7	AuNP/CSA-1-antibody/DSNB	FPMNP, AuNPs	*Salmonella typhimurium*	$10–10^7$ cells/mL	$10–10^7$ cells/mL	10 cell/mL	10 cell/mL		SERS	Signal enhancement,	Cottage cheese	Chattopadhyay et al. (2019)

(continued)

Table 11.2 (continued)

No.	Antibodies used	Nanomaterial use	Analyte	Linear range	Linear range (ng/mL)	LOD	LOD (ng/mL)	Response time (min)	Sensor type	Nanomaterial role	Sample	Reference
	(Raman probe) FPMNP/CSA-1-Antibody (capture antibody)									magnetic separation	(Indian paneer), egg white, and canned mixed fruit juice	
8	MNP-anti 2,4-D Antibody	HAu@AgNFs, MNPs	2,4-D (pesticide)	0.001–100 μg/mL	$1–1 \times 10^5$	0.11 ng/mL	0.11		SERS	Signal enhancement, Surface support	Tea and milk	Xu et al, (2020a, b)

Abbreviations: *CYP-OVA* Cyproheptadine hydrochloride-ovalbumin antigen, *FB1* Fumonisin B1, *AFB1* Aflatoxin B1, *OTA* Ochratoxin A, *AuNFs* Gold nanoflowers (blue), *AuNS* Gold nanospheres (red), *HAuPB* Hyperbranched Au plasmonic blackbodies (black), *AuNC* Gold nanocacti (purple), *CPAOZ* 3-[(4-Carboxyphenyl) monomethyl] amino-2-oxazolidinone, *SEA* Staphylococcal enterotoxin A, *ICTS* Immunochromatographic test strip, *AME* Alternariol monomethyl ether, *S. typhimurium Salmonella* Typhimurium, *gtAm-GOx* Glucose oxidase-labeled goat anti-mouse IgG, *AMD* Amantadine, *BLG* β-lactoglobulin, *2,4-D* 2,4-Dichlorophenoxyacetic acid, *MOF* Metal organic framework, *PtNPs* Platinum nanoparticles, *PV* Parvalbumin, *GOx* Glucose oxidase, 8 *Pt@ZIF-8*; Platinum loaded zeolitic imidazolate framework, *MP* Magnetic particles, *HAu@AgNFs* Au@Ag bimetallic nanoflowers, *CSA* Common structure antigen, *FPMNP* Functionalized polymer magnetic nanoparticles, *MNP* Magnetic nanoparticles, *AuNF@Ag* Flower-like gold-silver core-shell bimetallic nanoparticles, *DTNB* 5,5-Dithiobis-2-nitrobenzoic acid, *DSNB* 5,50-Dithiobis(succinimidyl-2-nitrobenzoate), *MBA* 4-Mercaptobenzoic acid

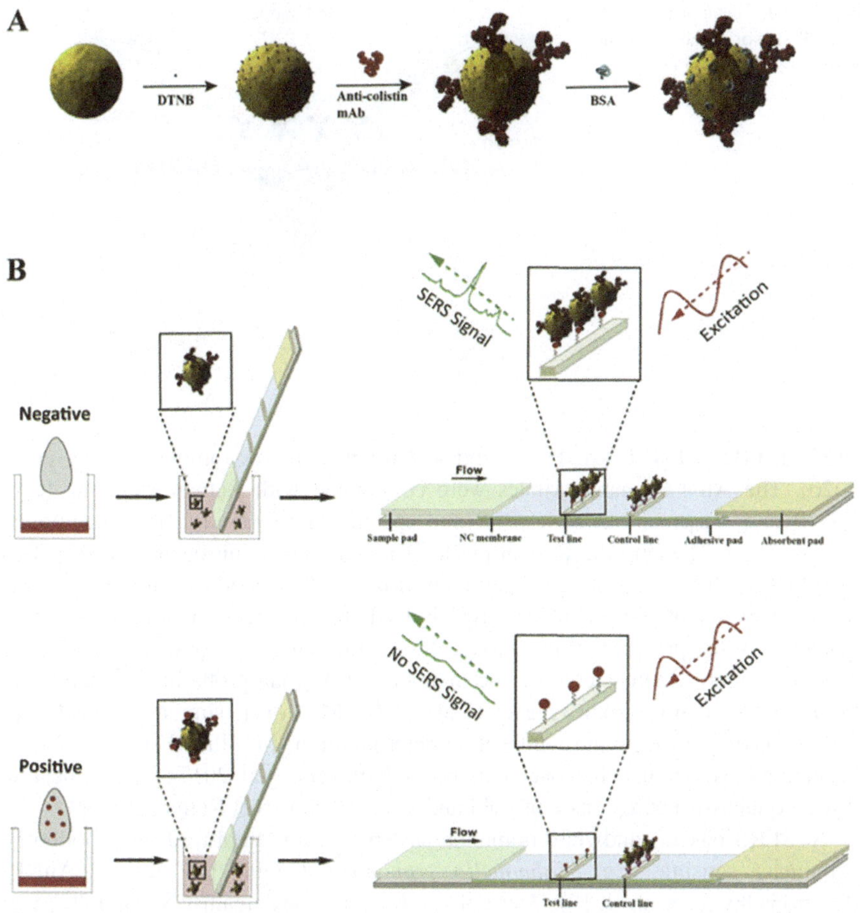

Fig. 11.9 Sensing of colistrin with DTNB-based Raman probe. (**a**) synthesis of Raman probe (**b**) Illustration of signal generation on LFIA strip in the absence/presence of colistrin. Adapted with permission from Li et al. (2019)

sensor was developed on nitrocellulose paper and LFIA strip is consisted of test line and a control line. A capture antigen and goat anti-mouse antibody was immobilized on test and control line, respectively. 5,5-dithiobis-2-nitrobenzoic acid (DTNB) (another Raman reporter) conjugated anti-colistin mAb was labeled with AuNPs and was used as Raman probe. Further, the target evaluation was performed by following competitive immunoassay where Raman probe was added in the microtiter plate. The Raman probe is attached to LFIA strip and generates Raman signals only when the test sample was devoid of colistrin (Fig. 11.9). The sensor could detect colistrin with LOD 0.10 ng/mL (Li et al. 2019).

Similarly, an Au@Ag core shell nanoparticles-based SERS LFIA was constructed to simultaneously sense six different mycotoxins, i.e., aflatoxin B1

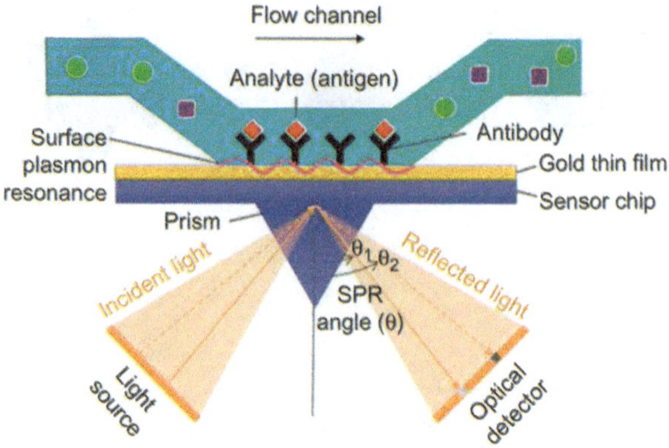

Fig. 11.10 Basic principle of an SPR sensor. Adapted with permission from Yu et al. (2015)

(AFB1), FB1, ZEN, OTA, DON, and T-2 toxin in maize samples (Zhang et al. 2020). The Au@Ag nanoparticles were conjugated with Raman reporters, (e.g., DTNB and 4-mercaptobenzoic acid) and specific antibodies for abovementioned mycotoxins to develop the Raman probe. The test line on nitrocellulose strip was modified by mycotoxin antigens and control line with secondary antibody. Performance analysis of nitrocellulose strip showed that the Raman probes peak was quenched in the presence of mycotoxin due to hindrance in its interaction with the capture antigen, whereas control line interact with Raman probe independently for the presence of mycotoxin (Zhang et al. 2020). Moreover, similar approach was exploited for penicillin and tetracycline determination (in milk) (Fan et al. 2020), brombuterol (BB) (in swine meat and urine sample) (Fu et al. 2017) using bimetallic flowered nanostructures made of gold and silver to construct Raman probes.

A SERS-based sandwich immunoassay was also employed for foodborne pathogens sensing, e.g., *Salmonella typhimurium* with the aid of AuNPs (Chattopadhyay et al. 2019). The CSA-1-Ab (antibody against *Salmonella*) was immobilized on AuNPs conjugated DSNB. Further, functionalized polymer magnetic nanoparticle-assisted CSA-1-Ab as capture probe. The target was sandwiched between capture probe and Raman probe and an immunocomplex was obtained. The presence of AuNPs was speculated for the enhancement in the Raman signals, while magnetic core facilitated facile analyte separation. This Raman probe was efficient enough to detect *Salmonella typhimurium* lower down to 10 cells/mL (Chattopadhyay et al. 2019). Magnetic structures have also assisted SERS-based immunoassay for pesticide (Xu et al. 2020) and mycotoxin detection (Ding et al. 2020).

SPR is the collection of oscillations generated by conduction band electrons of metals (Li and Zhang 2017). Basically, SPR is a physical phenomenon in which incident light photons excite the metal (mainly gold) surface electrons, which cause resonant oscillation of conduction electrons (Zhu and Gao 2019). In brief, when incident monochromatic beam falls onto the metal surface, it causes generation of evanescent waves. These waves interact with the plasmons (free electrons) of metal surface at a special angle of incidence (SPR angle) (Fig. 11.10) (Poltronieri et al.

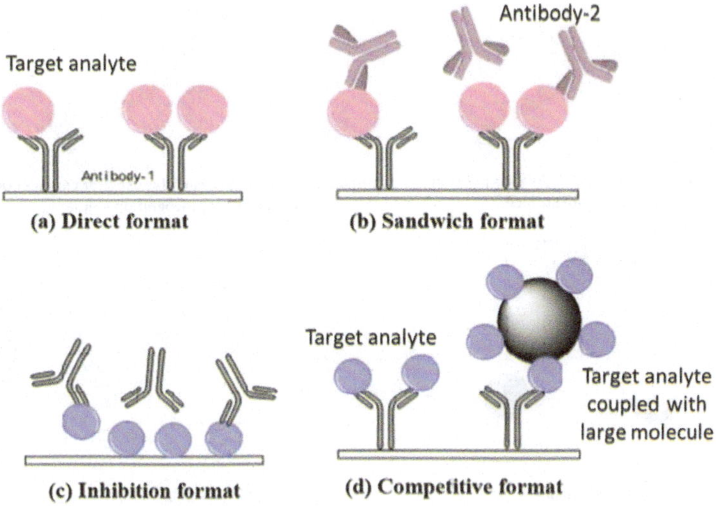

Fig. 11.11 Detection formats used in SPR immunosensing. Adapted with permission from Xu et al. (2010)

2014; Yu et al. 2015; Radhakrishnan and Poltronieri 2017). This SPR angle is highly sensitive for any change occurring at the metal-dielectric interface (Radhakrishnan and Poltronieri 2017). Thus, change in the SPR angle can be correlated with the amount of analyte reacted with the metal surface. In case of SPR immunosensor, the metal surface can be modified with the antibodies to sense specific antigen.

In general, SPR immunosensors is considered as a non-destructive optical analysis technique, which works on the measurement of change in SPR angle after interaction of thin-layered biomolecules, particularly Ag–Ab, on the sensor metal chip surface (Xu et al. 2010; Yu et al. 2015). The change in refractive index (SPR angle shift) can be used to calculate the change in mass of surface film due to formation of Ag–Ab complex on sensor surface, which can be used further to quantify the levels of target analyte (Karunakaran et al. 2015; Yu et al. 2015; Radhakrishnan and Poltronieri 2017). The change in angle is analyzed as a plot of resonance signal (proportional to change in mass) versus time (Xu et al. 2010; Yu et al. 2015).

As in other immunoassays, SPR-based immunosensing can be performed by (1) direct, (2) competitive, (3) sandwich, and (4) inhibition immunoassays (as shown in Fig. 11.11) (Xu et al. 2010; Yu et al. 2015; Vezocnik et al. 2017). Briefly, in direct format detection, the biorecognition element (such as antibody) is immobilized onto the SPR sensor surface (such as metal film) and target analyte in solution is allowed to flow as shown in Fig. 11.11a. Thus, the formed complex produces a change in refractive index; the alteration in the refractive index can be measured with SPR sensor to know the levels of target analyte. In sandwich type format (Fig. 11.11b), an additional second antibody is used to bind for the capturing of target analyte. This improves the specificity and LOD of the immunosensor

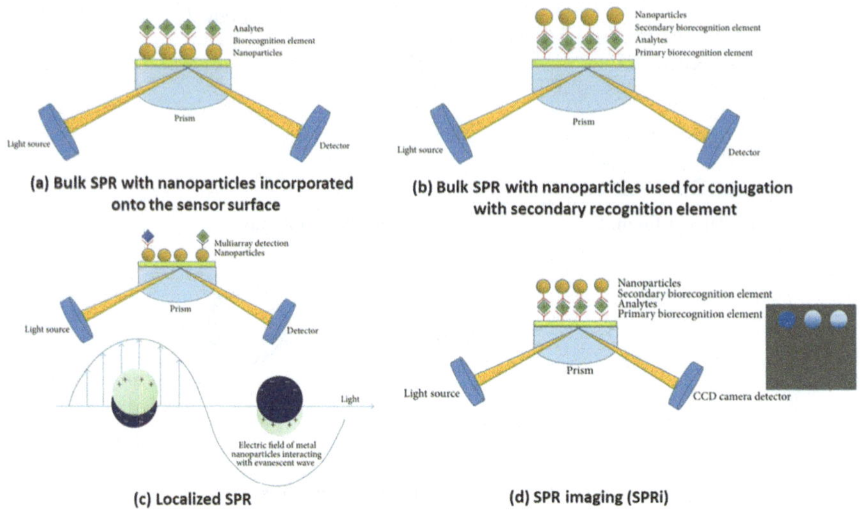

Fig. 11.12 Configurations of LSPR immunosensors. Adapted from Antiochia et al. (2016)

(Xu et al. 2010). Inhibition and competitive type immunoassays are usually used for small analytes (e.g., with molecular weight <5 kDa). The general mechanism for inhibition and competitive SPR immunoassays is shown in Fig. 11.11c and d. In both inhibition and competitive type of format, the sensing signals usually decreases with increasing analyte levels (Xu et al. 2010; Yu et al. 2015).

In SPR sensors, metal surfaces (e.g., gold) has been employed to generate the sensing signals. The sensing capability of SPR immunosensors has been further enhanced with the introduction of nanomaterials (e.g., metal nanoparticles, in particular AuNPs and AgNPs, CNTs, and MNBs (Pollet et al. 2011; Antiochia et al. 2016; Zhu and Gao 2019). In general, nanomaterials-based SPR immunosensors can be explored in three different configurations: (1) bulk SPR, (2) LSPR, and (3) SPR imaging (Fig. 11.12) (Antiochia et al. 2016).

1. *Bulk SPR*: In this case, the sensing of analyte molecules can be achieved by observing the small changes in the refractive index of a thin metal film due to conjugation of analyte molecule with transducer. The metal nanomaterials are employed to enhance the bulk SPR sensitivity, thus lowering the LOD. These can be arranged either to modify the sensing surface (e.g., metal surface) and/or secondary biorecognition element (e.g., secondary antibodies) to be used in sandwich immunoassays as shown in Fig. 11.12a and b (Antiochia et al. 2016).
2. *LSPR*: When charge density oscillations are confined to metal nanostructures, the LSPR phenomenon takes place as shown in Fig. 11.12c. LSPR not only deals with light scattering, but also absorbance, thus resulting in local electromagnetic field. This absorbance depends on type of nanomaterial and its morphology. Thus, sensitivity and selectivity could be improved depending on these factors (Antiochia et al. 2016).

3. *SPR imaging (SPRi)*: It is a combined phenomenon of SPR and spatially resolved measurement in which two-dimensional charge-couple device (CCD) detector is employed. The output of SPRi is an image consisting of spots indicating the light intensities when different concentration of analyte interacts with biorecognition element (Fig. 11.12d) (Antiochia et al. 2016).

The SPR immunosenors, being very sensitive and selective, have been utilized for sensing applications in diverse fields, e.g., biomedical, food, and environment (Ricci et al. 2007). In case of food analysis, the SPR immunosensors are employed for analysis of microorganisms, toxins, and allergens (Table 11.3) (Ricci et al. 2007; Mustafa and Andreescu 2018). For example, LSPR-based immunosensor was designed for label-free casein detection in the cow milk by employing gold-coated silica nanoparticles modified sensor surface (Hiep et al. 2007). (Note that casein is one of the abundant proteins found in the cow milk and can cause allergy in younger children). The biorecognition element for casein, for example, Ara h1antibody (anti-casein Ab) was attached onto the sensor surface via 4,4″-dithiodibutyric acid (DDA) and N-hydroxysuccinimide (NHS) linkers. This affinity SPR immunosensor exhibited the ability to detect 10^5–10^7 ng/mL casein (Hiep et al. 2007). In another SPR immunosensor, MNBs were used to analyze arachis hypogaea (Ara h1) (a peanut butter allergen). The magnetic beads were functionalized with polyclonal Ara h1 antibodies via carbodiimide reaction to give sandwich type immunoassay with 0.09×10^3 ng/mL LOD (Pollet et al. 2011).

Likewise, the nanomaterials are also capable of enhancing the SPR immunosensing of various food toxins (e.g., mycotoxins) (Xu et al. 2013; Hu et al. 2014). For instance, AuNPs-based competitive SPR immunoassay was used for the analysis of ochratoxins (a mycotoxins produced by fungi) (Yuan et al. 2009). This involvement of AuNPs was found to show LOD values of 0.3 and 0.5 ng/g in cereals (oats and corns) and beverages (wine and juices), respectively (Yuan et al. 2009). The use of AuNRs for the fabrication of competitive type SPR immunosensor was found excellent to detect AFB1 (Xu et al. 2013). The AuNRs-based sensor worked efficiently in the range of 0.5–20 ng/mL with 0.16 ng/mL LOD (Xu et al. 2013). Fascinatingly, the AuNP-enhanced SPRi strategy was efficient enough to detect multiple toxins present in the food samples, e.g., AFB1, OTA, and ZEN using competitive type immunoassay (Hu et al. 2014). The sensing surface of this microarray-type SPR sensor was fabricated by the uniform attachment of mycotoxin antigens on modified SPRi gold chip. The detection of mycotoxins (AFB1, OTA, and ZEN) was performed through competitive immunoassay by adding respective antibodies. After competitive binding, AuNPs-conjugated secondary antibody interacted and bound on the captured mAbs to further amplify the SPRi signal. The designed sensor achieved an LOD of 8×10^{-2}, 3×10^{-2}, and 1.5×10^{-2} ng/mL for AFB1, OTA, and ZEN, respectively (Hu et al. 2014).

As mentioned earlier, the nanomaterials-based SPR immunosensors are also applicable for the sensing of microorganisms in food. In a report on the SPR immunosensors, antibody functionalized carboxyl-modified Fe_3O_4 magnetic nanoparticles were found to be highly sensitive and selective detection of *Salmonella*

Table 11.3 Performance of nanomaterials-based SPR immunosensors in food sample analysis

No.	Antibodies used	Analyte	LOD	LOD (ng/mL)	Sensing Range	Sensing Range (ng/mL)	Response time (min)	Sensor type	Nanomaterials used	Nanomaterial role	Food/ sample tested	Reference
1.	Anti-casein	Casein	10 ng/mL	10	0.1–10 mg/mL	10^5–10^7		LSPR	Gold-capped silica NPs	To immobilize antibody for signal enhancement and better performance	Milk	Hiep et al. (2007)
2.	Anti-BZ	BZ	7 pg/mL	0.7×10^{-2}	0.1–80 ng/mL	0.1–80	4	SPR; Competitive type	AuNPs	Used in conjugation with secondary antibody to enhance SPR signals	Beverage	Gobi et al. (2008)
3.	Mouse anti-ochratoxin A mAb	OTA	0.41 ± 0.043 ng/mL	0.41 ± 0.043	0.058–0.4 ng/mL	0.058–0.4	<10	SPR; competitive type	AuNPs	Used on sensor surface for signal enhancement	White wine	Yuan et al. (2009)
			0.33 ± 0.028 ng/mL	0.33 ± 0.028							Red wine	
			0.058 ng/mL	0.058							Grape juice	
			0.084 ng/mL	0.084							Apple juice	
			0.33 ± 0.034	0.33 ± 0.034							Oats	
			0.5 ± 0.022 ng/mL	0.5 ± 0.022							Corn	
4.	Ara h1	Ara h1	0.09 µg/mL	0.09×10^3	10^2–2×10^3_ng/mL	10^2–2×10^3		SPR; sandwich	MNBs	Linker to antibody	Peanut butter, chocolate candy bar	Pollet et al. (2011)

5.	Anti-AFB1	AFB1	0.16 ng/mL	0.16	0.5–20 ng/mL	0.5–20		SPR; competitive type	AuNRs	To conjugate with competitive analyte	Peanut sample	(Xu et al. 2013)
6.	Primary: Anti-AFB1 Secondary: Anti-mouse IgG	AFB1	8 pg/mL	8×10^{-2}				SPRi	AuNPs	Used in conjugation with secondary antibody to enhance SPRi signals	Peanut extract	Hu et al. (2014)
7.	Primary: Anti-OTA Secondary: Anti-Mouse IgG	OTA	30 pg/mL	3×10^{-2}				SPRi	AuNPs	Used in conjugation with secondary antibody to enhance SPRi signals	Peanut extract	Hu et al. (2014)
8.	Primary: Anti-ZEN Secondary: Anti-Mouse IgG	ZEN	15 pg/mL	1.5×10^{-2}				SPRi	AuNPs	Used in conjugation with secondary antibody to enhance SPRi signals	Peanut extract	Hu et al. (2014)
9.	Anti-Salmonella polyclonal antibody (PAb)	*Salmonella enteritidis*	14 cfu/mL	14 cfu/mL				SPR	Magnetic nanoparticles	Used for functionalization of antibody	Egg shell	Liu et al. (2016)

(continued)

Table 11.3 (continued)

No.	Antibodies used	Analyte	LOD	LOD (ng/mL)	Sensing Range	Sensing Range (ng/mL)	Response time (min)	Sensor type	Nanomaterials used	Nanomaterial role	Food/ sample tested	Reference
10.	Anti-*E.coli*, b-anti-*E. coli*, anti-*Salmonella*, b-anti-*Salmonella*	*Escherichia coli* O157: H7 and *Salmonella* sp.	*E. coli*: 17 cfu/ mL and *Salmonella*: 11.7 cfu/mL					SPR	Streptavidin-AuNPs	Signal enhancement	hamburger	Vaisocherová-Lísalová et al. (2016)
			E. coli: 57 cfu/ mL and *Salmonella:* 7.4 cfu/mL								Cucumber	
11.	Magainin I	*Escherichia coli* O157: H7	5.0×10^2 cfu/ mL					Fiber optic SPR	AgNPs and rGO	Used on sensor surface for signal enhancement	Fruit and vegetable juice	Zhou et al. (2018a, b)

Abbreviations: *BZ* Benzaldehyde, *LSPR* Localized surface plasmon resonance, *AuNPs* Gold nanoparticles, *OTA* Ochratoxin A, *mAb* Monoclonal antibody, *AFB1* Aflatoxin B1, *ZEN* Zearalenone, *AgNPs* Silver nanoparticles, *rGO* Reduced graphene oxide *enteritidis* in egg shell (Liu et al. 2016). This magnetic nanoparticles-based sensor is tested on 1.4×10^1–1.4×10^9 cfu/mL of *S. enteritidis* and exhibited 14 cfu/ mL LOD, while in the absence of nanoparticles, the LOD was 1.4×10^4 cfu/mL for *S. enteritidis* (Liu et al. 2016). Likewise, poly(carboxybetaine acrylamide) was employed for the fabrication of SPR immunosensor to examine *E. coli* and *Salmonella* (Vaisocherová-Lísalová et al. 2016). This assay worked in three steps: (1) capturing of the microbial species (from the test samples), (2) addition and capturing of biotinylated secondary antibody (specific for the target analyte), and (3) signal enhancement with streptavidin-coated AuNPs. Streptavidin, being a biotin-binding protein, interacts with biotinylated secondary antibody. This sensor was evaluated by sensing of bacterial strains in cucumber and hamburger. The aforementioned sensor exhibited an LOD of 57 and 17 cfu/mL for *E. coli* whereas 7.4×10^3 and 11.7×10^3 cfu/mL for *Salmonella sp.* in cucumber and hamburger, respectively (Vaisocherová-Lísalová et al. 2016).

The modification of SPR immunosensor with optical fibers has also been explored to improve the applicability of these sensors in food analysis. The fiber optic SPR (FOSPR) sensor can show LOD of 5.0×10^2 cfu/mL for *E. coli* (Zhou et al. 2018a, b). To develop FOSPR, antimicrobial peptides, e.g., Magainin I and AgNPs-rGO were used as a biorecognition elements and signal amplifier, respectively. The developed sensor is assured to have high selectivity, efficiency, reproducibility, and sensitivity (Zhou et al. 2018a, b).

Not only this, SPR-based nano-immunosensors were also designed for rapid analysis of food characteristics such as flavor and fragrant. For example, a highly regenerable polyethylene glycol (PEG)-dialkanethiols and AuNPs-based SPR immunosensor was fabricated by Gobi et al. (2008) for trace level detection of benzaldehyde (BZ). (Note that BZ is a characteristic fragrant compound of peach). This sensor was based on competitive SPR immunosensing, where BZ analog was covalently bound to PEG monolayer sensor chip. The modified sensor chip was treated with solution mixture containing sample and fixed amount of BZ-Ab. The free BZ (present in sample) and its analog compete to bind with BZ-Ab. Thus, more the amount of BZ in the sample, more the formation of antigen (BZ)-antibody (BZ-Ab) complex and thus lower the sensor response. The sensor exhibited a good sensitivity for BZ and can be effective for the sensing of 0.1–80 ng/mL of BZ. It is worth mentioning that the use of secondary antibody functionalized AuNPs improved the strength of sensing signals by 15–18 folds, which was reflected in exceptionally low LOD value of 0.7×10^{-2} ng/mL for BZ detection (Gobi et al. 2008).

Overall, the use of nanomaterials in SPR immunosensing was found to show several fold enhancements in the sensing parameters. The signal enhancement due to metal nanoparticles occurs due to contribution of LSPR of metal nanoparticles (Wang et al. 2010). The key sensing characteristics of nanomaterials-based SPR immunosensors is shown in Table 11.3.

11.4.2.3 Fluorescent-Based Immunosensor

Another convenient way to immunosense food contaminant is to rely on the fluorescent nanomaterials. The fluorescent-based immunosensors work on principle of recognizing an analyte molecule through fluorescently active molecules (Kłos-Witkowska 2016). Accordingly, the detection of the analyte can be performed on the basis of variations in fluorescent intensity.

In food analysis, the nanomaterials-based fluorescent immunosensors mainly tested for the determination of food-borne pathogens (Table 11.4). To test the applicability of fluorescent mercaptopropionic acid-capped CdTe QDs (MPA-CdTe QDs) in immunosensing, E. coli O157:H7 was analyzed using H_2O_2-sensitive fluorescence properties of MPA-CdTe QDs (Chen et al. 2016). The immuosensing of bacteria via MPA-CdTe QDs was performed by employing rabbit anti-E. coli O157:H7 polyclonal antibody as the capture antibody. In addition, the biotinylated mouse anti-E. coli O157:H7 mAb was used as detection antibody (Fig. 11.13). The mAb was connected with enzyme CAT via streptavidin-biotin system. In the absence of E. coli O157:H7 in test sample, the MPA-CdTe QDs fluorescence was quenched due to H_2O_2. On the contrary, in the presence of target bacteria, CAT bound to the bacterial surface consumed H_2O_2 present in the solution, which led to no decrease in the fluorescence of MPA-CdTe QDs. It is worth mentioning that the change in the fluorescence signals of MPA-CdTe QDs was used to measure E. coli O157:H7 concentrations. This CAT-mediated fluorescence immunoassay exhibited a high sensitivity with an LOD of 5×10^2 CFU/mL for E. coli O157:H7. This test was also effective to analyze E. coli O157:H7 spiked milk samples. Moreover, this fluorescence-based immunoassay was ~140 times more efficient when compared with HRP-based colorimetric immunoassay.

A H_2O_2-induced decrease in fluorescence was also used to sense ochratoxin A (OTA) by MPA-CdTe QDs (Huang et al. 2016). The OTA was conjugated with CAT; CAT acts as competitive antigen against mouse anti-OTA mAb, which has relatively high affinity for OTA. After the addition of OTA-CAT, it is attached with mAb and the immobilized CAT consumed the available H_2O_2 (Huang et al. 2016). The absence (or decreased levels) of H_2O_2 led to no or less quenching of QDs fluorescence signals (Huang et al. 2016). In contrast, in the absence of OTA-CAT, the H_2O_2 is free to quench the fluorescence of QDs. This CdTe QDs-based method can work efficiently for the detection of 0.05–10 pg/mL OTA with LOD of 0.05 pg/mL. It is worth to mention that the LOD value of CdTe QDs-based method was ~300-times better than HRP-based conventional ELISA.

Similarly, a cow milk allergen called bovine β-lactoglobulin (BLG) (125–4000 and 0.48–62.5 ng/mL) was detected by fluorescent sandwich ELISA (sELISA) with fluorescence of thiolated CdTe QDs (He et al. 2018a, b). This sELISA rely on mAb1G9 BLG-specific IgE and antigen interaction. As discussed earlier, the sensing signals were generated on the basis of CAT-mediated fluorescence extinction thiolated CdTe QDs by H_2O_2. This QDs and CAT-based sELISA method exhibited an LOD value of 0.49 ng/mL (16-fold better LOD in comparison to HRP-based sELISA). In another study on CdSe/ZnS QDs-based fluorescence probe,

Table 11.4 Performance of nanomaterials-based fluorescent immunosensors in food sample analysis

No.	Analyte	Immunogenic molecule	Fluorophore	Nanomaterials Used	Linear range	Linear range (ng/mL)	LOD	LOD (ng/mL)	Sample	Response time (min)	Reference
1.	BLG	mAb1G9; specific to the IgE linear epitope for BLG	Thiolated CdTe QDs	CdTe QDs	125–4000 ng/mL and 0.48–62.5 ng/mL	125–4000 and 0.48–62.5	0.49 ng/mL	0.49	Cow milk		He et al. (2018a, b)
2.	α-Lactalbumin	mAbs	mAbs-CdSe/ZnS QDs	CdSe/ZnS QDs	0.1–1000 ng/mL	0.1–1000	0.1 ng/mL	0.1	Dairy products		Yang et al. (2014)
3.	OTA	Mouse anti-OTA mAb	H_2O_2-induced fluorescence quenching	MPA-modified CdTe QDs	0.05–10 pg/mL	5×10^{-5} – 10^{-2}	0.05 pg/mL	5×10^{-5}	Rice, wheat, and corn	15	Huang et al. (2016)
4.	OTA	OTA-specific aptamer	Fluorescein-labeled aptamer	AuNPs	25–300 nM (78–94 ng/mL)	78–94	22.7 nM (7 ng/mL)	7	Corn flour, beer, and coffee	30	Lv et al. (2017)
5.	OTA	OTA-specific aptamer	GQDs-aptamer bioconjugates	GQDs	0–1 ng/mL	0–1	13 pg/mL	13×10^{-3}	Red wine	50	Wang et al. (2017a, b)
6.	AFB1	Aptamers against AFB1	Aptamer-conjugated QDs	AuNPs	10–400 nM (30–125 ng/mL)	30–125	3.4 nM (1 ng/ml)	1	Rice and peanut samples	15	Sabet et al. (2017)
7.	*Escherichia coli* O157:H7	A rabbit anti-E. coli O157:H7 pAb and biotinylated mouse anti-E. coli O157:H7 biotin@mAb	CdTe QDs	CdTe QDs	1.18×10^3–1.18×10^6 CFU/mL		5×10^2CFU/mL		Milk	15	Chen et al. (2016)
8.	*V. parahaemolyticus*	IgY (extracted from the egg yolks of immunized hens)	CdSe/ZnS QDs	AuNPs	10–10^5 CFU/mL		10 CFU/ml		Shrimp, ground beef, and salted vegetables		Liu et al. (2017)
9.	*P. acidilactici* CFR K7	Bacterial cell wall	PD–AuNPs composite	PD–AuNPs composite	10^2–10^6 CFU/mL		10^2 CFU/mL		Water samples		Sahoo et al. (2012)

(continued)

Table 11.4 (continued)

No.	Analyte	Immunogenic molecule	Fluorophore	Nanomaterials Used	Linear range	Linear range (ng/mL)	LOD	LOD (ng/mL)	Sample	Response time (min)	Reference
10.	*P. aeruginosa*	Aptamers against *P. aeruginosa*	5-carboxyfluorescein-labeled complementary DNA (FAM-cDNA)	GOQDs	1.28×10^3–2.00×10^7 CFU/mL		100 CFU/mL		Drinking water, orange juice, and popsicle samples	120	Gao et al. (2018)

Abbreviation: *BLG* Bovine *β*-lactoglobulin, *QDs* Quantum dots, *mAbs* Monoclonal antibodies, *OTA* Ochratoxin A, *GQDs* Graphene quantum dots, *AFB1* Aflatoxin B1, *pAb* Polyclonal antibody, *mAb* Monoclonal antibody, *AuNPs* Gold nanoparticles, *GOQDs* Graphene oxide quantum dots, *MPA* Mercaptopropionic acid

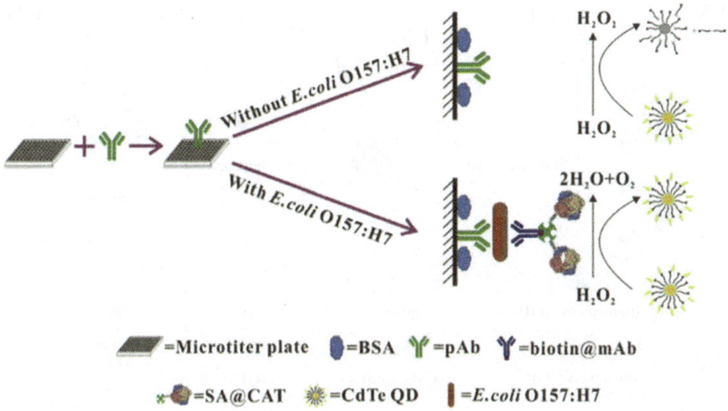

Fig. 11.13 Schematic of H_2O_2-sensitive MPA-CdTe QDs-based immunoassay for the of E. coli O157:H7 sensing. The bacteria captured by the sensing probe restricted the quenching of QDs by consuming the H_2O_2. On the basis of alteration in the fluorescence signals, the levels of E. coli O157:H7 were monitored. Adapted with permission from Chen et al. (2016)

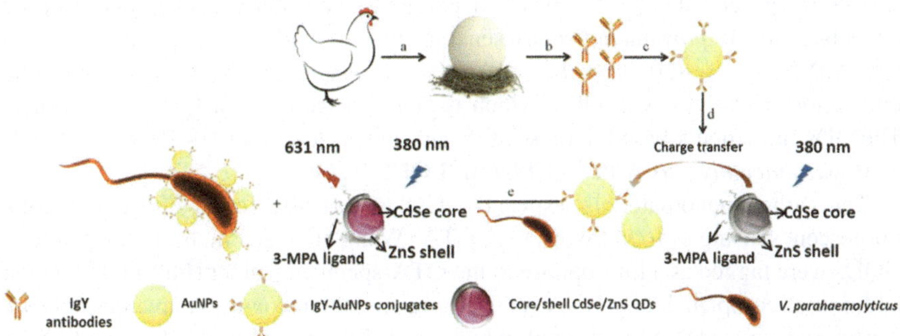

Fig. 11.14 Representation of steps followed for *V. parahaemolyticus* detection *by* AuNPs and CdSe/ZnS QDs-based immunosensor. (**a**) Immunization of specific pathogen-free (SPF) hen to generate anti-*V. parahaemolyticus*, (**b**) Extraction and purification of antibodies, (c) Conjugation of IgY antibodies on the surface of AuNPs, (d) fluorescence quenching of CdSe/ZnS QD upon interaction with AuNPs-IgY, and (**e**) Aggregation of AuNP-IgY in the presence of *V. parahaemolyticus* and restoration of fluorescence signal to achieve sensing of *V. parahaemolyticus*. Adapted with permission from (Liu et al. 2017)

0.1–1000 ng/mL of bovine α-lactalbumin (α -La) was examined successfully in dairy products with LOD value of 0.1 ng/mL (Yang et al. 2014).

The combination of two types of nanomaterials (e.g., AuNPs and QDs) are also impressive for the development of immunosensor for common pathogen *Vibrio parahaemolyticus* (*V. parahaemolyticus*) (Liu et al. 2017). In this sensor, the *V. parahaemolyticus*-specific antibody (IgY) immobilized AuNPs was used as a fluorescence quencher for CdSe/ZnS QDs (Fig. 11.14) (Liu et al. 2017). Note that the IgY was obtained from the egg of immunized hens and extracted IgY was

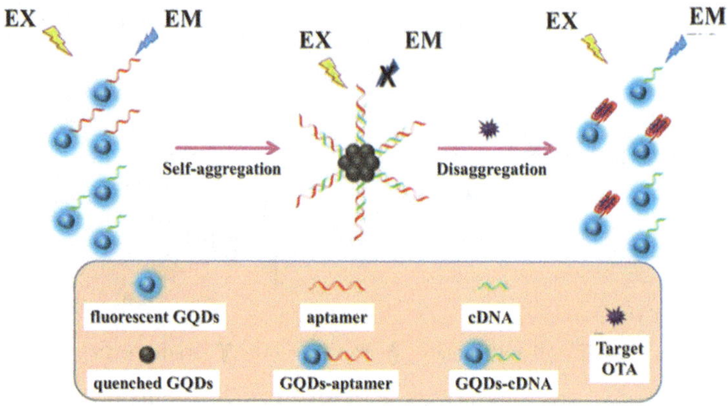

Fig. 11.15 Sensing mechanism of GQDs-based fluorescence aptasensor for OTA. Adapted with permission from Wang et al. (2017a, b)

attached on AuNPs through electrostatic self-assembly (Fig. 11.14). The sensing signals were generated on the basis of charge-transfer derived QDs fluorescence quenching for *V. parahaemolyticus* sensing. In this work, a quick aggregation of IgY-AuNPs was observed in the presence of *V. parahaemolyticus* and halted the interaction of AuNPs with QDs, which resulted in recovery of QDs fluorescence. The QDs and AuNPs-based fluorescence sensing was tested for 10–10,000 CFU/mL of *V. parahaemolyticus* with 10 CFU/mL LOD.

The carbon nanomaterials, especially GQDs was also found effective for the fluorescent sensing of food toxins, e.g., OTA (Wang et al. 2017a, b). In this context, GQDs were tagged as a fluorophore to the OTA-specific aptamer (Fig. 11.15). Upon addition of complementary DNA (cDNA), against the recognition aptamer, a DNA duplex (aptamer-cDNA) is formed, which caused aggregation of GQDs and hence quenching in the fluorescence signals of GQDs (Wang et al. 2017a, b). Further, the addition of OTA (specific for aptamer) caused unwinding of DNA duplex and formation of aptamer-OTA complex. The unwinding process led to the dissociation of aggregated GQDs and thus recovering their fluorescence. Hence, this sensor uniquely works with ON/OFF mechanism depending on the absence or presence of OTA, respectively. Moreover, this aptasensor works well for 0–1 ng/mL of OTA and exhibited an LOD of 13 pg/mL.

The use of a few drug molecules, e.g., Paracetamol (PD) (also known as p-hydroxyacetanilide) has also found to be useful for the AuNPs-based fluorescence immunosensing of pathogens (Sahoo et al. 2012). The PD is used as fluorophore and displays an emission at 435 nm after excitation by 320 nm light wavelength (Sahoo et al. 2012). (Note that in PD-AuNPs, PD functioned as stabilizing agent for AuNPs.) Interestingly, PD also exhibits the affinity to interact with the bacterial cell wall. In the presence of bacteria, the attachment of PD on bacterial cell wall (if present) and destabilizes/aggregates the AuNPs. This aggregation led to the fluorescence quenching of PD-AuNPs. Thus, the fluorescence intensity was decreased in the presence of bacterial cells. This method was found excellent for the analysis of six different strains of bacteria and exhibited sensitivity value of 100 CFU/mL. It was observed that Gram-positive (*Bacillus cereus* MTCC 1305, *Pediococcus acidilactici* CFR K7, and *Enterococcus faecalis* MTCC 439) and Gram-negative bacteria (*Enterobacter aerogenes* MTCC 2822, *Escherichia coli* MTCC 433, and *Pseudomonas aeruginosa* MTCC 2488) quenched the fluorescence of PD-AuNPs composite.

Like antibodies-based sensors, nanomaterials-based aptasensors have also employed for the testing of bacteria, e.g., *Pseudomonas aeruginosa* (Gao et al. 2018). The 5-carboxyfluorescein (FAM)-conjugated complementary DNA (FAM-cDNA) hybridized in the absence of *P. aeruginosa* with partial aptamer sequence, which led to the quenching of FAM fluorescence by graphene oxide QDs (GOQDs). In contrast, the fluorescence of FAM was recovered in of aptamer conjugated *P. aeruginosa*. The FAM-cDNA hybridized with the aptamer (present with *P. aeruginosa*) to release of FAM-cDNA from GOQDs, which led to the recovery of FAM fluorescence. On the basis of abovementioned FRET phenomenon, linear response for *P. aeruginosa* sensing was achieved for 1.28×10^3 – 2.00×10^7 CFU/mL. This GOQDs-based fluorescence sensor exhibited an LOD and response time of 100 CFU/mL and 2 h, respectively. This platform was also applied successfully to detect *P. aeruginosa* in water, juice, and popsicle (Gao et al. 2018).

In another report on OTA immunosensor, aptamer (against OTA)-conjugated AuNPs were tested for 25–300 nM of OTA to achieve an LOD of 22.7 nM (7 ng/mL) (Lv et al. 2017). To fabricate AuNPs-based aptasensor for OTA, FAM-modified aptamers were adsorbed onto the surface of AuNPs. The adsorption of aptamers on AuNPs makes them stable even in high salts concentration (e.g., NaCl). On the other hand, after attachment with AuNPs, FAM-aptamer fluorescence was quenched by AuNPs. On the contrary, the conformation of FAM-aptamer changes to binds with OTA, which caused the recovery of FAM fluorescence signals. The process of OTA interaction with FAM-aptamer led to the AuNPs aggregation under high salt levels. Thus, the sensing signals can be obtained from the FAM (fluorescence signals) and AuNPs (colorimetric signals) to determine the presence of OTA.

Another toxic allergen AFB1, readily produced by fungi has also been targeted through aptamer-conjugated QDs fluorescence immunosensors (Sabet et al. 2017). Upon addition of citrate-AuNPs into the test solution, the aptamers (conjugated with QDs) adsorbed on the surface of citrate-AuNPs, which caused quenching of QDs fluorescence (Sabet et al. 2017). After the addition of analyte molecules, i.e., AFB1 is added, the aptamer specifically binds with AFB1 leaves, which released AuNPs and hence recovery of QDs fluorescence. On the basis of QDs fluorescence recovery, 10–400 nM (30–125 ng/mL), of AFB1 was sensed by this system with LOD value of 3.4 nM (1 ng/mL) in rice and peanut samples. The optimum response time for this sensor was reported to 15 min. The performances of nanomaterials in the form of fluorescent sensors are shown in Table 11.4.

Overall, diverse materials from semiconductor to metal nanostructures (alone or in combination of other nanomaterials) have been tested for fabrication of fluorescent nanobiosensors for food contaminants. These sensors usually work on the decrease or increase in the nanoprobe fluorescence in the presence of target food contaminant.

11.4.3 Piezoelectric-Based Immunosensor

The QCM-based sensors usually work on the piezoelectric principle and exhibit great potentials in the development of label-free immunosensor (Cervera-Chiner et al. 2018; Cervera-Chiner et al. 2020). However, a scanty research is available on the utilization of nanomaterials-based piezoelectric immunosensor for food contaminants (Table 11.5). In majority of nanomaterials-based immunosensors, gold-based structures were majorly explored (Karczmarczyk et al. 2017; Haddada et al. 2018) . For example, in a study on piezoelectric sensor, Au- or Si-coated quartz crystal sensor chip was modified by AuNPs to detect SEA (Haddada et al. 2018). It was speculated that AuNPs increased the surface area of the working electrode and modify surface topography to improve accessibility of binding site of the biosensor. Anti-SEA was immobilized via affinity interaction on protein A-modified AuNPs. The AuNPs-based direct QCM sensor was found more sensitive (e.g., LOD = 8 ng/mL), when compared to the non-AuNPs-modified QCM-immunosensor (LOD = 20 ng/mL). Furthermore, the sandwich assay was performed with exhibited lower LOD, i.e., 1 ng/mL for SEA detection (Haddada et al. 2018). A similar observation was realized for the immunosensing of AFB1 by electrochemical QCM-based immunosensor developed by covalent (EDC/NHS linker) immobilization of anti-AFB1 on the AuNPs surface (Chauhan et al. 2016). The interaction of AFB1 with anti-AFB1 was monitored in the form of current and frequency change in QCM. The increase in the AFB1 levels increased the current and frequency of the QCM sensor. This QCM sensor exhibited a good LOD of 8 pg/mL for AFB1 detection (Chauhan et al. 2016).

The AuNPs were also involved in the amplification of sensing signals of QCM-based immunosensors. For instance, AuNPs-conjugated secondary antibody (Ab_2-AuNP) was used to amplify the weak signals of OTA QCM sensor (Karczmarczyk et al. 2017). In this case, the sensing of OTA was achieved by incubation of toxin containing samples with primary antibody (Ab1). In the presence of OTA in test sample, OTA-Ab1 complex was formed, which was removed in subsequent washing steps. As a result, the binding of secondary antibody (Ab_2)-AuNPs was less. Correspondingly, the sensor displayed a mass gain in comparison to the samples without OTA. The resultant change in the oscillation frequency was used to determine OTA with 0.16 ng/mL LOD (Karczmarczyk et al. 2017). Similarly, AuNP-conjugated Ab-based sandwich immunoassay displayed an LOD value of 150 CFU/mL for *Campylobacter jejuni* (a food-borne pathogen) (Masdor et al. 2016). In another study, QCM sensing of *C. jejuni* was performed with two anti-*C. jejuni*. The first anti-C. jejuni was conjugated on the surface of QCM sensor, while second anti-*C. jejuni* was linked with MNBs; the attachment of antibodies with magnetic beads was helpful for their easy removal from the sample (Wang et al.

Table 11.5 Applications of different nanomaterials in piezoelectric immunosensors for food analysis

No.	Antibodies used	Nanomaterials used	Analyte	Sensing range	sensing range (ng/mL)	LOD	LOD (ng/mL)	Response time (min)	Sensor type	Nanomaterial role	Type of Sample	Reference
1	Anti-SEA (rabbit)	AuNPs	SEA (mycotoxin)			1 ng/mL	1	10–15	QCM	Surface modification		Haddada et al. (2018)
2	Anti-AFB1	AuNPs	AFB1 (mycotoxin)	0.008–0.3 ng/mL and 1–10 ng/mL	0.008–0.3 and 1–10	8 pg /mL	0.008		QCM-CV	Surface modification	Spiked corn flakes	Chauhan et al. (2016)
3	Anti-OTA (primary antibody from rabbit), AuNP/anti rabbit (secondary antibody from goat)	AuNPs	OTA (mycotoxin)	0.2–40 ng/mL	0.2–40	0.16 ng/mL	0.16		QCM	Signal amplifier	Spiked red wine	Karczmarczyk et al. (2017)
4	Anti-*C. jejuni* (capture antibody), AuNP/anti-*C. jejuni* (detection antibody) (rabbit)	AuNPs	*C. Jejuni*	1×10^3–1×10^5 CFU/mL	1×10^3–1×10^5 CFU/mL	150 CFU/mL	150 CFU/mL	9	QCM	Signal amplification		Masdor et al. (2016)

(continued)

Table 11.5 (continued)

No.	Antibodies used	Nanomaterials used	Analyte	Sensing range	sensing range (ng/mL)	LOD	LOD (ng/mL)	Response time (min)	Sensor type	Nanomaterial role	Type of Sample	Reference
5	Anti-*C. jejuni* (mouse) (capture antibody), MNB/anti-*C. jejuni*, (rabbit), AuNP/anti-rabbit Ig G-(secondary from goat)	AuNPs and MNB	*C. Jejuni*	1–1000 CFU/mL	0–1000 CFU/mL	20–30 CFU/mL	20–30 CFU/mL	30	QCM	Signal amplification and immunoseparation	Spiked broiler carcass wash and ground Turkey	Wang et al. (2018a, b)
6	Nanolipsome/anti-AFB1 (rabbit)	Nanoliposome	Aflatoxin B1 (mycotoxin)	1.0 ng/kg–10 mg/kg	1.0 ng/kg–10×10^3 ng/kg	0.83 ng/kg	0.83 ng/kg		QCM	Signal generator	Spiked peanut	Tang et al. (2018)

Abbreviation: *QCM* Quartz crystal microbalance, *AFB1* Aflatoxin, *C. jejuni Campylobacter jejuni*, *OTA* Ochratoxin A, *MUA* 11-Mercaptoundecanoic acid, *SEA* Staphylococcus enterotoxin

2018a, b). The *C. jejuni* from sample was allowed to sandwich between these two anti-*C. jejuni* to form an immunocomplex on QCM electrode. Subsequently, anti-rabbit Ig G (from goat) conjugated with gold was allowed to bind with above immunocomplex. Both MNBs and AuNP-conjugated antibodies amplified the signal of QCM sensor by increasing mass change and hence the relative frequency shifts. The sensor could detect *C. jejuni* with LOD value of 20–30 CFU/mL (Wang et al. 2018a, b).

In another study, piezoelectric immunosensor was used to detect AFB1, where anti-AFB1 conjugated with glucose encapsulated nanoliposome for signal augmentation (Tang et al. 2018). The QCM gold chip surface was modified and activated with dextran and AFB1-BSA-concanavalin A (Con A), respectively. After the addition of anti-AFB1-nanoliposome on modified QCM electrode, it formed immunocomplex with AFB1. In the subsequent step, Triton X-100 was added to the aforementioned immunocomplex. The addition of Triton X-100 led to the bursting of nanoliposome to release encapsulated glucose. Further, due to strong affinity of Con A toward glucose than the dextran, AFB1-BSA-Con A established interactions with the glucose and detached from the QCM gold chip. However, in the presence of toxin, AFB1 competes with toxin present in the samples to interact with the antiAFB1-nanoliposome present on QCM surface. Consequently, on the basis of amount of AFB1 toxin present in the test sample, less amount of AFB1-BSA-Con A was detached from the QCM surface, which led to the increase in the frequency of QCM electrode. This QCM sensor exhibited an LOD value of 0.83 ng/kg for AFB1 (Tang et al. 2018).

All the abovementioned immunosensors (electrochemical, optical, and QCM-based) displayed excellent performances for the detection of food contaminants. In the context of futuristic sensors, the portability and applicability of these sensors can be further improved by incorporation of new technologies. Out of several key methodologies, the paper and smart phone-based sensor exhibited tremendous potential to be used in the futuristic immunosensors (Rateni et al. 2017; Lu et al. 2019; Mahato and Chandra 2019; Purohit et al. 2020a, b). The involvement of these new technologies in immunosensors will be highly beneficial for the development of point-of-care devices.

11.5 Conclusion and Future Prospective

The nanomaterials are found to enhance the sensing characteristic of an immunosensors to a great extent. A number of carbon nanomaterials and metal/metal oxides/metal sulfides have been examined successfully as a part of immunosensors for food contaminants sensing. These nanomaterials are found excellent in improving sensitivity, portability, response time, and procedural steps for the analysis of diverse food contaminants. The nanomaterials are mainly employed for the immunosensing of food contaminants via electrochemical, optical, and piezoelectric approach. All the explored sensing approaches are efficient enough to detect the food contaminants below the allowed limit. These nanomaterials, especially metal nanostructures are also capable of fabricating sensors as simple as paper strip. Such paper strips are excellent in performing the food contaminant analysis in a very simple manner. Likewise, easily readable signals were generated by the ECM-IS in response to food contaminants. Overall, the nanomaterials-based immunosensors can be the future of the easy, sensitive, and rapid sensing devices for food contaminations. Moreover, by manipulating the biorecognition element, these sensing devices can also be applicable for other potential sensing services, e.g., detection of diseases, pathogenic microorganisms, pollutants, explosives, and hazardous materials. Nonetheless, a number of gaps are still present in nanomaterials-based immunosensors, which may restrict the point-of-care use of nanomaterials-based immunosensor. The first and major aspect of immunosensors is the cost and stability of antibodies. Due to expensive procedures, the final cost of antibodies is very high, which can be suspected for increased cost of immunosensing. However, the cost can be reduced by increasing the efficient production of antibodies. Likewise, for long-term stability, low storage temperature is needed for immunosensors. The alternative molecules such as aptamers and synthetic antibodies can be helpful in resolving the antibodies-related issues in immunosensors.

The cost of nanomaterials can be counted as another hurdle in the fabrication of cost-effective immunosensors. However, it can be assumed that in future the increased production of nanomaterials can lower their cost. In addition to that, on the basis of types of food, a variety of components are present in food, which may hinder the final sensing signals of an immunosensor. So, these sensing devices should be examined on complex food samples to test their applicability on extended range of food products.

A very few nanomaterials-based immunosensing devices are present in the market; the future research should be oriented for the development of immunosensors as per the market need. It will increase the commercialization aspect of these sensors. Moreover, nanomaterials array-based immunosensor can be the focus of upcoming research to detect multiple analytes by a single device.

Acknowledgments This work acknowledges the support from the Department of Science and Technology, New Delhi, India, in the form of an INSPIRE Faculty Award (DST/INSPIRE/04/2017/002953) and from the Science and Engineering Research Board (SERB), Government of India, under the Early Career Research (ECR) award (File No. ECR/2018/000748).

References

Abdallah ZB, Grauby-Heywang C, Beven L, Cassagnere S, Morotė F, Maillard E, Sghaier H, Bouhacina TC (2019) Development of an ultrasensitive label-free immunosensor for fungal aflatoxin b1 detection. Biochem Eng J 150:107262

Agache I, Miller R, Gern JE, Hellings PW, Jutel M, Muraro A, Phipatanakul W, Quirce S, Peden D (2019) Emerging concepts and challenges in implementing the exposome paradigm in allergic diseases and asthma: a practall document. Allergy 74(3):449–463

Ahmed Adam MA, Tabana YM, Musa KB, Sandai DA (2017) Effects of different mycotoxins on humans, cell genome and their involvement in cancer. Oncol Rep 37(3):1321–1336

Ahmed MU, Zourob M, Tamiya E (2019) Immunosensors. Royal Society of Chemistry

Alves RC, Pimentel FB, Nouws HP, Correr W, González-García MB, Oliveira MBP, Delerue-Matos C (2015) Detection of the peanut allergen ara h 6 in foodstuffs using a voltammetric biosensing approach. Anal Bioanal Chem 407(23):7157–7163

An X, Shi X, Zhang H, Yao Y, Wang G, Yang Q, Xia L, Sun X (2020) An electrochemical immunosensor based on a combined amplification strategy with the go–cs/ceo 2–cs nanocomposite for the detection of aflatoxin m 1. New J Chem 44(4):1362–1370

Angulo-Ibanez A, Eletxigerra U, Lasheras X, Campuzano S, Merino S (2019) Electrochemical tropomyosin allergen immunosensor for complex food matrix analysis. Anal Chim Acta 1079: 94–102

Antiochia R, Bollella P, Favero G, Mazzei F (2016) Nanotechnology-based surface plasmon resonance affinity biosensors for *in vitro* diagnostics. Int J Anal Chem 2016:2981931. https://doi.org/10.1155/2016/2981931

Aranda PR, Messina GA, Bertolino FA, Pereira SV, Baldo MAF, Raba J (2018) Nanomaterials in fluorescent laser-based immunosensors: review and applications. Microchem J 141:308–323

Aydin EB, Aydin M, Sezgintürk MK (2019) Advances in electrochemical immunosensors. In: Advances in clinical chemistry. Elsevier, pp 1–57

Banica F-G (2012) Chemical sensors and biosensors: fundamentals and applications. John Wiley & Sons

Baniukevic J, Kirlyte J, Ramanavicius A, Ramanaviciene A (2013) Application of oriented and random antibody immobilization methods in immunosensor design. Sensors Actuators B Chem 189:217–223

Bansal SA, Kumar V, Karimi J, Singh AP, Kumar S (2020) Role of gold nanoparticles in advanced biomedical applications. Nanoscale Adv 2(9):3764–3787. https://doi.org/10.1039/D0NA00472C

Berhanu AL, Gaurav, Mohiuddin I, Malik AK, Aulakh JS, Kumar V, Kim K-H (2019) A review of the applications of schiff bases as optical chemical sensors. TrAC Trends Anal Chem 116:74–91. Available from https://www.sciencedirect.com/science/article/pii/S0165993619301153. https://doi.org/10.1016/j.trac.2019.04.025

Bhardwaj J, Devarakonda S, Kumar S, Jang J (2017) Development of a paper-based electrochemical immunosensor using an antibody-single walled carbon nanotubes bio-conjugate modified electrode for label-free detection of foodborne pathogens. Sensors Actuators B Chem 253:115–123

Bhardwaj H, Singh C, Kotnala R, Sumana G (2018) Graphene quantum dots-based nano-biointerface platform for food toxin detection. Anal Bioanal Chem 410(28):7313–7323

Bhardwaj SK, Bhardwaj N, Kumar V, Bhatt D, Azzouz A, Bhaumik J, Kim K-H, Deep A (2021) Recent progress in nanomaterial-based sensing of airborne viral and bacterial pathogens. Environ Int 146:106183. Available from https://www.sciencedirect.com/science/article/pii/S0160412020321383. doi https://doi.org/10.1016/j.envint.2020.106183

Bhatnagar D, Kumar V, Kumar A, Kaur I (2016) Graphene quantum dots fret based sensor for early detection of heart attack in human. Biosens Bioelectron 79:495–499. Available from https://www.sciencedirect.com/science/article/pii/S0956566315307387. doi: https://doi.org/10.1016/j.bios.2015.12.083

Bratakou S, Nikoleli GP, Siontorou CG, Nikolelis DP, Karapetis S, Tzamtzis N (2017) Development of an electrochemical biosensor for the rapid detection of saxitoxin based on air stable lipid films with incorporated anti-stx using graphene electrodes. Electroanalysis 29(4):990–997

Campuzano S, Yáñez-Sedeño P, Pingarrón JM (2020) Electrochemical affinity biosensors based on selected nanostructures for food and environmental monitoring. Sensors 20(18):5125

Capoferri D, Della Pelle F, Del Carlo M, Compagnone D (2018) Affinity sensing strategies for the detection of pesticides in food. Foods 7(9):148

Cervera-Chiner L, Juan-Borrás M, March C, Arnau A, Escriche I, Montoya Á, Jiménez Y (2018) High fundamental frequency quartz crystal microbalance (hff-qcm) immunosensor for pesticide detection in honey. Food Control 92:1–6

Cervera-Chiner L, March C, Arnau A, Jiménez Y, Montoya Á (2020) Detection of ddt and carbaryl pesticides in honey by means of immunosensors based on high fundamental frequency quartz crystal microbalance (hff-qcm). J Sci Food Agric 100(6):2468–2472

Chandra P, Prakash R (2020) Nanobiomaterial engineering. Springer

Chang H, Lv J, Zhang H, Zhang B, Wei W, Qiao Y (2017) Photoresponsive colorimetric immunoassay based on chitosan modified agi/tio2 heterojunction for highly sensitive chloramphenicol detection. Biosens Bioelectron 87:579–586

Chattopadhyay S, Sabharwal PK, Jain S, Kaur A, Singh H (2019) Functionalized polymeric magnetic nanoparticle assisted Sers immunosensor for the sensitive detection of S. typhimurium. Anal Chimica Acta 1067:98–106

Chauhan R, Singh J, Solanki PR, Manaka T, Iwamoto M, Basu T, Malhotra B (2016) Label-free piezoelectric immunosensor decorated with gold nanoparticles: kinetic analysis and biosensing application. Sensors Actuators B Chem 222:804–814

Chekin F, Singh SK, Vasilescu A, Dhavale VM, Kurungot S, Boukherroub R, Szunerits S (2016) Reduced graphene oxide modified electrodes for sensitive sensing of gliadin in food samples. ACS Sensors 1(12):1462–1470

Chen Y, Huang X-h, H-s S, Wang Y (2011) Research progress of piezoelectric immunosensors. In: 2011 Symposium on piezoelectricity, acoustic waves and device applications (SPAWDA). IEEE, pp 286–289

Chen R, Huang X, Li J, Shan S, Lai W, Xiong Y (2016) A novel fluorescence immunoassay for the sensitive detection of escherichia coli o157:H7 in milk based on catalase-mediated fluorescence quenching of cdte quantum dots. Anal Chim Acta 947:50–57. Available from https://www.sciencedirect.com/science/article/pii/S0003267016312004. doi: https://doi.org/10.1016/j.aca.2016.10.017

Cho I-H, Lee J, Kim J, Kang M-s, Paik JK, Ku S, Cho H-M, Irudayaraj J, Kim D-H (2018) Current technologies of electrochemical immunosensors: perspective on signal amplification. Sensors 18(1):207

Choi H, Hwang BK, Kim B-S, Choi SH (2020) Influence of pathogen contamination on beef microbiota under different storage temperatures. Food Res Int 132:109118. Available from https://www.sciencedirect.com/science/article/pii/S0963996920301435. doi: https://doi.org/10.1016/j.foodres.2020.109118

Cristea C, Florea A, Tertiş M, Săndulescu R (2015) Immunosensors. In: Biosensors-micro and nanoscale applications. IntechOpen

Daliri F, Aboagye AA, Kyei-Baffour V, Elahi F, Chelliah R, Daliri EB-M (2019) Immunosensors for food safety: current trends and future perspectives. 한국식품위생안전성학회지 34(6):509–518

Ding Y, Shang H, Wang X, Chen L (2020) A Sers-based competitive immunoassay for highly sensitive and specific detection of ochratoxin a. Analyst 145(18):6079–6084

Duffy G, Moore E (2017) Electrochemical immunosensors for food analysis: a review of recent developments. Anal Lett 50(1):1–32

Falleh H, Ben Jemaa M, Saada M, Ksouri R (2020) Essential oils: a promising eco-friendly food preservative. Food Chem 330:127268. Available from https://www.sciencedirect.com/science/article/pii/S0308814620311304. doi: https://doi.org/10.1016/j.foodchem.2020.127268

Fan R, Tang S, Luo S, Liu H, Zhang W, Yang C, He L, Chen Y (2020) Duplex surface enhanced raman scattering-based lateral flow immunosensor for the low-level detection of antibiotic residues in milk. Molecules 25(22):5249

Felix FS, Angnes L (2018) Electrochemical immunosensors–a powerful tool for analytical applications. Biosens Bioelectron 102:470–478

Filik H, Avan AA (2019) Nanostructures for nonlabeled and labeled electrochemical immunosensors: simultaneous electrochemical detection of cancer markers: a review. Talanta 205:120153

Fu X, Chu Y, Zhao K, Li J, Deng A (2017) Ultrasensitive detection of the β-adrenergic agonist brombuterol by a Sers-based lateral flow immunochromatographic assay using flower-like gold-silver core-shell nanoparticles. Microchim Acta 184(6):1711–1719

Gamella M, Bueno-Díaz C, Montiel VR-V, Povedano E, Reviejo A, Villalba M, Campuzano S, Pingarrón J (2020) First electrochemical immunosensor for the rapid detection of mustard seeds in plant food extracts. Talanta 219:121247

Gao R, Zhong Z, Gao X, Jia L (2018) Graphene oxide quantum dots assisted construction of fluorescent aptasensor for rapid detection of pseudomonas aeruginosa in food samples. J Agric Food Chem 66(41):10,898–10,905

García-Díaz M, Patiño B, Vázquez C, Gil-Serna J (2019) A novel niosome-encapsulated essential oil formulation to prevent aspergillus flavus growth and aflatoxin contamination of maize grains during storage. Toxins 11(11):646

Gobi KV, Matsumoto K, Toko K, Miura N (2008) Highly regenerable and storageable all-chemical based peg-immunosensor chip for spr detection of ppt levels of fragrant compounds from beverage samples. Sens & Instrumen Food Qual 2(4):225. https://doi.org/10.1007/s11694-008-9033-5

Goud KY, Kailasa SK, Kumar V, Tsang YF, Lee SE, Gobi KV, Kim K-H (2018) Progress on nanostructured electrochemical sensors and their recognition elements for detection of mycotoxins: a review. Biosens Bioelectron 121:205–222. Available from https://www.sciencedirect.com/science/article/pii/S0956566318306262. doi: https://doi.org/10.1016/j.bios.2018.08.029

Guo M, Sun L, Liu L, Song S, Kuang H, Cui G (2018) Ultrasensitive immunochromatographic strip for detection of cyproheptadine. Food Agric Immunol 29(1):941–952

Guo R, Huang F, Cai G, Zheng L, Xue L, Li Y, Liao M, Wang M, Lin J (2020a) A colorimetric immunosensor for determination of foodborne bacteria using rotating immunomagnetic separation, gold nanorod indication, and click chemistry amplification. Microchim Acta 187(4):1–9

Guo Y, Girmatsion M, Li H-W, Xie Y, Yao W, Qian H, Abraha B, Mahmud A (2020b) Rapid and ultrasensitive detection of food contaminants using surface-enhanced raman spectroscopy-based methods. Crit Rev Food Sci Nutr 1–14

Haddada MB, Salmain M, Boujday S (2018) Gold colloid-nanostructured surfaces for enhanced piezoelectric immunosensing of staphylococcal enterotoxin a. Sensors Actuators B Chem 255: 1604–1613

Han E, Li X, Zhang Y, Zhang M, Cai J, Zhang X (2020) Electrochemical immunosensor based on self-assembled gold nanorods for label-free and sensitive determination of staphylococcus aureus. Anal Biochem 611:113982

Hassan FI, Niaz K, Khan F, Maqbool F, Abdollahi M (2017) The relation between rice consumption, arsenic contamination, and prevalence of diabetes in south asia. EXCLI J 16:1132

He S, Li X, Gao J, Tong P, Chen H (2018a) Development of a h2o2-sensitive quantum dots-based fluorescent sandwich elisa for sensitive detection of bovine β-lactoglobulin by monoclonal antibody. J Sci Food Agric 98(2):519–526. Available from https://onlinelibrary.wiley.com/doi/abs/10.1002/jsfa.8489. https://doi.org/10.1002/jsfa.8489

He S, Li X, Wu Y, Wu S, Wu Z, Yang A, Tong P, Yuan J, Gao J, Chen H (2018b) Highly sensitive detection of bovine β-lactoglobulin with wide linear dynamic range based on platinum nanoparticles probe. J Agric Food Chem 66(44):11830–11838

Hiep HM, Endo T, Kerman K, Chikae M, Kim D-K, Yamamura S, Takamura Y, Tamiya E (2007) A localized surface plasmon resonance based immunosensor for the detection of casein in milk. Sci Technol Adv Mater 8(4):331–338. https://doi.org/10.1016/j.stam.2006.12.010

Hong J, Wang Y, Zhu L, Jiang L (2020) An electrochemical sensor based on gold-nanocluster-modified graphene screen-printed electrodes for the detection of β-lactoglobulin in milk. Sensors 20(14):3956

Hu W, Chen H, Zhang H, He G, Li X, Zhang X, Liu Y, Li CM (2014) Sensitive detection of multiple mycotoxins by spri with gold nanoparticles as signal amplification tags. J Colloid

Interface Sci 431:71–76. Available from http://europepmc.org/abstract/MED/24992296. https://doi.org/10.1016/j.jcis.2014.06.007

Hu M, Hu X, Zhang Y, Teng M, Deng R, Xing G, Tao J, Xu G, Chen J, Zhang Y (2019) Label-free electrochemical immunosensor based on aunps/zn/ni-zif-8-800@ graphene composites for sensitive detection of monensin in milk. Sensors Actuators B Chem 288:571–578

Huang X, Zhan S, Xu H, Meng X, Xiong Y, Chen X (2016) Ultrasensitive fluorescence immunoassay for detection of ochratoxin a using catalase-mediated fluorescence quenching of cdte qds. Nanoscale 8(17):9390–9397

Iarossi M, Schiattarella C, Rea I, De Stefano L, Fittipaldi R, Vecchione A, Velotta R, Ventura BD (2018) Colorimetric immunosensor by aggregation of photochemically functionalized gold nanoparticles. ACS Omega 3(4):3805–3812

Iglesias-Mayor A, Amor-Gutiérrez O, Costa-García A, de la Escosura-Muñiz A (2019) Nanoparticles as emerging labels in electrochemical immunosensors. Sensors 19(23):5137

Jung Y, Jeong JY, Chung BH (2008) Recent advances in immobilization methods of antibodies on solid supports. Analyst 133(6):697–701

Kabir E, Raza N, Kumar V, Singh J, Tsang YF, Lim DK, Szulejko JE, Kim K-H (2019) Recent advances in nanomaterial-based human breath analytical technology for clinical diagnosis and the way forward. Chem 5(12): 3020–3057. Available from https://www.sciencedirect.com/science/article/pii/S2451929419303730. doi: https://doi.org/10.1016/j.chempr.2019.08.004

Karczmarczyk A, Haupt K, Feller K-H (2017) Development of a qcm-d biosensor for ochratoxin a detection in red wine. Talanta 166:193–197

Karunakaran C, Pandiaraj M, Santharaman P (2015) Chapter 4 - Immunosensors. In: Karunakaran C, Bhargava K, Benjamin R (eds) Biosensors and bioelectronics. Elsevier, pp 205–245

Kausaite-Minkstimiene A, Ramanaviciene A, Kirlyte J, Ramanavicius A (2010) Comparative study of random and oriented antibody immobilization techniques on the binding capacity of immunosensor. Anal Chem 82(15):6401–6408

Kempahanumakkagari S, Kumar V, Samaddar P, Kumar P, Ramakrishnappa T, Kim K-H (2018) Biomolecule-embedded metal-organic frameworks as an innovative sensing platform. Biotechnol Adv 36(2):467–481. Available from https://www.sciencedirect.com/science/article/pii/S0734975018300144. doi: https://doi.org/10.1016/j.biotechadv.2018.01.014

Kłos-Witkowska A (2016) The phenomenon of fluorescence in immunosensors. Acta Biochim Pol 63(2):215–221

Kolok AS, Ali JM, Rogan EG, Bartelt-Hunt SL (2018) The fate of synthetic and endogenous hormones used in the us beef and dairy industries and the potential for human exposure. Curr Environ Health Reports 5(2):225–232. https://doi.org/10.1007/s40572-018-0197-9

Kour R, Arya S, Young S-J, Gupta V, Bandhoria P, Khosla A (2020) Recent advances in carbon nanomaterials as electrochemical biosensors. J Electrochem Soc 167(3):037555

Krishna VD, Wu K, Su D, Cheeran MC, Wang J-P, Perez A (2018) Nanotechnology: review of concepts and potential application of sensing platforms in food safety. Food Microbiol 75:47–54

Kukkar D, Vellingiri K, Kumar V, Deep A, Kim K-H (2018) A critical review on the metal sensing capabilities of optically active nanomaterials: limiting factors, mechanism, and performance evaluation. TrAC Trends Anal Chem 109:227–246. Available from https://www.sciencedirect.com/science/article/pii/S016599361830284X. doi https://doi.org/10.1016/j.trac.2018.09.009

Kukkar D, Kukkar P, Kumar V, Hong J, Kim K-H, Deep A (2021) Recent advances in nanoscale materials for antibody-based cancer theranostics. Biosens Bioelectron 173:112787. Available from https://www.sciencedirect.com/science/article/pii/S0956566320307740. doi https://doi.org/10.1016/j.bios.2020.112787

Kumar V, Chopra A, Arora S, Yadav S, Kumar S, Kaur I (2015) Amperometric sensing of urea using edge activated graphene nanoplatelets. RSC Adv 5(18):13,278–13,284. https://doi.org/10.1039/C4RA12594K

Kumar V, Kim K-H, Park J-W, Hong J, Kumar S (2017a) Graphene and its nanocomposites as a platform for environmental applications. Chem Eng J 315:210–232. Available from https://

www.sciencedirect.com/science/article/pii/S1385894717300098. doi: https://doi.org/10.1016/j.cej.2017.01.008

Kumar V, Mahajan R, Kaur I, Kim K-H (2017b) Simple and mediator-free urea sensing based on engineered nanodiamonds with polyaniline nanofibers synthesized in situ. ACS Appl Mater Interfaces 9(20):16,813–16,823. https://doi.org/10.1021/acsami.7b01948

Kumar V, Kukkar D, Hashemi B, Kim K-H, Deep A (2019) Advanced functional structure-based sensing and imaging strategies for cancer detection: possibilities, opportunities, challenges, and prospects. Adv Funct Mater 29(16):1807859. Available from https://onlinelibrary.wiley.com/doi/abs/10.1002/adfm.201807859. https://doi.org/10.1002/adfm.201807859

Kumar V, Kaur I, Arora S, Mehla R, Vellingiri K, Kim K-H (2020a) Graphene nanoplatelet/graphitized nanodiamond-based nanocomposite for mediator-free electrochemical sensing of urea. Food Chem 303:125375. Available from https://www.sciencedirect.com/science/article/pii/S030881461931489X. doi: https://doi.org/10.1016/j.foodchem.2019.125375

Kumar V, Vaid K, Bansal SA, Kim K-H (2020b) Nanomaterial-based immunosensors for ultrasensitive detection of pesticides/herbicides: current status and perspectives. Biosens Bioelectron:112382

Kumar S, Nehra M, Khurana S, Dilbaghi N, Kumar V, Kaushik A, Kim K-H (2021) Aspects of point-of-care diagnostics for personalized health wellness. Int J Nanomedicine 16:383

Lara S, Perez-Potti A (2018) Applications of nanomaterials for immunosensing. Biosensors 8(4): 104

Li H, Zhang L (2017) Photocatalytic performance of different exposed crystal facets of biocl. Curr Opin Green Sustain Chem 6:48–56. Available from http://www.sciencedirect.com/science/article/pii/S2452223617300494. doi: https://doi.org/10.1016/j.cogsc.2017.05.005

Li Y, Chen Q, Xu X, Jin Y, Wang Y, Zhang L, Yang W, He L, Feng X, Chen Y (2018) Microarray surface enhanced raman scattering based immunosensor for multiplexing detection of mycotoxin in foodstuff. Sensors Actuators B Chem 266:115–123

Li Y, Tang S, Zhang W, Cui X, Zhang Y, Jin Y, Zhang X, Chen Y (2019) A surface-enhanced raman scattering-based lateral flow immunosensor for colistin in raw milk. Sensors Actuators B Chem 282:703–711

Liang Y, Huang X, Chen X, Zhang W, Ping G, Xiong Y (2018) Plasmonic elisa for naked-eye detection of ochratoxin a based on the tyramine-h2o2 amplification system. Sensors Actuators B Chem 259:162–169

Liu X, Hu Y, Zheng S, Liu Y, He Z, Luo F (2016) Surface plasmon resonance immunosensor for fast, highly sensitive, and in situ detection of the magnetic nanoparticles-enriched salmonella enteritidis. Sens Actuat B Chem 230:191–198. Available from https://www.sciencedirect.com/science/article/pii/S0925400516301939. doi https://doi.org/10.1016/j.snb.2016.02.043

Liu Y, Zhao C, Fu K, Song X, Xu K, Wang J, Li J (2017) Selective turn-on fluorescence detection of vibrio parahaemolyticus in food based on charge-transfer between cdse/zns quantum dots and gold nanoparticles. Food Control 80:380–387

Lu Y, Shi Z, Liu Q (2019) Smartphone-based biosensors for portable food evaluation. Curr Opin Food Sci 28: 74-81. Available from https://www.sciencedirect.com/science/article/pii/S2214799319300700. DOI 10.1016/j.cofs.2019.09.003

Lv X, Zhang Y, Liu G, Du L, Wang S (2017) Aptamer-based fluorescent detection of ochratoxin a by quenching of gold nanoparticles. RSC Adv 7(27):16,290–16,294

Mahato K, Chandra P (2019) Paper-based miniaturized immunosensor for naked eye alp detection based on digital image colorimetry integrated with smartphone. Bios Bioelectron 128:9–16. Available from https://www.sciencedirect.com/science/article/pii/S0956566318309606. doi https://doi.org/10.1016/j.bios.2018.12.006

Mahato K, Kumar S, Srivastava A, Maurya PK, Singh R, Chandra P (2018) Electrochemical immunosensors: Fundamentals and applications in clinical diagnostics. In: Handbook of immunoassay technologies. Elsevier, pp 359–414

Mahato K, Purohit B, Kumar A, Chandra P (2020) Clinically comparable impedimetric immunosensor for serum alkaline phosphatase detection based on electrochemically engineered au-nano-dendroids and graphene oxide nanocomposite. Biosens Bioelectron 148:111815. Available from https://www.sciencedirect.com/science/article/pii/S0956566319308942. doi doi:https://doi.org/10.1016/j.bios.2019.111815

Makaraviciute A, Ramanaviciene A (2013) Site-directed antibody immobilization techniques for immunosensors. Biosens Bioelectron 50:460–471

Malvano F, Pilloton R, Albanese D (2020) Label-free impedimetric biosensors for the control of food safety–a review. Int J Environ Anal Chem 100(4):468–491

Man Y, Ren J, Li B, Jin X, Pan L (2018) A simple, highly sensitive colorimetric immunosensor for the detection of alternariol monomethyl ether in fruit by non-aggregated gold nanoparticles. Anal Bioanal Chem 410(28):7511–7521

Masdor NA, Altintas Z, Tothill IE (2016) Sensitive detection of campylobacter jejuni using nanoparticles enhanced qcm sensor. Biosens Bioelectron 78:328–336

Mistry KK, Layek K, Mahapatra A, RoyChaudhuri C, Saha H (2014) A review on amperometric-type immunosensors based on screen-printed electrodes. Analyst 139(10):2289–2311

Mohamad A, Teo H, Keasberry NA, Ahmed MU (2019) Recent developments in colorimetric immunoassays using nanozymes and plasmonic nanoparticles. Crit Rev Biotechnol 39(1):50–66

Mollarasouli F, Kurbanoglu S, Ozkan SA (2019) The role of electrochemical immunosensors in clinical analysis. Biosensors 9(3):86

Mustafa F, Andreescu S (2018) Chemical and biological sensors for food-quality monitoring and smart packaging. Foods (Basel, Switzerland) 7(10):168. Available from https://pubmed.ncbi.nlm.nih.gov/30332833. https://www.ncbi.nlm.nih.gov/pmc/articles/PMC6210272/. https://doi.org/10.3390/foods7100168

Narayanan J, Sharma MK, Ponmariappan S, Shaik M, Upadhyay S (2015) Electrochemical immunosensor for botulinum neurotoxin type-e using covalently ordered graphene nanosheets modified electrodes and gold nanoparticles-enzyme conjugate. Biosens Bioelectron 69:249–256

Neng J, Zhang Q, Sun P (2020) Application of surface-enhanced raman spectroscopy in fast detection of toxic and harmful substances in food. Biosens Bioelectron 112480

Nerín C, Aznar M, Carrizo D (2016) Food contamination during food process. Trends Food Sci Technol 48:63–68. Available from https://www.sciencedirect.com/science/article/pii/S0924224415301370. doi: https://doi.org/10.1016/j.tifs.2015.12.004

Niu X, Cheng N, Ruan X, Du D, Lin Y (2019) Nanozyme-based immunosensors and immunoassays: recent developments and future trends. J Electrochem Soc 167(3):037508

Njobeh PB, Dutton MF, Koch SH, Chuturgoon A, Stoev S, Seifert K (2009) Contamination with storage fungi of human food from Cameroon. Int J Food Microbiol 135(3):193–198. Available from https://www.sciencedirect.com/science/article/pii/S0168160509003870. doi https://doi.org/10.1016/j.ijfoodmicro.2009.08.001

Pal M, Lee S, Kwon D, Hwang J, Lee H, Hwang S, Jeon S (2017) Direct immobilization of antibodies on zn-doped fe3o4 nanoclusters for detection of pathogenic bacteria. Anal Chim Acta 952:81–87

Pan M, Gu Y, Yun Y, Li M, Jin X, Wang S (2017) Nanomaterials for electrochemical immunosensing. Sensors 17(5):1041

Patra S, Roy E, Madhuri R, Sharma PK (2017) A technique comes to life for security of life: the food contaminant sensors. In: Nanobiosensors. Elsevier, pp 713–772

Pei X, Zhang B, Tang J, Liu B, Lai W, Tang D (2013) Sandwich-type immunosensors and immunoassays exploiting nanostructure labels: a review. Anal Chim Acta 758:1–18

Peng J, Song S, Liu L, Kuang H, Xu C (2015) Development of sandwich elisa and immunochromatographic strip for the detection of peanut allergen ara h 2. Food Anal Methods 8(10):2605–2611

Pohanka M (2018) Overview of piezoelectric biosensors, immunosensors and DNA sensors and their applications. Materials 11(3):448

Pollet J, Delport F, Janssen KP, Tran DT, Wouters J, Verbiest T, Lammertyn J (2011) Fast and accurate peanut allergen detection with nanobead enhanced optical fiber spr biosensor. Talanta 83(5):1436–1441. https://doi.org/10.1016/j.talanta.2010.11.032

Poltronieri P, Mezzolla V, Primiceri E, Maruccio G (2014) Biosensors for the detection of food pathogens. Foods (Basel, Switzerland) 3(3):511–526. Available from https://pubmed.ncbi.nlm.nih.gov/28234334. https://www.ncbi.nlm.nih.gov/pmc/articles/PMC5302249/. https://doi.org/10.3390/foods3030511

Pottathara YB, Thomas S, Kalarikkal N, Grohens Y, Kokol V (2019) Nanomaterials synthesis: design, fabrication and applications. Elsevier

Pu Y, Cai F, Wang D, Wang J-X, Chen J-F (2018) Colloidal synthesis of semiconductor quantum dots toward large-scale production: a review. Ind Eng Chem Res 57(6):1790–1802

Purohit B, Kumar A, Mahato K, Chandra P (2020a) Smartphone-assisted personalized diagnostic devices and wearable sensors. Curr Opin Biomed Eng 13:42–50. Available from https://www.sciencedirect.com/science/article/pii/S2468451119300376. doi https://doi.org/10.1016/j.cobme.2019.08.015

Purohit B, Vernekar PR, Shetti NP, Chandra P (2020b) Biosensor nanoengineering: design, operation, and implementation for biomolecular analysis. Sens Int 1:100040. Available from https://www.sciencedirect.com/science/article/pii/S2666351120300401. doi https://doi.org/10.1016/j.sintl.2020.100040

Radhakrishnan R, Poltronieri P (2017) Fluorescence-free biosensor methods in detection of food pathogens with a special focus on listeria monocytogenes. Biosensors 7:63. https://doi.org/10.3390/bios7040063

Rateni G, Dario P, Cavallo F (2017) Smartphone-based food diagnostic technologies: a review. Sensors 17(6):1453

Ricci F, Volpe G, Micheli L, Palleschi G (2007) A review on novel developments and applications of immunosensors in food analysis. Anal Chim Acta 605(2):111–129

Sabet FS, Hosseini M, Khabbaz H, Dadmehr M, Ganjali MR (2017) Fret-based aptamer biosensor for selective and sensitive detection of aflatoxin b1 in peanut and rice. Food Chem 220:527–532. Available from https://www.sciencedirect.com/science/article/pii/S0308814616316211. doi https://doi.org/10.1016/j.foodchem.2016.10.004

Sahoo AK, Sharma S, Chattopadhyay A, Ghosh SS (2012) Quick and simple estimation of bacteria using a fluorescent paracetamol dimer–au nanoparticle composite. Nanoscale 4(5):1688–1694. https://doi.org/10.1039/C2NR11837H

Saito M, Kitsunai M, Ahmed MU, Sugiyama S, Tamiya E (2008) Label-free electrochemical detection for food allergen using screen printed carbon electrode. Electrochemistry 76(8):606–609

Savas S, Altintas Z (2019) Graphene quantum dots as nanozymes for electrochemical sensing of yersinia enterocolitica in milk and human serum. Materials 12(13):2189

Shukla S, Haldorai Y, Bajpai VK, Rengaraj A, Hwang SK, Song X, Kim M, Huh YS, Han Y-K (2018) Electrochemical coupled immunosensing platform based on graphene oxide/gold nanocomposite for sensitive detection of cronobacter sakazakii in powdered infant formula. Biosens Bioelectron 109:139–149

Suman P, Chandra P (2020) Immunodiagnostic technologies from laboratory to point-of-care testing. Springer Singapore, Singapore

Sun X, Ye Y, He S, Wu Z, Yue J, Sun H, Cao X (2019) A novel oriented antibody immobilization based voltammetric immunosensor for allergenic activity detection of lectin in kidney bean by using aunps-pei-mwcnts modified electrode. Biosens Bioelectron 143:111607

Suri CR, Boro R, Nangia Y, Gandhi S, Sharma P, Wangoo N, Rajesh K, Shekhawat GS (2009) Immunoanalytical techniques for analyzing pesticides in the environment. TrAC Trends Anal Chem 28(1):29–39. Available from https://www.sciencedirect.com/science/article/pii/S0165993608002331. doi https://doi.org/10.1016/j.trac.2008.09.017

Talib NAA, Salam F, Sulaiman Y (2018a) Development of highly sensitive immunosensor for clenbuterol detection by using poly (3, 4-ethylenedioxythiophene)/graphene oxide modified screen-printed carbon electrode. Sensors 18(12):4324

Talib NAA, Salam F, Yusof NA, Ahmad SAA, Azid MZ, Mirad R, Sulaiman Y (2018b) Enhancing a clenbuterol immunosensor based on poly (3, 4-ethylenedioxythiophene)/multi-walled carbon nanotube performance using response surface methodology. RSC Adv 8(28):15522–15532

Tang Y, Tang D, Zhang J, Tang D (2018) Novel quartz crystal microbalance immunodetection of aflatoxin b1 coupling cargo-encapsulated liposome with indicator-triggered displacement assay. Anal Chim Acta 1031:161–168

Thakur M, Ragavan K (2013) Biosensors in food processing. J Food Sci Technol 50(4):625–641

Tian J, Huang J, Zhao Y, Zhao S (2012) Electrochemical immunosensor for prostate-specific antigen using a glassy carbon electrode modified with a nanocomposite containing gold nanoparticles supported with starch-functionalized multi-walled carbon nanotubes. Microchim Acta 178(1):81–88

Tsagkaris AS, Pulkrabova J, Hajslova J (2021) Optical screening methods for pesticide residue detection in food matrices: advances and emerging analytical trends. Foods 10(1):88

Tsekenis G, Chatzipetrou M, Massaouti M, Zergioti I (2019) Comparative assessment of affinity-based techniques for oriented antibody immobilization towards immunosensor performance optimization. J Sens 2019

Turner AP (1997) Immunosensors: the next generation. Nat Biotechnol 15(5):421–421

Vaid K, Dhiman J, Kumar S, Kim K-H, Kumar V (2020a) A novel approach for effective alteration of morphological features of polyaniline through interfacial polymerization for versatile applications. Nano 10(12):2404

Vaid K, Dhiman J, Sarawagi N, Kumar V (2020b) Experimental and computational study on the selective interaction of functionalized gold nanoparticles with metal ions: sensing prospects. Langmuir 36(41):12,319–12,326. https://doi.org/10.1021/acs.langmuir.0c02280

Vaisocherová-Lísalová H, Víšová I, Ermini ML, Špringer T, Song XC, Mrázek J, Lamačová J, Scott Lynn N Jr, Šedivák P, Homola J (2016) Low-fouling surface plasmon resonance biosensor for multi-step detection of foodborne bacterial pathogens in complex food samples. Biosens Bioelectron 80:84–90. https://doi.org/10.1016/j.bios.2016.01.040

Vezocnik V, Hodnik V, Anderluh G (2017) Surface plasmon resonance analysis of food toxins and toxicants. pp 195–216

Vikrant K, Tsang DCW, Raza N, Giri BS, Kukkar D, Kim K-H (2018) Potential utility of metal–organic framework-based platform for sensing pesticides. ACS Appl Mater Interfaces 10(10): 8797–8817. https://doi.org/10.1021/acsami.8b00664

Wang B, Park B (2020) Immunoassay biosensing of foodborne pathogens with surface plasmon resonance imaging: a review. J Agric Food Chem 68(46):12,927–12,939

Wang J, Munir A, Zhu Z, Zhou HS (2010) Magnetic nanoparticle enhanced surface plasmon resonance sensing and its application for the ultrasensitive detection of magnetic nanoparticle-enriched small molecules. Anal Chem 82(16):6782–6789. https://doi.org/10.1021/ac100812c

Wang M, Kang H, Xu D, Wang C, Liu S, Hu X (2013) Label-free impedimetric immunosensor for sensitive detection of fenvalerate in tea. Food Chem 141(1):84–90

Wang X, Niessner R, Tang D, Knopp D (2016) Nanoparticle-based immunosensors and immunoassays for aflatoxins. Anal Chim Acta 912:10–23

Wang Q-L, Li J, Ding L-S, Xie J, Qing L-S (2017a) A simple nano-Sio 2-based elisa method for residue detection of 2, 4-dichlorophenoxyacetic acid in bean sprouts. Food Anal Methods 10(5): 1500–1506

Wang S, Zhang Y, Pang G, Zhang Y, Guo S (2017b) Tuning the aggregation/disaggregation behavior of graphene quantum dots by structure-switching aptamer for high-sensitivity fluorescent ochratoxin a sensor. Anal Chem 89(3):1704–1709

Wang H, Wang L, Hu Q, Wang R, Li Y, Kidd M (2018a) Rapid and sensitive detection of campylobacter jejuni in poultry products using a nanoparticle-based piezoelectric immunosensor integrated with magnetic immunoseparation. J Food Prot 81(8):1321–1330

Wang Y, Zhang L, Peng D, Xie S, Chen D, Pan Y, Tao Y, Yuan Z (2018b) Construction of electrochemical immunosensor based on gold-nanoparticles/carbon nanotubes/chitosan for sensitive determination of t-2 toxin in feed and swine meat. Int J Mol Sci 19(12):3895

Wang L, Huo X, Zheng L, Cai G, Wang Y, Liu N, Wang M, Lin J (2020a) An ultrasensitive biosensor for colorimetric detection of salmonella in large-volume sample using magnetic grid separation and platinum loaded zeolitic imidazolate framework-8 nanocatalysts. Biosens Bioelectron 150:111862

Wang Y, Qi Q, Zhou J, Li H, Fu L (2020b) Graphene oxide and gold nanoparticles-based dual amplification method for immunomagnetic beads-derived elisa of parvalbumin. Food Control 110:106989

Welch NG, Scoble JA, Muir BW, Pigram PJ (2017) Orientation and characterization of immobilized antibodies for improved immunoassays. Biointerphases 12(2):02D301

Wu Y, Zhou Y, Huang H, Chen X, Leng Y, Lai W, Huang X, Xiong Y (2020) Engineered gold nanoparticles as multicolor labels for simultaneous multi-mycotoxin detection on the immunochromatographic test strip nanosensor. Sensors Actuators B Chem 316:128107

Xu X, Ye Z-Z, Wu J, Ying Y-B (2010) Application and research development of surface plasmon resonance-based immunosensors for protein detection. Chinese J Anal Chem 38(7): 1052–1059. Available from http://www.sciencedirect.com/science/article/pii/S1872204009600591. doi: https://doi.org/10.1016/S1872-2040(09)60059-1

Xu X, Liu X, Li Y, Ying Y (2013) A simple and rapid optical biosensor for detection of aflatoxin b1 based on competitive dispersion of gold nanorods. Biosens Bioelectron 47:361–367. Available from http://europepmc.org/abstract/MED/23603134. https://doi.org/10.1016/j.bios.2013.03.048

Xu N, Wang Y, Pan L, Wei X, Wang Y (2017) Dual-labelled immunoassay with goldmag nanoparticles and quantum dots for quantification of casein in milk. Food Agric Immunol 28(6):1105–1115

Xu Y, Kutsanedzie FY, Hassan MM, Zhu J, Li H, Chen Q (2020) Functionalized hollow au@ ag nanoflower Sers matrix for pesticide sensing in food. Sensors Actuators B Chem 324:128718

Xu Z, Long L-L, Chen Y-q, Chen M-L, Cheng Y-H (2021) A nanozyme-linked immunosorbent assay based on metal–organic frameworks (mofs) for sensitive detection of aflatoxin b1. Food Chem 338:128039

Yan L, Dou L, Bu T, Huang Q, Wang R, Yang Q, Huang L, Wang J, Zhang D (2018) Highly sensitive furazolidone monitoring in milk by a signal amplified lateral flow assay based on magnetite nanoparticles labeled dual-probe. Food Chem 261:131–138

Yang A, Zheng Y, Long C, Chen H, Liu B, Li X, Yuan J, Cheng F (2014) Fluorescent immunosorbent assay for the detection of alpha lactalbumin in dairy products with monoclonal antibody bioconjugated with cdse/zns quantum dots. Food Chem 150:73–79

Yao S, Li J, Pang B, Wang X, Shi Y, Song X, Xu K, Wang J, Zhao C (2020) Colorimetric immunoassay for rapid detection of staphylococcus aureus based on etching-enhanced peroxidase-like catalytic activity of gold nanoparticles. Microchim Acta 187(9):1–8

Yu H-W, Halonen MJ, Pepper IL (2015) Chapter 12 - Immunological methods. In: PepperC IL, Gerba P, Gentry TJ (eds) Environmental microbiology, 3rd edn. Academic Press, San Diego, pp 245–269

Yu W, Zhang T, Ma M, Chen C, Liang X, Wen K, Wang Z, Shen J (2018) Highly sensitive visual detection of amantadine residues in poultry at the ppb level: a colorimetric immunoassay based on a Fenton reaction and gold nanoparticles aggregation. Anal Chim Acta 1027:130–136

Yu W, Sang Y, Wang T, Liu W, Wang X (2020) Electrochemical immunosensor based on carboxylated single-walled carbon nanotube-chitosan functional layer for the detection of cephalexin. Food Sci Nutr 8(2):1001–1011

Yuan J, Deng D, Lauren D, Aguilar M-I, Wu Y (2009) Surface plasmon resonance biosensor for the detection of ochratoxin a in cereals and beverages. Anal Chim Acta 656:63–71. https://doi.org/10.1016/j.aca.2009.10.003

Zhang S, Shen Y, Shen G, Wang S, Shen G, Yu R (2016a) Electrochemical immunosensor based on pd–au nanoparticles supported on functionalized pdda-mwcnt nanocomposites for aflatoxin b1 detection. Anal Biochem 494:10–15

Zhang X, Li C-R, Wang W-C, Xue J, Huang Y-L, Yang X-X, Tan B, Zhou X-P, Shao C, Ding S-J (2016b) A novel electrochemical immunosensor for highly sensitive detection of aflatoxin b1 in corn using single-walled carbon nanotubes/chitosan. Food Chem 192:197–202

Zhang L, Salmain M, Liedberg B, Boujday S (2019) Naked eye immunosensing of food biotoxins using gold nanoparticle-antibody bioconjugates. ACS Appl Nano Mater 2(7):4150–4158

Zhang W, Tang S, Jin Y, Yang C, He L, Wang J, Chen Y (2020) Multiplex Sers-based lateral flow immunosensor for the detection of major mycotoxins in maize utilizing dual raman labels and triple test lines. J Hazard Mater 393:122348

Zheng L, Cai G, Wang S, Liao M, Li Y, Lin J (2019) A microfluidic colorimetric biosensor for rapid detection of escherichia coli o157: H7 using gold nanoparticle aggregation and smart phone imaging. Biosens Bioelectron 124:143–149

Zhou C, Zou H, Li M, Sun C, Ren D, Li Y-X (2018a) Fiber optic surface plasmon resonance sensor for detection of E. coli o157:H7 based on antimicrobial peptides and agnps-rgo. Biosens Bioelectron 117. https://doi.org/10.1016/j.bios.2018.06.005

Zhou Y, Huang X, Zhang W, Ji Y, Chen R, Xiong Y (2018b) Multi-branched gold nanoflower-embedded iron porphyrin for colorimetric immunosensor. Biosens Bioelectron 102:9–16

Zhu Z (2017) An overview of carbon nanotubes and graphene for biosensing applications. Nano-micro Lett 9(3):1–24

Zhu X, Gao T (2019) Chapter 10 - spectrometry. In: Li G (ed) Nano-inspired biosensors for protein assay with clinical applications. Elsevier, pp 237–264

Role of Analytical Techniques in Food Quality Control and Safety

12

Vikas Kumar, Jaspreet Kaur, Kartik Sharma, Satish Kumar, and Rakesh Sharma

Abstract

Food quality control and analysis have many attributes that are discussed in this chapter. Quality and safety are the major parameters in any food industry, the importance of which is discussed in this chapter. Food analysis involves various steps along with different methods, the selection of which depends on various factors such as composition of food product which are mentioned in this chapter. The brief overview of different analytical techniques including sample preparation techniques, general analysis techniques, determinative and separation techniques, biological techniques, rheological techniques, radiochemical and electrochemical techniques, and their selection methods are also discussed in this chapter.

V. Kumar (✉)
Food Science and Technology, Punjab Agricultural University, Ludhiana, Punjab, India

J. Kaur
Food Technology and Nutrition, School of Agriculture, Lovely Professional University, Phagwara, Punjab, India

K. Sharma
Department of Biotechnology, Council of Scientific and Industrial Research—Institute of Himalayan Bioresource Technology (CSIR-IHBT), Palampur, Himachal Pradesh, India

S. Kumar
College of Horticulture and Forestry, Thunag, Mandi. Dr YS Parmar University of Horticulture and Forestry, Nauni, Solan, Himachal Pradesh, India

R. Sharma
Department of Food Science and Technology, Dr. YS Parmar University of Horticulture and Forestry, Solan, Himachal Pradesh, India

P. Chandra, P. S. Panesar (eds.), *Nanosensing and Bioanalytical Technologies in Food Quality Control*, https://doi.org/10.1007/978-981-16-7029-9_12

Keywords

Food quality · Food safety · Food analysis · Quality control · Analytical techniques · HACCP · GMP

12.1 Quality and Safety

Food is any product or substance, which when eaten by human or animals provides them energy in the form of nutrition. It is considered to be the most important and basic need of any living being, which comes prior to other basic needs like clothing and shelter. Food, when consumed serves various functions in body, such as it is responsible for body's maintenance, growth, repairment as well as reproduction (Rajput et al. 2019). Thus, quality along with the safety of food is among the crucial parameters for underdeveloped, developing, and developed economies. The quality of food determines its characteristics which are desirable and are acceptable among the customers. These characteristics can be chemical, physical, sensory (e.g. smell, taste) or convenience (e.g. steps in preparation). Therefore, it can be undoubtedly stated that the term food quality applicable over broader range than food safety. Hazard free or food safety is one of the most important parameter of the food quality system. Setting up the measures to achieve consistent quality is not just an option in any industry but every batch must reach up to the quality standards as set by the industries, thereby maintaining the consistency in quality. These systems somehow are mandatory by law while some are implemented voluntary by the members of food chain in industry (Sikora and Strada 2005).

Nowadays, consumers are exposed to diversified range of food products due to increased international trade of different food products, as a result of which, the consumers are getting prone to various risks associated with food safety. Thus, FSMS (Food Safety Management System) which includes GMP (Good Manufacturing Practices), HACCP (Hazard Analysis Critical Control Points), and GHP (Good Hygiene Practices) needs to put proper vigilance over the food hygiene and its supply. In recent years, the trend for implementation of these systems has come into existence in different countries. Despite this, the public health issues are still prevailing and are continuously arising along with other food-borne diseases. These food-borne diseases generally produce symptoms related to gastro-intestinal, the severity of which may result in organ failure or may cause cancer (Oldewage-Theron and Egal 2016). These food-borne diseases have emerged as a global health challenge for public due to various reasons. While the former diseases are cured from time to time but the new threats emerge continuously. Also, due to increased length of food supply chain and its globalization, various food items are available throughout the world, which in turn is responsible for spreading of pathogens to various geographical regions, and therefore the consumers get exposed to such unfamiliar food-borne diseases in such a new environment. Besides this, changes in microorganisms in different environment have led to evolution of novel pathogens which are resistant to antibiotics. The food usually prepared outside the home in poor

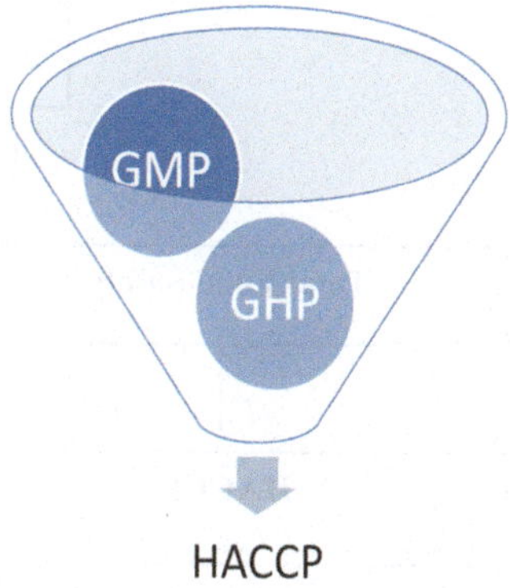

Fig. 12.1 Diagram illustrating the relationship between GHP, GMP, and HACCP (Sikora and Kołożyn-Krajewska 2001)

hygienic conditions poses greater risk for causing the food-borne diseases (WHO 2008). Thus, there is an urgent need to establish the reliable sanitary-surveillance system for identification of risk analysis, potential hazards and to control the spread of food-borne disease (Camino Feltes et al. 2017). However, majority of government across the globe are concerned for the persisting issues of food-borne diseases and are putting their efforts to solve this problem.

Protection of the consumer is the most important parameter and ultimate goal of any food industry in terms of its quality control and in order to ensure these protocols in terms of consistency and reliability. Several sets of laws, rules, and regulations have been made which covers different acts that affect the market in one form or the other. As discussed above, these include HACCP, GMP, GHP, along with various federal laws, regulations, and regular inspections related to factory, import, and export. These systems comprise of systematic approach which assures the particular traits of food product at particular stage of manufacturing, production as well as distribution.

GMPs or good manufacturing practices are the set of regulations that are to be followed and must be fulfilled during the manufacturing process that assures the safety of the food being produced. In a similar way, GHPs or good hygienic practices are the set of regulations that are to be followed to maintain proper hygienic conditions which should be monitored during all the steps of food chain and this in turn assures the safety of food. The prerequisites of good manufacturing practices and good hygienic practices when incorporated together forms another broader term related to food safety assurance system, called HACCP or hazard analysis and critical control point, the relation of which can be illustrated through Fig. 12.1. HACCP is a systematic method which assures food safety. It functions by identification, evaluation, and controlling the food hazards. It acts as a tool for the safety and management of product, which is further, linked with various management systems as given in Fig. 12.2. The complete HACCP system includes overall 12 stages, which is further composed of 7 principles and 5 preliminary tasks, which are as follows:

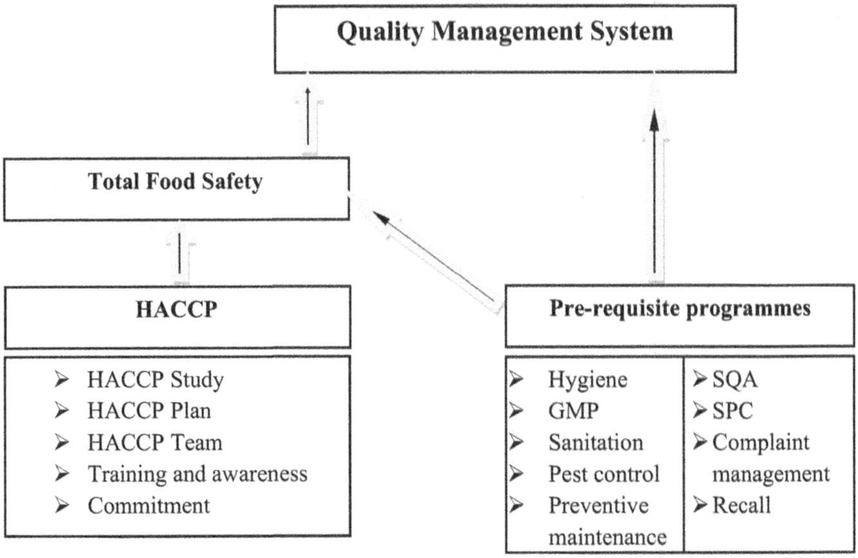

Fig. 12.2 Food safety within the QMP (Quality management programme)

12.1.1 Principles of HACCP

- Conduct hazard analysis.
- Determination of CCP, i.e. critical control point.
- Establishing the critical limits.
- Establishing the monitoring protocols.
- Establishing the corrective actions.
- Establishing the verification protocols.
- Documenting and record keeping.

12.1.2 Preliminary Tasks of HACCP

- Assembling of HACCP team.
- Food description and distribution.
- Description regarding the intended use of food to the consumer.
- Developing the flow diagram describing the process.
- Verification of flow diagram.

Several parameters are taken into consideration and are evaluated by various instruments to ensure the proper quality of the food products, viz. physico-chemical parameters, rheological parameters, phytochemical parameters, and packaging materials.

12.2 History of Utilization of Analytical Techniques in Food Quality Control and Safety

Since 2500 years back, prevention of meat and meat products was done by Egyptian and Mosaic laws. At the same time around 2000 years back, countries like India had established the regulations for pro by Egyptian and Mosaic laws. At the same time around 2000 years back, countries like India had established the regulations for prohibiting adulterations in fats and grains. It was mentioned in the books of *Old Testament* to restrict the consumption of the meat of such animals which were not slaughtered and died of unnatural cause. Regulated weights were used for measurement purposes. As per the records mentioned by Lasztity et al. (2004), wines were inspected and control on beers was done just to ensure the purity and quality of these commodities. Roman government had strict control over the supplies of food and helps the customers to prevent themselves from fraudulence and bad quality. However, it was still observed that during scarce conditions, there was resultant increase in the demand and enhanced fraudulent practices were observed (Adamson 2004).

During the middle ages, tradesmen with specific speciality were given responsibilities to control and supervise the quality of products. For instance, specific troops were deployed in 1419 to prohibit the blending of wines collected from various geographical regions. Different countries opt different ways for controlling the food quality in one form or the other, just in order to assure the safety to the consumers. Later in eighteenth century, concept of chemistry came into focus, where the chemistry was used as analytical tool against the adulteration practices. Robert Boyle proposed the use of specific gravity's principle to detect the adulteration in various foods (Adamson 2004).

In the mid of the nineteenth century, periodic standardization was done in an organized manner for various analytical techniques. This was the time when the industries were not limited to specific regions but started spreading themselves to far off places. This was basically the period of 'Industrial revolution' when the industries started expanding themselves in various fields without following proper hygiene and sanitization. The society changes from rural to urban, domestic factories got converted to food factory system thereby placing strains on production and distribution of food. Adhering to such changes within such a short span, poverty came into existence leading to the development of various health issues among the population. Unfortunately, the awareness regarding the adulteration, hygiene, and quality was limited and was not taken into that much consideration. In 1858, municipal services were set up in Amsterdam to control beverages and foodstuffs. This was later followed by England in 1860 and with this the first modern food law was introduced in the world, i.e. 'An act to prevent Adulteration of Foods and Drinks'. The scientific approaches were made with this act to tackle the problems related to food and analyst was appointed to check the purity of drinks and foods (FAO 1999). Few years later, municipal services were established to control the drinking water. Various laws were then established in different countries like Belgium, Austria, Hungary, Italy, etc. Various institutions were established working on food quality and inspection. The efficacy of such institution can be measured by the reports of food and drug inspection, which records for controlling 176,000 samples, of which 11,000 were adulterated. The major activities based on quality and safety was observed in industrialized nations and countries like Canada, the USA, and Australia enacted these laws.

During the twentieth century, there occurred substantive development in India. Various laws related to adulteration of food and quality control were amended to ensure consistency in the quality of food articles sold in country from 1919 to 1941. Prevention of Food Adulteration Act was enacted in India and was then amended later on, which is still applicable. Other nations such as Portugal and Spain also amended their laws with significant differences and efforts are still being made after a number of amendments to harmonize these laws with proper functionality. It is quite obvious that there is still need of adjustments in various regulations as conditions at present varies from the past, when these regulations were made. Also, soon after the industrial revolution, there felt the need of developing such mechanisms that detect frauds and maintain financial accountability, the consequence of which, investors started depending on the financial organizations associated with the joint stock market. The concept of auditing was made mandatory after the crash of stock market in 1929. Along with establishments of modern standards, BRC, i.e. British Retail Consortium Global Standards were introduced which aimed to protect the consumer's health. It offers consistency, upgrades the supplier's standard, and helps in preventing the product from failure, thereby reducing the units of audits to be performed in manufacturing units.

12.3 Importance of Quality and Safety in Food Industry

Quality is considered to be the most important parameter in any product's market success. Earlier, the quality of food defines the lack of defects in it. Food safety is another important parameter in food industry that comes after assurance of food quality. It is the first ever demand and expectation of the consumer that the food he eats should be of good quality and must be safe. It is considered as integral part of food security. FAO defines food security as a situation where all people have constant physical, economic, and social access to food, which accomplish the nutritional requirement of an individual (World Health Organization 2020). For end users, i.e. customers, food safety is considered as the valuable parameter of food quality. Customers from the industrialized countries demand the product with consistent and high quality throughout the time. Various strategies have been opted by various governing bodies and food industries to ensure food safety due to increased risk of contamination. Examples of such contamination include unauthorized or illegalized food ingredients in various food supplements, melamine in different milk products, mercury-tainted milk powder, carbendazim's presence in juices, especially that of orange juice, etc. The food is considered unsafe mainly due to the three hazards, viz.

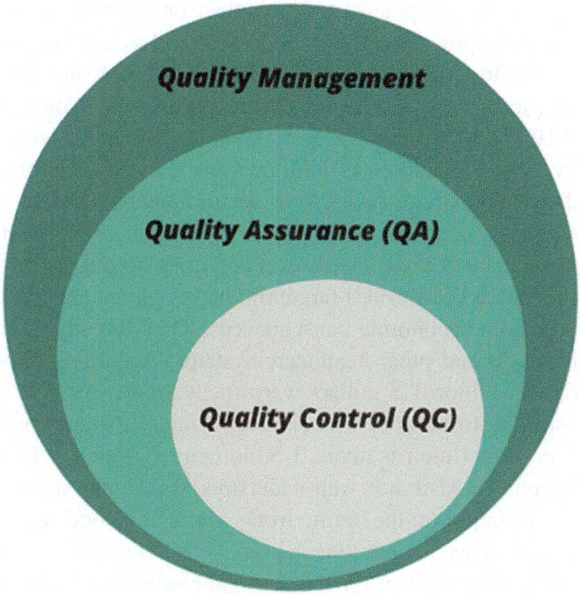

Fig. 12.3 Relation between QC, QA, and QM

(a) Physical hazard—includes extraneous matter such as insect matter, wood, metal, and glass.
(b) Chemical hazard—includes the presence of toxic chemicals such as herbicides, pesticides, and insecticides.
(c) Biological hazard—includes the presence of harmful microbes such as Salmonella and Listeria.

However, it is solely the responsibility and duty of manufacturer to eliminate the foreign particles from the produce he produces, which can be done by practising strict good manufacturing practices. The ultimate aim of following all these protocols is to ensure the safe delivery of products from farm to fork. The relation between quality control, quality management, and quality assurance is given in Fig. 12.3.

The major objective of HACCP is to prevent the human from the risks associated with food and to prevent the adulteration of food by various control measures in various steps of production as various health hazards are usually related with each step. It is applicable to every single step in production process, beginning from the cultivation of animal or plant (i.e. primary production process) including processing, manufacturing by the industry till its consumption by the customer. Various programmes like good manufacturing practices and other prerequisite SOPs (Standard Operating Procedures) should be established and followed before the implementation of HACCP (de Oliveira et al. 2016). Beside this, various other factors are also responsible for variability in the quality of finished products, for instance, defects in equipment, technologies, or methods used in process. Proper use of statistical process is a must to ensure and assure the good quality of the product.

The loopholes in food quality and safety, if not handled and managed properly may lead to social, economic and environmental consequences. As per the reports of WHO 2002, waterborne and foodborne diseases have consumed around 2.2 million lives, out of which 1.9 million were children. Outbreak of these diseases causes damage to trade, tourism which ultimately leads to unemployment and loss of earning among the population, resulting in social disturbance in the society. The prevalence of these diseases in countries highlights the major food safety concerns (WHO 2007). Besides targeting the people's health, these foodborne diseases also pose some economic consequences. They directly or indirectly impose burden on hospitals and other healthcare systems. According to a study in the USA, around 6.5–35 billion US dollars were spent in year 1995 to tackle 3.3–12 million cases suffering from foodborne diseases. Annual cost expenditure on healthcare system in European Union is about 3 billion euros, which persists solely due to *Salmonella* infections. Also, it is well understood that due to increased international food trade, the distance of the farm, from where the food was produced to the end user, i.e. customer has no longer been same. Thus, utilization of resources, energy, and exhaustion of gases (GHS) during the processes including consumption, transportation, and other factors is unavoidable. Therefore, the concept of food miles (distance from farm to end user), food chain should be sustainable so as to reduce the indirect burden on the environment. Food spoilage is another issue which arises due to poor quality and safety of food. Usually, every food product has limited shelf life and is perishable. High-quality food needs rapid minimal temperature conditions during the processing, the temperature abuse of which may lead to microbial growth, thereby causing spoilage of the food and onset of foodborne illness. According to International Institute of Refrigeration, around 300 MT of the produce is wasted annually because of improper refrigeration. Such big wastage of food along with resources is a big environmental persisting issue. Thus, considering the proper importance of food quality and safety in mind and by following the strict hygienic practices, these social, economic, and environmental loses can be overcome.

12.4 Steps in Food Analysis

Food analysis is said to be completed after completion of a number of steps, viz. sample preparation, performing analytical procedures, statistical data analysis, and reporting. The first step in food analysis starts from sample collection followed by sample preparation, which is further followed by performing of various analytical procedures. The selection of the sample should be done in such a way that it must represent the whole population. 'Population' refers to whole material, the properties of which have to be analysed, while 'sample' refers to some fraction of the selected population. The sample may be either one or more selected from the variable region of same population. The amount of sample is usually increased by keeping in mind the size of population. Further 'laboratory sample' is the one which is subpart of the sample obtained from the population due to its larger size. This fraction is actually used for the laboratory analysis. The analyst should perform this task with very precise and accurate measurements in order to obtain the accurate results. Also, this fact cannot be denied that the samples provide only the estimate of true value of whole population.

12.4.1 Sample Preparation

Before getting the sample ready for analysis, there are many other terms that need to be focused on. These includes sample size, sample location, and sample collection. Sample size is mainly dependent upon the expected disparity in properties of a population so that the sample may be able to represent whole of the population. Subsamples are taken in cases where the population size is too large and is performed till the ratio of good and bad sample lies under the predefined value, on the basis of which the population can be rejected or accepted. In case of homogenous population, sample location doesn't interfere with the result as all the subsamples possess similar properties. However, in case of heterogeneous populations, the location of collection of subsamples extremely matters. In such cases where there is heterogeneous population of large size, random sampling is preferred as there are less chances of human bias in it. Other parameters of sample location includes systematic sampling, in which the samples are taken from systematic time or location, say every eighth box from the batch to be analysed or sample to be collected from conveyor belt after every single minute; judgement sampling, in which subsamples are taken from the population by experience and judgement of the analyst. The collection of sample can be done either manually or by sampling devices. Picking the samples from conveyor belt or taking out samples from the sack using special containers or cups is a commonly practised example of manual sampling (Lazzaro and Pike 2014).

12.4.1.1 Homogenizing the Sample

Once the sample is collected by above-mentioned means, making the sample homogeneous is the very next step. The samples collected from the population are usually heterogeneous in majority of the cases and therefore, there is higher probability of variation in properties of samples collected from different location from same population. It is therefore compulsory to have the samples in homogeneous before they are analysed. Homogenization can be performed by the use of various mechanical devices depending upon the type of food (liquid, semi-solid, or solid). Usually mixers, slicers, blenders, and grinders are employed as mechanical devices for homogenization purposes. Other methods for homogenization include chemical methods where strong acids, base, and detergents are used; enzymatic method involves the use of enzymes like lipases, proteases, cellulases, etc.

12.4.1.2 Reducing the Sample Size

The homogenized sample is used to draw the manageable portion which is known as laboratory sample. This laboratory sample represents the properties of population and is used for further analysis.

12.4.1.3 Preservation of Sample

After the collection of sample, it is necessary to preserve the sample so as to maintain itself in its original form. Delaying in preservation of sample may lead to significant changes in sample, which may be physical, chemical, microbial, or enzymatic.

- Physical changes include loss or gain of moisture due to evaporation or condensation, disturbance in structural properties. The extent of physical changes can be reduced by means of adjusting the temperature where the sample is kept and by controlling the forces it experiences.
- Every sample contains loads of microorganisms in them, which exceeds above the safe levels, if not kept properly, leading to spoilage of the sample. Various treatments such as heat treatments, drying, freezing, and chemical preservatives can be used alone or in combination to limit the extent of microorganisms growing in food.
- Many food samples contain enzymes that may lead to various changes in properties of food before analysis is performed. Such enzymes include lipases, cellulases, and proteases. This ultimately ends up by providing erroneous data. Therefore, these enzymes must either be eliminated or inactive soon after the collection of samples, which can be done by the use of chemical preservative, heat treatment, freezing, drying, or combination of these, depending upon the type of sample.
- Samples rich in fat content may be prone to lipid peroxidation. Various factors such as elevated temperature, light exposure, pro-oxidants, and oxygen increase the rate of these reactions leading to spoilage of the sample at much faster rate. Such samples with higher content of unsaturated lipid can be stored under inert gas packaging such as nitrogen in dark rooms at refrigerated temperatures (Nielsen 1998).

12.4.1.4 Labelling the Sample and Its Identification

Samples should be labelled carefully from the very first day it is obtained so that if the problem persists during later stages, its origin can be identified. Following information is to be labelled before keeping the sample to storage:

(a) Description of sample
(b) Time when the sample was collected
(c) Location of sample
(d) Name of the person responsible for sample collection
(e) Selection method of sample
(f) Unique coding of sample

The analyst performing the tests should maintain a notebook in which detailed documentation of sample selection, its preparation procedures, and other results are to be recorded. As stated above, the samples must be labelled with unique code, the details of which should be properly recorded in the notebook by the analyst, so that in case if problem arises in future, the sample can be easily identified.

12.4.2 Analytical Procedures

Various analytical methods such as mass spectroscopy, liquid chromatography, gas chromatography, infrared spectroscopy, polymerase chain reaction, and many other analytical methods are widely used to assess the quality of food products (Tang et al. 2019). The detailed description regarding various analytical techniques and their selection for analysis of food is discussed later in this chapter.

12.4.3 Data Analysis and Reporting

Number of measurements is done on same sample to obtain the best value of data which indicates the value's reliability. Various techniques are used which enable us to gather the information of the laboratory sample, such as measure of central tendency, measure of spread data, sources and propagation of errors, rounding of significant figures, standard curves, and regression analysis. Measure of central tendency is the most commonly practised parameter. It gives the mean value that represents the overall properties for the number of measurements. Though it is not sure that which value is nearest to the true value, therefore we measure the mean using all the values and represent the result in the form of mean. Mean of the data is considered as the best estimated experimental value derived from measurements. Mean is calculated using the equation:

$$\overline{x} = \frac{x_1 + x_2 + x_3 + \dots\dots\dots\dots + x_n}{n} = \frac{\sum x_i}{n}$$

where $\overline{x}$ = mean, $x_1 + x_2 + x_3 + \dots\dots\dots\dots + x_n$ = individually measured values, n = number of measurements.

Median is another method that is used for determination which depicts the mid value of numbers within the group. Usually few values of the experiment lies above the mid value while other lies below it. Median is not commonly used as mean is considered as superior estimator.

The measure of spread of data depicts the closeness of the repeated measurements. Standard deviation is used as measure of spread in experimental measurements. Following equation is used to determine the standard deviation during experimental measurement:

$$\text{S.D.} = \sqrt{\frac{\sum_{i=1}^{n} (x_i - \overline{x})^2}{n - 1}}$$

where SD is standard deviation, $\sum$ is to sum, X_i is each score in distribution, $\bar{x}$ is sample mean, n is number of cases in sample.

Sources of errors—Accuracy as well as precision are considered to be very important in all analytical determinations. Also, it is impossible to expect that analytical technique will be entirely error free. The best we can expect during analytical measurements is less variation and consistent data. Errors can be classified as determinate or systematic error, indeterminate or random error, blunder or gross error. Systematic error produces result which deviates consistently from expected value in one or the other direction. It is not only difficult but also time consuming to identify the source of such errors as it may happens due to inaccurate instruments or devices. For instance, an impaired pipette consistently delivering the incorrect amount of reagent and gives the result with high precision but would be ultimately inaccurate. Sometimes the quality of chemicals is the issue creator, poor quality, or impure chemicals give improper results. Such systematic errors can be overcome by timely calibration of the instruments, running blank samples, etc. indeterminate errors or the so-called random errors always there in an analytical measurement. As the name suggests, these errors fluctuate randomly and are unavoidable. For instance, detection of end point during titration, using of pipette, reading analytical balance, all these causes random errors. Although it is difficult to avoid such errors, but fortunately these are usually small. Gross errors produce the results with values that are too far from the true value. Using wrong reagent, incorrect techniques cause such errors. These blunders or gross errors can be identified and corrected easily (Garfield 1991).

Rejecting data—Sometimes it is observed that while performing an analytical experiment, one of the observation is deviated too much from the mid value and other value which might be due to blunder in the protocol. In such cases, that particular value can be rejected and is considered to be incorrect. For such type of cases, *Q-test* is usually performed in order to decide either to accept or reject that particular value.

$$Q = \frac{X_{BAD} - X_{NEXT}}{X_{HIGH} - X_{LOW}}$$

where X_{BAD} is questionable value, X_{NEXT} is next closet value to X_{BAD}, X_{HIGH} is the highest value of data set, X_{LOW} is the lowest value of data set.

It is to be noted that sample can be rejected if *Q-value* is greater than given value in *Q*-test table for number of samples that are to be analysed (Nielsen 1998).

12.5 Selection of Analytical Methods for Food Analysis

Food analysis or the analysis of food is an interdisciplinary approach as it includes the impact of various spheres including health, economic impact as well as societal impact. The chief objective of analysing any food material is to ensure that the food which is supposed to be consumed by the consumer is appropriate and acceptable in terms of its chemical constituents, quality aspect, organoleptic properties, safety, and also the nutritional value. There are major factors which can play in significant role in affecting the molecular and chemical composition of food products such as geographical distribution, genetic origin, environmental conditions, farming practices, breeding, soil fertility, water quality, processing conditions, presence of adulterants as well as any type of contaminant can affect the food material. Therefore, the proper analysis of food products is very essential as it can pose a significant effect on the health of consumers. Till date, it has not become possible to establish a single perfect method for analysing each and every component of food so the different analytical methods are used in association with each other to come to a final conclusion. Taking this into consideration, researchers and scientists are trying to develop a reliable, powerful, and relatively inexpensive analytical tool in order to analyse the quality and quantity of food products rapidly and at the same time, the results should be accurate too. Mass spectroscopy (MS), liquid chromatography (LC), gas chromatography (GC), infrared spectroscopy (IR), capillary electrophoresis (CE), enzyme-linked immunosorbent assay (ELISA), polymerase chain reaction (PCR), and nuclear magnetic resonance spectroscopy (NMR) are some of the most widely and extensively used analytical methods to assess and analyse the quality of food products (Tang et al. 2019).

Selection of a particular type of analytical food for carrying out food analysis largely depends upon certain inherent characteristics such as specificity, selectivity, precision, and accuracy. The food analyst must address any kind of interferences among the different properties of food, ensure high degree of specificity, variability of test results, percent recovery of the sample to be measured, and must compare the method being employed with the standard or traditional method of food analysis in terms of accuracy. Moreover, there are some other factors also which affect the selection of a specific type of analytical method to be used such as sample size, reagents, equipment, cost of method, usefulness in terms of time requirement, reliability as well as the need of method. The food analyst must ensure that the sample size must fit his needs as it should neither be too small not too large. At the same time, the sample should fit the equipment or glassware intended to be used for the procedure. The stability of equipment and reagent is equally important while selecting a suitable analytical method as it must be addressed whether the equipment being used is able to withstand temperature and pressure conditions or not. Cost of the analytical method to be employed must be taken into consideration as it should be apt in terms of reagents, equipment as well as personnel. Thus, these points are quite advantageous in order to evaluate the suitability of any particular analytical method being considered (Nielsen 2017).

Objective of the assay or measurement plays a major part in the selection of appropriate analytical method. For instance, in case of a rapid processing measurement, a less reliable and less precise method can be employed rather than using an official or stand test method. On the other hand, the official, reference, definitive, or primary methods are employed in settings or analytical laboratories which are very well equipped and have trained personnel. It is a common practice among the food industries and companies that they use the rapid and unofficial analytical methods as they are less time consuming and then validate their results against the standard or official analytical test procedures. For example, the moisture content can be determined using a calibrated moisture balance rather than hot air oven method which is more time consuming (Wetzel and Charalambous 1998).

Another major factor affecting the selection of analytical method include the characteristics of food constituents as the chemical components of food matrix such as carbohydrate, protein, and lipids affect the performance of several analytical methods. For instance, it has been reported in research studies that different types of disturbances are observed in the high-sugar and high-fat food products as compared to the low-sugar or low-fat food materials. Therefore, it is possible that the use of multiple analytical techniques may be required rather than a single technique for a specific food component owing to the complexity of food materials. The nature of food matrix or food system determines the digestion trials as well the extraction steps required for the accurate and precise analytical test results. A schematic layout of the food matrix (triangle scheme) based on the chemical composition of food products (carbohydrates, lipid, and protein) was suggested by AOAC International (Association of Official Analytical Chemists) in which foods were categorized into different levels and were rated as 'high', 'low,' or 'medium'. These key nutrients, viz. carbohydrates, fats, and proteins are known to have a significant effect on the performance of any specific analytical method. By doing so, nine different combinations of high, medium, and low levels of carbohydrates, fat, and proteins were created which helps in determining the suitability of an analytical method based upon the composition of food matrix (Ikins et al. 1993; Lovett 1997; Ellis et al. 1997; Devries and Silvera 2001; Sharpless et al. 2004; Nielsen 2017).

Specificity, accuracy, sensitivity, and precision are the important characteristics while selection of any analytical method. At the same time, the variability of the data obtained from analytical method must be addressed considerably in order to detect and accept the differences related to a specific characteristic to the consumer. Sampling must be carefully done so to ensure that the number of selected samples to be analysed is a representative of the whole population. The equipment used for analysis must be appropriately standardized in order to obtain valid, accurate, comparable, and reproducible test results. Moreover, any limitations or drawbacks related to the performance of equipment must be addressed properly (Latimer Jr 1997; Nielsen 2017).

Method validity can also be determined by analysing the control materials or samples (check samples) against the samples to be tested which is an important part of quality control. These check samples services are provided by numerous government and non-government organizations (NIST, IRMM, BCR, AACC, etc.) in order to evaluate the dependability and reliability of any analytical method. The results so obtained after the analysis are then evaluated statistically followed by its comparison with other laboratory results to assess the degree of precision and accuracy. For instance, control samples of cereals-based samples are available for the analysis of ash content, moisture, protein, sugars, minerals, vitamins, total fibre (soluble and insoluble), etc. Similarly, another organization, viz. AOCS (American Oil Chemists' Society) offer check samples for oilseeds, marine oils, toxins (aflatoxin), minerals, specialty oils, and trace minerals for determining the suitability of fatty acid composition as well as for nutritional labelling. The data so obtained is then analysed by the trained researchers from different countries to determine the degree of accuracy of work done by the personnel (Ambrus 2008; Nielsen 2017).

It is not necessary to obtain the standard reference materials from outside organizations only, instead the check samples can be prepared internally in the laboratory as well because standard reference materials or the check samples are an important tool to ensure the reliability and accuracy of any data obtained using analytical methods. For doing so, the control sample of appropriate type must be selected, gathered, mixed, and prepared so as to ensure uniformity. Moreover, the packaging and storage must be done carefully and a routine analysis of the control samples must be done along with the test samples. It is important that the sample which is intended to be used as a reference or control must be similar to the food matrix of test sample being considered for analysis (Nielsen 2017).

12.6 Brief Overview of Various Analytical Techniques

Currently, a large number of analytical techniques or methods are being used for the analysis of food products for quality control. The various methods widely employed in food processing industries and laboratories include techniques of sample preparation, biological, separation, spectroscopic, rheological, thermal, radiochemical, electrochemical, and enzymatic analysis. Sample preparation technique involves headspace, microwave-assisted extraction, supercritical fluid extraction, solid-phase extraction, purge and trap, pressurized liquid extraction, flow injection analysis, and microextraction. PCR, biosensors, recombinant DNA technology, microbiological analysis, immunological assay, and others can be used as biological techniques. Similarly, for the separation of different food components from a food matrix or system, techniques like LC, GC, SDS/PAGE, supercritical fluid chromatography, capillary electrophoresis, and LC-GC can be used. Techniques including mass spectrometry, NMR, fluorescence, infrared, X-ray, ultraviolet, atomic spectroscopy, light scattering, electron spectroscopy, and circular dichroism can be employed for the qualitative and quantitative analysis of food materials. Creep, oscillatory shear, rheometry, viscometry, stress relaxation, normal stress, etc. can be used to determine the rheological properties of food products. Differential thermal analysis, DSC, thermogravimetry, and thermochemical techniques are used for the thermal analysis. Radiochemical and electrochemical techniques involve radioimmunoassay, isotropic method, radiochemical, radiometric, radioisotope, radiotracer, radiolabelling, biosensors, voltammetry, potentiometry, amperometry, polarography, conductometry, and coulometry. AACC International has also listed the procedures for the quantitative analysis of food materials including ash content, moisture, acidity, amino acid composition, crude fat, fibre, nitrogen, reducing sugar, total sugar, vitamin, mineral, and physicochemical tests (Cifuentes 2012). A brief overview of the various techniques is represented in Table 12.1.

Table 12.1 An overview of various qualitative and quantitative analytical techniques

Sr. no	Analytical technique	Application	Reference
A	**Sample preparation techniques**		
1.	Subsampling	• The most important and potential source of error in analysing any food material is the sample selection. • The test sample withdrawn from the whole lot of sample is the representable sample. • The sample must be drawn based upon the relationship of test sample with the whole lot of food.	Nielsen (1998), Cifuentes (2012), Moldoveanu and David (2021)
2.	Compositing	• The admixture of either two or more than two portions of any food material after subsampling is known as compositing. • The average of normal variation between the two different samples is done as a result of compositing. • It is essential that the individual samples selected by using subsampling technique should have same size, volume, and weight so as to make the sample homogenous and uniform in nature.	Nielsen (1998)
3.	Chopping, grinding, mixing	• The type of equipment used for the physical and mechanical processing of food products depends largely upon the food product to be treated as well as the moisture content of food. • The various equipment involved in this technique include mechanical choppers, grinders, mixers, mill, and blenders. • The food analyst must ensure that the mechanical process should prevent any changes in the food product as it can result in inaccurate and biased analytical results.	Nielsen (1998)
4.	Freezing and thawing	• Freezing is done to prevent any change in a food prior to analysis or to reserve storage. • The composition of the food should not be changed while	Nielsen (1998)

(continued)

Table 12.1 (continued)

Sr. no	Analytical technique	Application	Reference
		thawing; therefore, it should be done gradually, without giving heat and in a closed container so as to reserve the moisture content of food.	
5.	Microwave-assisted extraction (MAE)	• This method involves the extraction of compounds by using microwaves and can result in getting a higher yield in short span of time. • This technique is based on the mechanism in which the energy absorption takes place as soon as the microwave is passed through the solvent and gets converted into thermal energy. • Dipole rotation and ionic conduction accompany the heating caused by microwaves. • MAE is superior as it can be employed at the same temperature by the use of less solvent as compared to the conventional extraction techniques. • This technique can be used to extract various phenolic and biological compounds such as essential oils, organic acids, and fatty acids.	Proestos and Komaitis (2008), Sonar and Rathod (2020)
6.	Solid-phase extraction (SPE)	• SPE technique is a method of sample preparation in which a fused silica fibre is coated with a suitable stationary phase as a result of which the analyte present in the sample gets extracted and concentrated onto the fibre coating. • This technique is cost-effective in terms of solvent and disposal cost as well as saves time. • This method is generally used in combination with other techniques like gas chromatography and mass spectroscopy and is used to extract organic compounds (volatile as well as non-volatile)	Kataoka et al. (2000), Hanhauser et al. (2020)

(continued)

Table 12.1 (continued)

Sr. no	Analytical technique	Application	Reference
		from the biological, environmental, and food samples. • It is applied for the analysis of different aromatic and flavouring compounds present in the food samples. • This technique can also be employed in order to monitor the water quality for the identification of heavy metals and other contaminants in water samples.	
7.	Supercritical fluid extraction (SFE)	• SFE technique involves the use of supercritical CO_2 for the selective separation of desirable compounds without causing any degradation or toxicity of the food product. • It is used as an effective technology for the quantitative as well as qualitative analysis of the constituents which are naturally occurring and heat-labile in nature. • This technique can be used for the separation of high-quality essential oils (such as lemon oil, rosemary oil, lavender oil) and its derivatives, extraction of edible fats and oils, antioxidants, pesticides as well as for the detoxification of shellfish. • It is a promising technique for the extraction of microalgal compounds and thermolabile molecules and at the same time reduces energy costs by preserving the natural properties as well as qualities of compounds considered to be bioactive in nature.	Mohamed and Mansoori (2002), Molino et al. (2020)
8.	Flow injection analysis (FIA)	• FIA technique is widely used by researchers in order to analyse the sulphite content present in foods and beverages. • It is a cheap, accurate, simple, and quick analytical method by using relatively less amount of	Claudia and Francisco (2008), Bezerra et al. (2020)

(continued)

Table 12.1 (continued)

Sr. no	Analytical technique	Application	Reference
		reagent, small sample volume, less toxicity, and simple instrumentation. • It has been suggested through research studies that the implementation of FIA is accompanied by an increase in the throughput of analyte, decreased losses, minimized generation of waste as well as less chances of contamination. • The procedure involves two phases, viz. the former being extraction of sulphating agent and latter involves the injection of extracted sulphating agent into the liquid extract and its detection in the FIA system.	
9.	Pressurized liquid extraction (PLE)	• This technique is also known as accelerated solvent extraction (ASE), pressurized hot solvent extraction (PHSE), pressurized fluid extraction (PFE), high pressure solvent extraction (HPSE), subcritical solvent extraction (SSE) and high pressure, high temperature solvent extraction (HPHTSE). • PLE involves the extraction of compounds using solvents at high temperature and pressure. • This technique is considered to be a green and sustainable technique for extracting bioactive compounds from its natural as well as synthetic sources. It is also used in the detection of various bioactive compounds present in different food samples. • Various contaminants including polycyclic aromatic hydrocarbons, polychlorinated compounds, alkylphenols, pesticides, metals, drug residues, natural toxins, and other matrix components like polyphenols, essential oils, fat matter, pharmacologically active	Carabias-Martínez et al. (2005), Alvarez-Rivera et al. (2020)

(continued)

Table 12.1 (continued)

Sr. no	Analytical technique	Application	Reference
		compounds can be extracted by using PLE technique as an analytical method in food analysis.	
B	**General analysis techniques**		
1.	Heating and drying	• Various heating devices used in laboratory include hot plates, mantles, steam baths, water baths, burners, convection oven, vacuum oven, desiccator, etc. • In order to ensure uniform heating of sample, ovens are most extensively used. • Steam bath followed by oven heating must be used for extremely wet food materials as steam bath treatment allows evaporation of moisture prior to drying.	Nielsen (1998), Cifuentes (2012)
2.	Ashing and digestion	• The removal of organic matter to obtain inorganic residue by employing heat treatment using muffle furnace is known as ashing. • It is used to determine the total inorganic residues as well as for the analysis of trace minerals present in the sample. • The major drawback associated with dry ashing is that it is generally difficult to extract the metal completely from the ignited residues. • Another method involving the destruction of organic matter to estimate the inorganic residues is by acid digestion in which acids like H_2SO_4 and HNO_3 are used either individually or in conjunction. • Acid digestion technique used for the analysis of metal residue is accompanied with high chances for the sample to become contaminated. • Grossbier and Schoenfuss (2021) conducted a research study to do the comparative analysis between conventional	Nielsen (1998), Grossbier and Schoenfuss (2021)

(continued)

Table 12.1 (continued)

Sr. no	Analytical technique	Application	Reference
		digestion and microwave-accelerated digestion (MAD) method for the determination of equivalency. It was concluded that the rate of MAD was much faster than the conventional one for the sake of mineral analysis.	
3.	Extraction	• Extraction involves the partitioning of material between two phases. • Liquid–liquid extraction is based upon the extractant solubility in the two liquids. On the other hand, liquid–solid extraction is bit complicated due to the physical occlusion of extractant inside the solid material (inert in nature). • Extraction of fat content from a meat sample is one example of extraction technique where Soxhlet unit is utilized. The finely divided solid sample is mixed with extractant (solvent) which provides a good siphoning action. • Multiple extractions are often necessary for the quantitative analysis of food sample as a single extraction may not be suitable for partitioning of a substance.	Nielsen (1998)
4.	Distillation	• Distillation (simple, steam, and fractionation) is an analytical method used for the purification of a substance. • Simple distillation involves a liquid solution which is heated to a temperature when vapours begin to form followed by condensation of vapours and its collection. • A steam distillation unit consists of steam generator, sample flask and a condenser. Fractional distillation involves the process of separating a liquid mixture consisting of two or more constituents with close boiling points.	Nielsen (1998)

(continued)

Table 12.1 (continued)

Sr. no	Analytical technique	Application	Reference
		• It allows the separation of components on the basis of their volatility.	
5.	Titration	• Titration is one of the most extensively used analytical techniques for the analysis of check samples in laboratories. • It allows the quantitative determination of the concentration of an unknown or known analyte by using a titrant or titrator which is a standard solution of known volume and concentration.	Nielsen (1998), Rohindra and Lata (2020)
C	**Determinative and separation techniques**		
1.	Paper chromatography	• Paper chromatography is an important method used in food analysis for the extraction, identification, and isolation of synthetic food colours from various food products such as soft drinks, candies, and jellies. • It is based upon the principle of partition of compounds in which they get distributed between stationary phase (paper fibres) and mobile phase (developing solvent). • The components are identified and separated on the basis of Rf value of standard solution and samples.	Bachalla (2016)
2.	Gas-liquid chromatography	• This analytical technique is extensively used for the qualitative and quantitative analysis of food constituents, additives, flavouring compounds, aromatic components, contaminants, pesticides, preservatives, pollutants, natural toxins, transformation products, drugs, packaging materials, etc. • Gas chromatography helps to analyse the semi-polar, non-polar, volatile, semi-volatile chemicals, sterols, oils, fatty acid chains, off-flavours, etc. present in the food materials.	Lehotay and Hajšlová (2002), Cortes (2020)

(continued)

Table 12.1 (continued)

Sr. no	Analytical technique	Application	Reference
		• The time consumed in the analysis can be shortened by increasing the flow of carrier gas, increasing column diameter, heating the column, shortening the column length, or by reducing the viscosity of carrier gas. It ensures the required selectivity.	
3.	High performance liquid chromatography (HPLC)	• HPLC is employed for the separation of a mixture of compounds for the identification, quantification followed by purification of individual constituents of the mixture. • It plays an important role in food industries for quality role as it is used to analyse and separate food additives, preservatives, toxins, contaminants, and other food components. • It is based on the principle of column chromatography in which a high pressure of the mobile phase is applied to pump it through a packed column.	Nollet and Toldra (2019), Akash and Rehman (2020)
4.	Spectrophotometry	• Spectrophotometry is based upon the principle of reflectance of light by the sample material when it is exposed to a source of polychromatic light. • It helps in the detection of impurities, quantitative estimation of concentration of a component, characterization of proteins, structure elucidation of organic compounds, detection of functional groups in food constituents, determination of food dyes, quality evaluation of agricultural commodities, and many more.	Polesello et al. (1983)
5.	Refractometry	• Refractometry technique is used to determine the nature of food products. It is a method for the qualitative analysis of an unknown compound based upon the refractive index of the compound being considered, and	Bradley (2010)

(continued)

Table 12.1 (continued)

Sr. no	Analytical technique	Application	Reference
		it varies with the concentration of compound, wavelength of light as well as temperature. • It is used to determine the moisture of condensed milk, liquid sugar products, total soluble solids of fruits and fruit products.	
6.	Microscopy	• Electron microscopy works on the principle at which imaging is performed at room temperature and under high-vacuum since gas molecules may scatter electrons that reduce image resolution. • The wavelength range of visible light should be from 400–800 nm which adheres to the resolution of optical microscopy techniques. • High resolution imaging can be achieved using electron microscopy especially in food powders.	Burgain et al. (2017)
7.	Capillary electrophoresis (CE)	• CE is a technique used for the determination of free amino acids using optical detection, which is mainly based on the derivatizing agent, 4-Chloro-7-nitrobenzo-2-oxa-1,3-diazole (NBD-Cl), which employs laser-induced fluorescence (LIF) detector. • Other derivatizing reagents such as 6-aminoquinolyl-N-hydroxysuccimidyl carbamate (AQC), 9-fluorenylmethyl chloroformate (FMOC-Cl), naphthalene dicarboxaldehyde (NDA), o-phthalaldehyde (OPA), phenylisothiocyanate (PITC), and dansyl chloride are used for this detection. • It is used for the analysis of food additives, herbicide, animal nutrition, and detergents.	Omar et al. (2017)
8.	Supercritical fluid chromatography	• SFC allows the use of higher flow rates with lower pressure falls through the column. It leads	Bernal et al. (2013)

(continued)

Table 12.1 (continued)

Sr. no	Analytical technique	Application	Reference
		to better efficiency in a short duration of time as well as reduced utilization of organic solvents, which is carried out in subcritical conditions. • CO_2 is the most frequently used supercritical fluid owing to its correlating properties such as non-toxic, non-explosive, considered generally recognized as safe (GRAS) reagent and easily achievable experimental conditions, such as temperature 31 °C and pressure 73 bar.	
9.	Mass spectroscopy (MS)	• MS is based upon the principle of generation of ions from organic and inorganic compounds followed by their separation and detection both quantitatively as well as qualitatively. • It is used for the analysis of pesticides, lipids, drugs, mycotoxins, caffeine, contaminants, volatile compounds, toxins, phthalates, metabolites, amino acids, plasticizers, etc.	Nollet and Munjanja (2019)
10.	Fluorescence	• Fluorescence spectroscopy technique is used for the monitoring of food standards and analyses the food quality. • It is a quick, easy, rapid, and non-destructive method of analysing the quality of foods including dairy products, meat and seafood, eggs, vegetable oils, honey, wines, beers as well as detecting the various contaminants present in food products. • It is known to be a consolidated technique for the quantification of dissolved organic matter in a fast matter.	Sádecká and Tóthová (2007), Carstea et al. (2020)
11.	Nuclear magnetic resonance (NMR) spectroscopy	• NMR spectroscopy is an analytical technique which works on the principle of magnetic properties of substances.	Hatzakis (2019)

(continued)

Table 12.1 (continued)

Sr. no	Analytical technique	Application	Reference
		• It is used for the structural characterization of apple allergens and analysis of organic acids, conjugated linolenic acid, amino acids, antioxidants, carbohydrates, sugars, pectins, coffee, flowers components, phytochemicals, tea, fish, honey, spices, etc.	
12.	Infrared spectroscopy	• It is based upon the principle of absorption of electromagnetic radiation between 780 and 2500 nm wavelength. • The physical properties and chemical composition of food components (carbohydrates, water, proteins, fat) can be determined using IR spectroscopic technique. • It finds its application in the analysis of cereals and cereal products, dairy products, meat, fish, fruits, vegetables, confectionery, beverages and to assess the authenticity of food products.	Osborne (2006)
13.	X-ray spectroscopy	• X-ray spectroscopy is a non-destructive analytical technique for mineral analysis in cement industry, geology, petroleum, chemical, medical, and food industries. • Estimation of minerals like chromium, lead, titanium, iron, and zinc can be done by using this non-invasive analytical method.	Sosa et al. (2018)
14.	Ultraviolet spectroscopy	• This technique is extensively used in food industries for quality control and helps to detect adulterants, contaminants, identify the origin of food materials, variety of wine, analysis of food matrices (milk, coffee, wine, oil, meat), and differentiation between decaffeinated and caffeinated coffee.	Power et al. (2019)

(continued)

Table 12.1 (continued)

Sr. no	Analytical technique	Application	Reference
15.	Circular dichroism (CD)	• It is a spectroscopic technique involving conformational study of proteins, nucleic acids, and biomolecules. • CD data involves the study of structural composition of proteins, unfolding of proteins as a result of temperature or addition of chemical denaturants, stability of proteins, and effect of mutations on the structure of protein molecules.	Martin and Schilstra (2008)
D	**Biological techniques**		
1.	Biosensors	• Biosensors are widely used in food industries which involves the interaction of a biological element with the sample under observation (test sample) as a result of which a biological response is obtained. It is further transcribed into electrical signals with the help of a transducer. • This technique is used for the detection of toxic compounds, organophosphates, ammonia, methane, pathogenic organisms, water-soluble vitamins, antibodies, and chemical compounds. • It is used to test the quality of water and to measure carbohydrates, proteins, alcohols, phenols, acids, gases, inorganic compounds, amides, and certain heterocyclic compounds. • It plays an important role in the quantitative detection of ultra-low concentration of biomarkers in a very sensitive, robust, reliable, and selective manner.	Scott (1998), Mello and Kubota (2002), Purohit et al. (2020), Chandra et al. (2012), Choudhary et al. (2016), Deka et al. (2018), Mahato et al. (2018)
2.	PCR (polymerase chain reaction)	• It is a precise method of amplification of a desired fragment of DNA from a mixture of molecules of DNA. • PCR is used to detect genetically modified organisms, food toxic components	Klančnik et al. (2012)

(continued)

Table 12.1 (continued)

Sr. no	Analytical technique	Application	Reference
		(pathogens and contaminants), identify various species of dairy and meat products. It helps in quality control and protects human health against harmful consequences of toxicological compounds and hence ensures microbiological safety.	
3.	Microbiological analysis	• Microbiological analysis of food products is based upon accuracy of test, recovery of target organism, limit of detection of method, comparability of results, microbial growth inhibition, and competitive growth. • It assists in the Hazard Analysis Risk in management of food safety, verifying HACCP plans as well as to assess the storage stability or shelf life of food products.	Stannard (1997)
4.	Immunological technique	• Immunosensors are employed for the detection of any toxic substance present in food products which can pose deleterious effects on human health. • Immunological technique is used to detect veterinary drugs, anabolic steroids, pathogenic bacteria, aflatoxins, mycotoxins, GMOs, and pesticides present in the food samples. • SPR immunosensor helps to analyse food materials to determine the presence of food-borne pathogens and toxic compounds in food commodities.	Ricci et al. (2007)
E	**Rheological techniques**		
1.	Oscillatory shear	• It is a common rheological technique used for the determination of viscoelastic properties performed on rheometers either stress-controlled or strain-controlled. • This testing can be divided into two types, viz. small	Melito et al. (2012)

(continued)

Table 12.1 (continued)

Sr. no	Analytical technique	Application	Reference
		(SAOS) and large amplitude oscillatory shear (LAOS), where the former testing is performed in the linear viscoelastic region (LVR) and latter beyond the region of linearity. • LAOS has been used to study the mechanical properties of dispersed systems, such as emulsions, suspensions, and foams.	
2.	Rheometry	• Rheometry describes the relationship between the stress acting on a given material and the resulting deformation and/or flow that takes place in a specific period of time. • Stress is the measurement of force per unit of surface area and is expressed in Pascals (Pa). while strain represents a dimensionless quantity of relative deformation of a material that took place. • It is widely used to analyse rheological properties of food gels such as gelatin, jellies, and cooked egg whites; baked products, in starch and dairy products.	Tabilo-Munizaga and Barbosa-Cánovas (2005)
3.	Viscometry	• Capillary viscometer works on the principle of the drop of the pressure along the capillary which is transformed into a shear stress at the wall and the volumetric flow rate to shear rate. • It is used to analyse the viscosity and textural properties of food such as in wheat dough, soups, butter, honey, and sauces.	Campanella et al. (2002)
4.	Stress relaxation	• Stress relaxations provide information on permanent cross-linking, effects of different chemicals and enzymatic additives on baking quality as well as distinguish products from different origins. • It is an objective method for	Bhattacharya (2010)

(continued)

Table 12.1 (continued)

Sr. no	Analytical technique	Application	Reference
		the quality assessment of bread, wheat, and pulse-based doughs. Practical applications include design of product formulation, dough handling systems, and for the purpose of sheeting/ flattening. • The desirable characteristic for flattening purpose is that dough needs to possess a low value of residual stress for the preparation of chips and flakes.	
F	**Radiochemical and electrochemical techniques**		
1.	Radioimmunoassay	• Radioimmunoassay evaluates the quality and wholesomeness of food which requires a sample containing the antigen of interest, a complementary antibody, and a radiolabelled version of the antigen. When the radiolabelled antigen is added, it competes with the sample antigen and displaces it from the antibody. • The solution containing antigen–antibody complex is denser, so centrifuging the mixture is essential which allows separation, resulting in a pellet containing the bound sample antigen/radiolabelled antigen.	Grange et al. (2014)
2.	Radiometric	• Radiometry is a temperature measurement technique which is based on the principle where microwave frequency range, the thermal noise power emitted by a dissipative body is directly proportional to its temperature. So, the temperature inside a dissipative material can be determined using a radiometric system. • Quality control has one of the main concerns which is temperature control which is measured by infrared, optic fibre, and thermocouples measurements.	Cresson et al. (2008)

(continued)

Table 12.1 (continued)

Sr. no	Analytical technique	Application	Reference
3.	Voltammetry	• Stripping voltammetry is the most sensitive electroanalytical technique. The analyte is accumulated on a working electrode by controlled potential electrolysis followed by the dissolution of the deposit when a linear ramp is applied to the electrode. It results in the production of a detectable current at the electrode surface. • Anodic stripping voltammetry (ASV) was the first technique to be developed. It was mainly applied to trace analysis of heavy metal ions using a hanging mercury drop electrode. • It is generally used for the determination of food contaminants (toxic metals, pesticide, fertilizers, and veterinary drugs residuals), trace essential elements, food additive dyes, etc.	Alghamdi (2010)
4.	Potentiometry	• The basic principle of potentiometry is voltage measurement at null current. When an electrode is placed in a solution, it tends to send its ions into the solution and those ions in the solution react with the electrode. • Its advantages include low cost, ease of commercial production, and the possibility of obtaining selective sensors. • They are used to monitor cheese fermentation, evaluation of the impact of micro-oxygenation and oak chip maceration on wine composition, monitor changes during beer brewing, etc.	Pomeranz and Meloan (1994), Sliwinska et al. (2014)
5.	Amperometry	• Amperometry is the electroanalytical technique that involves the application of a constant reducing or oxidizing potential to an indicator electrode as well as measurement	Adeloju (2005), Scampicchio et al. (2008)

(continued)

Table 12.1 (continued)

Sr. no	Analytical technique	Application	Reference
		of the resulting steady-state current. • Amperometric biosensors include electronic tongue, which works on the principle of an electrochemical conversion occurring at an electrode and the resulting current is measured. • Biosensors are used for the analysis of complex mixtures such as wine and must.	
6.	Conductometry	• Conductometric sensors work on the principle of changes in conductivity, which result from the interactions with the volatile odorants, leading to the changes in the sensor's electrical resistance. • The three types of conductometric sensors used most commonly in electronic noses include metal oxide semiconductors (MOS), conductive polymer (CP) sensors, and metal oxide semiconductor field-effect transistors (MOSFET). • Some of the food applications include the monitoring of spoilage (like red wine), the dehydration of tomatoes, determination of the meat freshness, classification of fruits based upon ripeness and detection of aflatoxins in corn.	Sliwinska et al. (2014)
G	**Some other techniques**		
1.	Microfluidics	• Microfluidics is a new and emerging technology that is utilized in 3D printing which offers various advantages in terms of less generation of waste, consumption of reagent, cost, and other factors. • Different methods including micromilling, micromachining, hot embossing, etc. are used to develop microfluidic devices. • This analytical technique is used for the chemical and	Nielsen et al. (2020)

(continued)

Table 12.1 (continued)

Sr. no	Analytical technique	Application	Reference
		biological analyses and is considered to be one of the most advanced, well developed, and emerging technology in the field of 3D printing. • There has been an emerging research to study the application of 3D printers for the development of microfluidic devices. But it was concluded that the creation of small interior fluidic features is quite difficult with the use of 3D printers. On the other hand, surface feature devices are beneficial for the development of microfluidic structures.	
2.	Smartphone-based food analysis	• Smartphone-based food analysis technique has gained much attention recently different sectors, viz. food industry, agriculture sector, healthcare, and environmental monitoring. • Owing to the numerous useful components of smartphones such as Bluetooth, Wi-Fi, battery, processor, cellular data, camera, video, and visual display; it allows the rapid, low-cost, and easy measurement as well as detection of components. • Unlike laboratory equipment and tests, the smartphone-based technology can't be used alone. • This technology can be used along with other methodologies like fluorescent imaging, microsphere fluorescent immunoassay, colorimetric assays, microfluidics, lateral flow immunoassay, colorimetric imaging, voltammetry assay for the detection of bacteria in water, antibodies in milk, allergens in food samples, aflatoxin in maize, fluoride, and catechols in water..	Rateni et al. (2017)

(continued)

Table 12.1 (continued)

Sr. no	Analytical technique	Application	Reference
3.	Paper-based diagnostics	• This technology is highly appreciated identification of the causative agent of any underlying disease in the healthcare sector and also for the detection of contaminants in food samples. • The utilization of paper-based biosensors has gained commercial attraction due to its suitability for the analysis of biofluids being portable in nature, specific, user-friendly, rapid, easily transportable, highly sensitive, affordable, and easily available. • Various types of paper-based biosensors used are dipstick type paper-based biosensor, paper-based lateral flow assay, μ-PAD, and smart accessories-based paper bio-analyser. • This technology is used for the detection of glucose, uric acid, protein, pH, lactate, etc.	Mahato et al. (2017)
4.	Bluetooth devices	• Automated dietary monitoring assessment is a nutritional approach which helps in the analysis of food in terms of nutrition, assists dietary recall and nutritional assessment. • Nowadays, Bluetooth devices are being used widely for monitoring an individual's dietary behaviour. The advanced automated dietary monitoring systems help in the estimation of calorie (energy) consumption based on the number of detected bites by an individual. • Gao et al. (2016) conducted a research study in order to study the sound pattern of different food samples for the detection of eating behaviour. Food samples were categorized into four groups, viz. very soft, soft, hard, and very hard.	Gao et al. (2016)

12.7 Conclusion

With compliance to food and trade laws, the analysis of food products is very important in order to avoid contamination of foodstuffs, study the chemical composition, food processing, and quality control. For the sake of achieving high standard in terms of food safety, various analytical techniques are compulsorily and indispensably employed in food industries. It has been concluded in many researches that food industries often face serious challenges with respect to adulteration and high capital cost of food control systems. Therefore, it is highly desired to develop rapid, effective, and sensitive analytical techniques for quality control and food analysis.

References

Adamson MW (2004) Food in medieval times. Greenwood Publishing Group, Westport, CT, pp 64–67

Adeloju SB (2005) Amperometry, encyclopedia of analytical science, 2nd edn. Elsevier Academic Press, pp 70–79

Akash MSH, Rehman K (2020) High performance liquid chromatography. In: Essentials of pharmaceutical analysis. Springer, Singapore, pp 175–184

Alghamdi AH (2010) Applications of stripping voltammetric techniques in food analysis. Arab J Chem 3(1):1–7

Alvarez-Rivera G, Bueno M, Ballesteros-Vivas D, Mendiola JA, Ibañez E (2020) Pressurized liquid extraction. In: Liquid-phase extraction. Elsevier, pp 375–398

Ambrus A (2008) Quality assurance, Ch. 5. In: Tadeo JL (ed) Analysis of pesticides in food and environmental samples. CRC, New York, p 145

Bachalla N (2016) Identification of synthetic food colors adulteration by paper chromatography and spectrophotometric methods. Int Arch Integr Med 3(6):182–191

Bernal JL, Martín MT, Toribio L (2013) Supercritical fluid chromatography in food analysis. J Chromatogr A 1313:24–36

Bezerra MA, Santelli RE, Lemos VA, dos Santos Alves JP, Braz BF, Santos LB (2020) Strategies to make methods based on flow injection analysis greener. CLEAN–Soil, Air, Water 48(7–8): 2000007

Bhattacharya S (2010) Stress relaxation behaviour of moth bean flour dough: product characteristics and suitability of model. J Food Eng 97(4):539–546

Bradley RL (2010) Moisture and total solids analysis. In: Food analysis. Springer, Boston, MA, pp 85–104

Burgain J, Petit J, Scher J, Rasch R, Bhandari B, Gaiani C (2017) Surface chemistry and microscopy of food powders. Prog Surf Sci 92(4):409–429

Campanella OH, Li PX, Ross KA, Okos MR (2002) The role of rheology in extrusion. In: Engineering and food for the 21st century. CRC Press, pp 423–444

Camino Feltes MM, Arisseto-Bragotto AP, Block JM (2017) Food quality, food-borne diseases, and food safety in the Brazilian food industry. Food Qual Saf 1(1):13–27

Carabias-Martínez R, Rodríguez-Gonzalo E, Revilla-Ruiz P, Hernández-Méndez J (2005) Pressurized liquid extraction in the analysis of food and biological samples. J Chromatogr A 1089(1–2):1–17

Carstea EM, Popa CL, Baker A, Bridgeman J (2020) In situ fluorescence measurements of dissolved organic matter: a review. Sci Total Environ 699:134361

Chandra P, Koh WCA, Noh HB, Shim YB (2012) In vitro monitoring of i-NOS concentrations with an immunosensor: the inhibitory effect of endocrine disruptors on i-NOS release. Biosens Bioelectron 32:278–282

Choudhary M, Yadav P, Singh A, Kaur S, Ramirez-Vick J, Chandra P, Arora K, Singh SP (2016) CD 59 targeted ultrasensitive electrochemical immunosensor for fast and noninvasive diagnosis of oral cancer. Electroanalysis 28:2565–2574

Cifuentes A (2012) Food analysis: present, future, and foodomics. Int Sch Res Notices 2012:1–16

Claudia R-C, Francisco J-C (2008) Determination of preservatives in meat products by flow injection analysis (FIA). Food Addit Contam Part A Chem Anal Control Expo Risk Assess 25:1167–1178

Cortes HJ (2020) Multidimensional chromatography: techniques and applications. CRC Press, Boca Raton

Cresson PY, Ricard C, Dubois L, Vaucher S, Lasri T, Pribetich J (2008) Temperature measurement by microwave radiometry. In: 2008 IEEE instrumentation and measurement technology conference. IEEE, pp 1344–1349

Deka S, Saxena V, Hasan A, Chandra P, Pandey LM (2018) Synthesis, characterization and in vitro analysis of α-Fe2O3-GdFeO3 biphasic materials as therapeutic agent for magnetic hyperthermia applications. Mater Sci Eng C 92:932–941

Devries JW, Silvera KR (2001) AACC collaborative study of a method for determining vitamins A and E in foods by HPLC (AACC method 86-06). Cereal Foods World 46(5):211–215

Ellis C, Hite D, Van Egmond H (1997) Development of methods to test all food matrixes unrealistic, says OMB. Inside Lab Manag 1(8):33–35

FAO (1999) FAO trade-related technical assistance and information. Rome. Food and Agriculture Organization of the United Nations (2005). The state of food insecurity in the world

Gao Y, Zhang N, Wang H, Ding X, Ye X, Chen G, Cao Y (2016) iHear food: eating detection using commodity bluetooth headsets. In: 2016 IEEE first international conference on connected health: applications, systems and engineering technologies (CHASE). IEEE, pp 163–172

Garfield FM (1991) Quality assurance principles for analitical laboratories. https://scholar.google.com/scholar_lookup?title=Quality+Assurance+Principles+for+Analytical+Laboratories&author=Garfield+F.M.&author=Klesta+E.&author=Hirsch+J.&publication_year=2000&pages=122. Accessed 1 April 2021

Grange RD, Thompson JP, Lambert DG (2014) Radioimmunoassay, enzyme and non-enzyme-based immunoassays. Br J Anaesth 112(2):213–216

Grossbier DT, Schoenfuss TC (2021) Using microwave-accelerated digestion instead of dry ashing during sodium analysis of low-moisture, part-skim mozzarella. JDS Commun 2(1):13–15

Hanhauser E, Bono MS Jr, Vaishnav C, Hart AJ, Karnik R (2020) Solid-phase extraction, preservation, storage, transport, and analysis of trace contaminants for water quality monitoring of heavy metals. Environ Sci Technol 54(5):2646–2657

Hatzakis E (2019) Nuclear magnetic resonance (NMR) spectroscopy in food science: a comprehensive review. Compr Rev Food Sci Food Saf 18(1):189–220

Ikins W, DeVries J, Wolf WR, Oles P, Carpenter D, Fraley N, Ngeh-Ngwainbi J (1993) A food matrix organizational system applied to collaborative studies. The Referee AOAC Int 17:1–7

Kataoka H, Lord HL, Pawliszyn J (2000) Applications of solid-phase microextraction in food analysis. J Chromatogr A 880(1–2):35–62

Klančnik A, Kovač M, Toplak N, Piskernik S, Jeršek B (2012) PCR in food analysis. In: Hernandez-Rodrigues P, Ramirez Gomez AP (eds) Polymerase chain reaction. Intech, Rijeka, pp 195–220

Lasztity R, Petro-Turza M, Foldesi T (2004) History of food quality standards. In: Lasztity R (ed) Food quality and standards, Encyclopedia of life support systems (EOLSS), developed under the auspices of the UNESCO. Eolss Publishers, Oxford

Latimer GW Jr (1997) Check sample programs keep laboratories in sync. Inside Lab Manag 1(4): 18–20

Lazzaro S, Pike E (2014) The importance of sample preparation in food analysis. http://epgp.inflibnet.ac.in/epgpdata/uploads/epgp_content/S000015FT/P000065/M002606/ET/14619144641ET.pdf. Accessed 1 April 2021

Lehotay SJ, Hajšlová J (2002) Application of gas chromatography in food analysis. TrAC Trends Anal Chem 21(9–10):686–697

Lovett RA (1997) US food label law pushes fringes of analytical chemistry. Inside Lab Manag 1(4): 27–28

Mahato K, Srivastava A, Chandra P (2017) Paper based diagnostics for personalized health care: emerging technologies and commercial aspects. Biosens Bioelectron 96:246–259

Mahato K, Kumar S, Srivastava A, Maurya PK, Singh R, Chandra P (2018) Electrochemical immunosensors: fundamentals and applications in clinical diagnostics. In: Vashist SK, JHT L (eds) Handbook of immunoassay technologies. Academic Press, London

Martin SR, Schilstra MJ (2008) Circular dichroism and its application to the study of biomolecules. Methods Cell Biol 84:263–293

Melito HS, Daubert CR, Foegeding EA (2012) Validation of a large amplitude oscillatory shear protocol. J Food Eng 113(1):124–135

Mello LD, Kubota LT (2002) Review of the use of biosensors as analytical tools in the food and drink industries. Food Chem 77(2):237–256

Mohamed RS, Mansoori GA (2002) The use of supercritical fluid extraction technology in food processing. Food Technol Mag 20(7):134–139

Moldoveanu S, David V (2021) Modern sample preparation for chromatography. Elsevier

Molino A, Mehariya S, Di Sanzo G, Larocca V, Martino M, Leone GP, Marino T, Chianese S, Balducchi R, Musmarra D (2020) Recent developments in supercritical fluid extraction of bioactive compounds from microalgae: role of key parameters, technological achievements and challenges. J CO2 Utilization 36:196–209

Nielsen SS (1998) Food analysis. http://154.68.126.6/library/Food%20Science%20books/batch1/Food%20Analysis%20Fourth%20Edition.pdf. Accessed 1 April 2021

Nielsen SS (2017) Introduction to food analysis. In: Food analysis. Springer, Cham, pp 3–16

Nielsen AV, Beauchamp MJ, Nordin GP, Woolley AT (2020) 3D printed microfluidics. Annu Rev Anal Chem 13:45–65

Nollet LM, Munjanja BK (eds) (2019) Ambient mass spectroscopy techniques in food and the environment. CRC Press

Nollet LM, Toldrá F (eds) (2019) Food analysis by HPLC. CRC press

Oldewage-Theron WH, Egal AA (2016) Food quality and food safety. In: Temple NJ, Steyn N (eds) Community nutrition for developing countries. AU Press, Edmonton, p 430

de Oliveira CAF, da Cruz AG, Tavolaro P, Corassin CH (2016) Food safety: good manufacturing practices (GMP), sanitation standard operating procedures (SSOP), Hazard analysis and critical control point (HACCP). In: Barros-Velázquez J (ed) Antimicrobial food packaging. Academic Press. https://repositorio.usp.br/item/002737353. Accessed 1 April 2021

Omar MMA, Elbashir AA, Schmitz OJ (2017) Capillary electrophoresis method with UV-detection for analysis of free amino acids concentrations in food. Food Chem 214:300–307

Osborne BG (2006) Near-infrared spectroscopy in food analysis. In: Meyers RA (ed) Encyclopedia of analytical chemistry: applications, theory and instrumentation. Wiley

Polesello A, Giangiacomo R, Dull GG (1983) Application of near infrared spectrophotometry to the nondestructive analysis of foods: a review of experimental results. Crit Rev Food Sci Nutr 18(3): 203–230

Pomeranz Y, Meloan CE (1994) Food analysis: theory and practice, 3rd edn. Chapman & Hall, New York

Power AC, Chapman J, Chandra S, Cozzolino D (2019) Ultraviolet-visible spectroscopy for food quality analysis. In: Zhong J, Wang X (eds) Evaluation technologies for food quality. Woodhead Publishing, Duxford, pp 91–104

Proestos C, Komaitis M (2008) Application of microwave-assisted extraction to the fast extraction of plant phenolic compounds. LWT-Food Sci Technol 41(4):652–659

Purohit B, Vernekar PR, Shetti NP, Chandra P (2020) Biosensor nanoengineering: design, operation, and implementation for biomolecular analysis. Sens Int:100040

Rajput H, Rehal J, Goswami D, Mandge HM (2019) Methods for food analysis and quality control. Apple Academic Press. https://www.researchgate.net/profile/Rajput-2/publication/330982949_Methods_for_Food_Analysis_and_Quality_Control/links/5cbd8f0692851c8d22fc3e98/Methods-for-Food-Analysis-and-Quality-Control.pdf. Accessed 1 April 2021

Rateni G, Dario P, Cavallo F (2017) Smartphone-based food diagnostic technologies: a review. Sensors 17(6):1453

Ricci F, Volpe G, Micheli L, Palleschi G (2007) A review on novel developments and applications of immunosensors in food analysis. Anal Chim Acta 605(2):111–129

Rohindra D, Lata RA (2020) Volumetric analysis-titration for beginners. Pacific Studies Press

SádeCká J, TóThoVá J (2007) Fluorescence spectroscopy and chemometrics in the food classification—a review. Czech J Food Sci 25(4):159–173

Scampicchio M, Ballabio D, Arecchi A, Cosio SM, Mannino S (2008) Amperometric electronic tongue for food analysis. Microchim Acta 163(1–2):11–21

Scott AO (1998) Biosensors for food analysis. Woodhead Publishing

Sharpless KE, Greenberg RR, Schantz MM, Welch MJ, Wise SA, Ihnat M (2004) Filling the AOAC triangle with food-matrix standard reference materials. Anal Bioanal Chem 378(5):1161–1167

Sikora T, Kołożyn-Krajewska D (2001) Quality assurance and. Food Health Safety 6(55):15–18, 25. http://yadda.icm.edu.pl/baztech/element/bwmeta1.element.baztech-article-BPG8-0029-0004. Accessed 1 April 2021

Sikora T, Strada A (2005) Safety and quality assurance and management systems in food industry: an overview. The food industry in Europe. Agricultural University of Athens, Athens, pp 85–95

Sliwinska M, Wisniewska P, Dymerski T, Namiesnik J, Wardencki W (2014) Food analysis using artificial senses. J Agric Food Chem 62(7):1423–1448

Sonar MP, Rathod VK (2020) Microwave assisted extraction (MAE) used as a tool for rapid extraction of Marmelosin from Aegle marmelos and evaluations of total phenolic and flavonoids content, antioxidant and anti-inflammatory activity. Chem Data Collections 30:100545

Sosa P, Guild G, Burgos G, Bonierbale M, Zum Felde T (2018) Potential and application of X-ray fluorescence spectrometry to estimate iron and zinc concentration in potato tubers. J Food Compos Anal 70:22–27

Stannard C (1997) Development and use of microbiological criteria for foods. Food Sci Technol Today 11(3):137–177

Tabilo-Munizaga G, Barbosa-Cánovas GV (2005) Rheology for the food industry. J Food Eng 67(1–2):147–156

Tang F, Vasas M, Hatzakis E, Spyros A (2019) Magnetic resonance applications in food analysis. Annu Rep NMR Spectrosc 98:239–306

Wetzel DL, Charalambous G (eds) (1998) Instrumental methods in food and beverage analysis. Elsevier

WHO (2002) WHO global strategy for food safety: safer food for better health. https://apps.who.int/iris/bitstream/handle/10665/42559/9241545747.pdf?sequence=1&isAllowed=y. Accessed 1 April 2021

WHO (2007). WHO Food safety and foodborne illness. World Health Organization. https://www.who.int/news-room/fact-sheets/detail/food-safety. Accessed 1 April 2021

World Health Organization (WHO) (2008) Foodborne disease outbreaks: guidelines for investigation and control. https://www.who.int/foodsafety/publications/foodborne_disease/outbreak_guidelines.pdf. Accessed 1 April 2021

World Health Organization (2020) The state of food security and nutrition in the world 2020: transforming food systems for affordable healthy diets, vol 2020. Food & Agriculture Org

Electronic Noses and Tongue-Based Sensor Systems in Food Science 13

Gurlovleen Kaur, Ranjeeta Bhari, and Kuldeep Kumar

Abstract

Food safety is a multifaceted term and also a predominant element in food and beverage industries. It ensures safe handling, storage, and preparation of food and drinks to prevent food-related illness and diseases. The worldwide significance of food safety is continuously growing. With the advent of new technologies, innovative methods are being developed for the identification, assessment, and monitoring of the foodborne hazards. Great technological advances are being made for the development of electronic nose (e-nose) and electronic tongue (E-tongue)-based sensors for food safety.

Keywords

Sensor · Food safety · Electronic nose · Electronic tongue · Food science

13.1 Introduction

The most important part in the food and liquid drinks manufacturing is ensuring the standardization of products. The quality of the food, beverages, and various chemical products can be measured by means of different analytical devices. Many industries use the analytical tools that can sense the smell and the taste in the food which is produced due to the specific and nonspecific molecular interactions. The conventional methods, namely plasma atomic emission spectrometry, high

G. Kaur · K. Kumar (✉)
Department of Biotechnology, M. M. Modi College, Patiala, Punjab, India

R. Bhari
Department of Biotechnology, Punjabi University, Patiala, Punjab, India

P. Chandra, P. S. Panesar (eds.), *Nanosensing and Bioanalytical Technologies in Food Quality Control*, https://doi.org/10.1007/978-981-16-7029-9_13

performance liquid chromatography (HPLC), gas chromatography (GC), and electrophoresis are employed for the odor and flavor analysis, but they are costly as well as inappropriate for the real-time quality control (Ghasemi-Varnamkhasti et al. 2012). Among all the analytical devices, Electronic nose (e-nose) and Electronic tongue (e-tongue) are being increasingly used to monitor the quality of food and drinks. Electronic nose (e-nose) can recognize the odor and electronic tongue (e-tongue) can sense the taste of the sample.

13.2 Electronic Nose (e-nose)

The word "electronic nose" came into being in the late 1990s (Craven et al. 1996). Electronic nose is defined as an instrument that is composed of a multisensor array which is accountable for detection of one or more chemical components. It can also be defined as "an instrument which contains an array of electronic chemical sensors with the partial specificity and a suitable pattern recognition system, able of recognizing simple and complex odor." Due to the prerequisite of the authentic solution, compactness, cost-effectiveness, and rapidity, the idea of the electronic nose has become popular in the food and water industry, agricultural system, security systems, pharmaceuticals, and many more other areas. This was all possible due to the improvement of technology in sensors and is encouraged from sensation of smell. It is also known as an artificial nose, odor detector, multisensor array, aroma sensor, mechanical nose, smell sensing system, electronic olfactometry, and flavor sensor (Dymerski et al. 2011).

The olfactory systems of the living beings make them conscious of their environment, potential dangers, and help them in recognition as well as classification of food. But the challenge here is to automatically identify and classify odors as there is natural intercommunication of the smells in the chemical mixtures. This natural communication is mainly of three types: compensation, masking, and synergism (Fig. 13.1). Compensation is a process in which one component of the odor counteracts another component. Masking is defined as a process in which combination of one pleasant odor occurs with an unpleasant one. Synergism is a process in which interaction of two or more separate components produces a shared odor which is quite stronger than any of those individual constituents.

Electronic nose is similar to human nose in various aspects (Table 13.1). Human nose brings the smell to the epithelium layer with the help of lungs whereas electronic nose utilizes a pump to sense the odor. In human nose, hair and mucus membrane act as a filter and in electronic nose filtration is provided by an inlet sampling system. There is an olfactory epithelium in human nose that contains millions of sensing cells which interacts with the odors while electronic nose acts in a different manner by sensing odors by making use of range of sensors that interact in a diverse way with the odor molecules. In human nose, the distinctive patterns of the chemical response are perceived by the human receptors which are further converted into the nerve impulse. This nerve impulse is passed on to neurons and finally reaches the olfactory cortex of the brain for ultimate interpretation.

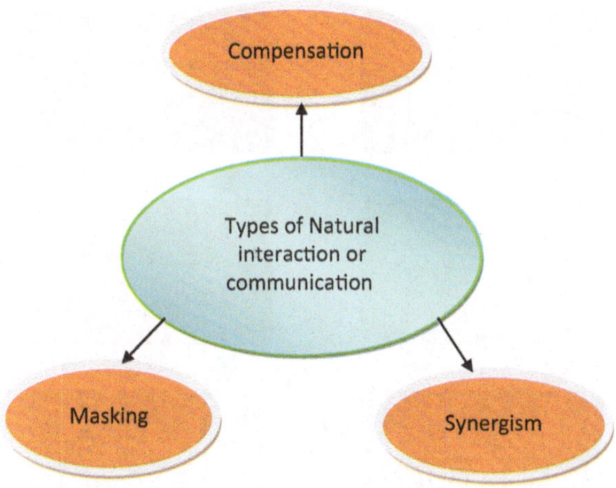

Fig. 13.1 Types of natural intercommunication between the odors that occurs during the automatic identification of chemical mixtures

Table 13.1 Comparative study of human and electronic noses

Properties	Human nose	Electronic nose
Odor identification	Epithelium layer	Pump
Filter	Mucus and Hair	Inlet sampling system
Type of sensors	Olfactory epithelium receptors	Metal oxide semiconductor sensors (MOS), Optical sensors, Conducting polymer sensors (CP), Piezoelectric sensors, etc.
Type of signal	Chemical signals	Electronic signals
Final interpretation	Olfactory cortex in Brain	Pattern recognition by Computer

Whereas, chemical sensors are present in electronic nose that reacts to the sample and which results in generation of electrical signals. Final interpretation of electrical signals takes place by the pattern classification algorithm, and this exclusive pattern of signals is read by computer. Different gas mixtures can be characterized with the help of electronic nose as well as by human nose. But there are also a few differences between them. The evaluation by the human sensory system is subjective. There is also individual variability as the same individual can present dissimilar valuation information in diverse experiments. So, the sensory panels are not consistent. Furthermore, it also depends upon the physical and mental state of an individual and the evaluation can be imprecise (Dutta et al. 2003).

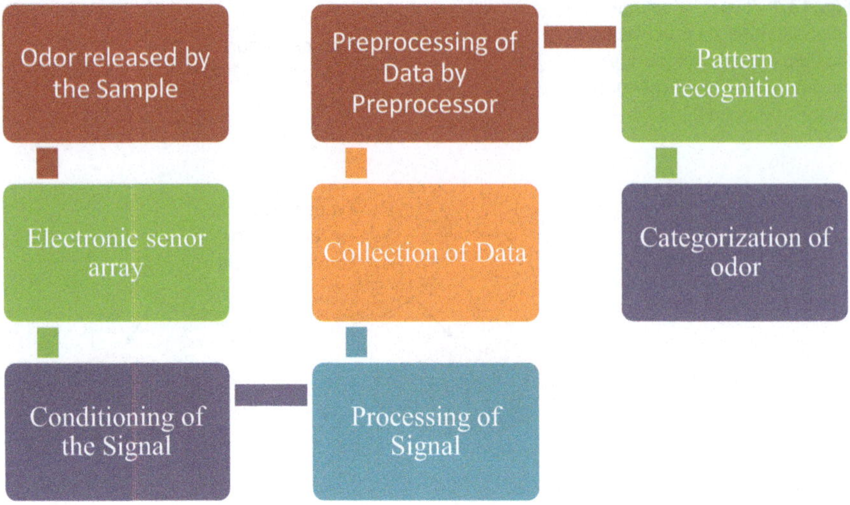

Fig. 13.2 Sequence of events that occur during the detection of odor by the electronic nose device

13.2.1 Components of the Electronic Nose

An Electronic nose consists of hard and software constituents, briefly described in Fig. 13.2.

Firstly, sensor array absorbs the released gas mixture. Its results in the change in the current, frequency, voltage according to the type of compounds reaching the sensor array and input signals are detected. Different type of sensors are grouped to form sensor arrays so preprocessing of signals is must which helps to identify the physical changes accurately. Then dataset is formed by digitalizing the obtained signals. Thus, suitable manipulation which includes filtration as well as amplification of the sensed signals occurs. Now the signal can be straightforwardly used for the subsequent stages. Next step is data gathering in which the analysis of the processed signals is done according to their specific properties. Adequate data is obtained from signals and preprocessing of data is done as per necessity of the working pattern recognition algorithm. Last of all, pattern recognition phase classifies the odor.

13.2.1.1 Sensors and Chemicals

Sensing system has developed speedily which resulted in the advancement of variety of sensors as well as in intricate microarray sensors. It is a major reactive component of the device. In 1980, sensors array with distinct sensitivity and selectivity were assembled together to improve the instrument (Ampuero and Bosset 2003; Bajpai et al. 2018; Chandra 2016; Purohit et al. 2020a, b). Qualitative and quantitative information is gathered from each sensor of the sensor array which creates the fingerprint of the sample. Hence, dataset or collected library is created which resembles a digital signature or fingerprints of specific odors. So, a particular sensor is required for the detection of specific odors and targeted chemicals are detected in

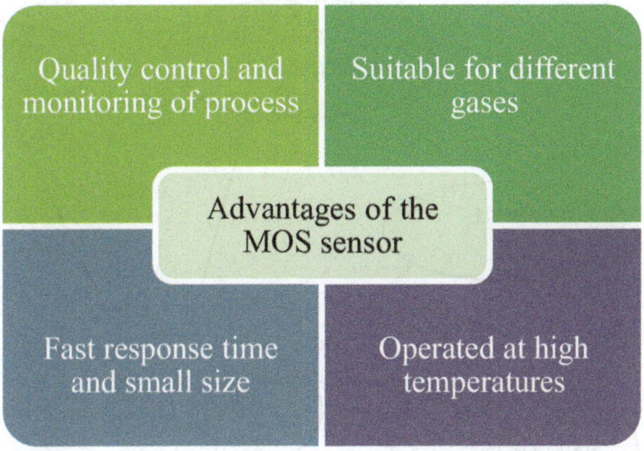

Fig. 13.3 Properties of a suitable sensor array used in the electronic nose device

the sample by sensor arrays as main purpose of the electric nose is to sense various chemicals (Boeker 2014). The properties of the suitable sensor are mentioned in Fig. 13.3.

Types of Sensors

Metal Oxide Sensor (MOS)

MOS sensors form major type of sensors which are normally used in electronic nose device because of its appropriateness for broad variety of gases (Fig. 13.4) (Wang et al. 2010).

According to the response to distinct gases, MOS sensors are divided into two types:

1. n-type
   In n-type sensors, reaction occurs in between the surface of sensor and oxygen molecules of air. It results in the entrapment of the free electrons on the surface of the sensor and the resultant potential barrier is formed between the grains. It inhibits the mobility of carrier which results in the production of huge resistance areas.
2. p-type
   The p-type sensors are most frequently used sensors in electronic nose instrument due to high sensitivity as well as high selectivity. Their reaction with oxidizing gases removes the electrons which lead to production of holes. Main characteristic of the p-type sensors is absorption of oxygen and surface reactivity which greatly increases the functioning of the sensor. It enhances the speed and reduces the reliance of signals on humidity (Kim and Lee 2014).

Conducting Polymer Based Sensor

They are extensively used sensors in electronic nose devices because of their capability for regulating the conductivity in response to organic compounds

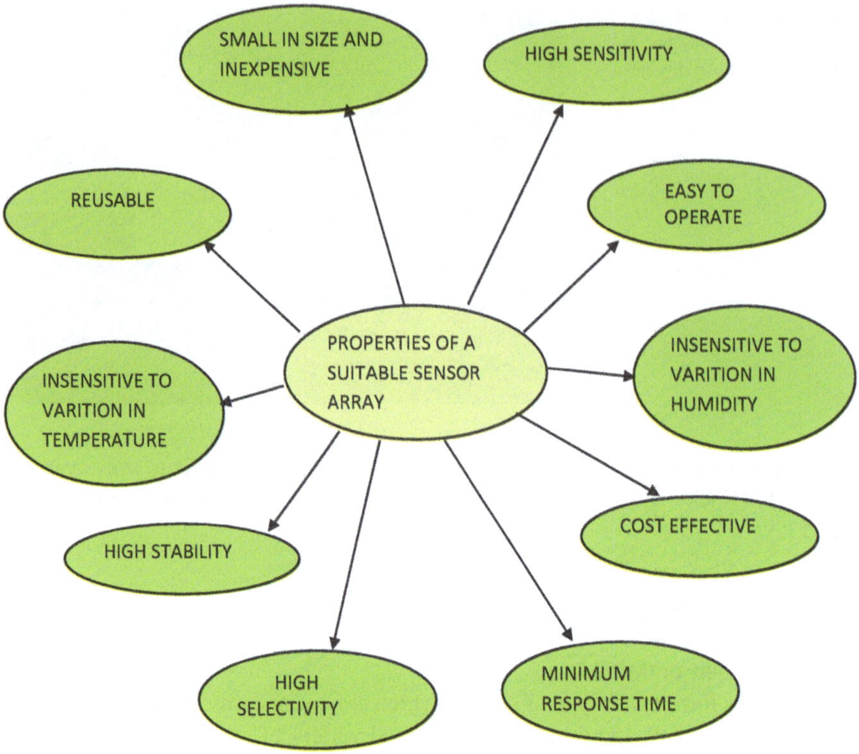

Fig. 13.4 Advantages of the metal oxide sensors in the field of food safety

Fig. 13.5 Advantages of conducting polymer sensors in the field of food safety

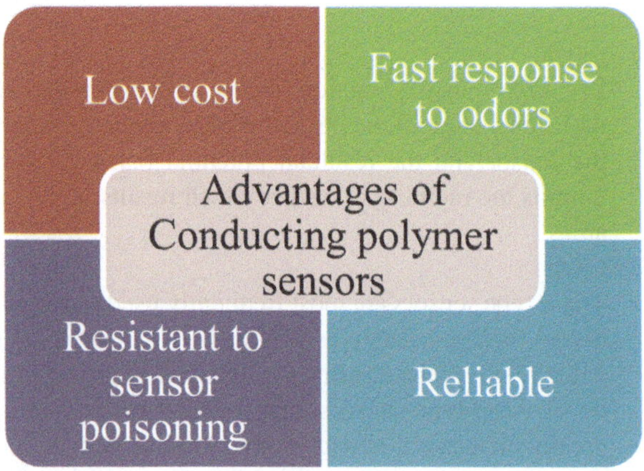

(Fig. 13.5). Active layer is the critical component of this sensor, and it can be manufactured by using different techniques such as electrochemical deposition, vapor deposition polymerization, Langmuir-Blodgett method (LB) , layer-by-layer

self-assembly method (LbL), thermal evaporation, and many more. Different polymers are used for the manufacturing of this sensor, for example, polyaniline and polythiophene (Bai and Shi 2007). The principle behind this sensor is that when an analyte interacts with the material, the alteration in the material takes place which changes resistance of sensor under ambient temperature conditions and detection of the various gases takes place.

Optical Sensors

Optical sensors are one of the most attractive sensors that have many applications and usually volatile molecules are excited by the light source. Optical sensors are used for the measurement of light polarization, fluorescence, absorbance, reflectance, optical layer thickness, chemiluminescence, and colorimetric dye response (Esfahani and Covington 2017). The odors in the immediate environment can be detected by these optical changes (Fig. 13.6). Different detectors are used for the detection of output signals which include photodiodes, charged-coupled device, and many more (Chodavarapu et al. 2007).

Optical sensors are mainly of two types:

1. Colorimetric sensors
   The optical sensors are sensors in which the process of detection is based on the variation of color. Chemically reactive dyes are used to make the thin films of this sensor.
2. Fluorescence sensors
   The fluorescence sensors are the sensors in which detection is based on the light emitted by the sample. They are relatively more sensitive than the colorimetric sensors and can be used for the detection of particular compounds in the sample mixture.

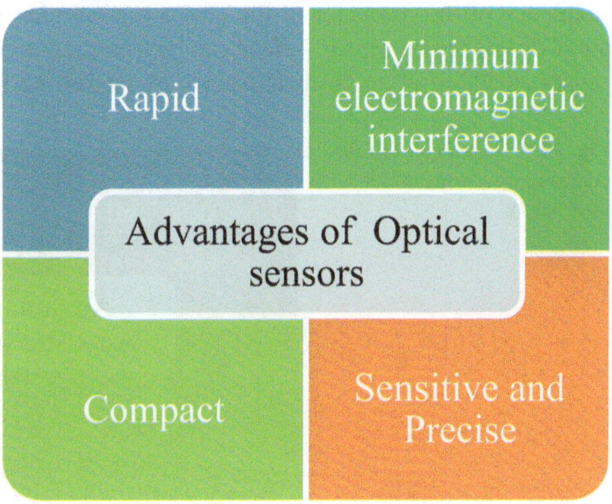

Fig. 13.6 Advantages of the optical sensors for food safety

But there are few limitations such as high cost of the associated electronic and software system. Also its operation is pretty complex. Also they have short life span which further increases the cost of the instrument.

Acoustic Wave-Based Sensors

They are a type of piezoelectric sensors. It includes tube acoustic wave device, bulk acoustic wave (BAW) sensor, flexural plate device, fiber acoustic sensor, transverse device, and surface acoustic wave (SAW) sensor (Fig. 13.7). Above all most widely used sensors are BAW and SAW. They are comparatively small sized more sensitive and are cost-effective as compared to others. Moreover, they respond to just about all the gases (Cheeke and Wang 1999). A surface acoustic wave sensor generates bigger change in frequency as they typically operate at high frequencies (100–1000 MHz). Interdigital transducer is the central part of the SAW sensor and is used as an output as well as input transducers. If the gas molecules are absorbed, it will result in the change of atmosphere of the transducer which results in the transformation of the vibration frequency. So, by comparison of the output and input interdigital signal weight information of the gas molecules can be provided.

However, the limitation of these sensors is the inadequate performance of the sensors in the liquid medium. Due to the high operating frequencies, the signal-to-noise ratio of surface acoustic wave is high (Fig. 13.8).

Quartz Crystal Microbalance Sensor (QCM)

QCM sensors are also a type of piezoelectric sensors. They have various applications in food safety, medicine, security, and environment monitoring (Fig. 13.9). The properties of the various chemicals can be determined by this sensor as it can operate in gas as well as in liquid environments (Turner et al. 2017). Surface of the sensor is covered by the receptive coating. Released gas from the surroundings is received by the sensitive coating of crystal which raises the total mass of the crystal. It leads to reduction in the frequency as there is change in the mass of gold surface. So, frequency of the resonator of the quartz crystal microbalance changes and even a small difference on sensor surface can be measured accurately. Hence, QCM sensor senses small variation on their surface by measuring the frequency change on the

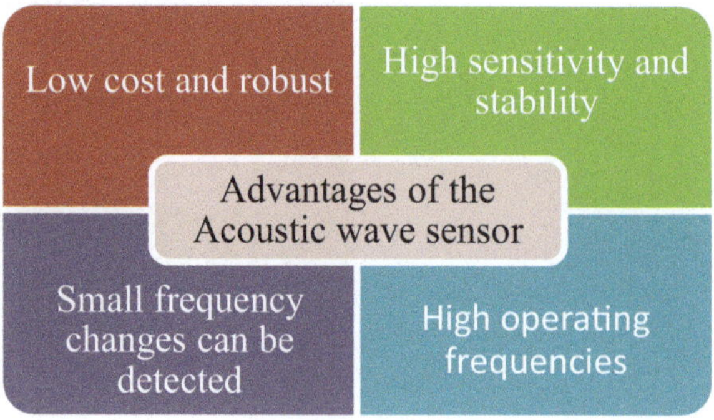

Fig. 13.7 Advantages of the Acoustic wave sensors for food safety

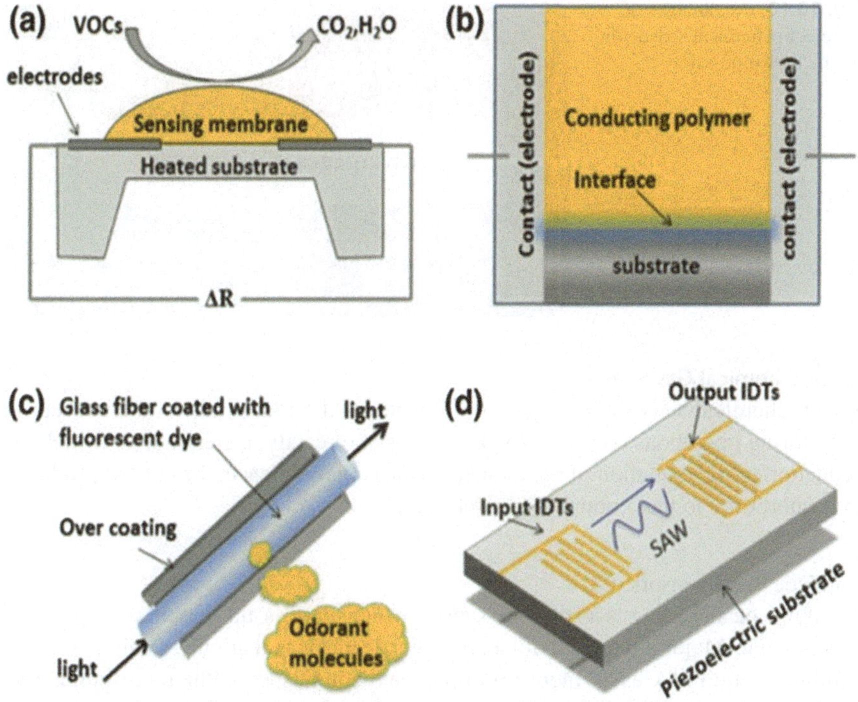

Fig. 13.8 Schematic diagram of (**a**) MOS sensor, (**b**) CP sensor, (**c**) optical fiber and (**d**) SAW sensor (Adopted from Zou et al. 2015)

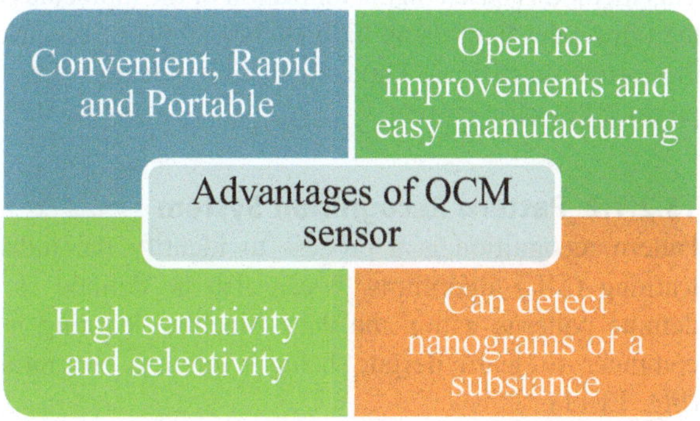

Fig. 13.9 Advantages of Quartz crystal microbalance sensor in the field of food safety

quartz crystal resonator (Huang et al. 2017). Surface layers of the quartz crystal microbalance are coated by different coatings such as acidized-multiwalled carbon nanotubes, biomimetic peptide-based sensing materials, molecularly imprinted polymers (MIPs), gold films, multiwall carbon nanotubes, calixarenes, and many more.

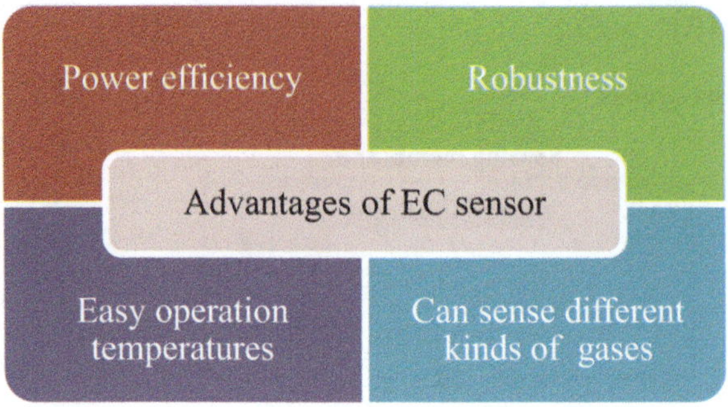

Fig. 13.10 Advantages of the electrochemical sensors in the field of food safety

Electrochemical Gas Sensors (EC)

Electrochemical sensors are mostly employed for the industrial, food and security monitoring operations (Fig. 13.10). They consist of catalytic electrode on which the reduction or the oxidation of the molecules takes place. Hence, the current produced is proportional to the amount of gas released.

Catalytic Bead Sensors (CB)

Catalytic bead sensor is a type of sensor which detects the gas based up on its combustion. Major advantage of catalytic bead sensor is that only small amount of sample is required. Gas of interest is burned and combustion enthalpy is found. This method is chiefly utilized for detection of combustible gases (Liu et al. 2012).

Photo Ionization Detector Sensors (PID)

This method is based on the ionization of the molecules of the target gas by means of the ultraviolet light using high energy photons. Electric current is produced due to the formation of ions. Photo ionization detectors are used for the detection of the produced electric current.

13.2.1.2 Pattern Recognition System

Pattern recognition is a process to identify the individual patterns by machine learning (ML) algorithms. It can also be defined as an ability to recognize the definite patterns within the data. Machine learning is defined as the ability of a computer to learn despite being explicitly programmed for a particular work (Fig. 13.11).

Machine learning algorithms are mainly divided into four types:

- Supervised learning
  Supervised learning is generally used for the regression, forecasting as well as for the classification. ML algorithm is provided with dataset having well-known labeled inputs and outputs. Vigorous mapping model obtained from input and output enables the system to calculate outputs for new original input data.
- Semi-supervised learning

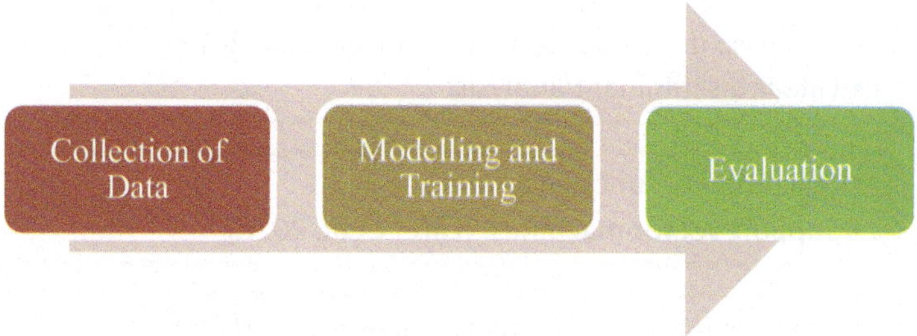

Fig. 13.11 Components of the machine learning algorithm

Semi-supervised learning dataset consists of labeled along with unlabeled data. Computational complexity can be reduced by using the semi-supervised machine learning algorithm, and it is used to solve problems having huge quantity of unlabeled data as a result it helps to save time.

- Unsupervised learning
  Dataset used for the unsupervised learning contains simply inputs. Dataset is investigated by the machine learning algorithm and learns inbuilt structures from the input data as well as try to recognize the particular patterns by finding out the relationships and correlations among the data points. Unsupervised learning is used for the problems such as association, dimensionality reduction, and clustering.
- Reinforcement learning
  Reinforcement learning is a type of ML algorithm which is based upon the learning from the past experiences and is considered as an adaptive method. It works automatically and determines the ultimate performance of the system while maximizing performance for the specific task. System evaluates a new input sample and accurate output is produced (Fig. 13.12).

(a) Stepwise discriminant analysis (SDA)
SDA is a widely used tool as it is easy to operate and has linear decision boundary. Moreover, it is a rapid method for the classification applications. But there are few disadvantages such as prominent computational time for training and is based on the Gaussian assumption.

(b) Linear discriminant analysis (LDA)
The tool LDA is utilized to observe associated features that distinguish and classify more than one substances and is employed for dimensionality reduction. This method is often used for pattern recognition, machine learning, and statistics (Wei et al. 2017). It is directly linked to PCA as both of them find out the linear arrangement of the variables which specifically describes the data. But the difference between the two is that LDA gives superior results for bigger multiple class datasets, and it is also a supervised technique.

(c) Support vector machines (SVMs)
It is a supervised learning method and rapid, memory efficient technique utilized for multi-class as well as for binary classification. It functions great with the

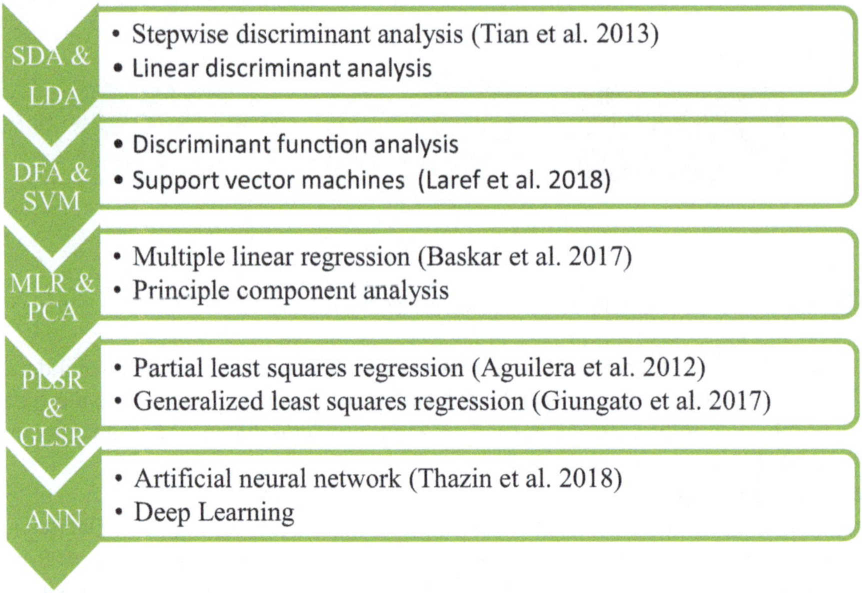

Fig. 13.12 Various tools used in the pattern recognition algorithms of electronic nose

linear as well as for the nonlinear data and is efficient in high dimensional spaces. Moreover, it is effectual even in the conditions where the sum total of dimensions is larger than the amount of data samples. It can also be utilized for the regression problems (Leal et al. 2019). There are few limitations of the SVM such as large computational time is needed with data in huge amount.

(d) Principle component analysis (PCA)
PCA, an unsupervised tool, is broadly utilized to lessen the data dimensionality and also provides the position of uncorrelated constituents. Moreover, it is used to measure the probability evaluation of high dimensional data as well as maintain the variance arrangement of data up to rotation. The key inadequacy of this method is that high computational time is required with the huge amount of data (Romani et al. 2012).

(e) Artificial neural networks (ANNs)
They are essentially the computational tool which is influenced from the operational principle of the biological neural systems. ANNs has adaptive nature as they are proficient in machine learning and in pattern recognition. The information processing unit is a crucial component of the artificial neural networks and consists of processing elements which are interconnected with each other. Their working is similar to that of neurons of human neural system and similarly work together to unravel particular problems. An artificial neural network is divaricated into three distinct layers (Fig. 13.13). First layer is an input layer that consists of input signals, second layer is hidden layer that usually consists of more than one layers and final layer, an output layer that provides the output

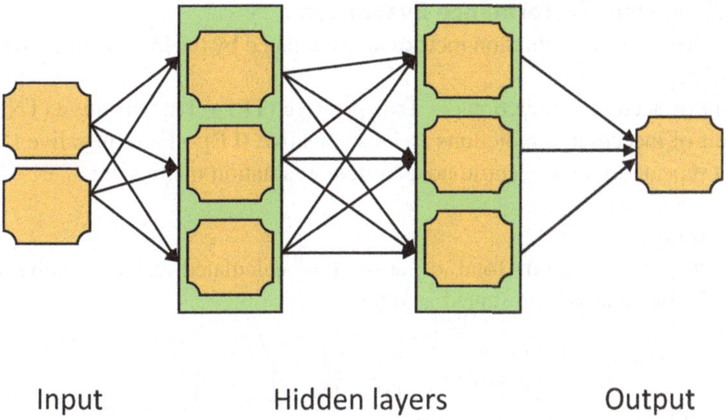

Fig. 13.13 Different layers of an artificial neural network (ANN)

signals. There is change in the number of hidden layers according to the nature of work to be performed. Signal processing nodes of the hidden layers as well as of output layers are connected to each other to form a network and coupling weight sets the power of the interconnections. The whole process is interlinked as activation of the hidden layer is determined by input layer along with the weights connecting the input and hidden layers. Likewise, output layer activation depends upon the hidden layers as well as the weights linking them. ANN is chiefly used for nonlinear data classification as ANN model is nonlinear.

(f) Deep learning

It is defined as a capability of a device to detect selected information automatically, and then its classification is done in the raw database. It is a set of advance methods that are frequently used for learning of neural networks (Lecun et al. 2015). It has different industrial plus medical applications, for example, classification as well as recognition of object and image, recognition of speech and face, and for the analysis of image and of video. There are various deep learning methods such as CNN (Convolutional neural network), Boltzmann machines, RNN (Recurrent neural networks). But CNNs have been commonly used among others, and it is a layered structure and consists of diverse layers such as an input layer, numerous convolutional layer, then pooling layer, after that nonlinear activation layers, last but not least fully connected layer furthermore, an output layer. Key advantage is that, there is no need to preprocess the input data and selection as well as feature extraction is carried out automatically (Qi et al. 2017). It is employed in the e-nose technology for identification along with classification of the gases and liquors. They are also used to carry out the rapid gas recognition and prevent the misidentification of the aromas by using dimension reduction along with the clustering (Tang et al. 2015).

13.2.1.3 System Performance Evaluation

System performance evaluation metrics are calculated by the inserting the (Karakaya et al. 2020):

Result of accurate predictions = True positive (TP) + True negative (TN)
Result of inaccurate predictions = False positive (FP) + False negative (FN)
Most repeatedly used electric nose system evaluation metrics are:

(a) Accuracy
Accuracy is the sum total of accurately calculated values among all the predictions and is represented as under:

$$\text{Accuracy} = \frac{\text{TP} + \text{TN}}{\text{TP} + \text{TN} + \text{FN} + \text{FP}}$$

(b) Precision
Precision is to be defined as value which provides data regarding rightly calculated affirmative label among entire positive values. It is evaluated as given:

$$\text{Precision} = \frac{\text{TP}}{\text{TP} + \text{FP}}$$

(c) Sensitivity
Sensitivity is to be defined as number of correctly categorized definite positive values. It is found as under:

$$\text{Sensitivity} = \frac{\text{TP}}{\text{TP} + \text{FN}}$$

(d) Specificity
Specificity is to be defined as the rate of accurately determining the actual negative values and is represented as under:

$$\text{Specificity} = \frac{\text{TN}}{\text{TN} + \text{FP}}$$

(e) F1-score.
F1-score is to be calculated as harmonic mean of sensitivity and precision. It is found as under:

$$\text{F1-score} = \frac{2 \times \text{Precision} \times \text{Sensitivity}}{\text{Sensitivity} + \text{Precision}}$$

13.3 Electronic Tongue (e-tongue)

Electronic tongue (e-tongue) is a diagnostic tool that includes an arrangement of wide ranging chemical sensors that are extremely stable and have cross-sensitivity to various components of solution along with the suitable technique of multivariate calibration for the data processing. As per the IUPAC definition, electronic tongue is defined as the “a multisensor system, which consists of a number of low selective

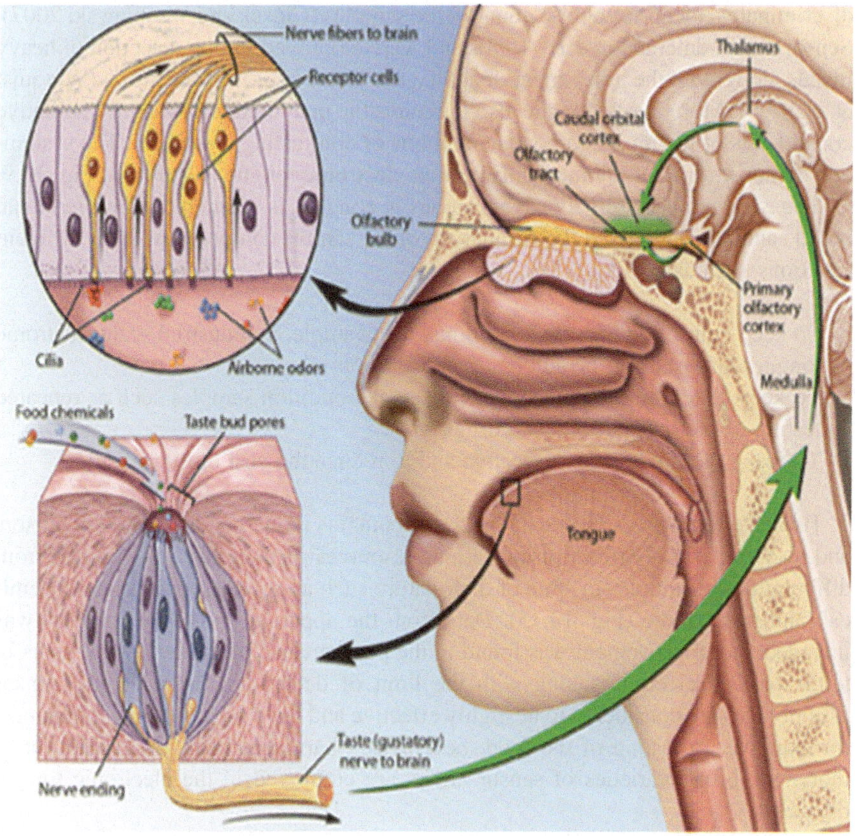

Fig. 13.14 The working of the gustatory system of the human beings (Adapted from the Brainstem.org)

sensors and uses advanced mathematical procedures for signal processing based on the pattern recognition (PARC) and/or multivariate data analysis" (Vlasov et al. 2005). They can also be defined as the analytical devices that can artificially or unnaturally reproduce the sensation of taste plus have collection of sensors which are joined to the chemometric processing unit and are used to distinguish the complex liquid solutions. The postulation of the electronic tongue was initiated from the functioning of the gustatory system of human beings (Fig. 13.14). The five tastes that include sweet, salty, bitter, umami, and sour are detected by the human tongue by means of the gustatory receptor cells that are clustered together and are known as gustatory buds. Therefore, different tastes are detected by these receptor cells and the information is further transmitted to the brainstem nuclei by the cranial nerves. Finally, cerebral cortex of brain analyzes the information and distinct tastes are interpreted as shown in Fig. 13.8. Identical goal is achieved by the electronic tongue which employs the chemometric technique along with the artificial intelligence and discriminates, identifies or else quantify the samples (Ciosek and Wróblewski 2007). Sensor arrays emerged in early 1990s and were employed for the detection of heavy metals along with the assessment of spoilage and flavor of the food stuffs (Winquist et al. 1998). Moreover, e-tongue also measures the quantitative as well as qualitative constituents of the multicomponent solutions of contrasting characters. The systematic demonstration of the mechanism of an electronic tongue is shown in Fig. 13.9.

The e-tongue is extremely advantageous in conditions where human professional board cannot be employed for the analysis of the sample components. It has different functions such as:

- To keep control on process conditions, for example, at industrial scale electronic tongues are used for the automatic process control
- To keep check on the poisonous and severe condition samples such as repeated testing of pharmaceuticals and drugs
- To make the process cost-effective and to reduce the cost of the product

Hence, e-tongue is a physical object that contains numerous incorporated sensors and also consists of collection of various resources of data that are obtained from different liquid samples in spite of their nature such as spectroscopic, electrochemical, and many more (Fig. 13.15). Data from the apparently unrelated sensors was included and improvement was found in the performance of the separate sensors in terms of the selectivity along with the limit of detection. These diverse sensor assemblages are considered to be highly effective and have much wider applications, for example, in testing of the food, beverages, pharmaceuticals, drugs, and many more. So, broad varieties of sensor arrays are employed in the electronic tongue devices.

13.3.1 Types of Chemical Sensors

Varieties of chemical sensors are utilized in the electronic tongue devices (Table 13.2). First are electrochemical which are further classified into amperometric, potentiometric, impedimetric, are conductometric. Next are optical sensors which include surface plasmon resonance, bioluminescence, reflectance, and another is gravimetric. A perfect matrix is mainly made up of both selective as well as of cross-sensitive sensors. Cross-sensitivity is to be defined as capability of the sensor in detection of various analytes present in the sample and to respond consistently to the different analytes. The most frequently employed sensor among the above mentioned sensors are electrochemical. Voltammetric and potentiometric are mainly used sensors in the electronic tongue devices. A voltammetric sensor produces more complex data as compared to potentiometric sensors but it provides the superior information.

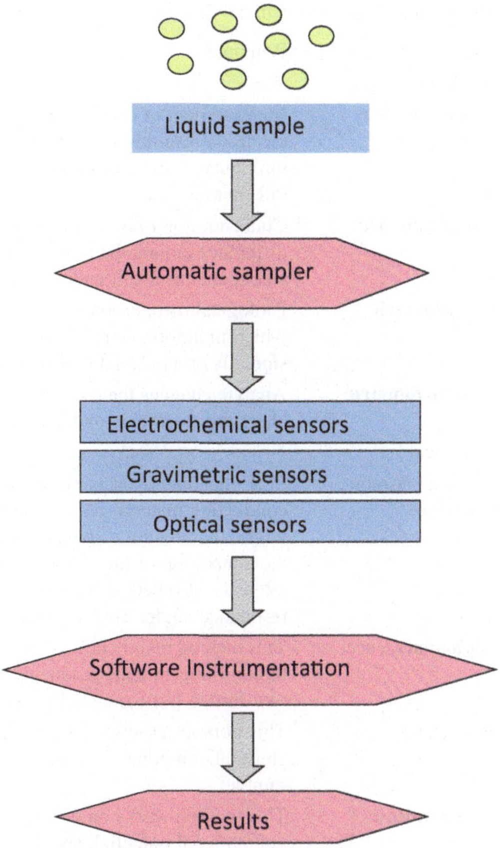

Fig. 13.15 The systematic demonstration of the working of an electronic tongue

13.4 Applications of Electronic Nose and Electronic Tongue in Food Science

13.4.1 Coffee Analysis

Coffee is an expensive commodity and is consumed by almost all the individuals throughout the world. Hence, improvement in its taste is of utmost importance. Therefore steps starting from the selection of grains, drying process, treatment, roasting, grinding as well packaging is carefully controlled. So, proficient quality control devices are required during the production for the monitoring of the flavor, aroma, adulterants, and other bioactive molecules present in the coffee. Different bioactive components present in the coffee are chlorogenic acid, caffeine, and polyphenols (Table 13.3).

Different electrochemical techniques such as differential pulse voltammetry (DPV), square wave voltammetry (SWV), potentiometry were employed for

Table 13.2 Types of chemical sensors

Types of sensors	Principle
1. Electrochemical	
Amperometric	Measurement of current among a working electrode and a counter electrode resulting from the oxidation or reduction of a electro-sensitive biological element at working electrode which provides definite analytical information.
Potentiometric	Concentration of an analyte is determined by measuring the potential difference among the working electrode and reference electrodes in electrochemical cell under zero current flow condition.
Impedimetric	Biological recognition element is immobilized onto the electrode surface which monitors the reaction and the output of the electrical impedance signal is proportional to analyte activity.
Conductometric	Analyte changes the concentration of ionic species in the medium which alters the electrical conductivity of medium which can be measured.
2. Optical	
Surface plasmon resonance	SPR phenomena takes place on surface of the metal or some another conducting material on boundary of media typically glass with liquid while illumination is done by polarized light at a definite angle, results in the generation of the surface plasmon's and therefore a decrease in intensity of reflected light occurs at particular angle that is identified as resonance angle. This result is proportional to mass on the surface.
Bioluminescence	This method mainly utilizes the recombinant technology and recombinant bioluminescent cells producing bioluminescent signals that are transmitted from analyte by an optical fiber.
Reflectance	This photoelectric sensor emits a visible or infrared light from the light emitting component and the sensor detects the light reflected from the sample.
3. Gravimetric	This method works on the principle that every mass has a related gravitational potential, and it is used to measure the gravitational acceleration.

detection of bioactive components in the coffee which helps to improve the quality. An additional sensor was reported in which electronic nose combined with gas chromatography (GC) technique to detect volatile compounds present in Arabica coffee beans (roasted) collected from the various countries such as Costa Rica, Brazil, Peru, Ethiopia, and many more. It was found that the coffee from different countries contains diverse volatile compounds which resulted in production of dissimilar odors. After the quantitative determination of volatile compounds, it was found that azines, alcohols, hydrazides, acids, ketones, and aldehydes are most abundant. As a result, highest amount of pyridine (azine) was present in the coffees from the Peru and Brazil that is 21.9% and 28.7%, respectively, and is responsible for the bitter odor. Moreover, santos coffee collected from Brazil contains about 30% aromatic compounds. Most of aromatic compounds were azines, then aldehydes (17.5%); hydrazine, and acids are about 12–13% and ketons are nearly 7%. Similarly, coffees from the Guatemala and Costa Rica, consist comparable percentage of volatile compounds. Hydrazides were found to be about 23–25%,

Table 13.3 Range of Bioactive components detected in Samples of Coffee

Bioactive component	Sample of Coffee	Concentration detected	Method employed	Reference
Chlorogenic acid	Coffee	3.7 mg/mL	Differential pulse voltammetry	de Santos et al. (2011)
Caffeine	Coffee	163 mg/L	Square wave voltammetry	Mersal (2012)
Caffeine	Coffee	82 mg/L	Differential pulse voltammetry	Khoo et al. (2013)
Caffeine	Nescafe sachet	229.5 mM	Differential pulse voltammetry (bare GCE)	Carolina Torres et al. (2014)
Caffeine	Coffee	64.1 ± 2.5 mg/L	Differential pulse adsorptive stripping voltammetry	Tyszczuk-Rotko and Beczkowska (2015)
Caffeine	Espresso coffees	325 ± 24 mg/100 mL (At 89 °C)	–	Buratti et al. (2016)
Chlorogenic acid	Espresso coffees	178 ± 15 mg/100 mL (At 89 °C)	–	Buratti et al. (2016)

then alcohols nearly 18%, aldehydes and azines (15–17%), and finally, ketones about 8–10% (Marek et al. 2020).

13.4.2 Detection of Aroma and Taste of Olive Oil

Olive oil is extracted from the olive fruits. Large amount of triglycerides are present in the olive oil along with the small proportion of hydrocarbons, fatty acids, diglycerides, monoglycerides, phenolic compounds, and sterols. Different types of phenolic compounds are found in the olive oil that is mentioned in Fig. 13.16. These small compounds are incredibly essential for the taste as well as for the aroma of olive oil. So, quantitative as well as qualitative analysis of these compounds is important for validation and to detect the adulterants in the oil (Boskou 2006). Several volatile compounds are also present in the olive oil that are associated with the aroma and they are hydrocarbons, ketones, alcohols, esters, aldehydes, acids, pyrazines, thiophene derivatives, ethers, furan derivatives, and thiols. Different flavors of the olive oil are due to different compounds present in them. For example, fatty flavor is due to the presence of hepatanal, 2-nonenal, 3-nonenal, 2-octenal; fruity flavor is due to ethyl isobutyrate, hexyl acetate, ethyl 2--methylpropanoate, ethyl 2-methylbutyrate; green flavor is due to 2-hexenal, cis-3-hexenal, 2-hexen-1-ol, cis-3-hexen-1-ol; grassy flavor is due to hexanal, 3-hexen-1-ol; soapy flavor is due to octanal, nonanal; sweet flavor is due to hexyl acetate, phenyl acetaldehyde; bitter flavor is due to 2-hexenal, 2-hexen-1-ol, and blackcurrant flavor is due to 4-methoxy-2-methyl-2- butanethiol (Apetrei et al. 2016).

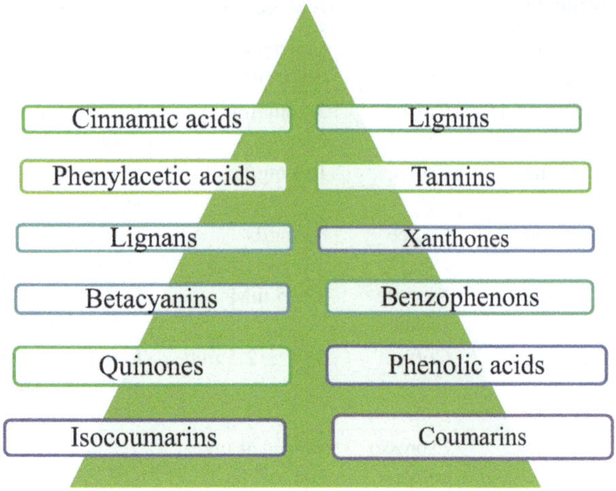

Fig. 13.16 Types of the phenolic compounds normally found in the olive oil

To overcome the shortcomings of gas chromatography (GC), mass spectrometry (MS) and human sensory evaluation; electronic noses and tongues are utilized for determination of aroma in addition to flavor of olive oil. For instance, deterioration of the olive oil occurs mainly by oxidation and lipolysis. So, electronic nose and electronic tongue were used along with multivariate analysis to keep check on oxidation of the olive oil in different storage conditions (Casio et al. 2007). Bitterness of extra virgin olive oils was determined by developing a multisensor system and is an accurate and cost-effective method. Voltammetric electrodes which are based up on the polypyrrole are employed as sensing units. This arrangement of the sensors was introduced to extra virgin olive oil emulsions. The redox properties of electro-active compounds such as antioxidants, which are found in the olive oil emulsions, are monitored by cyclic voltammetry. A comparison was done by evaluation of this method with the bitterness data obtained from the group of experts as well as by the physiochemical methods. In calibration, a fine correlation of 0.997 was attained along with root mean square error of calibration (RMSEC) was found to be 0.0762 and in the prediction, correlation of 0.995 was achieved along with root mean square error of prediction (RMSEP) of 0.1043 (Apetrei 2012). Also, quantitative as well as qualitative anaylsis of phenolic compounds present in olive oils was done via employing the voltammetric electronic tongue device. This device was constructed from an array of screen-printed electrodes altered by polypyrrole. Cyclic voltammetry (CV) technique was used for the analysis of the emulsions prepared from various samples of virgin olive oil. In calibration as well as in prediction excellent correlation coefficient of about 0.9976 and 0.9884 was obtained, in the range of 111.75 to 482.42 mg $\times$ kg^{-1} (Apetrei and Apetrei 2013). Another low-cost method known as fusion technique was discovered in which the Taguchi gas sensor (TGS) electronic nose and voltammetric electronic tongue were combined. This method was devised to overcome the shortcomings of the independently used electronic noses and electronic tongues. It was used for classification of the five different virgin olive oils collected from diverse geological areas of Morocco and the results outperformed the previously found results from the simple electronic tongue and nose system (Haddi et al. 2013).

A technique was developed to examine the quality of olives during the storage period. It was determined by observing the oxidative stability, free acidity, and peroxide values. The combination of electronic tongue and chemometric tools was utilized to monitor quality of extra virgin olives oils. This method was a success as analysis time and cost was reduced. And it was found that quality of the stored olives depends upon the storage time and storage lighting environment (Rodrigues et al. 2016). Another sensor was constructed by combining the electronic tongue, nose, and eye for categorization of extra virgin, olive as well as pomace olive oils along with the establishment of appropriate storage conditions for extra virgin olive oils. Data processing was done by principle component algorithm (PCA) and two categories of the olive oils were determined which are fresh and oxidized. Accuracy of 94% was accomplished. So, characterization and shelf-life assessment of olive oil was done (Buratti et al. 2018). Moreover, a method was developed for the post-harvest quality monitoring of olives which determines the quality of the produced olive oil. Electronic nose sensor along with the pattern recognition algorithm (PCA) was employed for prediction of status of olive oil by keeping a check on the harvested olive fruits. The experimentation was done on 82 samples and superior results were achieved with the multilayer perceptron-artificial neural network (MLP-ANN) with the accuracy of 90.2% (Gila et al. 2020). A new technique was reported to find out the ripe and green intensities of the olive fruits. Olives were classified into the ripened fruit or light greenly fruit or medium greenly fruit or intense greenly fruit by electronic nose device that is made up of nine (9) metal oxide sensor (MOS) (Teixeira et al. 2021).

13.4.3 Analysis of Meat Quality

Meat is among the most consumed foods all over the world and is a major source of proteins, vitamins, and minerals. But it is a perishable product and the deterioration rate of the meat is too rapid due to the presence of nutrients which results in change of its odor, flavor, and texture. So, it's safe storage and handling is of utmost importance. Electronic noses and electronic tongues are used for quality assessment, shelf-life examination, and verification of food products as well as for freshness assessment of foods (Gliszczynska-Swiglo and Chmielewski 2017). The quality of the meat can be analyzed by the e-nose devices by determining presence of volatile compounds (VOCs) such as alcohols, esters, methyl acetate, ethyl acetate, carboxylic acids, lactones, acetone, methyl ethyl ketone, furans, dimethyl sulfide, hydrogen sulfide, methyl mercaptan, propylene sulfide, pyridines, and dimethyl disulfide in the spoiled meat (Fig. 13.17).

Types of the sensors mainly employed for the meat sensing are metal oxide semiconductors and conducting polymers. The sensor was developed for detection of the *Pseudomonas aureofaciens* and assay time was low as 1–2 min along with the

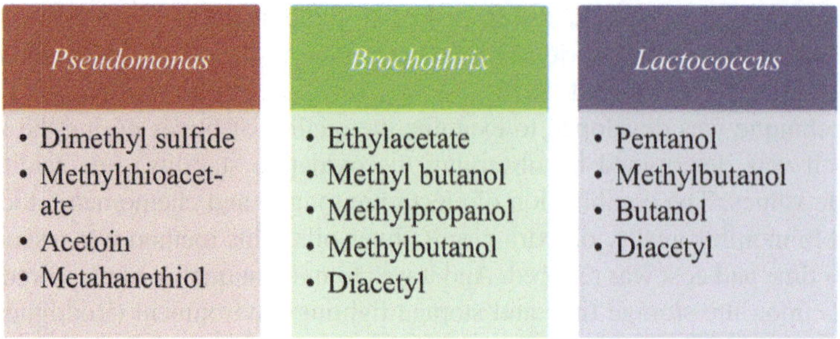

Fig. 13.17 Spoilage causing microbes and their related volatile compounds produced in the meat

detection limit that ranges from ppm (parts per million) to ppb (parts per billion) (Balasubramanian et al. 2016). Another commercial electronic nose device known as PEN3 was reported. It was used to investigate the meat samples of tilapia which were exposed to ozonated water of various concentrations. Principle component analysis (PCA) tool was employed for analysis of data and freshness of the meat was estimated by the reactive substances such as thiobarbituric acid and TVB-N. It was found that 10 min treatment of 5 mg/L of ozonated water results in the considerable reduction in the rate of declination of freshness of tilapia fillet (Yan et al. 2015). Another method was devised, in which eight (8) types of volatile compounds were detected in smoked chicken drumsticks. Electronic nose, electronic tongue along with the headspace solid-phase microextraction gas chromatography mass spectrometry (HS-SPME/GC-MS) were employed for assessing effect of different sugar smoking time periods on taste as well as aroma of the chicken drumsticks. The moisture content of drumsticks decreased to 65.23% from 71.2%, pH decreases from 6.66 to 5.36, and water activity decreased from 0.987 to 0.979 (Zhang et al. 2021).

Adulteration of the beef with the pork and turkey with chicken meat is a widespread problem. So, sensors based on the e-noses and e-tongues are employed to check the adulteration. A cost-effective method based on the colorimetric sensors was developed for the quantitative and qualitative detection of adulteration of beef meat with the pork meat. Three (3) samples containing pure pork, beef–pork combination and the pure beef were taken into considerationand were analyzed by the fisher LDA (linear discrimination analysis) as well as by the ELM (extreme learning machine). Prognosis of the adulteration level was made by the BP-ANNs (Back propagation artificial neural networks). It was found, ELM method is superior and identification rate of 91.27% was found in training set and 87.5% was found in prediction set was achieved, respectively (Han et al. 2020). An additional study based on the electronic tongue was done to find out the best possible dilution levels of the meat extract, and three systematized meat extraction methods were developed. This method aimed to study the small amount of adulterants present in meat, for example, adulteration of chicken in turkey and adulteration of pork in the beef. The optimum dilution factor was found to be 1% *w/v* of the liquid meat extract. Higher linear analysis accuracy (LDA) was achieved for instance recognition of 78.13% and

Table 13.4 Comparison of the flavor produced by the single-stage and double-stage Sous-vide during treatment with various temperature and time combinations

Stages of Sous-vide	Temperature	Time-period	Flavor detected by e-tongue	Compounds detected
Single-stage	70 °C	12 h	Unami	Adenosine-5-monophosphate (AMP), Guanosine-5-monophosphate
Single-stage	60 °C	3–12 h	Astringent and sour	Leucine, hypoxanthine, histidine, inosine
Double-stage	60 °C	6–12 h	Astringent and sour	Leucine, hypoxanthine, histidine, inosine

89.2% for the chicken and pork, respectively, as well as validation of 64.77% and 68.77% for the chicken and pork, correspondingly (Zaukuu et al. 2021).

Smart packaging of the meat products is a new trend in the meat industries. More and more advancements are being made in this field. One such innovation is the development of the electronic nose device for the quick assessment of the beef quality. Moreover, this device is rapid, cost-effective, and is easy to use as compared to the previous methods such as LDA (linear discriminant analysis), MLP (multi-layer perceptron), SVM (support vector machine), SVR (support vector regression), k-NN (k-nearest neighbor), and standard long short-term memory (LSTM) (Wijaya et al. 2021). The technique was used to examine the microbial population as well as the quality of the beef. Usually, electronic nose signal are contaminated with the noise, and the specialty of this method is to generate the contamination-free data. So, DWTLSTM (discrete wavelet transform and long short-term memory) was suggested for controlling the noise produced in the electronic nose signal during the examination of the beef quality. As expected, DWTLSTM method was far better than the conventional methods. Hence, this technique outperformed and average accuracy of 94.8% was achieved. Also, average of F-measure was found to be 85.05% and was effective for the calculation of the microbial population.

In another approach, e-tongue was utilized for the examination of non-volatile compounds present in beef muscles when exposed to different temperature and time mixtures (Table 13.4) (Ismaila et al. 2020).

An additional study was reported for the quick detection of PV that is peroxide value and IMF that is intramuscular fat of the pork meat. In this technique, information gathered from e-nose and hyperspectral imaging was combined and pork meat was exposed to different NaCl concentrations and temperatures (Aheto et al. 2020). A further method was invented for the monitoring of the spoilage of Salmon fish which is a highly perishable food product due to the changes in the pH, odor, temperature, and texture during the frost storage. An electronic nose device and the IoTMS (IoT-enabled monitoring system) were combined and were used for determination of the freshness as well as the quality of the Salmon fish. Experimentation was performed by combinations of the different temperature (0 °C, 4 °C, 6 °C) with changeable time periods (0, 3, 6, 9, 12 to 14 days). Electronic nose device information was clustered by the PCA that is a principle component analysis and the accuracy rate of more than the 90% was achieved (Feng et al. 2020).

In one another technique, electronic nose, electronic tongue was united with the microextraction gas chromatography–mass spectrometry (SPME-GC/MS) for checking the taste as well as the volatile compounds (VOCs) present in Harbin red sausages made traditionally and conventionally (Table 13.5). During the experimentation, four varieties of each traditional and conventional sausage were selected and 131 VOCs were detected. Out of 131, 50 volatile compounds (VOCs) were present in all the samples of sausages and the remaining 77 volatile compounds were present either in traditionally or conventionally prepared sausages. It was observed that number of volatiles found in traditional sausages was much more than the number of volatiles found in conventional sausages as mentioned in Table 13.5 (Yin et al. 2021):

13.4.4 Other Applications

Electronic noses and electronic tongues have many other functions, for example, assessment of freshness of the fruits (Jiang et al. 2018); categorization, and quality control of edible oils; detection of quality of beer, wine, and many other alcoholic drinks; qualitative and quantitative evaluation of tea (Xu et al. 2019); qualitative detection of spicy foods (Paup et al. 2019) and many more.

13.5 Conclusion

So, Electronic nose is a contrived olfaction technology which mimics the differentiation capacity of human olfactory system and is an odor examination technique. Whereas, Electronic tongue is a device that artificially reproduces the sensation of taste. Electronic nose and electronic tongue are devised of multisensor array along with an appropriate pattern recognition system. They are extremely beneficial devices and have many applications in food and beverage analysis, healthcare, drugs, pharmaceuticals, agriculture, forestry, military security system, civilian security system, indoor and outdoor monitoring, environmental monitoring, and medical diagnostics. Among all, most of them are from the food and beverage category. It includes the monitoring of quality from the procurement of raw materials to the final product fabrication and its packaging along with supervision of the storage conditions. More technical advancements are being made in this field to create miniature, portable, less arduous, more efficient, logical and easy-to-operate food analysis sensors.

Table 13.5 Types of volatile compounds and their concentration found in the traditional and conventional sausages

Volatile compounds	Examples of VOCs found in sausages	Concentration of VOCs in traditional sausages (ug/kg)	Concentration of VOCs in conventional sausages (ug/kg)
Alcohols	Hexanol, heptanol, octanol, ethanol, nonanol, benzyl alcohol	46.57 ± 2.60	72.53 ± 14.10
Aldehydes	Hexanal, 2-methyl-2- butenel, hetanal, octanal, nonanal, 2-heptenel, benzaldehyde	211.32 ± 44.65	80.56 ± 9.19
Ketones	2-heptanone, 3-hydroxy-2-butanone, 2-methyl-2-cyclopentane-1-one, 2-nonanone, 1-indanone, 3-ethyl-2-cyclopentane-1-one	404.28 ± 48.55	291.99 ± 23.97
Acids	Acetic acid, propionic acid, butyric acid, pentanoic acid, heptanoic acid, sorbic acid, palmitic acid, benzoic acid, lauric acid	156.61 ± 50.47	120.45 ± 31.75
Esters	Methyl benzoate, terpinyl acetate, butyrolactone	7.09 ± 0.91	34.81 ± 7.09
Phenols	Cresol, guaiacol, 2-methoxy-3-methylphenol, eugenol, 3-ethyl phenol, 2-methoxy-4-methylphenol, 3,4-dimethylphenol, 2-isopropylphenol	1795.40 ± 382.75	783.97 ± 68.93
Terpenes	Pinene, camphene, sabinene, terpinene, limonene, ocimene, cymene, styrene, terpinolene, elemane	1695.61 ± 283.40	2223.70 ± 198.34
Non-terpene hydrocarbons	Octane, toluene, 2-methylstyrene, naphthalene	496.30 ± 189.22	386.48 ± 98.97
Sulfur-containing compounds	Propylene sulfide, diallyl sulfide, dimethyl trisulfide, 3-methylthiophene	1842.20 ± 352.53	2397.50 ± 445.44
Furans	Furfural, furfuryl acetate, 2-acetylfuran, methyl-2-furoate, 5-methyl-2-acetylfuran	928.73 ± 236.59	226.70 ± 6.91
Pyridine	–	6.46 ± 2.67	2.41 ± 1.26
2-Picoline	–	1.99 ± 0.82	5.13 ± 1.82
3-Picoline	–	9.50 ± 4.83	4.54 ± 1.62
4-Picoline	–	0.81 ± 0.42	2.14 ± 0.85
2-Acetyl Pyrrole	–	1.02 ± 0.41	0.31 ± 0.16

References

Aguilera T, Lozano J, Paredes JA, Alvarez FJ, Suarez JI (2012) Electronic nose based on independent component analysis combined with partial least squares and artificial neural networks for wine prediction. Sensors 12(6):8055–8072

Aheto JH, Huang X, Tian X, Ren Y, Ernest B, Alenyorege EA, Dai C, Hongyang T, Xiaorui Z, Wang P (2020) Multi-sensor integration approach based on hyperspectral imaging and electronic nose for quantitation of fat and peroxide value of pork meat. Anal Bioanal Chem. https://doi.org/10.1007/s00216-019-02345-5

Ampuero S, Bosset J (2003) The electronic nose applied to dairy products: a review. Sens Actuators B 94(1):1–12

Apetrei C (2012) Novel method based on polypyrrole- modified sensors and emulsions for the evaluation of bitterness in extra virgin oils. Food Res Int 48(2):673–680

Apetrei IM, Apetrei C (2013) Voltammetric e- tongue for the quantification of total polyphenol content in olive oils. Food Res Int 54(2):2075–2082

Apetrei C, Ghasemi-Varnamkhasti M, Mirela Apetrei I (2016) Olive oil and combined electronic nose and tongue. In Electronic noses and tongues in food science, pp 277–289.

Bai H, Shi GQ (2007) Gas sensors based on conducting polymers. Sensors 7(3):267–307

Bajpai VK, Kamle M, Shukla S, Mahato DK, Chandra P, Hwang SK, Kumar P, Huh YS, Han YK (2018) Prospects of using nanotechnology for food preservation, safety, and security. J Food Drug Anal 26:1201–1214

Balasubramanian S, Amamcharla J, Shin JE (2016) Possible application of electronic nose systems for meat safety: an overview. In Electronic noses and tongues in food sciences, pp 59–71.

Baskar C, Nesakumar N, Rayappan JBB, Doraipandian M (2017) A framework for analysing E-nose data based on fuzzy set multiple linear regression: Paddy quality assessment. Sens Actuators A Phys 267:200–209

Boeker P (2014) On 'electronic nose' methodology. Sens Actuators B 204:2–17

Boskou D (ed) (2006) Olive oil chemistry and technology, 2nd edn. AOCS Press, Champaign, IL

Buratti S, Benedetti S, Giovanelli G (2016) Application of electronic senses to characterize espresso coffees brewed with different thermal profiles. Eur Food Res Technol 243(3):511–520

Buratti S, Malegori C, Benedetti S, Oliveri P (2018) E-nose, e-tongue and e-eye for edible olive oil characterization and shelf life assessment: a powerful data fusion approach. Talanta 182:131–141

Carolina Torres A, Barsan MM, Brett CMA (2014) Simple electrochemicalsensor for caffeine based on carbon and Nafion-modified carbon electrodes. Food Chem 149:215–220

Casio MS, Ballabio D, Benedetti S, Gigliotti C (2007) Evaluation of different storage conditions of extra virgin olive oils with an innovative recognition tool built by means of electronic nose and electronic tongue. Food Chem 101(2):485–491

Chandra P (2016) Nanobiosensors for personalized and onsite biomedical diagnosis. In Nanobiosensors for personalized and onsite biomedical diagnosis.

Cheeke JDN, Wang Z (1999) Acoustic wave gas sensors. Sens Actuators B 59(2–3):146–153

Chodavarapu VP, Shubin DO, Bukowski RM, Titus AH, Cartwright AN, Bright FV (2007) CMOS based phase fluorometric oxygen sensor system. IEEE Trans Circuits Syst I: Regul Pap 54(1): 111–118

Ciosek P, Wróblewski W (2007) Sensor arrays for liquid sensing—electronic tongue systems. Analyst 132:963–978

Craven MA, Gardner JW, Bartlett PN (1996) Electronic noses-development and future prospects. Trends Anal Chem 15(9):486–493

Dutta R, Hines EL, Gardner JW, Kashwan KR, Bhuyan M (2003) Tea quality prediction using a tin oxide-based electronic nose: an artificial intelligence approach. Sens Actuators B 94:228–237

Dymerski TM, Chmiel TM, Wardencki W (2011) Invited review article: an odor-sensing system-powerful technique for foodstuff studies. Rev Sci Instrum 82(11):111101

Esfahani S, Covington JA (2017) Low cost optical electronic nose for biomedical applications. Proceedings 1(4):589

Feng H, Zhang M, Liu P, Liu Y, Zhang X (2020) Evaluation of IoT-enabled monitoring and electronic nose spoilage detection for salmon freshness during cold storage. Food 9(11):1579

Ghasemi-Varnamkhasti M, Mohtasebi SS, Siadat M, Razavi SH, Ahmadi H, Dicko A (2012) Discriminatory power assessment of the sensor array of an electronic nose system for the detection of non-alcoholic beer aging. Czech J Food Sci 30:236–240

Gila DMM, Garcia JG, Bellincontro A, Mencarelli F, Ortega JG (2020) Fast tool based on electronic nose to predict olive fruit quality after harvest. Postharvest Biol Technol 160:111058

Giungato P, Laiola E, Nicolardi V (2017) Evaluation of industrial roasting degree of coffee beans by using an electronic nose and a stepwise backward selection of predictors. Food Anal Methods 10(10):3424–3433

Gliszczynska-Swiglo A, Chmielewski J (2017) Electronic nose as a tool for monitoring the authenticity of food: a review. Food Anal Methods 10:1800–1816

Haddi Z, Alami H, Bari NE, Tounsi M, Barhoumi H, Maaref M, Jaffrezic-Renault N, Bouchikhi B (2013) Electronic nose and tongue combination for improved classification of Moroccan virgin olive oil profiles. Food Res Int 54(2):1488–1498

Han F, Huang X, Aheto JH, Zhang D, Feng F (2020) Detection of beef adulterated with pork using a low-cost electronic nose based on colorimetric sensors. Foods 9(2):193

Huang XH, Bai QS, Hu JG, Hou D (2017) A practical model of quartz crystal microbalance in actual applications. Sensors 17(8):1785

Ismaila I, Hwang YH, Joo ST (2020) Low-temperature and long-time heating regimes on non-volatile compound and taste traits of beef assessed by the electronic tongue system. Food Chem 320:126656

Jiang H, Zhang M, Bhandaric B, Adhikari B (2018) Application of electronic tongue for fresh foods quality evaluation: a review. Food Rev Intl. https://doi.org/10.1080/87559129.2018.1424184

Karakaya D, Ulucan O, Turkan M (2020) Electronic nose and its applications: a survey. Int J Autom Comput. https://doi.org/10.1007/s11633-019-1212-9

Khoo WYH, Pumera M, Bonanni A (2013) Graphene platforms for the detection of caffeine in real samples. Anal Chim Acta 804:92–97

Kim HJ, Lee JH (2014) Highly sensitive and selective gas sensors using p-type oxide semiconductors: Overview. Sens Actuators B 192:607–627

Laref R, Losson E, Sava A, Adjallah K, Siadat M (2018) A comparison between SVM and PLS for E-nose based gas concentration monitoring. In Proceedings of IEEE International Conference on Industrial Technology, IEEE, Lyon, France, pp 1335–1339

Leal RV, Quiming AXC, Villaverde JF, Yumang AN, Linsangan NB, Caya MVC (2019) Determination of schizophrenia using electronic nose via support vector machine. In Proceedings of the 9th International Conference on Biomedical Engineering and Technology, ACM, Tokyo, Japan, pp 13–17

LeCun Y, Bengio Y, Hinton G (2015) Deep learning. Nature 521(7553):436–444

Liu X, Cheng ST, Liu H, Hu S, Zhang DQ, Ning HS (2012) A survey on gas sensing technology. Sensors 12(7):9635–9665

Marek G, Dobrzanski B, Oniszczuk T, Combrzynski M, Cwikla D, Rusinek R (2020) Detection and differentiation of volatile compound profiles in roasted coffee arabica beans from different countries using an electronic nose and GC-MS. Sensors 20(7):2124

Mersal GM (2012) Experimental and computational studies on the electrochemical oxidation of caffeine at pseudo carbon paste electrode and its voltammetric determination in different real samples. Food Anal Methods 5:520–529

Paup VD, Barnett SM, Diako C, Ross CF (2019) Detection of spicy compounds using the electronic tongue. J Food Sci. https://doi.org/10.1111/1750-3841.14709

Purohit B, Kumar A, Mahato K, Chandra P (2020a) Electrodeposition of metallic nanostructures for biosensing applications in health care. J Sci Res. https://doi.org/10.37398/jsr.2020.640109

Purohit B, Vernekar PR, Shetti NP, Chandra P (2020b) Biosensor nanoengineering: design, operation, and implementation for biomolecular analysis. Sensors Int 1:100040

Qi PF, Meng QH, Zeng M (2017) A CNN-based simplified data processing method for electronic noses. In Proceedings of ISOCS/IEEE International Symposium on Olfaction and Electronic Nose, IEEE, Montreal, Canada, pp 1–3

Rodrigues N, Dias LG, Veloso ACA, Pereira JA, Peres AM (2016) Monitoring olive oils quality and oxidative resistance during storage using an electronic tongue. LWT 73:683–692

Romani S, Cevoli C, Fabbri A, Alessandrini L, Dalla Rosa M (2012) Evaluation of coffee roasting degree by using electronic nose and artificial neural network for off-line quality control. J Food Sci 77(9):C960–C965

de Santos WJR, Santhiago M, Yoshida IVP, Kubota LT (2011) Novel electrochemical sensor for the selective recognition of chlorogenic acid. Anal Chim Acta 695(1-2):44–50

Tang CT, Huang CM, Tang KT, Chen H (2015) A scalable and adaptable probabilistic model embedded in an electronic nose for intelligent sensor fusion. In Proceedings of IEEE Biomedical Circuits and Systems Conference, IEEE, Atlanta, USA

Teixeira GG, Dias LG, Rodrigues N, Marx IMG, Veloso ACA, Pereira A, Peres AM (2021) Application of a lab-made electronic nose for extra virgin olive oils commercial classification according to the perceived fruitiness intensity. Talanta 226:122122

Thazin Y, Pobkrut T, Kerdcharoen T (2018) Prediction of acidity levels of fresh roasted coffees using e-nose and artificial neural network. In Proceedings of the 10th International Conference on Knowledge and Smart Technology, IEEE, Chiang Mai, Thailand, pp 210–215

Tian XJ, Wang J, Cui SQ (2013) Analysis of pork adulteration in minced mutton using electronic nose of metal oxide sensors. J Food Eng 119(4):744–749

Turner NW, Bloxham M, Piletsky SA, Whitcombe MJ, Chianella I (2017) The use of a quartz crystal microbalance as an analytical tool to monitor particle/surface and particle/particle interactions under dry ambient and pressurized conditions: A study using common inhaler components. Analyst 142(1):229–236

Tyszczuk-Rotko K, Beczkowska I (2015) Nafion covered lead film electrode for the voltammetric determination of caffeine in beverage samples and pharmaceutical formulations. Food Chem 172:24–29

Vlasov Y, Legin A, Rudnitskaya A, Di Natale C, D'amico A (2005) Nonspecific sensor arrays ("electronic tongue") for chemical analysis of liquids (IUPAC Technical Report). Pure Appl Chem 77(11):1965–1983

Wang CX, Yin LW, Zhang LY, Xiang D, Gao R (2010) Metal oxide gas sensors: sensitivity and influencing factors. Sensors 10(3):2088–2106

Wei ZB, Xiao XZ, Wang J, Wang H (2017) Identification of the rice wines with different marked ages by electronic nose coupled with smartphone and cloud storage platform. Sensor 17(11): 2500

Wijaya DR, Sarno R, Zulaika E (2021) DWTLSTM for electronic nose signal processing in beef quality monitoring. Sens Actuators B 326:128931

Winquist F, Krantz-Rülcker C, Wide P, Lundström I (1998) Monitoring of freshness of milk by an electronic tongue on the basis of voltammetry. Meas Sci Technol 9:1937

Xu M, Wang J, Zhu L (2019) The qualitative and quantitative assessment of tea quality based on E-nose, E-tongue and E-eye combined with chemometrics. Food Chem 289:482–489

Yan MY, Lu YQ, Chen DW (2015) Application of electronic nose in freshness evaluation of tilapia fillets as affected by ozone treatment. Food Sci Technol 36:265–269

Yin X, Lv Y, Wen R, Wang Y, Chen Q, Kong B (2021) Characterization of selected Harbin red sausages on the basis of their flavour profiles using HS-SPME-GC/MS combined with electronic nose and electronic tongue. Meat Sci 172:108345

Zaukuu JLZ, Gillay Z, Kovacs Z (2021) Standardized extraction techniques for meat analysis with the electronic tongue: a case study of poultry and red meat adulteration. Sensors 21(2):481

Zhang L, Hu Y, Wang Y, Kong B, Chen Q (2021) Evaluation of the flavour properties of cooked chicken drumsticks as affected by sugar smoking times using an electronic nose, electronic tongue, and HS-SPME/GC-MS. LWT 140:110764

Zou Y, Wan H, Zhang X, Ha D, Wang P (2015) Electronic nose and electronic tongue. In Bioinspired smell and taste sensors, pp 19–44. https://doi.org/10.1007/978-94-017-7333-1_2

Fungal β-D-Glucan Films for Electrochemical Biosensing in Food Analysis

14

Carlos Alberto Rossi Salamanca-Neto, Gabriel Junquetti Mattos, Aneli M. Barbosa-Dekker, Robert F. H. Dekker, and Elen Romão Sartori

Abstract

The exopolysaccharide botryosphaeran produced by the ascomycetous fungus, *Botryosphaeria rhodina* MAMB-05, and its chemical-derivative form (carboxymethyl-botryosphaeran) have emerged in the electroanalytical field in recent years as a platform for immobilizing the enzyme laccase on carbon-based electrodes. The bioelectrochemical devices fabricated have presented excellent performance towards the determination of phenolic compounds in food samples, with high sensitivity, selectivity, and long-term storage stability. Other applications have included analysis in the clinical, pharmaceutical, and environmental sectors.

Keywords

Botryosphaeria rhodina MAMB-05 · Botryosphaeran · Laccase · Carbon-based electrodes · Electroanalytical methods · Enzyme biosensor

C. A. R. Salamanca-Neto · G. J. Mattos · E. R. Sartori (✉)
Laboratório de Eletroanalítica e Sensores, Departamento de Química, Universidade Estadual de Londrina, Londrina, PR, Brazil
e-mail: elensartori@uel.br

A. M. Barbosa-Dekker · R. F. H. Dekker
Beta-Glucan Produtos Farmoquímicos EIRELI, Lote 24A, Bloco Zircônia, Universidade Tecnológica Federal do Paraná, Londrina, PR, Brazil

P. Chandra, P. S. Panesar (eds.), *Nanosensing and Bioanalytical Technologies in Food Quality Control*, https://doi.org/10.1007/978-981-16-7029-9_14

14.1 Introduction

The widespread interest and attention on phenolic compounds in foods are due to the effect of these compounds sequestering free radicals with consequent benefits to human health. Their concentration in foodstuffs is variable and related to sensory characteristics (flavor and taste), which can influence the nutritional value of teas, beverages (wines, fruit juices), and other products (Spanos and Wrolstad 1990; Herrmann 1989).

Enzyme biosensors based on amperometric or voltammetric detection have shown an incredible potential to replace the most complex analytical procedures used to determine phenolic compounds, ensuring selectivity, sensitivity, and quantitation (Rodríguez-Delgado et al. 2015). In the analytical procedures that use these types of devices, there is no need for tedious sample pretreatment or large amounts of reagents to be used, and the measurements are rapid, reproducible, direct, and reliable.

In developing electrochemical enzyme biosensors, the techniques of immobilizing enzyme on carbon-based electrodes are highly significant because the enzyme should maintain its catalytic activity and stability. Proper immobilization will aid good operational and storage stabilities, besides featuring high reproducibility of the device.

Recently, the exopolysaccharide botryosphaeran produced by the ascomycetous, filamentous, endophytic fungus *Botryosphaeria rhodina* MAMB-05 was successfully used to immobilize laccase on a glassy carbon electrode modified with multiwalled carbon nanotubes (MWCNTs) based upon non-covalent interactions between the enzyme and botryosphaeran (Coelho et al. 2019; Mattos et al. 2019). Another novelty that can be highlighted is the use of a derivative form of botryosphaeran (carboxymethyl-botryosphaeran) for the immobilization of laccases on the carboxymethyl-botryosphaeran in an aqueous solution, which subsequently is added onto the surface of a carbon black paste electrode (Gomes et al. 2020). Both approaches provided excellent analytical performance of the biosensors, as short time response, selectivity, stability, and also maintained the catalytic activity of the enzyme at the electrode/solution interface for more than 200 analytical measurements.

On this basis, this chapter presents a brief overview of the fundamentals of phenolic compounds, enzyme electrochemical biosensors, and the enzyme laccase when applied to the determination of phenolic compounds in food materials. A special focus is given to laccase immobilization on the surface of carbon-based electrodes by platform materials, botryosphaeran and its carboxymethylated form. The analytical applications of electrochemical (bio)sensors developed with these materials are also presented and discussed.

14.2 The Importance of Phenolic Compounds in Food Samples and Their Determination

Phenolic compounds are classified as secondary metabolites produced in plants. They present antioxidant activity due to their ability to donate hydrogen atoms or electrons acting in the defense mechanisms of plants against environmental damage for their survival, and are of nutritional interest, promoting several human health benefits (Jorge 2006). The consumption of antioxidant-rich vegetables and fruits avoids chronic cardiovascular and neurodegenerative diseases, some types of cancer, and also prevents premature aging (Swallah et al. 2020). They collectively act by inactivating the action of free radicals and reactive oxygen species before damaging vital biomolecules and body tissues, preventing or minimizing the triggering of oxidative reactions that cause oxidative stress which progressively leads to cell dysfunction and cell death (Lobo et al. 2010).

A phenolic compound is chemically defined as a molecule that has the presence of at least one aromatic ring with one or more hydroxyl groups directly bonded onto the cyclic structure (De Beer et al. 2002). The antioxidant action of phenolic compounds is related mainly to their chemical structure, which depends on the number and position of hydroxyl groups and the extent of conjugation in the molecule of interest. Flavonoids (e.g., quercetin) and phenolic acids (e.g., chlorogenic acid) are two of the main classes of phenolic compounds with known antioxidant activity (Tohma et al. 2017). Both classes protect food from oxidation and provide a beneficial effect on the health and well-being of consumers, and also contribute to the taste, bitterness, and aroma of the food (e.g., wines, green teas, fruit juices, and coffee-based beverages) (Spanos and Wrolstad 1990; Herrmann 1989).

The determination of the content of phenolic compounds is an effective way to determine food quality. Besides, it may be used to assess the market value, clonal variations, and seasonal quality variations of teas, wines, and fruit juices. Phenolic compounds may negatively affect the sensory characteristics of foods with an impact on their quality and they can cause browning reactions (Spanos and Wrolstad 1990; Herrmann 1989). Consequently, the food industry carries out strict quality control of the concentration of phenolic compounds in the composition of foods, especially in beverages, e.g., wines, spirits, and fruit juices.

A spectrophotometric method based upon a reagent proposed by Otto Folin and Vintila Ciocâlteu is one of the oldest methods extensively employed to quantify phenolic content in many types of samples (Musci and Yao 2017), including fruit juices and red/white wines. The Folin–Ciocâlteu reagent is composed of a mixture of phosphotungstic acid and phosphomolybdic acid in which the tungsten and the molybdenum are in the +6 oxidation state. This promotes electron-transfer reactions between the Folin–Ciocâlteu reagent and the phenolic compounds to form chromogens in an alkaline solution (pH of ~10, by adding sodium carbonate). The reagent, originally intense yellow, changes to blue during the redox reaction, which is directly proportional to the content of the reducing substances in the reaction. The Folin–Ciocâlteu assay presents sensitivity and precision, but lacks specificity due to the potential reducing substances added, or naturally present in the sample matrices that can respond in the same way with this reagent (Musci and Yao 2017).

Among the wide variety of known dosage procedures for the determination of phenolic compounds, those that employ electrochemical enzyme biosensors have great potential for the determination of these compounds in food samples (Rodríguez-Delgado et al. 2015; Della and Compagnone 2018). The incorporation of redox enzymes on the surface of carbon-based electrodes provides selectivity and sensitivity for analytical procedures applied to determine phenolic compounds in rather complex sample matrices, in a sustainable, environmentally friendly, rapid, and economical ways (Nguyen et al. 2019). Some of these devices may identify and determine individual phenolic compounds in complex matrix samples. The excellent performance of electrochemical enzyme sensors for analytical purposes requires the judicious choice of the redox enzyme(s), their form of immobilization (physical adsorption, ionic binding, covalent bonding), and the appropriate electrochemical transducer.

14.3 Electrochemical Enzyme Biosensors for the Determination of Phenolic Compounds in Food Samples

Novel electrochemical enzyme biosensors developed have evolved from the classic and successful glucose biosensor model. On the device originally developed by Clark and Lions in 1962 (Clark and Lyons 2006), the enzyme glucose oxidase (EC 1.1.3.4), obtained from *Aspergillus niger*, was immobilized in a semipermeable dialysis membrane combined with a Clark-type oxygen electrode for measuring glucose levels in the management of diabetes. This device measured either the oxygen reduction or hydrogen peroxide oxidation after the enzymatic reaction, with the main advantages of high sensitivity and low response times. Glucose oxidase, like other oxidases that liberate H_2O_2 in converting substrate to product, can also use other cofactors, such as nicotinamide adenine dinucleotide (NAD^+), flavin adenine dinucleotide (FAD), or nicotinamide adenine dinucleotide (NADH), to assist in transferring electrons, in which they can be externally supplied or immobilized into the electrochemical sensor (Bollella and Katz 2020).

A different approach to the electron-transfer method that has been used for the measurement of the enzymatic reaction was the replacement of oxygen by an electron mediator (i.e., ferrocene derivatives, ferrocyanide, conducting organic salts, and quinones) (Mohammad et al. 2013). Mediators are small electroactive molecules shuttling electrons between the redox center at the enzyme active sites and electrode surface (Bollella and Katz 2020). Briefly, the electrons generated in the enzymatic reaction are transferred to the mediator, which, in turn, is reduced. In the next step, the oxidation of the mediator on the electrode surface produces a current signal that is directly proportional to the concentration of the analyte (e.g., phenolic compound). An advantage of this type of biosensor is that the electrochemical detection of the phenolic compound occurs at a potential closer to zero, but the mediators can also facilitate the electron transfer from secondary redox reactions and

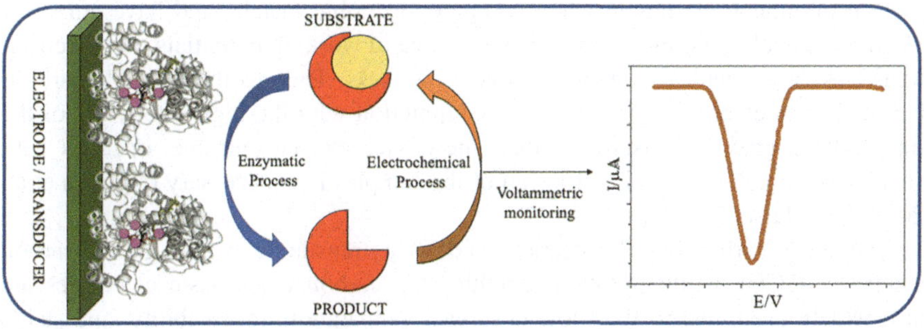

Fig. 14.1 Enzyme-based conversion of substrate into products and reverse electrochemical process monitored by voltammetry

present low chemical stability. Subsequently, electrochemical biosensors have also been developed in which the electron transfer takes place between the active center of the enzyme and the electrode surface without the use of mediators (Freire et al. 2003). With this setup it is possible to obtain a simplified and miniaturized device that operates at potentials closer to those of the enzyme, minimizing the effects of concomitant compounds from complex samples.

In general, electrochemical detection techniques can be categorized and based upon voltammetric, amperometric, impedimetric, potentiometric, or conductometric procedures. Comparatively, amperometric or voltammetric methods are those most used, and the ones with much interest due to their economical qualities, high sensitivity, fast response times, simple design, and construction (Nguyen et al. 2019).

Figure 14.1 presents a brief schematic illustration of an electrochemical enzymatic biosensor that employed voltammetry as an electrochemical detection technique. An electrochemical enzyme biosensor consists of a redox enzyme layer properly immobilized onto an appropriate electrode (transducer). In an aqueous solution, the phenolic compound diffuses into the enzyme layer, where the biochemical reaction will take place. As a result, the electrical parameters of the solution change due to the electron-transfer processes.

The detection system will measure the current flowing through the system of electrodes by varying the potential applied between the biosensor (working electrode) and the reference electrode. Generally, the system contains three electrodes immersed in a supporting electrolyte: biosensor (working electrode—where the electrochemical action takes place), reference electrode (all measured potentials are referenced), and the counter electrode (completes the circuit). The current measured is associated with oxidation or reduction of the reactant or the product of the enzymatic reaction which is correlated to the concentration of the analyte in solution. If measurements of the current are carried out at constant potential, an amperometric system is obtained.

Carbon-based materials are outstanding transducers for the immobilization of redox enzymes. Pyrolytic graphite, glassy carbon, and carbon black electrodes are a more economical alternative when compared to noble metals, and have attractive electrochemical properties, such as a wide range of working potentials, fast electron-transfer kinetics, and reasonably chemical inertness. The readily available carbon-based electrodes provide sensitivity in conjunction with the high selectivity of the immobilized enzyme, ensure stability, the device can be reusable, operates with small sample volumes, and a clean-up of the samples is unnecessary (Nguyen et al. 2019; Sassolas et al. 2012).

The most carbon-based material used as a transducer is the glassy carbon electrode (GCE), while pyrolytic graphite and carbon black-paste electrodes are also widely used, due to their low cost, easy preparation, renewability, and stable responses (Švancara et al. 2009). Carbon pastes usually are constituted of a mixture of carbon powder and a hydrophobic organic liquid binder (Adams 1963). Several strategies of modification of carbon-based electrodes may be used in the development of biosensors, as they provide a new sensor device with desired, often predefined, properties.

The modification may be realized with a nanostructured material including multi-walled carbon nanotubes (MWCNTs) to improve the performance of detection platforms that allow an enhancement of conductivity, and an extensive surface area on the sensor (Della and Compagnone 2018; Merkoçi 2006; Luong et al. 2008; Balasubramanian and Burghard 2006). By definition, MWCNTs are formed by two or more concentric cylindrical shells of graphene sheets with spacing between the layers of 0.34 nm. Smaller tubes are contained within the larger outer shell. Most commercially available MWCNTs are synthesized by chemical vapor deposition techniques during the pyrolysis of hydrocarbon gases at high temperatures (Jacobs et al. 2010). MWCNTs can be added to graphite paste, in an appropriate proportion, as well as by drop-casting them onto the surface of a GCE. GCEs modified with MWCNTs display a low detection limit, high sensitivity, and rapid sensor kinetics (Balasubramanian and Burghard 2006).

Another attractive alternative that has long been practiced for biosensor fabrication is the use of carbon black (CB) paste electrodes as transducers. CB is an inexpensive material, abundant, and does not require any prior treatment before use. This carbonaceous material is produced from heavy aromatic petroleum oils by several well-established manufacturing processes including oil furnace, e.g., thermal black, acetylene black, lamp black, channel black, and gas black (Long et al. 2013). Among these, the oil furnace process presents a highly efficient method that permits rigid control of the chemical and physical properties of CB (Cabot 1960). As the final product of this process, a fine black powder in fluffy form is produced that is composed of spherical primary particles in diameters ranging from 10 to 100 nm, and with surface areas from 25 to 1500 m^2/g, with a semi-graphitic structure of nearly pure elemental carbon (Cabot 1960). CB primary particles fuse, forming aggregates, which connect to form agglomerates. A more recent overview (Arduini et al. 2020) has highlighted interesting electroanalytical properties of CB in the design of sensors and biosensors besides cost-effectiveness, and current trends of this carbonaceous material.

Redox enzymes have also been immobilized onto nanostructured materials mainly by physical absorption or covalent interactions to develop sensitive biosensors.

14.4 Laccase as a Powerful Biocatalyst in Analysis of Phenolic Compounds in Foods

Laccases (EC 1.10.3.2) belong to the polyphenol multicopper oxidase class of enzymes and are a promising candidate to act as a biorecognition element in bioelectrochemical devices (Cannatelli and Ragauskas 2017). Their structure is based on a glycoprotein chain and a copper cluster in the active site, which acts as a cofactor on the biocatalysis of the oxidation process of (*ortho*- and *para*-) (poly)-phenols in the reduced form to the corresponding quinones, simultaneously reducing molecular oxygen (final electron acceptor) to water, as exemplified in the scheme below (Fig. 14.2) (Mogharabi and Faramarzi 2014). These enzymes are obtained from several biological sources, including bacteria, plants, insects, and fungi. Fungal laccases have gained much importance in the design of electrochemical biosensors, especially due to their higher oxidation potential compared to laccases from other sources, and their ability to catalyze electron-transfer processes without additional cofactors or chemical mediators (Moraes et al. 2019).

In view of the application of laccase-based biosensors in the food industry, some electrochemical biosensors have been developed using laccase from different sources, including the basidiomyceteous fungi, *Coriolus versicolor,* and *Ganoderma* spp., that biodegrade lignin a (poly)phenolic/aromatic plant cell wall biopolymer (Rodríguez-Delgado et al. 2015). These devices are employed to specifically determine phenolic compounds and ensure the quality of the final products, as discussed above. For this purpose, laccase has been immobilized on the surface of carbon-

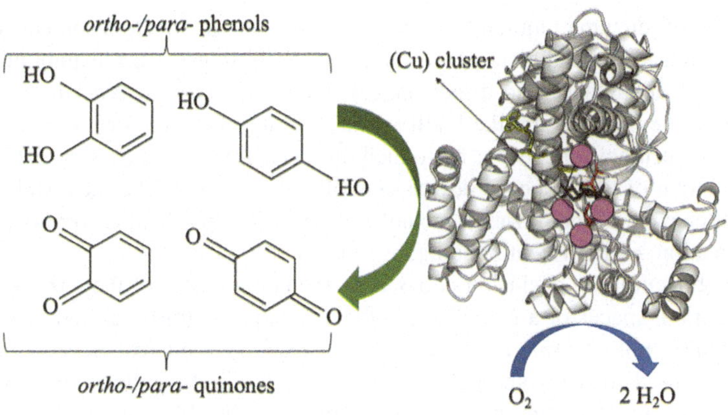

Fig. 14.2 The laccase-catalyzed oxidation reaction of *ortho-/para-* phenols to *ortho-/para-* quinones and electrochemical reduction of molecular oxygen to water

based electrodes by usual enzyme immobilization procedures, and the devices were applied to determine phenolic compounds in tea leaves at different stages of tea production (Ghindilis et al. 1992) and commercial fruit juices (Chawla et al. 2011).

The ascomyceteous ligninolytic fungus *Botryosphaeria rhodina* MAMB-05 produces an extracellular laccase constitutively, which can be induced to higher enzyme titer levels by veratryl alcohol, a laccase inducer (Vasconcelos et al. 2000). Recent studies reported in the literature describe the use of crude enzyme extracts obtained by submerged fermentation of fungi (Coelho et al. 2019; Mattos et al. 2019; Gomes et al. 2020; Moraes et al. 2019). The application of crude laccase extracts in the architecture of bioelectronic devices may result in selectivity problems. However, this procedure reduces the cost of preparing the biosensor on a large scale, in addition to being simpler and ensuring a longer lifetime of the biocatalytic activity when compared to the purified enzyme.

Laccase from *B. rhodina* MAMB-05 has been applied in electrochemical biosensing of several (poly)phenolic compounds, with simple and complex organic structures, including hydroquinone (Mattos et al. 2019), dopamine (Coelho et al. 2019), chlorogenic acid (Salamanca-Neto et al. 2020a), flavonoids (Gomes et al. 2020; Mattos et al. 2021), and also for indirect determination of spironolactone (Coelho et al. 2019) based on a mechanism of enzyme inhibition. Although these studies have used different electrochemical platforms in the transduction system, the mechanism of monitoring is derived from the same laccase-based reaction. The laccase activity based on a stereospecific and selective reaction promotes the formation of large amounts of the oxidized species (*ortho-/para*-quinones) at the interface electrode/solution, which results in a greater signal of current arising from the electrochemical reduction of those species.

14.5 Enzyme Immobilization Methods for Electrochemical Biosensing in Food Analysis

The type of enzyme immobilization procedure on an electrochemical support directly affects the analytical performance of the biosensor. Each immobilization method has advantages and disadvantages; thus, each type of procedure must be studied taking into account the lifetime and stability of the biorecognition element (enzyme), accuracy, precision, reproducibility, sensitivity, limits of detection, and quantification of the proposed device (Sassolas et al. 2012). The most well-known types of enzyme immobilization include gel occlusion, physical adsorption, microencapsulation, and covalent bonding, as illustrated in Fig. 14.3.

Immobilization by occlusion is based on the confinement of the enzyme within the interstitial spaces of a gel, allowing the diffusion of substrates and products in and out of the polymer matrix and retaining the enzyme of interest. Several materials have been used for enzyme occlusion, such as polyvinyl alcohol, polyacrylamide gels, anionic and cationic groups, among others. A disadvantage of the occlusion method is related to the possible loss of enzyme activity due to leaching, which is governed by the pore sizes of the gel (Fernández-Fernández et al. 2013).

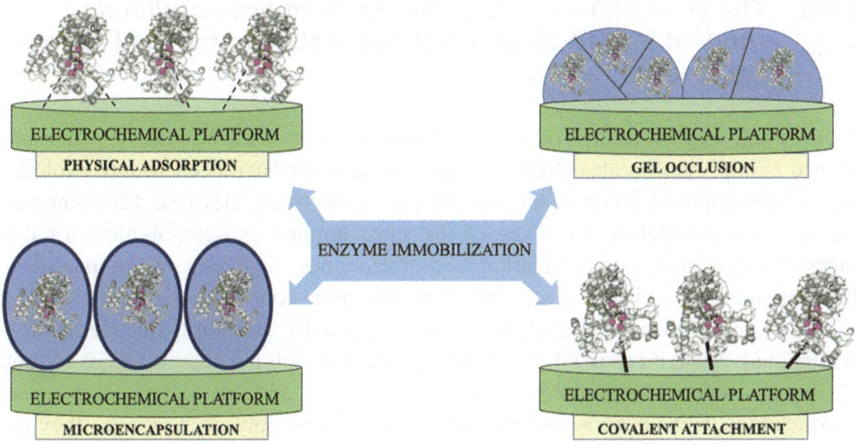

Fig. 14.3 Pictorial representation of different enzyme immobilization methods

Physical adsorption is the simplest method of enzyme immobilization which has the advantages of low cost and ease of immobilization based on weak physical forces/interactions between the enzyme and the immobilized matrix; they include van der Waals forces and hydrogen bonding. Compared to other methods of enzyme immobilization, the disadvantages of physical adsorption include low stability of the immobilized enzyme under severe conditions of temperature, pH, and ionic strength, which may lead to a fast washing-out of the biomaterial from the electrode surface. However, this method has been widely applied in biosensor projects since it remains the fastest and most applicable procedure for the immobilization of biocatalysts (Brady and Jordaan 2009).

Microencapsulation is based upon trapping the enzyme within a membrane deposited on the electrode surface. This material retains the enzyme, and its porosity allows the diffusion of the substrate and the products formed in the reaction, which are monitored on the interface electrode/solution. Examples of membranes include nylon, chitosan, cellulose acetate, and polycarbonates. A remarkable disadvantage of this method is related to the impediment of the mass transfer of the substrate and the reaction products caused by the membrane (Rochefort et al. 2008).

The covalent attachment of the enzyme is the most stable and efficient method in the immobilization of enzyme onto a support material. This procedure occurs through chemical bonds between functional groups on the enzyme (the part not essential for catalytic activity) and reactive groups on the support material that may include hydroxyls, carbonyls, amines, phenolic groups, imidazole, and thiols. The most widely applied support materials include insoluble polymers such as chitosan, cellulose, dextran, and sepharose. Examples of covalent binding agents used in the construction of electrochemical enzyme biosensors have included: 1-ethyl-3-[3-dimethylaminopropyl] carbodiimide (EDC), N-hydroxysuccinimide (NHS), epichlorohydrin, glyoxal, aminothiols, and glutaraldehyde (Casero et al. 2013).

14.6 The (1 → 3)(1 → 6)-β-D-Glucan from *Botryosphaeria rhodina* MAMB-05 as a Natural Stabilizer for the Enzyme Laccase

Some fungi and bacteria produce exocellular carbohydrate biopolymers that are named exopolysaccharides (EPSs). These biomaterials have been widely explored in the development of new enzyme-based bioelectronic devices. Chitosan and carboxymethyl-cellulose are some of the most applied exo-biopolymers for this purpose considering some remarkable characteristics, including biocompatibility, biodegradability, ability to form adherent thin films on support electrochemical platforms, renewable, abundant in nature, and usually are non-toxic (Hernández-Ibáñez et al. 2016; Fu et al. 2015). These organic materials are obtained by extraction of fungal mycelia, microalgae, plant cell walls, and crustacean and insect exoskeletons. EPSs have advantages over other natural polysaccharides as they are produced by fermentation by microbial species, and can be obtained from the fermentation broths by precipitation using alcohol with consequent high yields, and resulting in lower production costs.

Apart from laccase secreted extracellularly, *B. rhodina* MAMB-05 also produces an exopolysaccharide of the (1 → 3)(1 → 6)-β-D-glucan type (named botryosphaeran, BOT) when cultivated by submerged fermentation on glucose (or sucrose) as sole carbon source. This exocellular biopolymer consists of a backbone chain bound by β-(1 → 3)-linked D-glucose residues with approximately 22% side-branching through C-6 of glucose and gentiobiose residues via β-(1 → 6) bonds along the backbone chain (Crognale et al. 2007; Barbosa et al. 2003; Dekker et al. 2019) as shown in Fig. 14.4.

As aforementioned, the fungus *B. rhodina* MAMB-05 produces both botryosphaeran and laccase under conditions of fermentation optimized for the production of either laccase or botryosphaeran. An attractive characteristic of this system is that botryosphaeran constitutes a biofilm upon which laccase can assemble providing a natural biochemical environment for enzyme immobilization. This feature was of interest to develop a new method for laccase immobilization on bioelectrochemical devices (Coelho et al. 2019; Mattos et al. 2019; Gomes et al. 2020).

Botryosphaeran was employed in the construction of a laccase-based biosensor for monitoring chlorogenic acid content to discriminate different types of brewed

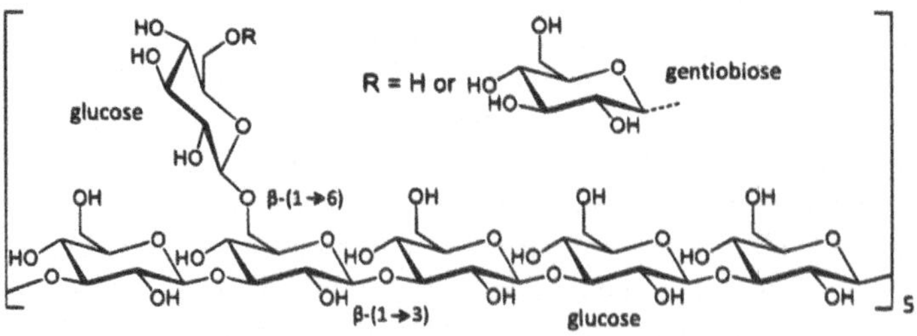

Fig. 14.4 Chemical structural representation of botryosphaeran

coffee (Salamanca-Neto et al. 2020a). A chemometric study based on statistical mixture design revealed that the presence of botryosphaeran enhances laccase activity. The absence of the biomolecule was expected to increase the analytical response for the target analyte due to lower resistance of charge transfer, but a decrease in the response was observed. On the other hand, the addition of the exopolysaccharide botryosphaeran to the composition of the biosensor promoted an increase in the current for chlorogenic acid even with an increase in the resistance to charge transfer. The biosensor response was due to the catalytic effect of the laccase in the oxidation of the polyphenolic compound to the *ortho*-quinone form, and this was followed by reduction by the transducer. The results indicated that the natural biochemical environment for laccase immobilization increased the catalytic effect of the enzyme with the need for small amounts of the biocatalyst on the chemical makeup of the biosensor. The biosensor based on laccase immobilized on botryosphaeran was successfully applied in the discrimination of specialty and traditional coffee beverages by using principal component analysis where the laccase-catalyzed reduction peak of chlorogenic acid was an important factor for the discrimination (Salamanca-Neto et al. 2020a).

14.7 Carboxymethylated-Derivative of Botryosphaeran and Its Application in Biosensing.

The derivatization of polysaccharides is performed by chemical modification methods where active functional groups are added to the polysaccharide chain. The different types of derivatization include ether and ester functional groups on the polysaccharides and are performed in cases where expected specific characteristics of the product are desired. For example, the addition of hydroxypropyl and methyl groups to cellulose provides solubility, increases viscosity in solution, and promotes stability against biodegradation; features that are very useful in the preparation of tablets in the pharmaceutical industry (Shokri and Adibki 2013). Carboxymethylation is the most commonly used chemical modification method of polysaccharides to increase water solubility and their biological activities. This reaction is based on the Williamson ether synthesis with several adaptations implied in low-cost procedures (Kagimura et al. 2015; Theis et al. 2019; Chakka and Zhou 2020). The modification of botryosphaeran by carboxymethylation resulted in increased water solubility and produced film characteristics of the natural exopolysaccharide, which have been used for the purpose of developing electrochemical sensor devices (Eisele et al. 2019; Salamanca-Neto et al. 2020b). The application of biosensing fabrication relies upon a covalent linkage of the enzyme to the carboxyl-terminal groups of the modified biopolymer using the cross-linking agents EDC and NHS. The covalent attachment of laccase to carboxymethyl-botryosphaeran (CMB) was carried out in a one pot-sequential reaction in phosphate buffer solution (pH 6.0), in which EDC was allowed to react with CMB, followed by the addition of NHS, and finally the laccase. An aliquot of the solution of laccase covalently bound to CMB was layered on the top surface of a carbon black paste electrode and allowed to cure. The reaction was accompanied by electrochemical impedance spectroscopy and revealed that this type of immobilization led to a very stable, reproducible, reusable, and sensitive biosensing platform. The biosensor device fabricated responded favorably to quercetin, and it was possible to selectively quantify this phenolic compound in complex matrices as red wine, green tea, fruit juices, pharmaceuticals, and human urine using square-wave voltammetry (Gomes et al. 2020).

14.8 Other Applications of Botryosphaeran and Its Carboxymethylated-Derivative in (Bio)Sensing

As stated above, both botryosphaeran and CMB were used in the construction of electrochemical biosensing devices for monitoring of chemical components in food products. On the other hand, botryosphaeran was also employed in the construction of laccase-based biosensors for monitoring hydroquinone (Mattos et al. 2019) and dopamine and spironolactone (Coelho et al. 2019), while CMB was explored in the construction of sensors for the phenolic compounds, paracetamol and dopamine (Eisele et al. 2019), and the antihistamine drug desloratadine (Salamanca-Neto et al. 2020b).

A carbon black paste electrode modified with gold nanoparticles was simultaneously covered by an aliquot of botryosphaeran and laccase in which the enzyme was immobilized by physical adsorption, eliminating the need for reagents acting as cross-linking agents. The fabricated biosensor provided a very sensitive response for hydroquinone over uric acid and other phenolic compounds. The biosensing device was applied to the determination of quercetin in dermatological cream, urine, and river water without any prior treatment being required (Mattos et al. 2019). The same procedure of layering botryosphaeran and laccase onto a glassy carbon electrode modified with MWCNTs was used to fabricate another biosensor device. This biosensor was employed in the determination of dopamine with satisfactory selectivity over other phenolic compounds. Additionally, using the same biosensor architecture, we found that the pharmaceutical spironolactone interacted with the dopamine-quinone, the oxidized form of dopamine catalyzed by laccase, forming an imine, which consequently reduced the reduction signal of dopamine. This behavior led to a novel bioanalytical method for the indirect determination of spironolactone by reducing the signal of the dopamine-quinone formed from the catalytic action of laccase (Coelho et al. 2019).

The first electrochemical use of CMB was published by Eisele et al. (Eisele et al. 2019). They employed CMB as an adherent layer of carbon black placed on the surface of a GCE. The properties of CMB promoted ionic interactions and hydrogen bonding between the electrode surface and the phenolic molecules, and such interactions resulted in an increase of the surface concentration of these species on the modified electrode surface assembly thereby increasing the voltammetric response. This kind of sensor was employed in the determination of paracetamol and dopamine in pharmaceutical formulations and synthetic cerebrospinal fluid.

While the botryosphaeran appears to act as a stabilizer of laccase, as it provides a natural environment for the enzyme favoring its catalytic effect, the derivatized form of the exopolysaccharide was successfully used to stabilize pristine multiwalled carbon nanotubes in an aqueous dispersion. In constructing MWCNTs modified electrode, it is first necessary to disperse the MWCNTs in solution. CMB could disperse the carbon nanostructured material in water without any previous functionalization. The dispersion resulted in a homogeneous distribution of the nanomaterial when layered onto the surface of a glassy carbon electrode, which was employed to determine the antihistaminic drug desloratadine in samples of pharmaceutical products and in rat serum (Salamanca-Neto et al. 2020b).

14.9 Conclusion and Future Perspectives

It is important to present novel biomaterials (e.g., botryosphaeran and its derivative forms) to design biosensors that have the potential for diverse applications. Both biomaterials are still little explored in the development of sensors and biosensors. In view of this, we hope that this brief overview will pique the interest of researchers to focus on these biotechnologically derived bio/sensors, which can contribute to reducing the expenditure of potentially toxic reagents used in alternative analytical procedures. The natural environment for laccase immobilization was responsible for its high catalytic activity, and higher stability of enzyme biosensors can be attained through different types of immobilization procedures (physical adsorption and covalent attachment). The carboxyl groups of CMB have strategically been used for the covalent attachment of laccase using cross-linking agents in an aqueous solution, allowing these devices capable of being produced in bulk for multiple bio/sensing applications.

Future work should focus on the application of botryosphaeran and its derivative forms (carboxymethyl-, sulfonated-, acetylated- and phosphorylated-botryosphaerans) to immobilize enzymes (laccases and other oxidoreductases) from other sources in developing novel carbon-based electrodes for determining drugs and other analytes.

References

Adams RN (1963) Carbon paste electrodes. Rev Polarogr 11:71–78. https://doi.org/10.5189/revpolarography.11.71

Arduini F, Cinti S, Mazzaracchio V et al (2020) Carbon black as an outstanding and affordable nanomaterial for electrochemical (bio)sensor design. Biosens Bioelectron 156:112033. https://doi.org/10.1016/j.bios.2020.112033

Balasubramanian K, Burghard M (2006) Biosensors based on carbon nanotubes. Anal Bioanal Chem 385:452–468. https://doi.org/10.1007/s00216-006-0314-8

Barbosa AM, Steluti RM, Dekker RF et al (2003) Structural characterization of Botryosphaeran: a (1→3;1→6)-β-D-glucan produced by the ascomyceteous fungus, *Botryosphaeria* sp. Carbohydr Res 338:1691–1698. https://doi.org/10.1016/S0008-6215(03)00240-4

Bollella P, Katz E (2020) Enzyme-based biosensors: tackling electron transfer issues. Sensors (Switzerland) 20:3517. https://doi.org/10.3390/s20123517
Brady D, Jordaan J (2009) Advances in enzyme immobilisation. Biotechnol Lett 31:1639–1650. https://doi.org/10.1007/s10529-009-0076-4
Cabot GL (1960) Carbon black. Ind Eng Chem 52:25A–26A. https://doi.org/10.1021/i650611a716
Cannatelli MD, Ragauskas AJ (2017) Two decades of laccases: advancing sustainability in the chemical industry. Chem Rec 17:122–140. https://doi.org/10.1002/tcr.201600033
Casero E, Petit-Domínguez MD, Vázquez L et al (2013) Laccase biosensors based on different enzyme immobilization strategies for phenolic compounds determination. Talanta 115:401–408. https://doi.org/10.1016/J.TALANTA.2013.05.045
Chakka VP, Zhou T (2020) Carboxymethylation of polysaccharides: synthesis and bioactivities. Int J Biol Macromol 165:2425–2431. https://doi.org/10.1016/j.ijbiomac.2020.10.178
Chawla S, Rawal R, Pundir CS (2011) Fabrication of polyphenol biosensor based on laccase immobilized on copper nanoparticles/chitosan/multiwalled carbon nanotubes/polyaniline-modified gold electrode. J Biotechnol 156:39–45. https://doi.org/10.1016/j.jbiotec.2011.08.008
Clark LC, Lyons C (2006) Electrode systems for continuous monitoring in cardiovascular surgery. Ann N Y Acad Sci 102:29–45. https://doi.org/10.1111/j.1749-6632.1962.tb13623.x
Coelho JH, Eisele APP, Valezi CF et al (2019) Exploring the exocellular fungal biopolymer botryosphaeran for laccase-biosensor architecture and application to determine dopamine and spironolactone. Talanta 204:475–483
Crognale S, Bruno M, Fidaleo M et al (2007) Production of β-glucan and related glucan-hydrolases by Botryosphaeria rhodina. J Appl Microbiol 102:860–871
De Beer D, Joubert E, Gelderblom WCA, Manley M (2002) Phenolic compounds: a review of their possible role as in vivo antioxidants of wine. S Afr J Enol Vitic 23:48–61. https://doi.org/10.21548/23-2-2155
Dekker RFH, Queiroz EAIF, Cunha MAA, Barbosa-Dekker AM (2019) Botryosphaeran—A fungal exopolysaccharide of the (1{\textrightarrow}3)(1{\textrightarrow}6)-$β$-D-glucan kind: structure and biological functions. In: Cohen E, Merzendorfer H (eds) Extracellular sugar-based biopolymers matrices. Chapter 11. Springer, Cham, pp 433–484
Della PF, Compagnone D (2018) Nanomaterial-based sensing and biosensing of phenolic compounds and related antioxidant capacity in food. Sensors (Switzerland) 18:462. https://doi.org/10.3390/s18020462
Eisele APP, Valezi CF, Mazziero T et al (2019) Layering of a film of carboxymethyl-botryosphaeran onto carbon black as a novel sensitive electrochemical platform on glassy carbon electrodes for the improvement in the simultaneous determination of phenolic compounds. Sensors Actuators B Chem 287:18–26
Fernández-Fernández M, Sanromán MÁ, Moldes D (2013) Recent developments and applications of immobilized laccase. Biotechnol Adv 31:1808–1825. https://doi.org/10.1016/J.BIOTECHADV.2012.02.013
Freire RS, Pessoa CA, Mello LD, Kubota LT (2003) Direct electron transfer: an approach for electrochemical biosensors with higher selectivity and sensitivity. J Braz Chem Soc 14:230–243. https://doi.org/10.1590/S0103-50532003000200008
Fu J, Li D, Li G et al (2015) Carboxymethyl cellulose assisted immobilization of silver nanoparticles onto cellulose nanofibers for the detection of catechol. J Electroanal Chem 738:92–99. https://doi.org/10.1016/J.JELECHEM.2014.11.025
Ghindilis AL, Gavrilova VP, Yaropolov AI (1992) Laccase-based biosensor for determination of polyphenols: determination of catechols in tea. Biosens Bioelectron 7:127–131. https://doi.org/10.1016/0956-5663(92)90017-H
Gomes A, Mattos GJ, Coldibeli B et al (2020) Covalent attachment of laccase to carboxymethyl-botryosphaeran in aqueous solution for the construction of a voltammetric biosensor to quantify quercetin. Bioelectrochemistry 135:107543. https://doi.org/10.1016/j.bioelechem.2020.107543
Hernández-Ibáñez N, García-Cruz L, Montiel V et al (2016) Electrochemical lactate biosensor based upon chitosan/carbon nanotubes modified screen-printed graphite electrodes for the

determination of lactate in embryonic cell cultures. Biosens Bioelectron 77:1168–1174. https://doi.org/10.1016/J.BIOS.2015.11.005

Herrmann K (1989) Occurrence and content of hydroxycinnamic and hydroxybenzoic acid compounds in foods. Crit Rev Food Sci Nutr 28:315–347. https://doi.org/10.1080/10408398909527504

Jacobs CB, Peairs MJ, Venton BJ (2010) Review: carbon nanotube based electrochemical sensors for biomolecules. Anal Chim Acta 662:105–127. https://doi.org/10.1016/j.aca.2010.01.009

Angelo PM, Jorge N (2006) Phenolic compounds in foods – a brief review. Rev Inst Adolfo Lutz 3221–2200

Kagimura FY, da Cunha MAA, Theis TV et al (2015) Carboxymethylation of (1→6)-β-glucan (lasiodiplodan): Preparation, characterization and antioxidant evaluation. Carbohydr Polym 127:390–399. https://doi.org/10.1016/j.carbpol.2015.03.045

Lobo V, Patil A, Phatak A, Chandra N (2010) Free radicals, antioxidants and functional foods: impact on human health. Pharmacogn Rev 4:118–126. https://doi.org/10.4103/0973-7847.70902

Long CM, Nascarella MA, Valberg PA (2013) Carbon black vs. black carbon and other airborne materials containing elemental carbon: physical and chemical distinctions. Environ Pollut 181: 271–286. https://doi.org/10.1016/j.envpol.2013.06.009

Luong JT, Male K, Hrapovic S (2008) Carbon nanotube-based electrochemical biosensing platforms: fundamentals, applications, and future possibilities. Recent Pat Biotechnol 1:181–191. https://doi.org/10.2174/187220807780809427

Mattos GJ, Moraes JT, Barbosa ECM et al (2019) Laccase stabilized on β-D-glucan films on the surface of carbon black/gold nanoparticles: a new platform for electrochemical biosensing. Bioelectrochemistry 129:116–123. https://doi.org/10.1016/J.BIOELECHEM.2019.05.002

Mattos GJ, Salamanca-Neto CAR, Barbosa ECM et al (2021) A photoelectrochemical enzyme biosensor based on functionalized hematite microcubes for rutin determination by square-wave voltammetry. Microchim Acta 188:28. https://doi.org/10.1007/s00604-020-04659-z

Merkoçi A (2006) Carbon nanotubes in analytical sciences. Microchim Acta 152:157–174. https://doi.org/10.1007/s00604-005-0439-z

Mogharabi M, Faramarzi MA (2014) Laccase and Laccase-mediated systems in the synthesis of organic compounds. Adv Synth Catal 356:897–927. https://doi.org/10.1002/adsc.201300960

Mohammad R, Ahmad M, Heng L (2013) An amperometric biosensor utilizing a ferrocene-mediated horseradish peroxidase reaction for the determination of Capsaicin (Chili Hotness). Sensors 13:10014–10026. https://doi.org/10.3390/s130810014

Moraes JT, Salamanca-Neto CAR, Švorc Ľ et al (2019) Laccase from *Botryosphaeria rhodina* MAMB-05 as a biological component in electrochemical biosensing devices. Anal Methods 11: 717–720. https://doi.org/10.1039/C8AY02805B

Musci M, Yao S (2017) Optimization and validation of Folin–Ciocalteu method for the determination of total polyphenol content of Pu-erh tea. Int J Food Sci Nutr 68:913–918. https://doi.org/10.1080/09637486.2017.1311844

Nguyen HH, Lee SH, Lee UJ et al (2019) Immobilized enzymes in biosensor applications. Materials (Basel) 12:1–34. https://doi.org/10.3390/ma12010121

Rochefort D, Kouisni L, Gendron K (2008) Physical immobilization of laccase on an electrode by means of poly(ethyleneimine) microcapsules. J Electroanal Chem 617:53–63. https://doi.org/10.1016/j.jelechem.2008.01.027

Rodríguez-Delgado MM, Alemán-Nava GS, Rodríguez-Delgado JM et al (2015) Laccase-based biosensors for detection of phenolic compounds. TrAC Trends Anal Chem 74:21–45. https://doi.org/10.1016/J.TRAC.2015.05.008

Salamanca-Neto CAR, Marcheafave GG, Scremin J et al (2020a) Chemometric-assisted construction of a biosensing device to measure chlorogenic acid content in brewed coffee beverages to discriminate quality. Food Chem 315:126306. https://doi.org/10.1016/j.foodchem.2020.126306

Salamanca-Neto CAR, Olean-Oliveira A, Scremin J et al (2020b) Carboxymethyl-botryosphaeran stabilized carbon nanotubes aqueous dispersion: a new platform design for electrochemical sensing of desloratadine. Talanta 210:120642. https://doi.org/10.1016/j.talanta.2019.120642

Sassolas A, Blum LJ, Leca-Bouvier BD (2012) Immobilization strategies to develop enzymatic biosensors. Biotechnol Adv 30:489–511. https://doi.org/10.1016/j.biotechadv.2011.09.003

Shokri J, Adibki K (2013) Application of cellulose and cellulose derivatives in pharmaceutical industries. In: van de Ven T, Godbout L (eds) Cellulose - medical, pharmaceutical and electronic applications. InTech Open. https://doi.org/10.5772/55178

Spanos GA, Wrolstad RE (1990) Influence of processing and storage on the phenolic composition of Thompson seedless grape juice. J Agric Food Chem 38:1565–1571. https://doi.org/10.1021/jf00097a030

Švancara I, Vytřas K, Kalcher K et al (2009) Carbon paste electrodes in facts, numbers, and notes: a review on the occasion of the 50-years jubilee of carbon paste in electrochemistry and electroanalysis. Electroanalysis 21:7–28. https://doi.org/10.1002/elan.200804340

Swallah MS, Sun H, Affoh R et al (2020) Antioxidant potential overviews of secondary metabolites (polyphenols) in fruits. Int J Food Sci 2020. https://doi.org/10.1155/2020/9081686

Theis TV, Queiroz Santos VA, Appelt P et al (2019) Fungal exocellular (1-6)-β-d-glucan: carboxymethylation, characterization, and antioxidant activity. Int J Mol Sci 20:2337. https://doi.org/10.3390/ijms20092337

Tohma H, Gülçin İ, Bursal E et al (2017) Antioxidant activity and phenolic compounds of ginger (Zingiber officinale Rosc.) determined by HPLC-MS/MS. J Food Meas Charact 11:556–566. https://doi.org/10.1007/s11694-016-9423-z

Vasconcelos AFD, Barbosa AM, Dekker RFH et al (2000) Optimization of laccase production by Botryosphaeria sp. in the presence of veratryl alcohol by the response-surface method. Process Biochem 35:1131–1138. https://doi.org/10.1016/S0032-9592(00)00149-7

Application of Nano-ELISA in Food Analysis 15

Long Wu

Abstract

ELISA is a widely applied technique with good reliability, sensitivity, and specificity. Compared to other immunoassays, ELISA has been intensively used in many fields like biology, toxicology, immunology, and medical diagnosis due to its simple operations and high reliability. Recently, ELISA has been widely used in food safety and control. Though ELISA has so many applications and superior advantages, it encounters a lot of restrictions, especially the relatively low sensitivity and stability. Based on this, abundant work has been done to improve the detection performances of conventional ELISA (c-ELISA), including the limit of detection (LOD), accuracy, and stability. Fortunately, combined with nanomaterials, various ELISA-based methods have been developed to address the limitations of c-ELISA. The nanomaterials-based ELISA (nano-ELISA) behaves additionally mechanical, electrical, magnetic, optical, and catalytic properties. Based on this, in this chapter, we summarize ELISA methods and provide an overall description of the history, principles, designs, and applications in analysis of food contaminants, which is expected to help facilitate the food safety and control in compliance with legislation and consumers' demands.

Keywords

ELISA · Nanomaterials · Detection · Food analysis · Biomarker

L. Wu (✉)
College of Food Science and Engineering, Hainan University, Haikou, Hainan, China

Key Laboratory of Tropical Fruits and Vegetables Quality and Safety for State Market Regulation, Hainan Institute for Food Control, Haikou, China

P. Chandra, P. S. Panesar (eds.), *Nanosensing and Bioanalytical Technologies in Food Quality Control*, https://doi.org/10.1007/978-981-16-7029-9_15

15.1 Introduction

It is known that ELISA is a popular biochemistry assay using a solid-phase enzyme immunoassay to analyze a target with a form of antibody-antigen recognition model, in a plate well or solutions (Voller et al. 1978; Butler 2000). As a golden standard in immunoassay, ELISA has been widely applied in laboratories and industries, which acts as a verification method, or a detection means in sample quality tests (Buss et al. 1997; Salomone et al. 2004; Török et al. 2015). Owing to its advantages such as good convenience, high specificity and feasibility, ELISA or ELISA-based method has been regarded as a powerful tool in analytical science (Gao et al. 2019a, b). Yet, c-ELISA suffers from inherent shortcomings like low efficiency, complicated operations, and single detection mode (usually antigen-antibody detection). Thus, it is vital to develop an effective method to solve the problem.

Since the extensive applications of ELISA, different ways have been tried to solve the existing problems. For instance, many studies have been carried out to enhance the LOD and detection accuracy of c-ELISA (He et al. 2016; Fadlalla et al. 2020). So far, not only the range of application of ELISA but also its detection performance has been developed a lot (Byer et al. 2008; Jaria et al. 2020). To be specific, from c-ELISA to avidin-biotin ELISA (ABS-ELISA), the immunoassay has been extended to in vivo detection with higher sensitivity (Peng et al. 2014). Besides that, it is an ultimate goal to achieve higher stability, better accuracy, simpler operations, and lower cost of the ELISA method.

Recently, a lot of work related to ELISA have been reported in the applications of food analysis. In particular, Zhang et al. summarized the application of ELISA in pesticide residues detection in food products, which emphasized the accuracy and universality of ELISA in pesticide residues analysis (Zhang et al. 2008). Besides, Aydin et al. gave a detailed description on the history, working principles, and different classifications of ELISA, as well as how to analyze peptide or protein using an ELISA method, including discussing what we can do with ELISA analytical errors (Aydin 2015). To be specific, Toh et al. introduced the application of aptamers in ELISA, that is aptamer-based ELISA, which uses aptamers to recognize the analytes or give signal outputs (Toh et al. 2015). After that, based on the localized surface plasmon resonance of nanomaterials such as nanogold or nanosilver particles, Satija et al. discussed the plasmonic-ELISA that used in visual detection applications (Satija et al. 2016). Later, Wang et al. reported the advances in ELISA for antibiotics detection in food matrices involving different immunosensors from electrochemical ELISA to fluorescence-ELISA (Wang et al. 2017). At present, nano-ELISA has been developed by integrating nanotechnology with ELISA, which brings much more convenience and superiority in food analysis. For example, Wu et al. introduced the development of c-ELISA in combination of nanomaterials (Wu et al. 2019a, b), especially the latest development of nano-ELISA and their applications in food safety. By modifying c-ELISA with nanomaterials, it shows superior performance with lower LOD and cost, higher stability, and accuracy. Thus, it is vital to develop nanomaterials (nanopolymer, nanoantibody, nanoprobes)-based ELISA and applied them in food analysis.

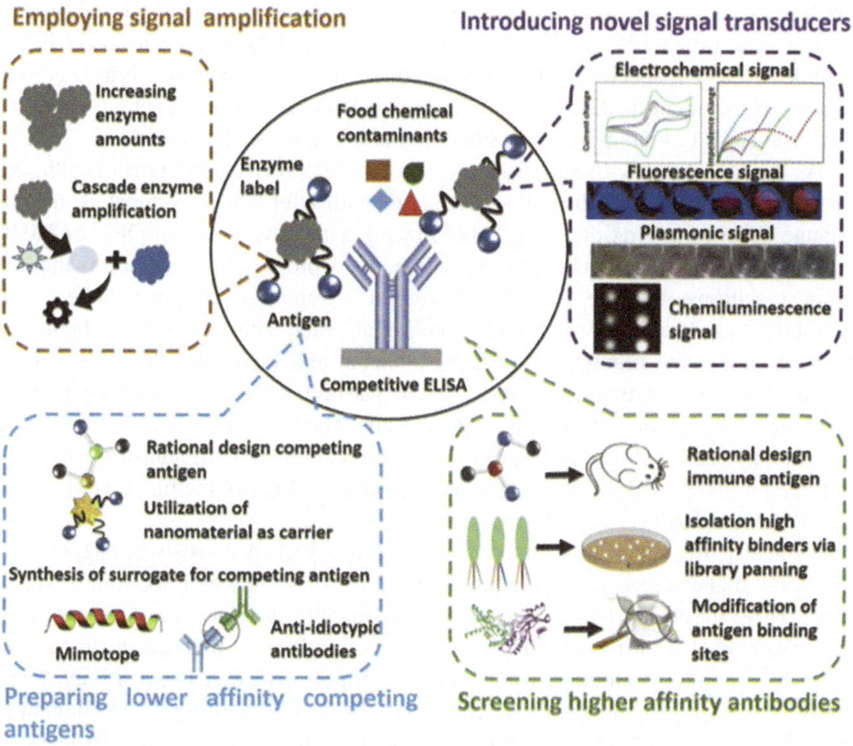

Fig. 15.1 Schematic presentation of emerging strategies for enhancing the sensitivity of c-ELISA. Reproduced from Ref. (Xiong et al. 2020) with permission from Elsevier

Based on the above background, the development of c-ELISA via nanomaterials (nano-ELISA) is given in this chapter, mainly referring to the four construction units of ELISA: substrate of sorbents, recognition models, enzyme labels, and chromogenic agents (Fig. 15.1). Also, advantages and disadvantages of c-ELISA and nano-ELISA are summarized and discussed. Different nanomaterials are described to improve or rebuild c-ELISA, which offer valuable guidance and strategies to design and construct nano-ELISA. In addition, different applications of newly developed nano-ELISA are described in food samples. Finally, challenges and perspectives on nano-ELISA are discussed, as well as their applications in food analysis and safety verification.

15.2 History and Development of ELISA

Before the development of the ELISA, radioimmunoassay was the only choice to carry out an immunoassay, which adopted radioactively labeled antigens or antibodies. At first, radioimmunoassay was proposed to achieve detection of endogenous plasma insulin (Yalow and Berson 1960). In the method, radioactive isotope like iodine-125 as labeled signal can indicate the existence of target in the sample, usually a specific antigen or antibody. The innovative idea brings convenience to the identification of certain targets. However, as radioactivity may do potential harm to human health, a safer alternative must be developed.

As the radioactivity of such label can pose potential risks on human health, the urgent affair is to replace the radioactive signal with safer labels. Fortunately, a color change occurs as peroxidase (e.g., HRP) reacts with substrates like OPD or TMB, which can be used as signal equals to radioactive isotope. The limitation is that the color variations must be initiated by certain enzyme, so enzymes linked with antibody was developed to provide labels that react with substrates, which was defined as enzyme immunoassay (EIA) (Nakane and Pierce Jr 1967). To conveniently remove the references through simple washing procedures, antibody/antigen must be anchored on the bottom of plate wells. Based on a sorbent substrate and enzyme-labeled antibody, the prototype of enzyme-linked immunosorbent assay (ELISA) was reported for the quantitative detection of IgG for the first time (Engvall and Perlmann 1971).

Typically, the signal transducer of conventional ELISA consists of enzyme and substrates that can react to produce color changes when analyte exists (Johnson et al. 1992; Hosseini et al. 2018). Based on the fundamental structures of ELISA, optical, electrochemical, and magnetic reporters are frequently used to generate signals for other ELISA-like techniques, which can outcompete c-ELISA in sensitivity, flexibility, and stability (Bouças et al. 2008; Phillips and Abbott 2008; Al Ghounaim et al. 2016). In technical terms, some of the assays cannot be classified as ELISAs, as they are not linked with enzyme or absorbed on a solid surface of well plates. However, they are instead linked to some nanozymes or anchored on other solid surfaces. Generally, their working principles are the same, so we accepted them as ELISAs. Till the year of 2012, De La Rica reported an ELISA using AuNP as a reporter to achieve the colorimetric detection of prostate-specific antigen and HIV-1 capsid antigen, which is the very beginning of nano-ELISA (De La Rica and Stevens 2012).

15.3 The Working Ways of ELISA

Conventionally, the specificity of antigen-antibody type reaction is used because it is easy to raise an antibody specifically against an antigen in bulk as a reagent. (Gaastra 1984). Taking food sample detection as an example, the specific substance to be detected (an analyte) is anchored on a solid substrate with specific recognition captured by antibody (a "sandwich" mode) (Fig. 15.2). After the analyte is anchored on the solid plate, a liquid sample is added onto a stationary solid phase with special binding properties, followed by multiple liquid reagents that are sequentially added, incubated, and washed. Finally, color development with some optical change can be observed in the final liquid in the well, which can be used to analyze the amount of analyte qualitatively and quantitatively.

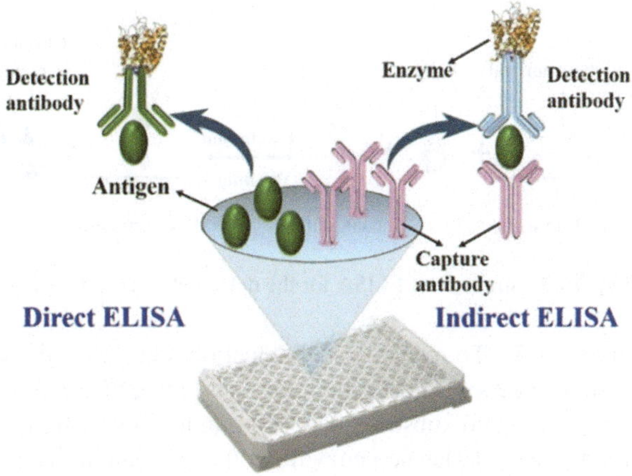

Fig. 15.2 Scheme illustration of the principle of direct ELISA and indirect ELISA

In addition, the analyte is also called the ligand because it will specifically bind or ligate to a detection reagent, thus ELISA falls under the bigger category of ligand binding assays (Ma and Shieh 2006). The ligand-specific binding reagent is usually coated and dried onto the transparent bottom and sometimes also an interface to generate a signal. In this regard, the recognition element-like aptamer can also be an alternative. On the other hand, for each washing step, the signal label and nonspecific or unbound components are washed away, but the reaction products immunosorbed on the solid phase (Heaney et al. 2020). That's to say, the ligand, immobilized antibody are parts of the plate, which is difficult to be developed into reusable ELISAs.

Generally, as a heterogenous assay, ELISA separates some components of the analytical reaction mixture by adsorbing certain components onto a solid phase which is physically immobilized (Kwong et al. 2002). In the most simple form of an ELISA, antigens from the sample to be tested are attached to a surface. Then, a matching antibody is applied over the surface so it can bind the antigen. This antibody is linked to an enzyme and then any unbound antibodies are removed. In the final step, a substance containing the enzyme's substrate is added. If there was binding, the subsequent reaction produces a detectable signal, most commonly a color change.

15.4 Structure of ELISA

As described above, from bottom to up, ELISA method consists of four main parts: solid substrates, sorbent antibody/antigen, enzyme labels, and chromogenic reagents (Fig. 15.3). To conduct the ELISA detection, the supporting substrate like microplate well provides a surface for antibody/antigen to bound to, then the enzyme-linked complementary biomolecules (a new specific antibody or a secondary antibody) will bind with the primary antibody/antigen and generate a bioconjugation (Waritani

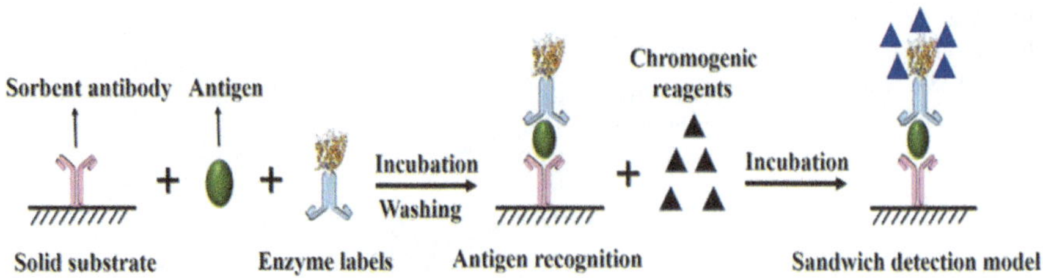

Fig. 15.3 Structure of ELISA for the detection of analytes (indirect ELISA method, **E**: enzyme)

et al. 2017). To realize visual detection, chromogenic reagents are used to generate visual detection signal (chemiluminescence, fluorescence color, etc.). In the ELISA structure, solid substrate acts as bounded foundation, biomolecules recognition as the framework, labeled enzyme as the reaction initiator, and reagent substrate as the signal indicator (Wu et al. 2019a, b).

In this part, to make it more understandable in designing the ELISA-based methods, the construction of c-ELISA will be introduced from the solid substrate to biomolecules recognition, then to the color development, as well as the washing and blocking steps. Based on the fundamental units, the improvement of c-ELISA with other new technology is introduced and discussed, especially the nanotechnology-directed ELISA method, also known as nano-ELISA.

15.4.1 Nanomaterials-Based Substrate

The most common solid substrate of ELISA is polystyrene used for most optical detection microplates (Hosseini et al. 2018). It can be colored blue by the addition of TMB/H_2O_2 for optical absorbance or luminol/H_2O_2 for luminescence detection or black by adding silver nanoparticles for biological assays (Wu et al. 2019a, b). Typically, a microplate has 6, 12, 24, 48, 96, 384, or 1536 sample wells, which is designed to allow low-volume and high-throughput assay for the samples. The plate well provides a solid surface for analytes, antibody, antigen, or other reagents to attach on, which are usually physically immobilized. The physical binding force between the bottom surface and the adsorbent provides a bridge for biomolecules conjugation, thus called "immunosorbent." In this regard, the solid substrate can be regarded as the foundation of signal recognition to be connected. So, an ideal substrate should meet the requirements of low cost, high light transmission, and weak-nonspecific adsorption.

As mentioned above, nonspecific adsorption is not welcome in the ELISA methods. However, most of the cases, the binding force of adsorbents comes from the physical adsorption of plate wells. So, the nonspecific adsorption is inevitable, which can pose great effects on the detection results, for example, false positive or negative, depending on the ELISA detection type. Thus, it is vital to develop appropriate substrate by using new materials to build the ELISA method. For instance, poly (dimethylsiloxane) (PDMS) membrane (Wang et al. 2013), nanofibers (Pan et al. 2015), molecularly imprinted polymers (MIPs) (Li et al. 2017) and magnetic nano-beads (Al Hamshary et al. 2020) are popular and reliable adsorbent substrate in ELISA.

15.4.2 Recognition Models

When adsorbents are bound to the surface of plate wells, the recognition between antibodies and antigens should be constructed for the detection of targets. To meet the requirements of practical assay, different models are designed with different detection intentions. For example, small molecules like pesticides are firstly conjugated with BSA and coated on the plate wells, then an antibody labeled with enzyme is applied to detect analyte, which is called direct ELISA. On the other hand, for the bigger molecules such as proteins, primary antibody firstly adsorbed on the plate wells, then the proteins are recognized and captured by the primary antibody, finally a second antibody labeled with enzyme is applied to give signals, which is called indirect ELISA. Thus, direct ELISA, indirect ELISA, sandwich ELISA, and competitive ELISA are the four main models used in immunoassay.

For the direct ELISA, several steps should be followed: (1) the antigen to be tested for is added to a microtiter plate and incubate for a certain time; (2) the block agent-like bovine serum albumin (BSA) or casein, is introduced to each well to cover any uncoated surface in the well; (3) the primary antibody labeled with enzyme is added to the well in order to specifically combine with the antigen; (4) a coloring substrate-like TMB/H_2O_2 is added to generate color variations. Obviously, the higher level of primary antibody present, more significantly the color changes. The major problem of the direct ELISA is that it lacks sensitivity using antigen immobilization. Taking serum detection as an example, when serum acts as the source of test antigen, all proteins in serum may adhere to the plate well, only a small amount of analyte in serum compete with other serum proteins. As a result, the color variations will not be that significant, which is the limitation for the detection sensitivity and accuracy.

Indirect ELISA can solve the above issue by using a labeled secondary antibody. Similar to the direct ELISA, antigen is bound by the primary antibody and then detected by a second antibody. A sandwich ELISA is distinct from an indirect ELISA by the recognition model of antibody-antigen-antibody. Despite from direct ELISA, the other ELISA methods are based on two antibodies. Nevertheless, antibody itself suffers from some disadvantages, such as high cost, hard to store long term, easy denaturation, and limited application conditions. Besides that, due to the steric effect of antibody, it cannot be easily applied to ELISA easily when come to the detection of small molecules. Thus, a special class of nucleic acid molecules that named aptamers has been developed to specifically recognize small molecules, which are poised to replace the monoclonal antibodies in therapeutics, diagnostics, and drug development (Toh et al. 2015; Lee and Zeng 2017; Wu et al. 2020; Her et al. 2017).

15.4.3 Nanozyme Labels

The enzyme acts as an amplifier to accelerate chemical reactions of substrate. Due to the high catalytic activity of enzyme, even though few enzyme-linked antibodies are bound, they will produce many signal molecules that can give special color. The more enzyme-labeled antibody is bound, the faster the color will develop. To a large extent, the quality of enzyme plays a crucial role in the construction of ELISA methods. Thus, the enzyme label should have the characteristics including high purity, good specificity, excellent stability, and long-term activity.

Among the natural enzymes, horseradish peroxidase and alkaline phosphatase, are the most commonly used antibody labels. Both can produce a colored, fluorimetric, or luminescent derivative when incubated with a proper substrate, allowing it to be detected and quantified. Compared to ALP, HRP is much better in ELISA applications as it is smaller, more stable, and less expensive. The enzyme label is directly related with the signal output, so it is the most concerned part in ELISA method. However, natural enzymes suffer from the limitations such as hard to be separated and purified, hard to store long term and mass produce, high cost, easy denaturation, and limited application conditions.

To solve the problem, nanozymes, a kind of nanomaterials with peroxide activity, have been introduced to replace natural enzymes. At the same time, due to the large specific surface area, nanozymes can also act as loading substrate to achieve signal amplification, thus enhancing the detection sensitivity of ELISAs. Compared with natural enzymes, nanozymes are easier to be modified and purified. Moreover, a variety of new nanomaterials have been found to mimic the activity of enzymes, such as metal–organic frameworks (MOFs), covalent organic frameworks COFs, and Prussian blue (PB). Given the advantages like low cost, recyclable utilization, high catalytic activity and stability, nanozymes are widely used in food safety. For instance, based on MOFs, an indirect competitive ELISA method was developed by replacing natural enzyme with MOFs nanozymes for the sensitive detection of AFB1 (Xu et al. 2021). To achieve high sensitivity, Tian et al. proposed a cascade reaction-based colorimetric aptasensor for the detection of OTA (Tian et al. 2019).

15.4.4 Enzymatic Markers

When the structure of ELISA is well constructed, enzymatic markers are needed to provide the readout signal for the targets. Usually, the markers are the catalytic substrate of the labeled enzyme, which can be oxidized by certain enzyme and produce the colored, fluorescent, or luminescent product. Most of the substrates (TMB, ABTS, OPD, etc.) are used in the presence of hydrogen peroxide (H_2O_2). For instance, the commonly used HRP enzyme could oxidize TMB when H_2O_2 exists, generating a blue color variation that is detectable. Based on the principle of electron donor and receptor, ABTS and OPD with H_2O_2 also behave a color change in the presence of peroxidase. Owing to its good stability, low toxicity and high sensitivity, TMB/H_2O_2 as a coloring system is widely applied in the colorimetric assays.

Till now, most of the ELISA methods adopt the TMB/H_2O_2 coloring system for signal output. For the visual detection, the system is only a single-color measurement, which rely on the gradation of color development to indicate the analyte. However, the single-color mode requires higher demand for the naked eye and is difficult to achieve multi-sample detection. In the view of visual detection, multi-color assay is easier to realize sensitive and accurate detection. Therefore, it is of great importance to develop new chromogenic mode or multi-color reagents.

To achieve multi-color signal output, metallic nanoparticles such as gold or silver are intensively studied. Based on their localized surface plasmon resonance properties, they behave specific optical performance such as red shift in spectra. For instance, based on the principle that TMB^{2+} can etch gold nanoparticles, Guo et al. proposed a dual-color response for prostate-specific antigen, with color variations from wine red to colorless and then to yellow (Guo et al. 2016a, b). On the basis of the principle of iodine etched gold nanorods, an ALP-based plasma ELISA strategy was developed for the sensitive detection of human immunoglobulin G (Zhang et al. 2017a, b). Both methods provide examples of nanoparticles as enzymatic markers.

15.4.5 Washing and Blocking Agents

Actually, to obtain accurate, stable, and sensitive ELISA results, washing and blocking steps are indispensable. As the ELISA method is constructed based on the immunosorbent strategy, the nonspecific adsorption and instable adhesion are inevitable, which could lead to false-positive or negative results. Thus, in each step of the establishment of ELISA, washing step is essential to remove excess antigen or antibody. The washing agent we used is a kind of buffer solution with 0.05% Tween-20 in PBS solutions. When antigen or antibody are attached to the well surface, washing buffer is needed to wash away the unbound biomolecules. Similarly, in the antibody-antigen recognition step, the washing is also acquired to remove excess biomolecules and remain the reacted ones. A relatively low concentration detergent can be used because high dosage of detergent would cause damage to biomolecules.

Usually in a direct or sandwich ELISA, a solution of nonreacting protein is added to each well to block any other surface in the plate wells. This kind of protein solution is known as blocking agent, which is prepared by introducing certain proteins in the washing buffer. The proteins can be all kinds of animal serums such as bovine serum albumin (BSA), rabbit serum, horse serum, or casein. After a known quantity of capture antibody is attached on the well surface, the blocking procedures are carried out to block any nonspecific-binding sites on the surface. The blocking step is usually performed once in the ELISA construction process. Typically, the blocking step can be completed by incubating certain volume of blocking agent in the wells for 2 h at 37 °C.

It was reported that various proteins as blocking agents can produce quantitative differences. So, many researches have been conducted to explore the effects of different proteins on ELISA performance. For example, Vogt Jr. et al. explored the influences of instantized dry milk, serum albumin, casein, gelatins on blocking nonspecific adsorption of ELISA, indicating that casein and instantized milk showed the best blocking property (Vogt Jr et al. 1987). Moreover, Xiao and Isaacs reported that different BSA preparations used as a blocking agent in an ELISA can give different amounts of nonspecific binding of ELISA reactants (Xiao and Isaacs 2012). The report reminded that critical controls are needed to ensure that ELISA reactants are appropriately bound to the blocking agent.

15.5 Applications of Nano-ELISA in Food

15.5.1 Detection of Biotoxins

Biotoxins are toxic secondary metabolites produced by living organisms, which are usually harmless to the organism itself, but after consumption, it will affect the health of humans or animals (Fletcher and Netzel 2020). Biotoxins can be divided into five categories according to their sources: mycotoxins, bacterial toxins, marine toxins, animal toxins, and phytotoxins. They can enter food through various ways, causing people to appear food poisoning and other phenomena, which are serious or even fatal. Therefore, how to efficiently and quickly detect these toxins is particularly important. On the other hand, nanomaterial-modified ELISA overcomes the relatively low stability and sensitivity of c-ELISA. Through its inherent nanostructure, stability and specificity are improved, and a more efficient and rapid detection method can be constructed (Fig. 15.4).

15.5.1.1 Detection of Mycotoxins

Mycotoxins, a kind of secondary organic metabolites, are produced by distinct fungi like *Aspergillus*, *Penicillium*, and *Fusarium* (Avery et al. 2019). In food, mycotoxins are toxic metabolites produced when fungi contaminate food, and they are also the most commonly seen toxins among different food toxins (Moretti et al. 2018). Generally, the most common mycotoxins include aflatoxin, ochratoxin, trichothecenes, zearalenone, and fumonisins. Humans and animals ingest food containing high level of mycotoxins at one time will cause acute poisoning, and long-term intake of food containing mycotoxins will also cause chronic poisoning, and even cause cancer and teratogenic effects.

Aflatoxins are common mycotoxins that contaminate food and agricultural products. The nano-ELISA based on direct competition and indirect competition can detect aflatoxins in grains and milk more sensitively, quickly, and specifically. Combining immunomagnetic beads (IMBs) with direct competition ELISA, Zhang et al. proposed a strategy using monoclonal antibody 5H3-modified immunomagnetic beads as capture probes, and AFB1-CMO-labeled horseradish peroxidase as probes. Competing with free aflatoxin, the total amount of aflatoxin in corn samples was determined (Zhang et al. 2017a, b). Subsequently, Zhou Xu et al. developed an indirect competitive ELISA based on an MOF material MIL-88, to construct AFB1 antigen, AFB1 antibody, and MIL-88-modified antibody solid-

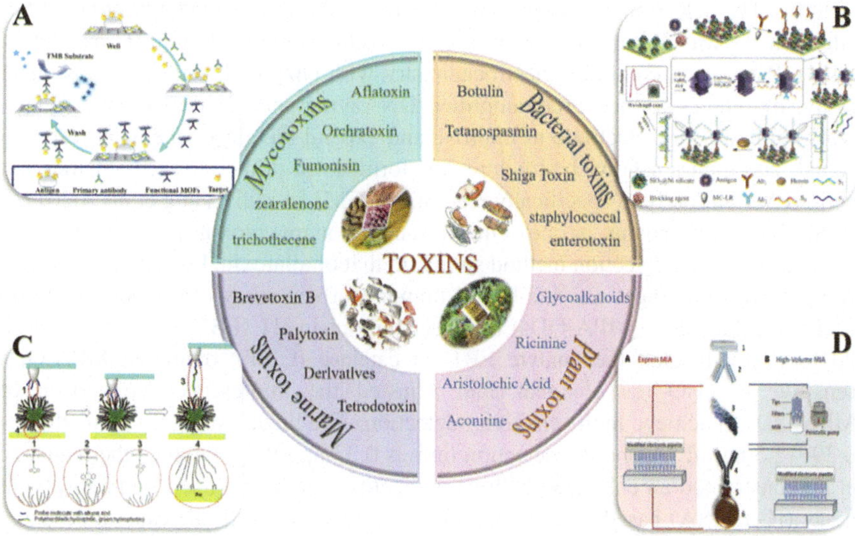

Fig. 15.4 Schematic illustration of nano-ELISA analytical techniques for biotoxins in food. Reproduced from Refs. with permission from Elsevier (A: Xu et al. (2020); B: Liu et al. (2019); C: Chen et al. (2011); D: Orlov et al. (2013))

phase antigen-tested antibody-enzyme label secondary antibody complex (Xu et al. 2020). By adding a substrate solution of TMB and H_2O_2 to observe the degree of color development, it can be used for the high-throughput determination of aflatoxin B1 in peanut milk and soy milk. The linear range of this method is 0.01–20 ng mL^{-1}, and the detection limit is 0.009 ng mL^{-1}.

In addition, the combination of various detection methods with ELISA is also a new trend in aflatoxin detection. Based on nanoenzymes, aptamers, and Fe_3O_4 magnetic nanoparticles (MNP), Long Wu has established a simple operation and separation of nanoenzymes and aptamer immunosorbent assay (NAISA) for aflatoxin B1 (AFB1) detection. In this work, mesoporous SiO_2/Au-Pt (m-SAP) was used as a signal marker with high catalase activity, and AFB1 in peanut was specifically recognized by an aptamer, and MNP was used to achieve magnetic separation. In order to verify the performance of NAISA, traditional ELISA (c-ELISA) and enhanced ELISA (e-ELISA) based on MNP and m-SAP nanozymes were applied to the detection of AFB1. The lowest detection limit of NAISA method is 5 pg mL^{-1}, which is 600 times and 12 times lower than c-ELISA (3 ng mL^{-1}) and e-ELISA (0.06 ng mL^{-1}), respectively (Wu et al. 2020). Zherdev immobilized the antibody on the surface of magnetic particles, changed the solid phase of ELISA, and constructed a microplate-based enzyme-linked immunoassay. The immobilized antibody reacts with the natural antigen and the labeled antigen in solution, thereby shortening the interaction time to 5 min without affecting the analysis results. The adsorption of immunoglobulins on the surface of magnetic nanoparticles increases their stability in water-organic media, thereby minimizing the degree of dilution required for samples. The detection of barley and corn extracts showed that the detection limit of aflatoxin B1 was 20 pg mL^{-1} with a total detection time of 20 min (Urusov et al. 2014). Pang et al. combines electrochemistry with enzyme-linked immunosorbent assay, introduces rolling circle amplified DNAzyme and covalent organic framework to modify the electrode to improve and expand the electrochemical response signal, and then a sandwich structure was formed via primer-antigen-aptamer and anti-AFM1 antibody to specifically recognize AFM1 in milk (Pang et al. 2020).

Similarly, fumonisins, as one of the most common mycotoxins in cereal products, have established a detection method similar to that of aflatoxin. Lu et al. developed a competitive fluorescent enzyme-linked immunosorbent assay (cFELISA) based on CdTe quantum dots (MPA-QDs) to detect fumonisin B1 (FB1) in corn (Lu et al. 2018). They labeled the analyte FB1 on catalase (CAT), outputted MPA-QDs sensitive to H_2O_2 as a signal, and adjusted the fluorescence conversion of MPA-QDs to achieve high-sensitivity detection. The linear range of this method is 0.39–12.5 ng mL^{-1}, and the detection limit is 0.33 ng mL^{-1}. On the other hand, Li et al. constructed an enhanced indirect competitive enzyme-linked immunosorbent assay (IC-ELISA) based on gold nanoparticles modified with mercaptoundecanoic acid (AuNPs-MUA) (Li et al. 2018a, b). Three hybridoma cell lines were obtained by immunization and cell cloning methods, which secreted monoclonal antibodies against fumonisin B1 (FB1); AuNPs-MUA was used as horseradish peroxidase (HRP)-goat antibody. The mouse IgA vector is used to observe the degree of color development by adding a substrate solution of TMB and H_2O_2, so as to quantitatively detect the total content of fumonisins (FB1, FB2, and FB3) in corn. The detection limit of this method is 0.078 ± 0.013 μg L^{-1}.

Ochratoxin A is a natural mycotoxin, which has been found in several food matrices. Due to its high toxicity, effective monitoring of its presence in food is particularly important. Zhu et al. prepared botryoid-shaped Au/Ag nanoparticles (BSNP) via a tailored galvanic reaction. In the presence of ascorbic acid, the silver nanoprism-BSA complex is used as a template to react with $HAuCl_4$. The formed BSNPs-HRP-IgG was used as a carrier of HRP-IgG to amplify the detection signal of indirect competitive ELISA against ochratoxin A. The linear range of this BSNPs-enhanced ELISA method is 0.016–0.05 ng mL^{-1} (Zhu et al. 2017). Karczmarczyk et al. used gold nanoparticles for QCM-D signal enhancement and successfully established an indirect competitive bioassay for the detection of ochratoxin A in red wine (Karczmarczyk et al. 2017). Based on the form of indirect competition, a specific rabbit PAb and a second goat anti-rabies PAb labeled with gold nanoparticles were used for signal amplification. The linear range of the method is 0.2–40 ng mL^{-1}, and the detection limit is 0.04 ng mL^{-1}.

Mak et al. reported an ultra-sensitive magnetic nanoparticle immunoassay that can detect more than one mycotoxin (Mak et al. 2010). The use of magnetic nanoparticles as the solid phase allows a significantly increased surface area for the immobilization of reactants and their uniform distribution in the entire volume of the reaction medium, thereby eliminating the diffusion limitation of traditional ELISA. The application of the magnetic field allows the reactants to be separated simply and quickly and simplifies the washing steps required for traditional microplate-based ELISA. Taking advantage of these advantages, an MNP-based immunoassay protocol was developed and implemented in the wells of ELISA microplates to detect AFB1, zearalenone, and HT-2 mycotoxins.

15.5.1.2 Detection of Bacterial Toxins

It is reported that more than half of food-borne diseases are caused by pathogenic bacteria, among which botulism caused by Clostridium botulinum, gastroenteritis, and staphylococcal poisoning caused by E. coli strains are the most common (Hernández-Cortez et al. 2017). In order to reasonably control-related food pollution, it is necessary for people to strictly manage food from the primary production to the final consumption. Therefore, it is necessary to construct feasible analytical techniques to monitor bacterial toxins in food samples.

Bhairab Mondal uses staphylococcal enterotoxin B (SEB)-binding body (SEB2) as capture, and unmodified gold nanoparticles (AuNPs) as colorimetric probes to construct a simple, sensitive, and specific detection of SEB (Bhairab et al. 2018). This method is based on the color change from red to purple caused by the conformational change of the aptamer in the presence of SEB, and the aggregation of AuNPs induced by salt, which can be monitored with the naked eye or UV-Vis spectrometer. The results show that AuNP can effectively distinguish SEB-induced conformational changes of nucleic acid aptamers at a certain high salt concentration, and the stability is effectively tested in artificially added milk samples. The linear range of this method is 50 μg mL^{-1} ~ 0.5 ng mL^{-1}, the lower limit of visual detection (LOD) reaches 50 ng mL^{-1} within a few minutes, and the spectrophotometric method increases to 0.5 ng mL^{-1}.

Orlov et al. determined two staphylococcal toxins in milk by a magnetic sandwich immunoassay (Orlov et al. 2013). The capture monoclonal antibody is fixed on the 3D fiber solid phase in the kit, and the recognition monoclonal antibody is conjugated to the magnetic nanoparticles through the biotin-streptavidin binding. When the interlayer is formed, the magnetic particles are captured by the cartridge and detected with extremely high sensitivity and selectivity by combining frequencies. The signal is read from the entire volume of the non-transparent 3D fiber structure used as the solid phase, providing a large reaction surface, rapid reagent mixing, and antigen immunofiltration directly during the measurement process. This method showed that the limits of detection (LOD) of Staphylococcal Enterotoxin A (SEA) and Toxic Shock Syndrome Toxin (TSST) were as low as 4 and 10 pg mL^{-1}, respectively.

15.5.1.3 Detection of Marine Toxins

Marine toxins are a kind of toxic natural active micromolecules that exist in marine organisms and have high toxicity. According to the different carrier, it can be divided into shellfish toxin, tetrodotoxin (TTX), and cigar toxin (CTX) (Wang et al. 2020a, b). After they enter the human body through food, they will act on the nervous or digestive system, causing food poisoning symptoms such as diarrhea, paralysis, and even lead to death (Grasso et al. 2019). These marine toxins not only affect food safety and human health, but also cause serious economic losses.

Therefore, in order to prevent the occurrence of food-borne marine toxin poisoning, the early diagnosis and detection of marine toxins is of great significance.

Campbell has developed a nano-array planar waveguide biosensor for detecting tetrodotoxin (TTX). The technology consists of a nanoprinted toxin-conjugate array constructed by an indirect competitive immunoassay, and it is used to analyze pufferfish samples under high flow conditions. By studying the matrix effect and the toxin recovery rate, the applicability to natural samples was studied. The biosensor can detect TTX in 0.4–3.29 μg g^{-1} puffer fish tissue (Reverte et al. 2017).

Liu et al. combined ELISA with nanozymes and established a sensitive colorimetric immunosensor to visually detect microcystin-LR (MC-LR) (Liu et al. 2019). The microchip is modified with flaky nickel silicate-coated silica nanospheres (SiO_2@Ni Silicate) to immobilize the antigen. The copper hydroxide nanozyme acts as a marker to capture the secondary antibody used for immune response and couples with the G-quadruplex/heme DNA enzyme to form a dual integrated mimic enzyme, which reflects the peroxidase activity of ABTS. Greatly improve the visual signal. The linear range of this method is 0.007–75 μg L^{-1} with the LOD of 6 ng mL^{-1}.

15.5.1.4 Detection of Phytotoxins

Phytotoxins are natural phytochemicals or secondary metabolites formed by plants, which can protect themselves from various threats, such as bacteria, fungi, insects, and natural enemies (Bucheli 2014). According to different chemical structures, they can be divided into three main chemical structures: alkaloids, terpenes, and phenols. Among them, furanocoumarin, lectin, carbohydrate alkaloid, and pyrrolidine nuclear alkaloid are the most studied (Gunthardt et al. 2018). In food, due to non-edible plant pollution, it can be specifically divided into two types: phytotoxins inherent in food crops and phytotoxins that enter food. For example, alkaloids and cyanogenic glycosides are phytotoxins inherent in potatoes and cassava (Mol et al. 2011). In order to effectively control food safety, it is of great necessity to construct a variety of detection methods for analyzing phytotoxins in food.

The ELISA method based on colloidal gold particles significantly shortened the measurement time of ricin. Xu et al. modified the gold-coated AFM tip with polyethylene glycol derivatives to add anti-ricin antibodies to identify ricin. The sensitivity is as high as the level of fg mL^{-1}, and the LOD is as low as 1 ng mL^{-1} (Chen et al. 2011). Christopher et al. used sensitive electrochemical biochip technology combined with ELISA to detect ricin. The capture antibody immobilized on the gold electrode facilitates the specific binding of ricin. The detection of bound ricin is achieved by applying an antibody-enzyme conjugate and measuring the current of the enzymatic redox reaction on the electrode. Among them, the high conversion of the enzymatic reaction facilitates signal amplification, and the built-in redox cycle program provides a second signal amplification, which can be very sensitive to identify ricin at about 50 °C within 20 min (Phlmann et al. 2017).

15.5.2 Detection of Pesticide Residues

Pesticide is a chemical agent used in agriculture to prevent plant diseases and insect pests and regulate plant growth (Xu et al. 2017a, b). There are many varieties of pesticides, which can be divided into nine categories according to their usage, including insecticides, acaricides, nematicides, molluscicides, rodenticides, fungicides, herbicides, defoliants, and plant growth regulators (Bhandari et al. 2019). The sources of pesticide residues in food mainly include three aspects: (1) the direct pollution of crops by pesticides; (2) the absorption of pesticides by crops from the polluted environment; (3) the accumulation of pesticides in organisms due to the effect of food chain. Foods containing residual pesticides can cause habitual headaches, dizziness, fatigue, sweating, depression, memory loss, weakness, and other hidden effects. Long-term consumption can cause cancer, arteriosclerosis, cardiovascular diseases, and other diseases (Samsidar et al. 2018; Silva et al. 2019). Therefore, so as to ensure the food safety of consumers, it is crucial to analyze pesticide residues in food samples.

So far, the applications of ELISA to mainly detect pesticide residues in food are insecticides, fungicides, and herbicides (Fig. 15.5). ELISA is a detection method based on the specific and reversible binding reaction of antigen and antibody. Highly selective antibodies can be obtained by preparing antigen hapten and its carrier conjugate, so that ELISA for detecting pesticide residues can be established, including indirect competition method, direct competition method, and labeled antigen competition method (Wu et al. 2019a, b). Therefore, nano-ELISA is becoming increasingly popular in food contaminants analysis. It can detect pesticide residues

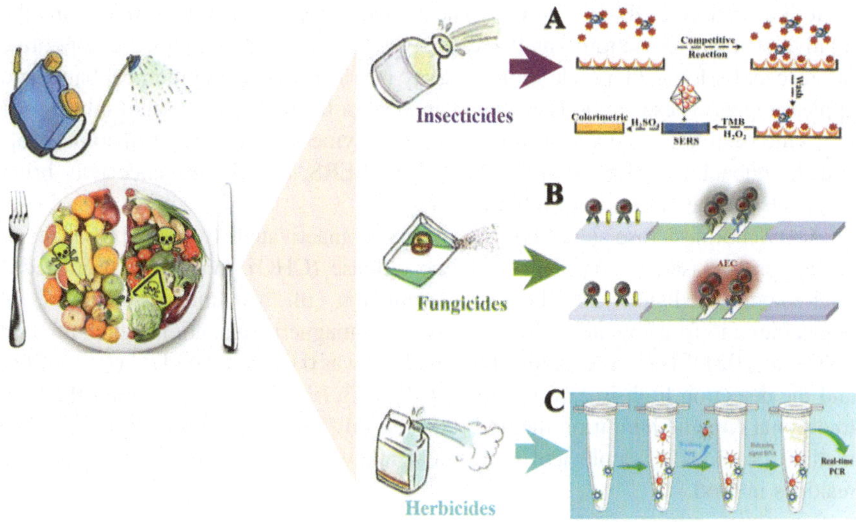

Fig. 15.5 Schematic illustration of nano-ELISA analytical techniques for pesticide residues in food samples. Reproduced from Refs. with permission from Elsevier and American Chemical Society (A: Yan et al. (2019); B: Wei et al. (2020); C: Guan et al. (2021))

in vegetables and fruits more sensitively, quickly, and specifically, and its detection level can reach ng or even pg level (Jia et al. 2009).

15.5.2.1 Insecticides

Insecticides were used to eliminate or reduce any kind of pests, such as insects and wheat aphids. Nowadays, different types of pesticides are widely used in agriculture to achieve high yields. The application of pesticides has ensured nearly one third of the world's crop production. Pesticides improve food production to meet the needs of a growing population (Nsibande and Forbes 2016). The prevention and control of plant diseases and insect pests is conducive to the prevention of harmful diseases of crops. Pesticide residues were found in many food samples, such as pyrethrins, dimethoate, imidacloprid, and triazophos. For example, In order to detect the residual content of triazophos in food, Yan et al. designed a biomimetic nano-ELISA based on molecularly imprinted polymers (MIP) and nanoenzyme markers for the detection of DDT (Yan et al. 2019). Wherein, MIP as a biomimetic antibody, Pt@BSA-hapten as a competitive probe to recognize the binding site of MIPs, and AuNP as a substrate for SERS enhancement. The detection limit of this method is 1 ng mL^{-1}. Based on competitive binding and biological barcode amplification, Zhang et al. designed an immunoassay method for the detection of triazophos (Zhang et al. 2018). The Au nanoparticles (AuNPs) are modified by monoclonal antibodies and single-stranded thiol-oligonucleotides labeled with 6-carboxyfluorescein, then bound to ovalbumin with antigenic haptens that coated on the bottom of the microplate to compete with triazophos in the sample. So, the fluorescence intensity of 6-FAM quenched by AuNPs was negative correlation to the triazophos concentration. The linear range of this method is 0.01–20 μg L^{-1}, and the limit of detection (LOD) is 6 ng L^{-1}. The recovery rate was 85.0%–110.3%, and the relative standard deviation was 9.4%–17.4%. This method performed competitive fluorescent biological barcode immunoassay in water, rice, cucumber, cabbage, and apple samples. These methods provided an idea for the rapid detection of other small molecule pesticide residues. Application the enzyme-like activity of nanomaterials and the quenching effect of fluorescent dyes, SERS, and fluorescence may bring prospects for the detection of other pollutants.

Acetylcholinesterase (AChE) activity was inactivated by the presence of acetamiprid. Based on AChEand choline oxidase (CHO), Wu et al. proposed a dual-enzyme-mediated Fe^{2+}/Fe^{3+} determination of acetamiprid residues in vegetables and fruits by an Fe^{2+}/Fe^{3+} conversion magnetic relaxation switch method (Wu et al. 2021). The linear range of this method was 0.01–1000 μg L^{-1} ($R^2 = 0.99$), and the detection limit is 2.66 ng mL^{-1} ($S/N = 3$, $n = 3$), which is better than the traditional enzyme inhibition method (0.89 μg mL^{-1}) increased by 335 times. This method provides a simple and convenient analytical tool for detecting pesticide residues in food.

15.5.2.2 Fungicides

Fungicides were a class of chemical reagents, which used to inhibit or kill pathogenic spores to protect crops, including bactericidal streptomycin, carbendazim, bordeaux mixture, zinc methylarsenate, and so on. For example, streptomycin (STR) is an antibiotic extracted from the culture broth of Streptomyces griseus (Yin et al. 2017). STR residues in food can have serious effects on human health, such as nephrotoxicity and ototoxicity. In order to detect aminoglycoside streptomycin, Wei et al. established a new type of lateral flow immunoassay (LFA) platform with Au@Pt as a marker with enzyme-like activity, with the limit of detection (LOD) of 0.1 ng mL^{-1} (Wei et al. 2020). This method was applied to the content of streptomycin in milk. LFA based on nanoenzymes is a promising tool for detecting pesticide residues in food.

15.5.2.3 Herbicides

Herbicides can inhibit the growth of weeds in the field and increase the yield of crops. For example, atrazine, 2,4-D, trifluralin, and glyphosate (GLYP) are the main pollutants of soil and water ecosystems. For example, atrazine molecules can be fully degraded (100%) by Fe_3O_4-TiO_2/rGO nanozyme under irradiation of natural sunlight. Based on competitive ELISA, Kwon et al. developed a peroxidase-like Pd@Pt nanoparticle-conjugated primary antibody as an enzyme marker to detect atrazine (Kwon et al. 2020). The method has high sensitivity, LOD is 0.5 ng mL^{-1}, and the recovery rate is between 99% and 115%, indicating that the immunoassay can detect atrazine and other small molecule herbicides and pesticides. Wang et al. developed a direct competitive enzyme-linked immunosorbent method (Wang et al. 2016). Using ovalbumin-2,4-D (OVA-2,4-D)-modified nano-silica (OVA-2,4-D-SiO_2 NPs) as a capture probe, horseradish peroxidase labeled with anti-2,4-D antibody as a probe, and competed 2,4-D in the samples. Thus, 2,4-D was detected to prevent the abuse of 2,4-D in the commercial production of bean sprouts. The linear range of this method was 1–350 ng mL^{-1}, and LOD was 0.079 ng mL^{-1}.

Further, in view of the wide application prospects and advantages of oligonucleotide functionalized AuNPs in biological analysis. Naiyu Guan et al. synthesized anti-GLYP antibody and double-stranded oligonucleotide bifunctional AuNP probe and established AuNP biological barcode immuno-PCR (AuNP-BB-iPCR) to detect GLYP (Guan et al. 2021). GLYP and OVA-GLYP coating to compete with functionalized antibodies, thereby releasing signal DNA and detecting by real-time PCR. The detection linear range of GLYP was 61.1 pg g^{-1}–31.3 ng g^{-1}, and LOD was 4.5 pg g^{-1}. This method was seven orders of magnitude lower than the conventional ELISA method (70 $\mu g \cdot g^{-1}$) established with the same antibody. The recoveries of soybean, rape, and corn samples were 99.8%, 102.6%, and 103.7%, and the relative standard deviations were all less than 12.9%. The detection time of AuNP-BB-iPCR (including food sample preparation) was 4 h, which can be used for sensitive detection of GLYP in food and environment.

15.5.3 Detection of Veterinary Drug Residues

Residues of veterinary drugs in food are mainly due to the residues in animal foods such as meat, eggs, and milk after the animals are used for drugs (Baynes et al.

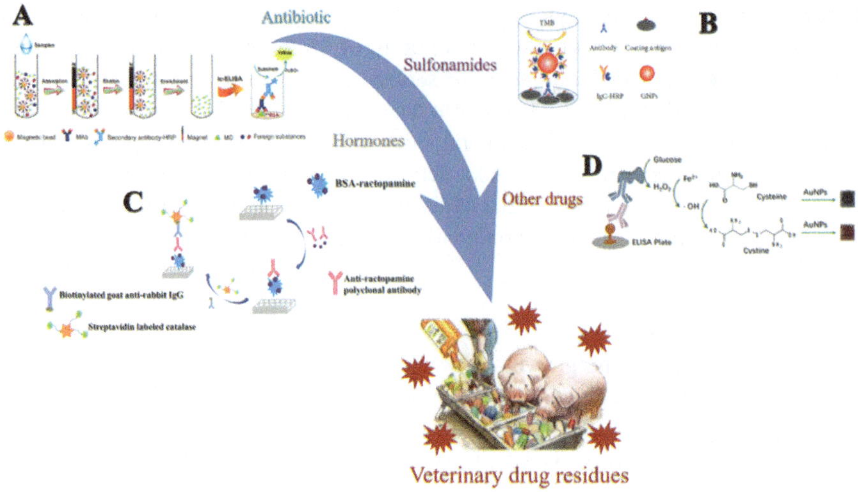

Fig. 15.6 Schematic illustration of nano-ELISA analytical techniques for veterinary drugs residues in food. Reproduced from Refs. with permission from Elsevier (A: Song et al. (2018); C: Han et al. (2018); B: Peng et al. (2013a, b); D: Yu et al. (2018))

2016). Veterinary drug residues in food are mainly antibiotics, sulfonamides, furans, antiparasites, and hormones. Therefore, it is necessary to detect veterinary drug residues in food. Using the enzyme-like activity of nanozymes, a new ELISA based on direct competition, indirect competition, and labeled antigen competition can be used to detect veterinary drug residues in animal foods more sensitively, quickly, and specifically. In this chapter, nano-ELISA detects pesticide residues in food mainly antibiotics, sulfonamides, and hormones (Fig. 15.6).

15.5.3.1 Antibiotics

Antibiotics are used to prevent and treat animal diseases and improve the performance of modern animal husbandry (English and Gaur 2010). Commonly used antibiotics include tetracyclines (TCs), maduramycin (MD), chloramphenicol, macrolides, etc. Illegal or excessive addition of antibiotics may cause animal poisoning or remain in animal muscle tissue (Ben et al. 2019). It is a potential hazard to human and environmental health. Therefore, the development and application of the nano-ELISA method is of great significance in food safety. For example, the use of nanomaterials (MBs and SiO_2) as absorption substrates greatly improves the sensitivity, accuracy, and stability of nano-ELISA. Antibody-functionalized MB (IMB) was synthesized based on the production of a specific anti-MD MAb. Song et al. established an IMBs-based indirect competitive ELISA (ic-ELISA) to detect MD in three chicken tissues (Song et al. 2018). The detection limits of MD in chicken muscle, skin and fat, and liver were 72, 74, and 173 μg kg^{-1}, respectively. The recovery rate was 80.0%–115.8%, and the coefficient of variation was less than 11.3%. Tao et al. proposed a competitive direct chemiluminescence immunoassay method based on MBs separation and AuNPs labeling technology for the detection of chloramphenicol (CAP) in milk (Tao et al. 2013). In two different extraction methods, the IC50 values of the chemiluminescent magnetic nanoparticle immunoassay (CL-MBs-nano-ELISA) were 0.017 and 0.17 μg L^{-1}, respectively.

Further, by nucleic acid aptamers replace antibodies and nanozymes replace biological enzymes, nano-ELISA improves the selectivity and sensitivity of traditional ELISA. For example, Sheng et al. reported an ultra-sensitive Apt-modified AuNPs (AuNPs-Apt) analysis method to detect tetracycline residues in honey based on the high selectivity of aptamers for analytes and the enhanced catalytic ability of AuNP (Sheng et al. 2020). TCs-BSA were coated on a microplate. Then, the TCs-BSA coating in the microplate competes with the free TCs in the sample for the limited AuNPs-Apt. As a kind of nanoenzyme, AuNPs show peroxidase activity, oxidized 3, 3′, 5, 5′-tetramethylbenzidine (ox-TMB) from colorless to blue, and measured at 652 nm.

15.5.3.2 Sulfonamides

Sulfa drugs are mainly used for antibacterial and anti-inflammatory, such as sulfadimethoxine (SDM), sulfamidine, and sulfadiazine. After long-term intake of animal food containing sulfa drug residues, the drug will continue to accumulate in the body, causing damage to the urinary system, causing crystaluria, hematuria, and tubular urine, etc. (Chang et al. 2020). Therefore, it was clear to regulate and control the useful dose of sulfa drugs at all over the world.

Due to its easy coupling of biomolecules and maintaining the biological activity of labeled molecules (antibodies and DNA), AuNPs make it widely used in the field of biosensors (Chandra et al. 2013; Chandra et al. 2010; Kumar et al. 2020; Mahato et al. 2019). The use of nanozymes to improve the sensitivity of traditional ELISAs brings prospects for the analysis of contaminants in food. For example, Peng et al. developed an ultra-sensitive nano-ELISA to detect SDM in chicken tissue, increasing the sensitivity of traditional ELISA by 20 times (Peng et al. 2013a, b). Using the biological coupling of AuNPs and enzyme-labeled antibodies as signal probes, a simple and sensitive detection of SDM residues in animal tissues is achieved. The sensitivity of nano-ELISA in the buffer was 5 pg·mL^{-1} and the LOD was 0.2 μg·kg^{-1} that can be obtained by simply extracting chicken liver with the buffer. The strategy was convenient and sensitive, which can be applied to improve the performance of the ELISA to detect small molecule contaminants.

15.5.3.3 Hormones

Hormonal drugs are mainly used to improve the reproduction and growth of animals. The hormones suitable for animals are sex hormones and corticosteroids, and sex hormones are the most commonly used. For example, testosterone, progesterone, and ractopamine hydrochloride. Excessive hormone drugs in foods will affect the normal physiological functions of consumers and have certain carcinogenicity, leading to health problems such as precocious puberty and abnormal growth of children (Hoga et al. 2018).

Ractopamine, also known as clenbuterol, is a common hormone veterinary drug residue in animal food. Pingli He et al. developed a colorimetric ELISA based on indirect competition with a linear range of 2–512 ng mL^{-1} and the LOD was 0.35 ng mL^{-1} (Han et al. 2018). Obtaining anti-ractopamine polyclonal antibodies by preparing antigenic ractopamine-glutaric acid-bovine serum albumin antigen. Based on Mn(VII)/Mn(II) conversion-induced change in low-field nuclear magnetic resonance of the transverse relaxation rate, Wang et al. report a magnetic immunosensor for the detection of food-borne pathogen and residue of veterinary drug (Wang et al. 2019a, b, c). This Mn-mediated magnetic immunosensor not only maintains the good stability of the traditional paramagnetic ion-mediated magnetic sensor, but also greatly improves the sensitivity of the sensor. And the LOD improve from ng mL^{-1} to pg mL^{-1}. This method provides a promising platform for sensitive, stable, and convenient biological analysis.

In addition, 19-nortestosterone (19-NT) was also a common hormonal veterinary drug residue in animal food. Peng et al. proposed a nano-ELISA method based on the coupling of AuNPs with goat anti-rabbit IgG and HRP for the highly sensitive detection of 19-NT in beef (Peng et al. 2013a, b). The AuNPs-IgG-HRP conjugate was simple prepare and stable. The sensitivity of this method in buffer was 0.01 ng mL^{-1}, which was 10 times that of c-ELISA. After simple pretreatment of beef samples, the LOD was 0.3 $mg{\cdot}kg^{-1}$.

15.5.3.4 Other Drugs

In addition, there are other types of veterinary drugs that are also easy to contaminate animal food, such as antiviral drugs and antiparasitic drugs. Antiviral drugs on the market mainly include amantadine (AMD), rimantadine, ribavirin, and other drugs represented by the symmetrical tricyclic amine structure (De Clercq 2001). Praziquantel (PZQ) is an antiparasitic drug for mammals and fish. Shen et al. developed an ic-ELISA to detect PZQ residues (Shen et al. 2019). The hapten PZQ-HS synthesizes the immunogen and the coating antigen with carrier protein by introducing an amino group into the benzene ring, thereby preparing a highly sensitive monoclonal antibody. Finally, an immunochromatographic test strip (ICS) for rapid detection of PZQ residues in mackerel was developed. Yu et al. developed an ic-ELISA that introduced Fenton reaction and gold nanoparticle aggregation (Yu et al. 2018). By Fenton reaction to form hydroxyl radicals, it significantly accelerates and controls the oxidation of cysteine, and amplifies the signal. At the same time, through the strong Au-S interaction and the AuNPs aggregation, led to a pronounced color change from red to dark purple in the solution, which could be easily distinguished with the naked eye, thereby detecting the residue of amantadine (AMD) in poultry. The detection limit of this method is 0.095 ng mL^{-1}.

15.5.4 Detection of Microorganism

Microbial contamination in food is the most important factor affecting food safety, and pathogenic microorganisms are the food safety issue that has the greatest impact

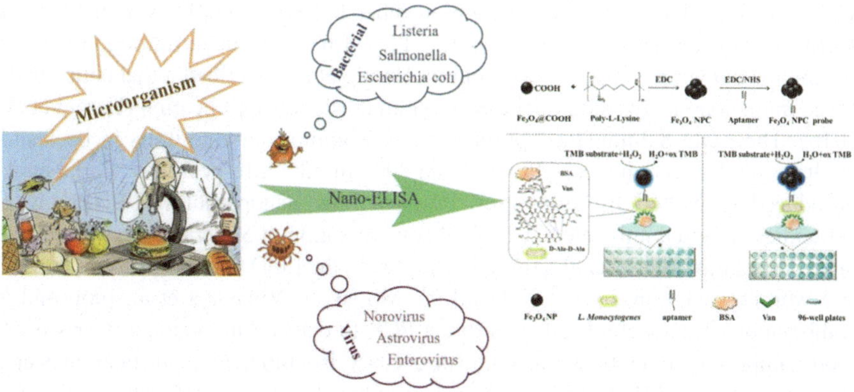

Fig. 15.7 Schematic illustration of nano-ELISA analytical techniques for food-borne pathogenic microorganisms. Reproduced from Refs. with permission from Elsevier (Zhang et al. 2016)

on consumer health. Therefore, microbial contamination in food has caused widespread concern in society. Common food-borne microorganisms are bacteria, fungi, and viruses. In particular, microorganisms that cause food spoilage and decay and food-borne pathogenic microorganisms have attracted great attention from society. They can multiply bacteria in food and even produce toxic metabolites, causing food poisoning or harmful infections. So far, the application of ELISA to detect microorganisms in food is mainly bacteria and viruses. However, traditional ELISA is limited by complicated procedures, relatively low detection limit, and large sample size. The nano-ELISA has been widely developed to enable simple and sensitive detection of microorganisms in food (Fig. 15.7).

15.5.4.1 Bacteria

Bacteria are the main pathogens in food, including *Escherichia coli*, *Salmonella*, *Listeria*, *Staphylococcus aureus Botox*, *Shigella*, *Streptococcus haemolyticus*, *Vibrio parahaemolyticus*, and so on. Among them, Salmonella has been the major cause of the food-borne contaminants in vegetables, egg, chicken pork, beef, or vegetable row crops (Vinayaka et al. 2019). The World Health Organization (WHO) reported that over 105 million diarrheal cases in the world every year are related to food-borne pathogens (Bull et al. 2020). Among them, Salmonella have drawn the most concerns because it is frequently found in food stuffs and may cause severe diseases (Wang et al. 2019a, b, c).

Based on gold nanoparticle-based enzyme-linked antibody-aptamer sandwich (nano-ELAAS), Wu et al. proposed a immunoassay for robust detection of Salmonella enterica serovar Typhimurium (STM) with high specificity (Wu et al. 2014). The STM in the sample is captured and pre-concentrated from the aptamer-modified magnetic particles, and then combined with the detector antibody. Then, a nanoprobe carrying a large number of reporter antibodies, and HRP is used for colorimetric signal amplification. The quantitative detection range of nano-ELISA was 1×10^3–1×10^8 cfu mL^{-1}, and the detection limit was 1×10^3 cfu mL^{-1}. The selectivity for STM in high concentration other bacterial samples was more than 10 times, and the detection time was less than 3 h. To achieve immunomagnetic separation and simple target concentration, Herzig et al. described a magnetic bead-based immunoassay for Salmonella with tyramide signal amplification (Herzig et al. 2016). The LOD Salmonella typhimurium and Salmonella enteritidis in beef and poultry samples were increased to 800 and 200 cfu mL^{-1}, respectively. Farka et al. introduced a method by combining Prussian blue nanoparticles (PBNPs) with antibodies and used it to detect microbial contamination of *Salmonella typhimurium* in powdered milk (Farka et al. 2018). Furthermore, the LOD was 6×10^3 CFU mL^{-1} and working range up to 10^6 CFU mL^{-1}. Moreover, Vinayaka et al. proposed a combination of magnetic beads and direct PCR to detect Salmonella vulgaris from food samples without bacterial culturing, DNA isolation, and purification steps (Vinayaka et al. 2019). Based on urease-induced silver metallization on gold nanorods (AuNR), Gao et al. reported an improved ELISA for the determination of Salmonella enterica Choleraesuis (Gao et al. 2019a, b). As aspect ratio (length divided by width) of AuNR decreased, AuNR solution showed a multi-color change, and it also behaved a significant blue shift in the absorption peak ($\Delta\lambda$max) of AuNR. Hence, the LOD of this method were as low as 1.21×10^2 cfu mL^{-1} for qualitative detection with naked eyes, and 1.21×10^1 cfu mL^{-1} for quantitative analysis.

Furthermore, *E. coli* O157:H7 (E. coli) is a highly infectious pathogen that spreads widely in food and water and poses a major challenge to public health (Scallan et al. 2011). Wei et al. synthesized anti-*E. coli* antibody-HRP-$Cu_3(PO_4)_2$ nanocomposites to replace HRP-conjugated antibody, and then the nanozyme label was applied in ELISA for *E. coli* detection (Wei et al. 2016). Based on DNA-based hybridization chain reaction (HCR) and biotin–streptavidin signal amplification, Guo et al. reported on a novel sandwich ELISA for the sensitive determination of E. coli (Guo et al. 2016a, b). The method behaved a detection range of 5×10^2 cfu mL^{-1} to 1×10^7 cfu mL^{-1}; and an LOD of 1.08×10^2 cfu mL^{-1} that is 185 times lower than that of traditional ELISA.

Listeria as a food-borne pathogen has posed great threats on human health all over the world (Li et al. 2018a, b). Based on nucleic acid hybridization reaction and magnetic signal readout, Qi et al. proposed a magnetic DNA sensor for sensitive detection of *Listeria* in ham sausage samples (Qi et al. 2021). The method can be realized in a one-step detection with LOD as low as 50 cfu mL^{-1} within 2 h, and the average recovery can reach 92.6%. Zhou et al. developed a sandwich enzyme-linked immunosorbent method, based on nanoparticle clusters (NPC), using vancomycin (Van) as the loading substrate to capture Listeria, Fe_3O_4 nanoparticle clusters (NPC)-modified aptamers are used as signal amplification nanoprobes to identify Listeria monocytogenes, thereby effectively detecting food-borne pathogens-Listeria monocytogenes in milk and other foods bacteria (Zhang et al. 2016). The linear range of this method was $5.4 \times 10^3 \sim 10^8$ cfu mL^{-1}, and the detection limit was 5.4×10^3 cfu mL^{-1}. Y. Wu et al. developed a sandwich pELISA for Cronobacter detection in powdered infant formula samples by mediating AuNP growth through DNA (Wu et al. 2019a, b). The catalase–hydrogen peroxide (Cat–H_2O_2) system was introduced to bridge the DNA-directed AuNP growth and pELISA, as such DNA can be cleaved into fragments by the hydroxyl radical generated from oxidation of H_2O_2 through Fenton reagents. The proposed pELISA can qualitatively detect Cronobacter species by the naked eye with a cut-off limit of 3×10^5 cfu mL^{-1}. This method linear range was 3×10^2 to 3×10^7 cfu mL^{-1}.

15.5.4.2 Virus

Viruses are also food-borne pathogenic microorganisms. Currently, common food-borne viruses mainly include hepatitis A and E viruses, rotavirus, astrovirus, enterovirus, and norovirus. Noroviruses is the family of Caliciviridae, which has been identified in human beings and several other animal species (Mauroy et al. 2009). Human norovirus (NoV) may cause viral gastroenteritis for human and viral food-borne outbreaks globally. Weerathunge et al. proposed a colorimetric sensor based on the combination of AuNPs and aptamers, which produces a blue color when norovirus was present. The concentration of the virus was detected by the shade of the color (Weerathunge et al. 2019). Khoris et al. proposed a colorimetric sensor based on peroxidase-like activity of silver ion-incorporated gold nanoparticles, which can make TMB develop color (Khoris et al. 2019). NoV-LPs were assayed with an LOD of 10.8 pg mL^{-1}, corresponding to at least 100-fold higher sensitivity compared to the traditional immunoassay.

15.5.5 Detection of Food Allergens

Food allergens refer to ingredients in food that can cause abnormal reactions in the body's immune system, which can cause a strong allergic reaction in the human body and harm to the body if you consume or contact with them (Sicherer 2001; Sathe et al. 2005; Ekezie et al. 2018). With the occurrence of food allergic diseases increasing year by year, the problem of food allergy has become a more prominent problem in the field of food safety (Monaci et al. 2018; Liu and Sathe 2018). Accurately determining the type and content of allergens in food and providing information about the background level of allergens in food and information about hidden or unlabeled allergens is of great significance for preventing and avoiding food allergies.

Food allergens can be divided into major allergens and minor allergens, which are generally proteins or glycoproteins with a relative molecular mass of 10,000–70,000 (Burgess et al. 2019). Major allergens are derived from peanuts, milk, nuts, crustaceans, and so on (Table 15.1). Those people who are allergic to food allergies can prevent them by avoiding the foods with allergens, which is also the most effective risk management measure (Chan et al. 2018). Therefore, manufacturers are required to correctly label allergic ingredients on food labels to prevent consumers from accidentally eating and causing allergies.

Allergic reactions will not only seriously affect the quality of life of patients, but also life-threatening. For example, more and more studies believe that allergic reactions can cause idiopathic inner ear disease—Meniere's disease (Sicherer and Sampson 2014). Therefore, it is important to analyze the inevitable allergens in food

Table 15.1 The common food sources of food allergens

Food sources	Allergens
Peanuts	Ara h1
	Ara h2
Wheat	Gliadin
Shrimp	Tropomyosin
Eggs	Ovalbumin
	Ovomucin
Milk	Casein
	Lactoglobulin
	Lactalbumin
	BSA

samples. The characteristic fragments present in trace amounts in allergens are usually lost due to pretreatment operations, making the analysis result lower than the actual value. ELISA is still the most extensive detection method at present, and the introduction of nanomaterials enables c-ELISA to overcome their own shortcomings and achieve more sensitive, stable, and low detection limits (Alves et al. 2016). On the other hand, modern analytical techniques have made significant progress in improving the accuracy, sensitivity, and stability of c-ELISA (Andjelkovic et al. 2017). Interestingly, the nano-ELISA has greatly developed the ELISA, which provides more ideas for the detection of food additives (Wu et al. 2019a, b).

15.5.5.1 Peanut Allergies

Peanuts, which are a type of nuts frequently encountered in daily life, are important food allergens and may cause extremely serious allergic reactions (Barnett et al. 1983; Chassaigne et al. 2007). According to British researchers, in the UK, about one in 200 people are sensitive to peanuts. Unfortunately, according to research investigations, peanut allergy is usually caused by childhood and is accompanied by life (Hebling et al. 2013; Turcanu et al. 2003). At the same time, it is also the one with the largest number of food allergy deaths, and 90% of deaths caused by food allergies are caused by peanuts. In the United States, about 100 people die every year from anaphylactic shock caused by peanut allergy (Hua et al. 2016; Vierk et al. 2007). The main peanut allergens are Arah1 (63.5 kDa), Arah2 (17 kDa), other related antigens are Arah3 (60 kDa), Arah4 (14 kDa), and the minor antigens are Arah 6, Arah7, and actin.

Although correct and decisive food labeling will enable consumers to avoid peanut allergens during food processing, cross-contamination, or ingredients containing hidden allergens during food processing will greatly weaken this task (Breiteneder and Radauer 2004). For example, the ELISA inhibition test confirmed that Ara h2 is the main antigen causing the cross-reaction of peanuts, hazelnuts, almonds, and so on. In addition, individuals who are allergic to eggs, milk, or walnuts are also allergic to peanuts, but no cross-reactions with walnuts have been found protein (Pele et al. 2007). Therefore, food manufacturers usually include precautionary words ("may contain traces of peanuts") on their packaging, not only to protect sensitive individuals, but also to protect themselves. From this perspective, a reliable, accurate, highly sensitive and selective method is needed to safely assess the presence of peanut allergens in food.

At present, the detection of allergens is mainly focused on the traditional ELISA, which is used to quantify the low-level food allergens in food ingredients and the preparation and processing of foods and beverages (Khedri et al. 2018). However, this method has many disadvantages, such as time-consuming and expensive. With the development of nanotechnology, more researchers have applied nanomaterials in ELISA for the detection of allergens, such as gold nanoparticles (AuNPs), grapheme oxide (GO), and even microfluidic technology (Xing et al. 2018; Arya and Estrela 2018; Tan et al. 2018; Cao et al. 2019). Alves et al. developed an electrochemical immunosensor based on AuNPs, using a screen-printed carbon electrode coated with AuNPs as the sensor to detect Ara h1. AuNPs are generated electrochemically on the surface of the electrode, and then two monoclonal antibodies are used to electrochemical detect the antibody–antigen interaction through the stripping analysis of the deposited silver by alkaline phosphatase (Alves et al. 2015). The immunosensor has been identified to be used in analyzing complex food matrices, with a detection limit of 3.8 ng mL^{-1}, RSD less than 8.7%, and recoveries above 96.6%. Based on the cyclic electrodeposition of alternate monolayers of graphene and AuNPs, Sun et al. reported a biosensor to detection Ara h1 on the surface of a GCE with a multilayer graphene–gold nanocomposite (Sun et al. 2015). Weng et al. applied microfluidic technology to the traditional enzyme-linked immunosorbent platform, combined with a designed light sensor to detect the proteins of Ara h1 and wheat gluten (Weng et al. 2016). Compared with commercial enzyme-linked immunosorbent kits, the developed microfluidic ELISA shortens the total detection time from several hours to 15–20 min, and reduces the sample consumption to 5–10 μL with higher sensitivity. At the same time, it shows that nanotechnology has a broader prospect in c-ELISA.

In addition to the Ara h1 protein, there are other allergen plant proteins that have been quantitatively and qualitatively analyzed by modified ELISA strategies. Glutathione-modified AuNPs (GSH-AuNPs) were used by Liu et al. to develop an electrochemical immunosensor for the detection of Ara h2 antibodies (Liu et al. 2010). The AuNPs were functionalized by 28 amino acid peptide fragments of the main IgE binding epitope of Ara h2 and coated on pyrolytic graphite on the surface of the electrode. Otherwise, Manfredi et al. developed a disposable amperometric biosensor based on AuNP-modified screen-printed carbon electrodes for rapid analysis of cytotoxic gliadin (Manfredi et al. 2016).

15.5.5.2 Crustaceans Allergens

As seafood is favored by more and more people, reports of such food allergies are gradually increasing. Among them, shrimp allergens have attracted much attention (Ho et al. 2014). It is reported that 0.6–2.8% of patients with allergic diseases are allergic to shrimp. People have detected at least 13 IgE-binding proteins in shrimp meat, but tropomyosin (TM) has been identified as the only major allergen with a relative molecular mass between 34,000 and 39,000. According to reports, TM is an important antigen of invertebrates such as shrimp, crab, oyster, and squid and has a highly conserved amino acid sequence.

TM is extremely stable, not only resistant to the effects of the digestive tract, but also resistant to heat, proteolysis, hydrolysis, and digestion. As a highly conserved muscle protein, TM, accounting for more than 90% of all food allergens, has cross-reactivity in many arthropods; therefore, it is also one of the most dangerous allergens (Moonesinghe et al. 2016; Wong et al. 2019). Angulo-Ibáñez et al. used carboxyl-functionalized magnetic microbeads in sandwich immunoassays, combined with disposable screen-printed electrodes, and developed an electrochemical immunoassay strategy based on H_2O_2 as the enzyme substrate and hydroquinone as the redox mediator to analyze shrimp TM (Angulo-Ibáñez et al. 2019). Specifically, the author used EDC/NHS to chemically activate carboxyl-functionalized MBs, covalently bind to the capture antibody with polyclonal rabbit anti-shrimp TM antibody, and use ethanolamine hydrochloride (ETA) to block unbound active sites to avoid specific adsorption. Next, functionalized MBs are used to specifically capture shrimp TM in all components of food extracts or standard solutions. Finally, a detection antibody and a secondary labeling antibody sandwich the bound TM to form complete sandwich immunity. Not surprisingly, MBs with sandwich immune complexes were brought to the surface of the working electrode, and then Abe was detected in the presence of H_2O_2 enzymatic substrates.

In addition, based on the less conserved sequence of TM in different phylogenetic species, this biosensor is currently being used to identify the adulteration of shellfish products using TM as a biomarker (Wang et al. 2019a, b, c). Jiang et al. embedded fluorescein isothiocyanate (FITC) on the SiO_2 modified Fe_3O_4 core, and then wrapped it in liposomes to form cationic magnetic fluorescent nanoparticles, which were used for electrochemical immunoassay for the detection of shrimp allergen myosin and fish allergen parvalbumin (Jiang et al. 2015). Interestingly, Wang et al. reported a novel immunomagnetic bead-derived ELISA method, which uses antibody-functionalized GO and AuNPs to amplify the detection signal of parvalbumin (Wang et al. 2020a, b).

15.5.5.3 Other Allergens

In addition, there are other types of food allergens that are also easy to hide in food or production lines, such as α-lactoglobulin, ovalbumin, ovomucoid, and casein. These allergens are also not to be ignored. They are easy to be contaminated on the production and processing lines but are most easily ignored. Especially for allergens in eggs, the positive rate is as high as 35% in children with food allergies, and as high as 12% in adults.

Maier et al. reported a sandwich-type immunoassay that can analyze ovalbumin and ovomucin at the same time, using AuNPs as an optical immune chip for signal sensor (Maier et al. 2008). Yang et al. used quantum dots (QDs) to covalently bind anti-α-lactic acid monoclonal antibodies to establish an immunoassay method for the detection of α-lactoglobulin in commercially available dairy products, which they called fluorescence enzyme-linked immunosorbent assay (FELISA). Compared with traditional competitive ELISA, best detection limit (0.1 ng mL^{-1}) and wide dynamic range (0.1–1000 ng mL^{-1}) of FELISA are better (Yang et al. 2014). He et al. used H_2O_2-mediated cadmium telluride sulfide QDs as an immunosensor for fluorescent signal output to detect β-lactoglobulin (He et al. 2018).

According to the foregoing, it can be seen that the application of nanomaterials such as AuNPs, QDs, and GO, to traditional ELISA has significantly improved the many shortcomings and defects of ELISA, and the accuracy and sensitivity have been greatly improved (Chen et al. 2016; Huang et al. 2016). Therefore, it is of great significance to accelerate the development of the combination of nanotechnology and traditional technology.

15.5.6 Food Additives

Food additives are micro-preparations added to food that can improve the color, flavor, and prevent food spoilage (Martins et al. 2019). It is mainly divided into natural extracts and artificial compounds and is an important part of the food industry. However, most of the food additives currently used is artificial compounds, and excessive use will cause varying degrees of harm to the human body (Carocho et al. 2017; Siegrist and Sütterlin 2017). Since the development of additive residue detection technology lags far behind the development of the food additive industry, illegal and excessive use of additives in the food production process is still very serious.

As necessities of the modern food industry, food additives have as many as 2000 varieties, including antioxidants, colorants, sweeteners, preservatives, and bleaching agents (Young et al. 1987). However, improper use of food additives may cause serious damage to human organs, such as diabetes and heart disease (Corkey 2012; Rangan and Barceloux 2009). Therefore, it is particularly important to construct effective strategies to detect illegal additives in food.

At present, the main analytical methods for food additives are chromatography or other liquid chromatography related to various detectors. Generally, chromatographic methods are a little time-consuming, complex pre-processing, and high cost, so it is particularly important to develop other technologies for additive detection (Zhang et al. 2008). For example, Han et al. reported a combined molecular imprinting technology with high-performance liquid chromatography for the selective detection of Sudan dyes in food (Han et al. 2007). However, the ELISA that has received much attention is rarely used in the detection of food additives. Compared with traditional detection methods, the most significant advantages of enzyme-linked immunosorbent assay are its sensitivity and specificity, simple preparation, and high throughput. Moreover, ELISA has been widely used in biology, agriculture, environment, and other fields, such as detecting proteins, microorganisms, antibiotics, and pesticides (Wu et al. 2019a, b).

In recent years, ELISA has attracted attention in the field of food additives due to the high specificity of the antigen–antibody interaction, making the sample analysis process simple and detecting one or more analytes at the same time. Berlina et al.

used Sudan dyes monoclonal antibody combined with AuNPs to establish a lateral flow immunoassay method to detect the food colorant Sudan (Berlina et al. 2017). The conditions of this experiment are optimized for the qualitative and quantitative control of Sudan in food matrix, and its detection limit is 2.5 ng mL^{-1}. In addition, indirect competitive chemiluminescence enzyme immunoassay (CLEIA) has also been applied to colorant detection. Based on polyclonal antibody and horseradish peroxidase-labeled antibody chemiluminescence system, a CLEIA analysis strategy for malachite green in seafood was established (Zhang et al. 2015). Secondly, the dye FB (Chr FB) can be detected by the horseradish peroxidase-luminol-H_2O_2 system, with p-iodophenol as an enhancer, combined with polyclonal antibodies to establish a chemiluminescence immunoassay method. This method can be used for rapid screening of Chr FB in yogurt candy, vitamin drinks, and bread. Compared with the traditional method, its sensitivity is improved by two orders of magnitude (Xu et al. 2017a, b). It can be seen that the analysis of food additives by ELISA method is very promising. The introduction of nanomaterials can improve the sensitivity of ELISA for trace detection in complex matrices; therefore, nano-ELISA needs to be developed in the field of food additive analysis.

15.6 Advantages and Disadvantages of c-ELISA and Nano-ELISA

Though c-ELISA in immunoassay behaves various disadvantages, it still suffers from many limitations. To improve c-ELISA, a lot of work been done to enhance the detection robustness and make it more convenient in operations. By introducing nanomaterials, c-ELISA has been greatly improved in sensitivity and flexibility, making nano-ELISA a powerful tool in many fields. As mentioned above, tons of functional nanomaterials are developed for ELISA to illustrate their potential application in food safety. The advantages of nano-ELISA are obvious, including higher sensitivity, faster response, richer detection strategies. However, many new problems appear when come to construct all kinds of nano-ELISA.

For instance, for the development of solid substrate, nano-fibers, MIPs, and magnetic nano-beads are commonly used in nano-ELISA, which can provide high specific surface area and easier operations. But their stability and reproducibility may become new problems for nano-ELISA. For the antibody recognition improvement, nanobody and aptamer and are the potential substitutes in ELISA, but the binding force with analytes needs to be further evaluated. For the enzyme labels, nanozymes (CuS nano-sheets, CuO, MnO_2, and CeO_2 nanoparticles) with peroxidase- or oxidase-like activity are widely used to replace natural enzymes, which is a good beginning for the development of enzymes, but how to take count of the advantages of natural enzymes (e.g. selectivity, biocompatibility) is another problem. From this point of view, it seems impossible to put all good things in one method. The detailed advantages and disadvantages of c-ELISA and nano-ELISA are listed in Table 15.2.

Table 15.2 Comparison between c-ELISA and nano-ELISA

c-ELISA		Nano-ELISA

Advantages	High selectivity of enzyme	Controllable catalytic activity of nanozyme
	Good substrate selectivity	Multienzyme mimetic activity
	Excellent biocompatibility	Low cost
	Clear standards	Easy to mass produce
	Mild reaction conditions	Long-term storage
	Stable structure	Robust to harsh environments
	Simple recognition model	Unique physicochemical properties
Disadvantages	High cost	Limited types of nanozymes
	Hard to be purified	Limited substrate selectivity
	Hard to store long term	Limited biocompatibility
	Limited application conditions	Limited stability
	Easy denaturation	Lack of standards

15.7 Perspectives and Challenges

The introduction of nanomaterial into ELISA (nano-ELISA) brings convenience to paper- and fiber-based ELISAs and the miniaturization of ELISA, which also provides a new avenue for other biosensors in electrochemistry, optics, and magnetism. Here, we mainly focused on the development and application of nano-ELISA. As a powerful visual method, ELISA has attracted intensive attention in many fields, and the improvement towards the c-ELISA is becoming more and more popular. Clearly, the performance of c-ELISA has been greatly improved, and the technique is intensively used in food safety. However, concerns still exist the developed ELISA, such as lack of stability and standards, time-consuming, and tedious operations. Hence, it is an alternative to construct facile ELISA by combining new nanomaterials and other techniques.

Up to now, nano-ELISA has been given sufficient attention in food analysis. A large number of studies have been reported on the development of c-ELISA. But most of the ELISA are enhanced in only a part, either the enzyme label or the recognition model. There is no doubt that some properties like sensitivity and simplicity can be greatly enhanced, but in most cases new problems appears. For example, the stability and specificity remain to be explored after applying nanozymes modified detection antibody in ELISA measurements. So, how to evaluate the whole performance of nano-ELISA is a critical issue. It is believed that in the near future the new developed nano-ELISA could have the golden standards like c-ELISA does.

On the other hand, nano-ELISA combined techniques such as electrochemical method, surface-enhanced Raman scattering technique, electrochemiluminescence, and smartphone are new arising analytical methods, which can realize rapid, sensitive, or even intelligent detection of analytes. Therefore, the combination of nano-ELISA with these techniques is a new trend in future applications. However, the nanomaterials also have problems in real applications due to their relatively low stability and bioconjugate efficiency. Thus, engineering stable and high-performance nanomaterials is still the challenge remained to be addressed. Overall, it remains a great challenge in developing stable and reliable ELISA methods in food safety, and the combination of nanomaterials and methodology can provide a good direction.

References

Al Ghounaim M, Longtin Y, Gonzales M, Merckx J, Winters N, Quach C (2016) Clostridium difficile infections in children: impact of the diagnostic method on infection rates. Infect Control Hospital Epidemiol 37(9):1087–1093

Al Hamshary AS, Bayoumi IR, Aly NS, Omar RE, Mohammed DA, Marei YM, Rashed GA (2020) Evaluation of Nano-based-ELISA for Serodiagnosis of human toxoplasmosis. Benha J Appl Sci 5(2 part (1)):1–7

Alves RC, Pimentel FB, Nouws HP, Marques RC, González-García MB, Delerue-Matos C (2015) Detection of Ara h 1 (a major peanut allergen) in food using an electrochemical gold nanoparticle-coated screen-printed immunosensor. Biosens Bioelectron 64:19–24

Alves RC, Barroso MF, González-García MB, Oliveira MBPP, Delerue-Matos C (2016) New trends in food allergens detection: toward biosensing strategies. Crit Rev Food Sci 56:2304–2319

Andjelkovic U, Gavrovic-Jankulovic M, Martinovic T, Josic D (2017) Omic methods as a tool for investigation of food allergies. TrAC Trends Anal Chem 96:107–115

Angulo-Ibáñez A, Eletxigerra U, Lasheras X, Campuzano S, Merino S (2019) Electrochemical tropomyosin allergen immunosensor for complex food matrix analysis. Anal Chim Acta 1079: 94–102

Arya SK, Estrela P (2018) Recent advances in enhancement strategies for electrochemical ELISA-based immunoassays for cancer biomarker detection. Sensors 18(7):2010

Avery SV, Singleton I, Magan N, Goldman GH (2019) The fungal threat to global food security. Fungal Biol 123(8):555–557

Aydin S (2015) A short history, principles, and types of ELISA, and our laboratory experience with peptide/protein analyses using ELISA. Peptides 72:4–15

Barnett D, Baldo BA, Howden ME (1983) Multiplicity of allergens in peanuts. J Allergy Clin Immunol 72(1):61–68

Baynes RE, Dedonder K, Kissell L, Mzyk D, Marmulak T, Smith G et al (2016) Health concerns and management of select veterinary drug residues. Food Chem Toxicol 88:112–122

Ben Y, Fu C, Hu M, Liu L, Wong MH, Zheng C (2019) Human health risk assessment of antibiotic resistance associated with antibiotic residues in the environment: a review. Environ Res 169: 483–493

Berlina AN, Zherdev AV, Xu C, Eremin SA, Dzantiev BB (2017) Development of lateral flow immunoassay for rapid control and quantification of the presence of the colorant Sudan I in spices and seafood. Food Control 73:247–253

Bhairab M, Shylaja R, Lavu PS, Bhavanashri N, Joseph KJF, i. M. (2018) Highly sensitive colorimetric biosensor for staphylococcal enterotoxin B by a label-free aptamer and gold nanoparticles. Front Microbiol 9:179

Bhandari G, Zomer P, Atreya K, Mol HGJ, Yang X, Geissen V (2019) Pesticide residues in Nepalese vegetables and potential health risks. Environ Res 172:511–521

Bouças RI, Trindade ES, Tersariol IL, Dietrich CP, Nader HB (2008) Development of an enzyme-linked immunosorbent assay (ELISA)-like fluorescence assay to investigate the interactions of glycosaminoglycans to cells. Anal Chim Acta 618(2):218–226
Breiteneder H, Radauer C (2004) A classification of plant food allergens. J Allergy Clin Immunol 113(5):821–830
Bucheli TD (2014) Phytotoxins: environmental micropollutants of concern? Environ Sci Technol 48(22):13,027–13,033
Bull FC, Al-Ansari SS, Biddle S, Borodulin K, Buman MP, Cardon G et al (2020) World Health Organization 2020 guidelines on physical activity and sedentary behaviour. Br J Sports Med 54(24):1451–1462
Burgess JA, Dharmage SC, Allen K, Koplin J, Garcia-Larsen V, Boyle R, Lodge CJ (2019) Age at introduction to complementary solid food and food allergy and sensitization: a systematic review and meta-analysis. Clin Exp Allergy 49(6):754–769
Buss H, Chan TP, Sluis KB, Domigan NM, Winterbourn CC (1997) Protein carbonyl measurement by a sensitive ELISA method. Free Radic Biol Med 23(3):361–366
Butler JE (2000) Solid supports in enzyme-linked immunosorbent assay and other solid-phase immunoassays. Methods 22(1):4–23
Byer JD, Struger J, Klawunn P, Todd A, Sverko ED (2008) Low-cost monitoring of glyphosate in surface waters using the ELISA method: an evaluation. Environ Sci Technol 42(16):6052–6057
Cao Y, Feng T, Xu J, Xue C (2019) Recent advances of molecularly imprinted polymer-based sensors in the detection of food safety hazard factors. Biosens Bioelectron 141:111447
Carocho M, Morales P, Ferreira IC (2017) Sweeteners as food additives in the XXI century: a review of what is known, and what is to come. Food Chem Toxicol 107:302–317
Chan ES, Abrams EM, Hildebrand KJ, Watson W (2018) Early introduction of foods to prevent food allergy. Allergy Asthma Clin Immunol 14(S2):57
Chandra P, Das D, Abdelwahab AA (2010) Gold nanoparticles in molecular diagnostics and therapeutics. Dig J Nanomater Biostruct 5(5):363–367
Chandra P, Singh J, Singh A, Srivastava A, Goyal RN, Shim YB (2013) Gold nanoparticles and nanocomposites in clinical diagnostics using electrochemical methods. J Nanoparticles 2013:1–12. https://doi.org/10.1155/2013/535901
Chang CP, Hou PH, Yang WC, Wu CF, Chang CC, Tsai MY et al (2020) Analytical detection of sulfonamides and organophosphorus insecticide residues in fish in Taiwan. Molecules 25(7)
Chassaigne H, Nørgaard JV, van Hengel AJ (2007) Proteomics-based approach to detect and identify major allergens in processed peanuts by capillary LC-Q-TOF (MS/MS). J Agric Food Chem 55(11):4461–4473
Chen G et al (2011) Single molecule interaction and conformation study based on atomic force microscopy. Doctoral dissertation, University of Georgia.
Chen R, Huang X, Li J, Shan S, Lai W, Xiong Y (2016) A novel fluorescence immunoassay for the sensitive detection of Escherichia coli O157:H7 in milk based on catalase-mediated fluorescence quenching of CdTe quantum dots. Anal Chim Acta 947:50–57
Corkey BE (2012) Diabetes: have we got it all wrong?: insulin hypersecretion and food additives: cause of obesity and diabetes? Diabetes Care 35(12):2432–2437
De Clercq E (2001) Antiviral drugs: current state of the art. J Clin Virol 22(1):73–89
De La Rica R, Stevens MM (2012) Plasmonic ELISA for the ultrasensitive detection of disease biomarkers with the naked eye. Nat Nanotechnol 7(12):821–824
Ekezie FGC, Cheng JH, Sun DW (2018) Effects of nonthermal food processing technologies on food allergens: a review of recent research advances. Trends Food Sci Technol 74:12–25
English BK, Gaur AH (2010) The use and abuse of antibiotics and the development of antibiotic resistance. Adv Exp Med Biol 659:73–82
Engvall E, Perlmann P (1971) Enzyme-linked immunosorbent assay (ELISA) quantitative assay of immunoglobulin G. Immunochemistry 8(9):871–874

Fadlalla MH, Ling S, Wang R, Li X, Yuan J, Xiao S et al (2020) Development of ELISA and lateral flow immunoassays for ochratoxins (OTA and OTB) detection based on monoclonal antibody. Front Cell Infect Microbiol 10:80

Farka Z, Cunderlova V, Horackova V, Pastucha M, Mikusova Z, Hlavacek A, Skladal P (2018) Prussian blue nanoparticles as a catalytic label in a Sandwich Nanozyme-linked immunosorbent assay. Anal Chem 90(3):2348–2354

Fletcher MT, Netzel GJT (2020) Food safety and natural toxins. Toxins (Basel) 12(4):236

Gaastra W (1984) Enzyme-linked immunosorbant assay (ELISA). In: Proteins. Humana Press, pp 349–355

Gao Y, Zhou Y, Chandrawati R (2019a) Metal and metal oxide nanoparticles to enhance the performance of enzyme-linked immunosorbent assay (ELISA). ACS Appl Nano Materials 3(1): 1–21

Gao B, Chen X, Huang X, Pei K, Xiong Y, Wu Y et al (2019b) Urease-induced metallization of gold nanorods for the sensitive detection of salmonella enterica Choleraesuis through colorimetric ELISA. J Dairy Sci 102(3):1997–2007

Grasso I, Archer SD, Burnell C, Tupper B, Rauschenberg C, Kanwit K, Record NR (2019) The hunt for red tides: Deep learning algorithm forecasts shellfish toxicity at site scales in coastal Maine. Ecosphere 10(12):e02960

Guan N, Li Y, Yang H, Hu P, Lu S, Ren H et al (2021) Dual-functionalized gold nanoparticles probe based bio-barcode immuno-PCR for the detection of glyphosate. Food Chem 338:128133

Gunthardt BF, Hollender J, Hungerbuhler K et al (2018) Comprehensive toxic plants-phytotoxins database and its application in assessing aquatic micropollution potential. J Agric Food Chem 66:7577–7588

Guo Q, Han JJ, Shan S, Liu DF, Wu SS, Xiong YH, Lai WH (2016a) DNA-based hybridization chain reaction and biotin-streptavidin signal amplification for sensitive detection of Escherichia coli O157:H7 through ELISA. Biosens Bioelectron 86:990–995

Guo L, Xu S, Ma X, Qiu B, Lin Z, Chen G (2016b) Dual-color plasmonic enzyme-linked immunosorbent assay based on enzyme-mediated etching of Au nanoparticles. Sci Rep 6(1):1–7

Han D, Yu M, Knopp D, Niessner R, Wu M, Deng A (2007) Development of a highly sensitive and specific enzyme-linked immunosorbent assay for detection of Sudan I in food samples. J Agri Food Chem 55(16):6424–6430

Han S, Zhou T, Yin B, He P (2018) Gold nanoparticle-based colorimetric ELISA for quantification of ractopamine. Microchim Acta 185(4)

He J, Wang Y, Zhang X (2016) Preparation of artificial antigen and development of IgY-based indirect competitive ELISA for the detection of kanamycin residues. Food Anal Methods 9(3): 744–751

He S, Li X, Gao J, Tonga P, Chen H (2018) Development of a H_2O_2-sensitive quantum dots-based fluorescent sandwich ELISA for sensitive detection of bovine β-Lactoglobulin by monoclonal antibody. J Sci Food Agric 98:519–526

Heaney JL, Campbell JP, Goodall M, Plant T, Shemar M, Hand C, Drayson MT (2020) Analytical validation of new ELISAs for the quantitation of polyclonal free light chains and comparison to existing assays for healthy and patient samples. J Immunol Methods 478:112713

Hebling CM, McFarland MA, Callahan JH, Ross MM (2013) Global proteomic screening of protein allergens and advanced glycation endproducts in thermally processed peanuts. J Agric Food Chem 61(24):5638–5648

Her J, Jo H, Ban C (2017) Enzyme-linked antibody aptamer assays based colorimetric detection of soluble fraction of activated leukocyte cell adhesion molecule. Sensors Actuators B Chem 242: 529–534

Hernández-Cortez C, Palma-Martínez I, Gonzalez-Avila LU, Guerrero-Mandujano A, Castro-Escarpulli G (2017) Food poisoning caused by bacteria (food toxins). IntechOpen

Herzig GP, Aydin M, Dunigan S, Shah P, Jeong KC, Park SH et al (2016) Magnetic bead-based immunoassay coupled with tyramide signal amplification for detection of S almonella in foods. J Food Saf 36(3):383–391

Ho MHK, Wong WHS, Chang C (2014) Clinical spectrum of food allergies: a comprehensive review. Clin Rev Allerg Immu 46:225–240

Hoga CA, Almeida FL, Reyes FGR (2018) A review on the use of hormones in fish farming: analytical methods to determine their residues. CyTA J Food 16(1):679–691

Hosseini S, Vázquez-Villegas P, Rito-Palomares M, Martinez-Chapa SO (2018) Advantages, disadvantages and modifications of conventional ELISA. In: Enzyme-linked immunosorbent assay (ELISA). Springer, Singapore, pp 67–115

Hua X, Goedert JJ, Pu A, Yu G, Shi J (2016) Allergy associations with the adult fecal microbiota: analysis of the American gut project. EBioMedicine 3:172–179

Huang X, Zhan S, Xu H, Meng X, Xiong Y, Chen X (2016) Ultrasensitive fluorescence immunoassay for detection of ochratoxin a using catalase-mediated fluorescence quenching of CdTe QDs. Nanoscale 8:9390–9397

Jaria G, Calisto V, Otero M, Esteves VI (2020) Monitoring pharmaceuticals in the aquatic environment using enzyme-linked immunosorbent assay (ELISA)—a practical overview. Anal Bioanal Chem 412(17):3983–4008

Jia CP, Zhong XQ, Hua B, Liu MY, Jing FX, Lou XH et al (2009) Nano-ELISA for highly sensitive protein detection. Biosens Bioelectron 24(9):2836–2841

Jiang D, Zhu P, Jiang H, Ji J, Sun X, Gu W, Zhang G (2015) Fluorescent magnetic bead-based mast cell biosensor for electrochemical detection of allergens in foodstuffs. Biosens Bioelectron 70: 482–490

Johnson AM, Roberts H, Tenter AM (1992) Evaluation of a recombinant antigen ELISA for the diagnosis of acute toxoplasmosis and comparison with traditional antigen ELISAs. J Med Microbiol 37(6):404–409

Karczmarczyk A, Haupt K, Feller KHJT (2017) Development of a QCM-D biosensor for Ochratoxin A detection in red wine. Talanta 166:193–197

Khedri M, Ramezani M, Rafatpanah H, Abnous K (2018) Detection of food-born allergens with aptamer-based biosensors. TrAC Trends Anal Chem 103:126–136

Khoris IM, Takemura K, Lee J, Hara T, Abe F, Suzuki T, Park EY (2019) Enhanced colorimetric detection of norovirus using in-situ growth of Ag shell on Au NPs. Biosens Bioelectron 126: 425–432

Kumar A, Purohit B, Mahato K, Mahapatra S, Srivastava A, Chandra P (2020) Bio-nano-interface engineering strategies of AuNPs passivation for next-generation biomedical applications. In: Chandra P, Pandey LM (eds) Biointerface engineering: prospects in medical diagnostics and drug delivery. Springer, Singapore. https://doi.org/10.1007/978-981-15-4790-4_10

Kwon EY, Ruan X, Wang L, Lin Y, Du D, Van Wie BJ (2020) Mesoporous Pd@Pt nanoparticle-linked immunosorbent assay for detection of atrazine. Anal Chim Acta 1116:36–44

Kwong LS, Hope JC, Thom ML, Sopp P, Duggan S, Bembridge GP, Howard CJ (2002) Development of an ELISA for bovine IL-10. Vet Immunol Immunopathol 85(3–4):213–223

Lee KH, Zeng H (2017) Aptamer-based ELISA assay for highly specific and sensitive detection of Zika NS1 protein. Anal Chem 89(23):12,743–12,748

Li L, Peng AH, Lin ZZ, Zhong HP, Chen XM, Huang ZY (2017) Biomimetic ELISA detection of malachite green based on molecularly imprinted polymer film. Food Chem 229:403–408

Li F, Li F, Aguilar ZP, Xiong Y, Xu H (2018a) Polyamidoamine (PAMAM) dendrimer-mediated biotin amplified immunomagnetic separation method coupled with flow cytometry for viable Listeria monocytogenes detection. Sensors Actuators B Chem 257:286–294

Li Z, Sheng W, Liu Q, Li S, Shi Y, Zhang Y, Wang S (2018b) Development of a gold nanoparticle enhanced enzyme linked immunosorbent assay based on monoclonal antibodies for the detection of fumonisin B1, B2, and B3 in maize. Anal Methods 10(28):3506–3513

Liu C, Sathe SK (2018) Food allergen epitope mapping. J Agric Food Chem 66(28):7238–7248

Liu H, Malhotra R, Peczuh MW, Rusling JF (2010) Electrochemical immunosensors for antibodies to Peanut allergen Ara h2 using gold nanoparticle-peptide films. Anal Chem 82:5865–5871

Liu W, Gan C, Chang W, Qileng A, Lei H, Liu Y (2019) Double-integrated mimic enzymes for the visual screening of microcystin-LR: copper hydroxide nanozyme and G-quadruplex/hemin DNAzyme. Anal Chim Acta 1054:128–136

Lu T, Zhan S, Zhou Y, Chen X, Huang X, Leng Y et al (2018) Fluorescence ELISA based on CAT-regulated fluorescence quenching of CdTe QDs for sensitive detection of FB1. Anal Methods 10(48):5797–5802

Ma H, Shieh KJ (2006) ELISA technique. Nat Sci 4(2):36–37

Mahato K, Nagpal S, Shah MA, Srivastava A, Maurya PK, Roy S, Jaiswal A, Singh R, Chandra P (2019) Gold nanoparticle surface engineering strategies and their applications in biomedicine and diagnostics. 3 Biotech. https://doi.org/10.1007/s13205-019-1577-z

Maier I, Morgan MRA, Lindner W, Pittner F (2008) Optical resonance enhanced absorption-based near-field immunochip biosensor for allergen detection. Anal Chem 80:2694–2703

Mak AC, Osterfeld SJ, Yu H, Wang SX, Davis RW, Jejelowo OA et al (2010) Sensitive giant magnetoresistive-based immunoassay for multiplex mycotoxin detection. Biosens Bioelectron 25(7):1635–1639

Manfredi A, Giannetto M, Mattarozzi M, Costantini M, Mucchino C, Careri M (2016) Competitive immunosensor based on gliadin immobilization on disposable carbon-nanogold screen-printed electrodes for rapid determination of celiotoxic prolamins. Anal Bional Chem 408:7289–7298

Martins FC, Sentanin MA, De Souza D (2019) Analytical methods in food additives determination: compounds with functional applications. Food Chem 272:732–750

Mauroy A, Scipioni A, Mathijs E, Saegerman C, Mast J, Bridger JC et al (2009) Epidemiological study of bovine norovirus infection by RT-PCR and a VLP-based antibody ELISA. Vet Microbiol 137(3–4):243–251

Mol HGJ, Dam RCJV, Zomer P, Mulder PPJJFA, Contaminants. (2011) Screening of plant toxins in food, feed and botanicals using full-scan high-resolution (Orbitrap) mass spectrometry. Food Addit Contam Part A Chem Anal Control Expo Risk Assess 28(10):1405–1423

Monaci L, De Angelis E, Montemurro N, Pilolli R (2018) Comprehensive overview and recent advances in proteomics MS based methods for food allergens analysis. TrAC Trends Anal Chem 106:21–36

Moonesinghe H, Mackenzie H, Venter C, Kilburn S, Turner P, Weir K, Dean T (2016) Prevalence of fish and shellfish allergy: a systematic review. Ann Allergy Asthma Immunol 117:264–272

Moretti A, Pascale M, Logrieco AF (2018) Mycotoxin risks under a climate change scenario in Europe. Trends Food Sci Technol 84:S0924224417304090

Nakane PK, Pierce GB Jr (1967) Enzyme-labeled antibodies for the light and electron microscopic localization of tissue antigens. J Cell Biol 33(2):307–318

Nsibande SA, Forbes PB (2016) Fluorescence detection of pesticides using quantum dot materials—a review. Anal Chim Acta 945:9–22

Orlov AV, Khodakova JA, Nikitin MP, Shepelyakovskaya AO, Brovko FA, Laman AG et al (2013) Magnetic immunoassay for detection of staphylococcal toxins in complex media. Anal Chem 85(2):1154–1163

Pan P, Wang Y, Zhu Y, Gao X, Ju Z, Qiu P et al (2015) Nontoxic virus nanofibers improve the detection sensitivity for the anti-p53 antibody, a biomarker in cancer patients. Nano Res 8(11): 3562–3570

Pang Y, Guo L, Shen X, Yang N, Yang CJEA (2020) Rolling circle amplified DNAzyme followed with covalent organic frameworks: Cascade signal amplification of electrochemical ELISA for alfatoxin M1 sensing. Electrochimica Acta 341:136055

Pele M, Brohée M, Anklam E, Hengel AJV (2007) Peanut and hazelnut traces in cookies and chocolates: relationship between analytical results and declaration of food allergens on product labels. Food Addit Contam 24(12):1334–1344

Peng C-F, Duan X-H, Pan Q-L, Liu L-Q, Xue F (2013a) Ultrasensitive nano-ELISA for detecting sulfadimethoxine in chicken tissue. J Chem 2013:1–5

Peng C-F, Liu C-L, Song S-S, Liu L-Q (2013b) Highly sensitive nano-ELISA for detecting 19-nortestosterone in beef. Food Agric Immunol 25(3):423–431

Peng Z, Ling M, Ning Y, Deng L (2014) Rapid fluorescent detection of Escherichia coli K88 based on DNA aptamer library as direct and specific reporter combined with immuno-magnetic separation. J Fluoresc 24(4):1159–1168

Phillips RW, Abbott D (2008) High-throughput enzyme-linked immunoabsorbant assay (ELISA) electrochemiluminescent detection of botulinum toxins in foods for food safety and defence purposes. Food Addit Contam 25(9):1084–1088

Phlmann C, Bellanger L, Drevinek M, Elner TJPT (2017) Multiplex detection of biothreat agents using an automated electrochemical ELISA platform. Toxins (Basel) 27:104–105

Qi X, Wang Z, Lu R, Liu J, Li Y, Chen Y (2021) One-step and DNA amplification-free detection of Listeria monocytogenes in ham samples: combining magnetic relaxation switching and DNA hybridization reaction. Food Chem 338:127837

Rangan C, Barceloux DG (2009) Food additives and sensitivities. Dis Mon 55(5):292–311

Reverte L, Campas M, Yakes BJ, Deeds JR, Katikou P, Kawatsu K et al (2017) Tetrodotoxin detection in puffer fish by a sensitive planar waveguide immunosensor b253(dec):967–976

Salomone A, Mongelli M, Roggero P, Boscia D (2004) Reliability of detection of citrus tristeza virus by an immunochromatographic lateral flow assay in comparison with ELISA. J Plant Pathol:43–48

Samsidar A, Siddiquee S, Shaarani SM (2018) A review of extraction, analytical and advanced methods for determination of pesticides in environment and foodstuffs. Trends Food Sci Technol 71:188–201

Sathe SK, Teuber SS, Roux KH (2005) Effects of food processing on the stability of food allergens. Biotechnol Adv 23(6):423–429

Satija J, Punjabi N, Mishra D, Mukherji S (2016) Plasmonic-ELISA: expanding horizons. RSC Adv 6(88):85,440–85,456

Scallan E, Hoekstra RM, Angulo FJ, Tauxe RV, Widdowson MA, Roy SL et al (2011) Foodborne illness acquired in the United States—major pathogens. Emerg Infect Dis 17(1):7–15

Shen X, Liu L, Xu L, Ma W, Wu X, Cui G, Kuang H (2019) Rapid detection of praziquantel using monoclonal antibody-based ic-ELISA and immunochromatographic strips. Food Agric Immunol 30(1):913–923

Sheng YM, Liang J, Xie J (2020) Indirect competitive determination of tetracycline residue in honey using an ultrasensitive gold-nanoparticle-linked aptamer assay. Molecules 25(9)

Sicherer SH (2001) Clinical implications of cross-reactive food allergens. J Allergy Clin Immunol 108(6):881–890

Sicherer SH, Sampson HA (2014) Food allergy: epidemiology, pathogenesis, diagnosis and treatment. J Allergy Clin Immunol 133:291–307

Siegrist M, Sütterlin B (2017) Importance of perceived naturalness for acceptance of food additives and cultured meat. Appetite 113:320–326

Silva V, Mol HGJ, Zomer P, Tienstra M, Ritsema CJ, Geissen V (2019) Pesticide residues in European agricultural soils—a hidden reality unfolded. Sci Total Environ 653:1532–1545

Song M, Xiao Z, Xue Y, Zhang X, Ding S, Li J (2018) Development of an indirect competitive ELISA based on immunomagnetic beads' clean-up for detection of maduramicin in three chicken tissues. Food Agric Immunol 29(1):590–599

Sun X, Jia M, Guan L, Ji J, Zhang Y, Tang L, Li Z (2015) Multilayer graphene–gold nanocomposite modified stem-loop DNA biosensor for peanut allergen-Ara h1 detection. Food Chem 172:335–342

Tan X, David A, Day J, Tang H, Dixon ER, Zhu H, Fan X (2018) Rapid mouse follicle stimulating hormone quantification and estrus cycle analysis using an automated microfluidic chemiluminescent ELISA system. ACS Sensors 3(11):2327–2334

Tao X, Jiang H, Yu X, Zhu J, Wang X, Wang Z et al (2013) An ultrasensitive chemiluminescence immunoassay of chloramphenicol based on gold nanoparticles and magnetic beads. Drug Test Anal 5(5):346–352

Tian F, Zhou J, Jiao B, He Y (2019) A nanozyme-based cascade colorimetric aptasensor for amplified detection of ochratoxin A. Nanoscale 11(19):9547–9555

Toh SY, Citartan M, Gopinath SC, Tang TH (2015) Aptamers as a replacement for antibodies in enzyme-linked immunosorbent assay. Biosens Bioelectron 64:392–403

Török K, Hajas L, Horváth V, Schall E, Bugyi Z, Kemény S, Tömösközi S (2015) Identification of the factors affecting the analytical results of food allergen ELISA methods. Eur Food Res Technol 241(1):127–136

Turcanu V, Maleki SJ, Lack G (2003) Characterization of lymphocyte responses to peanuts in normal children, peanut-allergic children, and allergic children who acquired tolerance to peanuts. J Clin Invest 111(7):1065–1072

Urusov A, Petrakova A, Vozniak M, Zherdev A, Dzantiev BJS (2014) Rapid immunoenzyme assay of aflatoxin B1 using magnetic nanoparticles. Sensors (Basel) 14(11):21,843–21,857

Vierk KA, Koehler KM, Fein SB, Street DA (2007) Prevalence of self-reported food allergy in American adults and use of food labels. J Allergy Clin Immunol 119(6):1504–1510

Vinayaka AC, Ngo TA, Kant K, Engelsmann P, Dave VP, Shahbazi MA et al (2019) Rapid detection of salmonella enterica in food samples by a novel approach with combination of sample concentration and direct PCR. Biosens Bioelectron 129:224–230

Vogt RV Jr, Phillips DL, Henderson LO, Whitfield W, Spierto FW (1987) Quantitative differences among various proteins as blocking agents for ELISA microtiter plates. J Immunol Methods 101(1):43–50

Voller A, Bartlett A, Bidwell DE (1978) Enzyme immunoassays with special reference to ELISA techniques. J Clin Pathol 31(6):507–520

Wang T, Zhang M, Dreher DD, Zeng Y (2013) Ultrasensitive microfluidic solid-phase ELISA using an actuatable microwell-patterned PDMS chip. Lab Chip 13(21):4190–4197

Wang Q-L, Li J, Li X-D, Ding L-S, Xie J, Qing L-S (2016) A simple nano-SiO2-based ELISA method for residue detection of 2,4-dichlorophenoxyacetic acid in bean sprouts. Food Anal Methods 10(5):1500–1506

Wang Z, Beier RC, Shen J (2017) Immunoassays for the detection of macrocyclic lactones in food matrices—a review. TrAC Trends Anal Chem 92:42–61

Wang S, Zheng L, Cai G, Liu N, Liao M, Li Y et al (2019a) A microfluidic biosensor for online and sensitive detection of Salmonella typhimurium using fluorescence labeling and smartphone video processing. Biosens Bioelectron 140:111333

Wang Z, Xianyu Y, Zhang Z, Guo A, Li X, Dong Y, Chen Y (2019b) Background signal-free magnetic bioassay for food-borne pathogen and residue of veterinary drug via Mn (VII)/Mn (II) interconversion. ACS Sensors 4(10):2771–2777

Wang Y, Rao Z, Zhou J, Zheng L, Fu L (2019c) A chiral assembly of gold nanoparticle trimer-based biosensors for ultrasensitive detection of the major allergen tropomyosin in shellfish. Biosens Bioelectron 132:84–89

Wang Q, Yang Q, Wu W (2020a) Ensuring seafood safe to spoon: a brief review of biosensors for marine biotoxin monitoring. Crit Rev Food Sci Nutr:1–13

Wang Y, Qi Q, Zhou J, Li H, Fu L (2020b) Graphene oxide and gold nanoparticles-based dual amplification method for immunomagnetic beads-derived ELISA of parvalbumin. Food Control 110:106989

Waritani T, Chang J, McKinney B, Terato K (2017) An ELISA protocol to improve the accuracy and reliability of serological antibody assays. MethodsX 4:153–165

Weerathunge P, Ramanathan R, Torok VA, Hodgson K, Xu Y, Goodacre R et al (2019) Ultrasensitive colorimetric detection of murine norovirus using NanoZyme Aptasensor. Anal Chem 91(5):3270–3276

Wei T, Du D, Zhu MJ, Lin Y, Dai Z (2016) An improved ultrasensitive enzyme-linked immunosorbent assay using hydrangea-like antibody-enzyme-inorganic three-in-one nanocomposites. ACS Appl Mater Interfaces 8(10):6329–6335

Wei D, Zhang X, Chen B, Zeng K (2020) Using bimetallic au@Pt nanozymes as a visual tag and as an enzyme mimic in enhanced sensitive lateral-flow immunoassays: application for the detection of streptomycin. Anal Chim Acta 1126:106–113

Weng X, Gaur G, Neethirajan S (2016) Rapid detection of food allergens by microfluidics ELISA-based optical sensor. Biosensors 6(2):24

Wong L, Tham EH, Lee BW (2019) An update on shellfish allergy. Curr Opin Allergy Clin Immunol 19:236–242

Wu W, Li J, Pan D, Li J, Song S, Rong M et al (2014) Gold nanoparticle-based enzyme-linked antibody-aptamer sandwich assay for detection of Salmonella typhimurium. ACS Appl Mater Interfaces 6(19):16,974–16,981

Wu L, Li G, Xu X, Zhu L, Huang R, Chen X (2019a) Application of nano-ELISA in food analysis: recent advances and challenges. TrAC Trends Anal Chem 113:140–156

Wu Y, Xiong Y, Chen X, Luo D, Gao B, Chen J et al (2019b) Plasmonic ELISA based on DNA-directed gold nanoparticle growth for Cronobacter detection in powdered infant formula samples. J Dairy Sci 102(12):10,877–10,886

Wu L, Zhou M, Wang Y, Liu J (2020) Nanozyme and aptamer-based immunosorbent assay for aflatoxin B1. J Hazard Mater 399:123154

Wu L, Zhou M, Liu C, Chen X, Chen Y (2021) Double-enzymes-mediated Fe(2+)/Fe(3+) conversion as magnetic relaxation switch for pesticide residues sensing. J Hazard Mater 403:123619

Xiao Y, Isaacs SN (2012) Enzyme-linked immunosorbent assay (ELISA) and blocking with bovine serum albumin (BSA)—not all BSAs are alike. J Immunol Methods 384(1–2):148–151

Xing B, Zhu W, Zheng X, Zhu Y, Wei Q, Wu D (2018) Electrochemiluminescence immunosensor based on quenching effect of SiO2@ PDA on SnO2/rGO/Au NPs-luminol for insulin detection. Sensors Actuators B Chem 265:403–411

Xiong Y, Leng Y, Li X, Huang X, Xiong Y (2020) Emerging strategies to enhance the sensitivity of competitive ELISA for detection of chemical contaminants in food samples. TrAC Trends Anal Chem 126:115861

Xu ML, Gao Y, Han XX, Zhao B (2017a) Detection of pesticide residues in food using surface-enhanced Raman spectroscopy: a review. J Agric Food Chem 65(32):6719–6726

Xu K, Long H, Xing R, Yin Y, Eremin SA, Meng M, Xi R (2017b) A sensitive chemiluminescent immunoassay to detect Chromotrope FB (Chr FB) in foods. Talanta 164:341–347

Xu Z, Long LL, Chen YQ, Chen ML, Cheng YHJFC (2020) A nanozyme-linked immunosorbent assay based on metal-organic frameworks (MOFs) for sensitive detection of aflatoxin B1. Food Chem 338:128039

Xu Z, Long LL, Chen YQ, Chen ML, Cheng YH (2021) A nanozyme-linked immunosorbent assay based on metal–organic frameworks (MOFs) for sensitive detection of aflatoxin B1. Food Chem 338:128039

Yalow RS, Berson SA (1960) Immunoassay of endogenous plasma insulin in man. J Clin Invest 39(7):1157–1175

Yan M, Chen G, She Y, Ma J, Hong S, Shao Y et al (2019) Sensitive and simple competitive biomimetic nanozyme-linked immunosorbent assay for colorimetric and surface-enhanced Raman scattering sensing of triazophos. J Agric Food Chem 67(34):9658–9666

Yang A, Zheng Y, Long C, Chen H, Liu B, Li X, Yuan J, Cheng F (2014) Fluorescent immunosorbent assay for the detection of alpha-lactalbumin in dairy products with monoclonal antibody bioconjugated with CdSe/ZnS quantum dots. Food Chem 150:73–79

Yin J, Guo W, Qin X, Zhao J, Pei M, Ding F (2017) A sensitive electrochemical aptasensor for highly specific detection of streptomycin based on the porous carbon nanorods and multifunctional graphene nanocomposites for signal amplification. Sensors Actuators B Chem 241:151–159

Young E, Patel S, Stoneham MD, Rona R, Wilkinson JD (1987) The prevalence of reaction to food additives in a survey population. J Roy Coll Phy Lond 21(4):241

Yu W, Zhang T, Ma M, Chen C, Liang X, Wen K et al (2018) Highly sensitive visual detection of amantadine residues in poultry at the ppb level: a colorimetric immunoassay based on a Fenton reaction and gold nanoparticles aggregation. Anal Chim Acta 1027:130–136

Zhang G, Wang X, Zhi A, Bao Y, Yang Y, Qu M et al (2008) Development of a lateral flow immunoassay strip for screening of sulfamonomethoxine residues. Food Addit Contam 25(4): 413–423

Zhang Y, Yang JY, Lei HT, Wang H, Xu ZL, Shen YD et al (2015) Development of chemiluminescent enzyme immunoassay for the determination of malachite green in seafood. Food Agric Immunol 26(2):204–217

Zhang L, Huang R, Liu W, Liu H, Zhou X, Xing D (2016) Rapid and visual detection of Listeria monocytogenes based on nanoparticle cluster catalyzed signal amplification. Biosens Bioelectron 86:1–7

Zhang S, Zhang D, Zhang X, Shang D, Xue Z, Shan D, Lu X (2017a) Ultratrace naked-eye colorimetric detection of Hg2+ in wastewater and serum utilizing mercury-stimulated peroxidase mimetic activity of reduced graphene oxide-PEI-Pd nanohybrids. Anal Chem 89(6): 3538–3544

Zhang X, Song M, Yu X, Wang Z, Ke Y, Jiang H, Wen KJFC (2017b) Development of a new broad-specific monoclonal antibody with uniform affinity for aflatoxins and magnetic beads-based enzymatic immunoassay. Food Control 79:309–316

Zhang C, Du P, Jiang Z, Jin M, Chen G, Cao X et al (2018) A simple and sensitive competitive bio-barcode immunoassay for triazophos based on multi-modified gold nanoparticles and fluorescent signal amplification. Anal Chim Acta 999:123–131

Zhu Y, Liu C-L, Xie Z-J, Immunology A (2017) Botryoid-shaped nanoparticles-enhanced ELISA for ochratoxin A. Food Agric Immunol 28(1/2):299–309

The manufacturer's authorised representative in the EU is Springer
Nature Customer Service Centre GmbH, Europaplatz 3, 69115 Heidelberg,
Germany. If you have any concerns regarding our products, please
contact ProductSafety@springernature.com

Printed and bound by CPI Group (UK) Ltd, Croydon, CR0 4YY
30/09/2026
02225324-0017